Lecture Notes in Computer Science 3260

Commenced Publication in 1973
Founding and Former Series Editors:
Gerhard Goos, Juris Hartmanis, and Jan van Leeuwen

Ignas Niemegeers
Sonia Heemstra de Groot (Eds.)

Personal Wireless Communications

IFIP TC6 9th International Conference, PWC 2004
Delft, The Netherlands, September 21-23, 2004
Proceedings

Volume Editors

Ignas Niemegeers
Delft University of Technology
Center for Wireless and Personal Communication
Mekelweg 4, 2628 DC Delft, The Netherlands
E-mail: I.Niemegeers@ewi.tudelft.nl

Sonia Heemstra de Groot
Twente Institute for Wireless and Mobile Communications
Institutenweg 30, 7521 PK Enschede, The Netherlands
E-mail: Sonia.Heemstra.de.Groot@ti-wmc.nl

Library of Congress Control Number: 2004111956

CR Subject Classification (1998): C.2, H.4, H.3, D.2, K.8

ISSN 0302-9743
ISBN 3-540-23162-5 Springer Berlin Heidelberg New York

Springer is a part of Springer Science+Business Media

springeronline.com

Printed in Germany

Preface

The IFIP TC-6 9th International Conference on Personal Wireless Communications, PWC 2004 is the main conference of the IFIP Working Group 6.8, Mobile and Wireless Communications.

The field of personal wireless communications is steadily growing in importance, from an academic, industrial and societal point of view. The dropping cost of WLAN and short-range technologies such as Bluetooth and Zigbee is causing the proliferation of personal devices and appliances equipped with radio interfaces. Together with the gradual deployment of powerful wireless infrastructure networks, such as 3G cellular systems and WLAN hotspots, the conditions are being created for affordable ubiquitous communication involving virtually any artifact. This enables new application areas such as ambient intelligence where a world of devices, sensors and actuators surrounding us use wireless technology to create systems that assist us in an unobtrusive way. It also allows the development of personal and personalized environments that accompany a person wherever he or she goes. Examples are Personal Area Networks (PAN) physically surrounding a person, and personal networks with a potentially global reach.

PWC 2004 reflects these developments, which are happening on a global scale. Researchers from all over the world, and in particular a large number from Asia, made contributions to the conference. There were 100 submissions. After a thorough reviewing process, 25 full papers and 13 short papers were retained for presentation in the technical sessions. The papers cover the whole range of wireless and mobile technologies: cellular systems, WLAN, ad hoc and sensor networks, host and network mobility, transport protocols for wireless systems, and the physical layer.

PWC 2004 was made possible by the enthusiasm, dedication and cooperation of many people. In particular we would like to thank the TPC members and the reviewers who were responsible for the high-quality reviewing process, and the Executive Committee for making the final selection of papers and the composition of the sessions. Of course, a special thanks to all the authors who showed their interest in the conference by submitting their papers. Also thanks to the organizations that supported the conference through various forms of sponsoring: KPN Mobile, Nokia, TNO Telecom and WMC. Last but not least, many thanks to the Organizing Committee that worked very hard to make sure that all organizational processes, the website, the electronic submissions, the financials, the social event and all the practical matters, were taken care of.

July 2004

Ignas G. Niemegeers,
Sonia Heemstra de Groot

Organization

Executive Committee

Conference Chair

Ignas Niemegeers	Delft University of Technology, The Netherlands

Conference Vice Chair

Sonia Heemstra de Groot	Twente Institute for Wireless and Mobile Communications, The Netherlands

Steering Committee

Piet Demeester	Ghent University, Belgium
Carmelita Görg	University of Bremen, Germany
Sonia Heemstra de Groot	Twente Institute for Wireless and Mobile Communications, The Netherlands
Petri Mähönen	RWTH Aachen, Germany
Ignas Niemegeers	Delft University of Technology, The Netherlands
Homayoun Nikookar	Delft University of Technology, The Netherlands

Organizing Committee

Nico Baken	Delft University of Technology, The Netherlands
Laura Bauman	IRCTR, Delft University of Technology, The Netherlands
Wendy Murtinu	Delft University of Technology, The Netherlands
Ignas Niemegeers	Delft University of Technology, The Netherlands
Homayoun Nikookar	Delft University of Technology, The Netherlands
Ad de Ridder	IRCTR, Delft University of Technology, The Netherlands
Nan Shi	Delft University of Technology, The Netherlands
Alex Slingerland	Delft University of Technology, The Netherlands
Mia van der Voort	IRCTR, Delft University of Technology, The Netherlands

Technical Program Committee

Arup Acharya	IBM T.J. Watson Research Center, USA
Dharma Agrawal	University of Cincinnati, USA
Paolo Bellavista	Università degli Studi di Bologna, Italy
Hans van den Berg	TNO Telecom/University of Twente, The Netherlands
Mauro Biagi	INFO-COM, University of Rome "La Sapienza", Italy
Chris Blondia	University of Antwerp, Belgium
Andrew Campbell	Columbia University, USA
Augusto Casaca	INESC-INOV, Portugal
Claudio Casetti	Politecnico di Torino, Italy
Piet Demeester	Ghent University, Belgium
Sudhir Dixit	Nokia, USA
Laura Marie Feeney	Swedish Institute of Computer Science, Sweden
Zhong Fan	Toshiba Research, UK
Luigi Fratta	Politecnico di Milano, Italy
Rajit Gadh	UCLA, USA
Silvia Giordano	SUPSI, Switzerland
Carmelita Görg	University of Bremen, Germany
Sonia Heemstra de Groot	Twente Institute for Wireless and Mobile Communications, The Netherlands
Geert Heijenk	University of Twente, The Netherlands
Georgios Karagiannis	University of Twente, The Netherlands
Mohan Kumar	University of Texas at Arlington, USA
Anthony Lo	Delft University of Technology, The Netherlands
Antonio Loureiro	Federal University of Minas Gerais, Brazil
Hiroyuki Morikawa	University of Tokyo, Japan
Luis Muñoz	University of Cantabria, Spain
Ignas Niemegeers	Delft University of Technology, The Netherlands
Homayoun Nikookar	Delft University of Technology, The Netherlands
Stephan Olariu	Old Dominion University, USA
Niovi Pavlidou	Aristotle University of Thessaloniki, Greece
Ramjee Prasad	Aalborg University, Denmark
Jan Slavik	Testcom, Czech Republic
Ivan Stojmenović	University of Ottawa, Canada
Rahim Tafazolli	University of Surrey, UK
Samir Tohmé	École Nationale Supérieure des Télécommunications (ENST), France
Christian Tschudin	University of Basel, Switzerland
Jörg Widmer	Swiss Federal Institute of Technology, Lausanne (EPFL), Switzerland
Magda El Zarki	University of California at Irvine, USA
Djamal Zeghlache	Groupe des Écoles des Télécommunications-Institut National des Télécommunications (GET-INT), France

Referees

Imad Aad
Arup Acharya
Dharma Agrawal
S. Aust
Paolo Bellavista
Hans van den Berg
Pavel Bezpalec
Mauro Biagi
Ramón Agüero Calvo
Augusto Casaca
Claudio Casetti
Dave Cavalcanti
Y. Che
Antonio Corradi
Piet Demeester
Hongmei Deng
Sudhir Dixit
M. Elkadi
Paul Ezchilvan
Zhong Fan
Laura Marie Feeney
N. Fikouras
Luigi Fratta
Susan Freeman
Rajit Gadh
Silvia Giordano
Carmelita Görg
Antonio Grilo
Meetu Gupta
Sonia Heemstra
Geert Heijenk
Jeroen Hoebeke
J. Hrad
Z. Irahhauten
Martin Jacobsson
Gerard Janssen
Georgios Karagiannis
A. Könsgen
M. Koonert
K. Kuladinithi
Dirk Kutscher
E. Lamers
Benoit Latre
Tom Van Leeuwen
Xi Li
Anthony Lo
Weidong Lu
Petri Mähönen
Peter Martini
Ruben Merz
Hiroyuki Morikawa
Anindo Mukherjee
Luis Muñoz
Ignas Niemegeers
Homayoun Nikookar
Mario Nunes
Stephan Olariu
Evgeny Osipov
Niovi Pavlidou
Paulo Pereira
Liesbeth Peters
Neeli Prasad
Ramjee Prasad
Vassilis Prevelakis
Janne Riihijarvi
Hartmut Ritter
R. Schelb
Vivek Shah
David Simplot-Ryl
Jan Slavik
Alex Slingerland
Ivan Stojmenović
A. Timm-Giel
Christian Tschudin
U. Türke
Einar Vollset
Boris Šimák
Natasa Vulic
Haitang Wang
L. Wang
Shu Wang
Jörg Widmer
Magda El Zarki
Djamal Zeghlache
Tomas Zeman

Sponsoring Institutions

Table of Contents

Self-Organization of Wireless Networks: The New Frontier (Keynote Speech)

Sudhir Dixit

Nokia Research, Burlington, MA, USA
sudhir.dixit@nokia.com

Abstract. The size of the internet will increase with the mainstream adoption of the broadband mobility connecting a myriad of devices and sensors at homes and businesses and the use of IPv6. All this will add to the spatio-temporal complexity of the network topology and dynamics. We present a brief overview of the role that self-organization can play in this new era of complexity. Issues of QoS, scalability, robustness, and reachability, among others (e.g., heterogeneity) will dominate the research in the future. First, we present the definition, scope, and applicability of self-organization. Then we briefly articulate the need for self-organization, and some recent breakthrough advances in this emerging area of research. This is followed by some near- and long-term scenarios where self-organization can be applied, and some results that we have obtained. We conclude the talk with a discussion on the key challenges that lie ahead.

I. Niemegeers and S. Heemstra de Groot (Eds.): PWC 2004, LNCS 3260, p. 1, 2004.

The Impacts of Signaling Time on the Performance of Fast Handovers for MIPv6*

Seung-Hee Hwang[1], Youn-Hee Han[2], Jung-Hoon Han[1], and Chong-Sun Hwang[1]

[1] Dept. of Computer Science and Engineering, Korea University, Seoul, Korea,
{shhwang,frajung,hwang}@disys.korea.ac.kr
[2] i-Networking Lab., Samsung Advanced Institute of Technology, Yongin,
Kyungki-do, Korea, yh21.han@samsung.com

Abstract. A Fast Handover protocol (FMIPv6) in IETF working group is proposed to reduce the handover latency in Mobile IPv6 standard protocol. The FMIPv6 proposes some procedures for fast movement detection and fast binding update to minimize the handover latency. Additionally, to reduce the lost packets caused by a handover, this protocol introduces buffers in access routers. However, the handover latency or the amount of lost packets are affected by the time to send signals such as Fast Binding Update message for the fast handover. In this paper, we inspect the impacts of the signaling time on packet loss and handover latency in FMIPv6 through the numerical analysis, we propose the optimal signaling time to improve the performance of FMIPv6 in terms of the handover latency and lost packets.

1 Introduction

In mobile network, a mobile user should communicate with its correspondent nodes via its IP address regardless of its location. However, the IP address is some location-dependent, so its IP address may be changed at its location change, and its communication also may be disconnected at its new location. To solve this problem, Mobile IP is proposed [1].

In MIPv6, MN has a home IP address (HoA) for identification and a temporal IP address for routing information. When MN moves to a new subnet, that is, it may disconnect with the current link and connect with a new link in link layer, and it should obtain a new temporal address called Care-of-Address (CoA) through stateless or stateful (e.g. DHCPv6) address auto-configuration [2] according to the methods of IPv6 Neighbor Discovery [8]. Then MN should register the binding its new CoA with its HoA to its home agent (HA) and its CNs. Therefore MN can maintain the connectivity with CNs regardless of its movement.

On the other hand, when MN in MIPv6 conducts these procedures which are called a *handover*, there is a period the MN is unable to send or receive packets;

* This works was supported by the Korea Rearch Foundation Grant (KRF-2003-041-D00403)

I. Niemegeers and S. Heemstra de Groot (Eds.): PWC 2004, LNCS 3260, pp. 2–13, 2004.

that is, the *handover latency* is defined as a duration from reception of last packet via previous link to reception of first packet via new link. This handover latency results from standard Mobile IPv6 procedures, namely movement detection, new Care of Address configuration and confirmation, and Binding Update, as well as link switching delay, and these procedures are time consuming tasks, so it is often unacceptable to real-time traffic such as VoIP.

To reduce the handover latency, Fast handovers for Mobile IPv6 (FMIPv6) [3], has been also proposed in IETF. FMIPv6 supports a fast handover procedure allowing starting handover in advance a movement [3]. In this proposal, MN obtains the new CoA (NCoA) before actual movement to new subnet through newly defined messages: *Router Solicitation for Proxy* (RtSolPr) and *Proxy Router Advertisement* (PrRtAdv). It also register its NCoA to previous AR (PAR) to indicate to forward the packets to its NCoA, so as soon as MN moves to the new subnet and connect with a new link, it can receive the forwarded packets from PAR. If feasible, buffers may exist in PAR and NAR for protecting packet loss. Therefore this proposal reduces the service disruption duration as well as the handover latency [3]. However, the various signaling in FMIPv6 makes it complicated to analyze the performance, which should be investigated, so we analyze the signaling time of FMIPv6 on the performance in terms of handover latency, packet loss, and required buffer size.

This paper is organized as follows: we will describe FMIPv6 protocol in Setion 2; we will expain the analytic models and calculate the performance functions for the handover latency, the number of lost packets, and the required buffer size, and then we will show the numerical results in Section 3; we will inspect of signaling time on the performance of FMIPv6 and propose the optimal signaling time for more effective FMIPv6 in Section 4; and we will conclude this paper with some words in Section 5.

2 FMIPv6

FMIPv6 is proposed to reduce the handover latency of MIPv6 by providing a protocol to replace MIPv6 movement detection algorithm and new CoA configuration procedure. Providing FMIPv6 is operated over IEEE 802.11 network, the new AP is determined by the scanning processes. The new associated subnet prefix information is obtained through the exchange of the *Router Solicitation for Proxy* (RtSolPr) and *Proxy Router Advertisement* (PrRtAdv) messages. Although the sequential L2 handover processes of scanning, authentication, and re-association are performed autonomously by firmware in most existing IEEE 802.11 implementations, these processes should not be executed autonomously in FMIPv6 to exchange RtSolPr and PrRtAdv messages, and *Fast Binding Update* (FBU) and *Fast Binding Acknowledgement* (FBAck) messages.

In FMIPv6 operated over IEEE 802.11 network, MN firstly performs a scan to see what APs are available. The result of the scan is a list of APs together with physical layer information, such as signal strength. And then, MN selects one or more APs by its local policy. After the selection, MN exchanges RtSolPr and PrRtAdv to get the new subnet prefix. In fact, there may or may not some delay

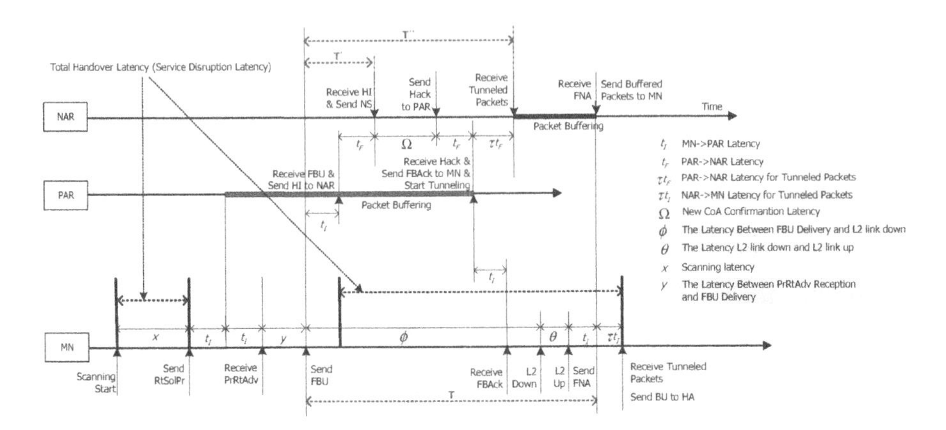

Fig. 1. Origin FMIPv6 handover procedure and timing diagram in case of receiving FBAck via the previous link

between scanning and sending RtSolPr; that is, MN can execute scanning at any time and may not do scanning for RtSolPr delivery, we assume that MN may scan available APs and select one or more APs just before sending RtSolPr in this paper. After receiving PrRtAdv, MN itself configures its prospective new CoA (NCoA) based on the new subnet prefix. And then, MN sends FBU to PAR to tell a binding of previous CoA (PCoA) and NCoA. At this point, the MN should wait FBAck message, if feasible, when it still presents on the previous subnet. If the MN receives FBAck in the previous subnet, it should move to NAR as soon as possible. If the MN does not receive FBAck in the current subnet and becomes under unavoidable circumstances (e.g., signal strength is very low) forcing it to move to NAR, MN should move to NAR without waiting more FBAck.

Fig.1 and Fig.2 describe FMIPv6 handover procedure and its timing diagram in the case that MN receives FBAck on the previous subnet. They also show the different time for buffering. When PAR in Fig.1 receives RtSolPr with buffering option from MN, it should start buffering according to [3], whereas PAR starts to store the packets destined for MN's PCoA into its buffer after receiving FBU from MN in Fig.2. In fact, if PAR starts buffering after receiving RtSolPr and finishes after processing FBU according to [3], that may cause needless buffering before reception of FBU, since the PAR delivers and also stores the packets in its buffer for the duration from receiving RtSolPr to receiving FBU 1. Therefore we assume PAR starts buffering at the reception of FBU like that in Fig.2. From this time, MN cannot receive packets and thus the service disruption time is measured. Then PAR sends *Handover Initiation* (HI) with NCoA to NAR. On receiving HI, NAR should confirm the NCoA and respond with *Handover Acknowledge* (HAck) message to PAR. At this time, the tunnel between PAR and the new location of MN is setup and the buffered packets are tunneled to NCoA. NAR must intercept the tunneled packets and store them into its buffer until it receives *Fast Neighbor Advertisement* (FNA) message from MN. FNA is the first message delivered from MN when it completes re-association with new

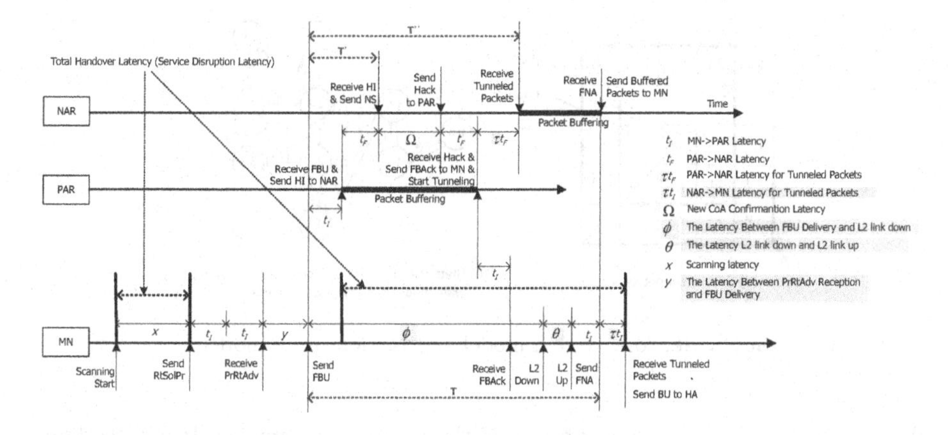

Fig. 2. FMIPv6 handover procedure and timing diagram in case of receiving FBAck via the previous link

AP. The reception of FNA allows NAR to release the buffered packets to MN. It means the end of service disruption time.

On the other hand, in the case the MN does not receive FBAck on the previous subnet, some different procedures are performed. MN moves to new AP area earlier than the reception of FBAck and, at this time, MN sends FNA immediately after attaching to new AP. It is noted that this FNA should encapsulate FBU in order to allow NAR to first check if NCoA is valid. When receiving such a FNA, NAR may not receive the tunneled packets delivered from PAR. In this case, NAR just forwards the tunneled packets to MN when the tunneled packets arrive at NAR.

3 Performance Analysis

3.1 Packet Level Traffic Model

IP traffic is characterized as connectionless transmission. Each packet has its destination address and is routed individually to the destination. In terms of IP traffic, we define a session or *session time* as a duration that packets with same source and same destination are been generating continuously, and *idle time* as a duration that packets are not generated until a new session starts. Generally today's Internet traffic is characterized as self-similar by nature. Therefore, recent research describes that a session time follows the Pareto distribution (or the Weibull distribution) which presents the self-similar property [6]. On the other hand, a session is arrived by the Possion process with a rate λ_c. In this paper, let a session time be t_{st}. For t_{st}, its probability density function $f_{st}(t)$ is defined from [6] as follows:

$$f_{st}(t) = \begin{cases} \frac{\alpha k^{\alpha}}{t^{\alpha+1}} & , t \geq k \\ 0 & , otherwise \end{cases} \tag{1}$$

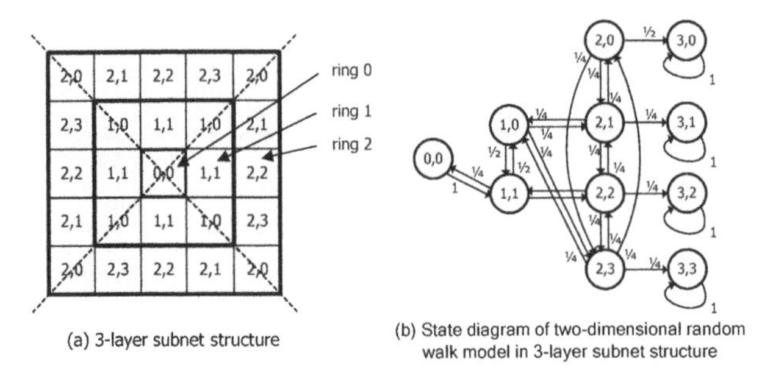

(a) 3-layer subnet structure

(b) State diagram of two-dimensional random walk model in 3-layer subnet structure

Fig. 3. Network system model and random walk model

where α is the shape parameter and k is the location parameter. If $\alpha \leq 2$, then the distribution has infinite variance, and if $\alpha \leq 1$, then it has infinite mean. The mean is as follows.

$$E_{st}[t] = \int_{0}^{\infty} t f_{st}(t) dt = \frac{\alpha \cdot k}{\alpha - 1}, \quad \alpha > 1 \tag{2}$$

For all examinations, we will use $\lambda_c = 0.002$ (so, the mean inter-session arrival time is 500 seconds.), $\alpha = 1.1$, and $k = 30$ as being similar to [6]. (so, the mean session time is 330 seconds.)

3.2 Network System Model and Mobility Model

We assume that each subnet also consists of more than one wireless AP areas. We assume that the homogeneous network of which all AP areas in a subnet have the same shape and size. We describe a two-dimensional random walk model for mesh planes in order to compute the domain and subnet residence time density functions. The mesh plane is drown as Fig.3 in which each 25 small squares and the entire square represents each AP areas and one subnet area, respectively.

A subnet is referred to as an *n-layer subnet* if it overlays with $N = 4n^2 - 4n + 1$ AP areas. For instance, Fig.3 (a) shows $3 - layer$ subnet, and the number of overlayed AP areas is $4 \times 3^2 - 4 \times 3 + 1 = 25$. The AP area in the center in Fig.3 (a) is called as *ring* 0, and the set of AP areas that surround ring 0 is called as *ring* 1, and so on. In general, the set of AP areas which surround the ring $x - 1$ is called as *ring* x. Therefore, an n-layer subnet consists of AP areas from ring 0 to ring $n - 1$. Especially, the AP areas that surround the ring $n - 1$ are referred to as *boundary neighbors*, which are outside of the subnet.

We assume that MN resides in an AP area for a period and moves to one of its four neighbors with the same probability, i.e., with probability 1/4. According to this equal moving probability assumption, we classify the AP areas in a subnet into several *AP area types*. An AP area type is represented as the form $<x, y>$, where x indicates that the AP area is in the ring x and y represents the $y + 1$th type in the ring x. Each type in each ring is named sequentially from 0, and the number 0 is assigned to APs in a diagonal line. For example, in Fig.3 (a), The

AP type $< 2, 1 >$ represents that this AP is in the ring 2 and it is the AP of 2nd type in ring 2.

In the random walk model, a state (x, y) represents that the MN is in one of the AP areas of type $<x, y>$. The absorbing state (n, j) in $n - layer$ subnet represents that an MN moves out of the subnet from state $(n-1, j)$, where $0 \leq j \leq 2n-3$ (For example, $j \in \{0, 1, 2, 3\}$ for 3-layer subnet). The state diagram of the random walk for 3-layer subnet is shown in Fig.3 (b).

Let t_p and t_s be i.i.d. random variables representing the AP area residence time and the subnet residence time, respectively. Let $f_p(t)$ and $f_s(t)$ be the density function of t_p and t_s, respectively. We assume that the AP area residence time of an MN has the Gamma distribution with mean $1/\lambda_p$ $(=E[t_p])$ and variance ν. The Gamma distribution is selected for its flexibility and generality. The Laplace transform of the Gamma distribution is $f_p^*(t) = \left(\frac{\gamma\lambda_p}{t+\gamma\lambda_p}\right)^\gamma$, where $\gamma = \frac{1}{\nu\lambda_p^2}$. Also, we can get the Laplace transform $f_s^*(t)$ of $f_s(t)$ and its expected subnet residence time $E[t_s]$ from [10,11].

From [10,11], during a session time t_{st}, the probabilities $\Pi_p(K)$ and $\Pi_s(K)$ that the MN moves across K AP areas and K subnets, respectively, can be derived as follows:

$$\Pi_p(K)=\begin{cases} 1 - \frac{E_{st}[t]}{E[t_p]}(1 - f_p^*(\frac{1}{E_{st}[t]})) & ,K = 0 \\ \frac{E_{st}[t]}{E[t_p]}(1 - f_p^*(\frac{1}{E_{st}[t]}))^2(f_p^*(\frac{1}{E_{st}[t]}))^{K-1} & ,K \geq 1 \end{cases} \tag{3}$$

$$\Pi_s(K)=\begin{cases} 1 - \frac{E_{st}[t]}{E[t_s]}(1 - f_s^*(\frac{1}{E_{st}[t]})) & ,K = 0 \\ \frac{E_{st}[t]}{E[t_s]}(1 - f_s^*(\frac{1}{E_{st}[t]}))^2(f_s^*(\frac{1}{E_{st}[t]}))^{K-1} & ,K \geq 1 \end{cases} \tag{4}$$

3.3 Performance Functions

At first, we introduce some parameters used for performance functions as follows:

- η: packet delivery delay in wireless path between AP and MN.
- ϵ: packet delivery delay per hop in wired path.
- Ω: NCoA confirmation latency in FMIPv6.
- τ: additional weight for tunnelled packets.
- a: #hops between AP and AR.
- b: #hops between PAR and NAR.
- $t_I(=\eta + \epsilon a)$: packet delivery delay between MN and AR.
- $t_F(=\epsilon b)$: packet delivery delay between two ARs.
- ϕ: latency between FBU transmission and L2-down trigger.
- θ: AP area switching latency (new AP re-association and authentication latency).
- x: AP Scanning latency.
- y: latency between PrRtAdv reception and FBU transmission. MN receives the packets from PAR in this duration.

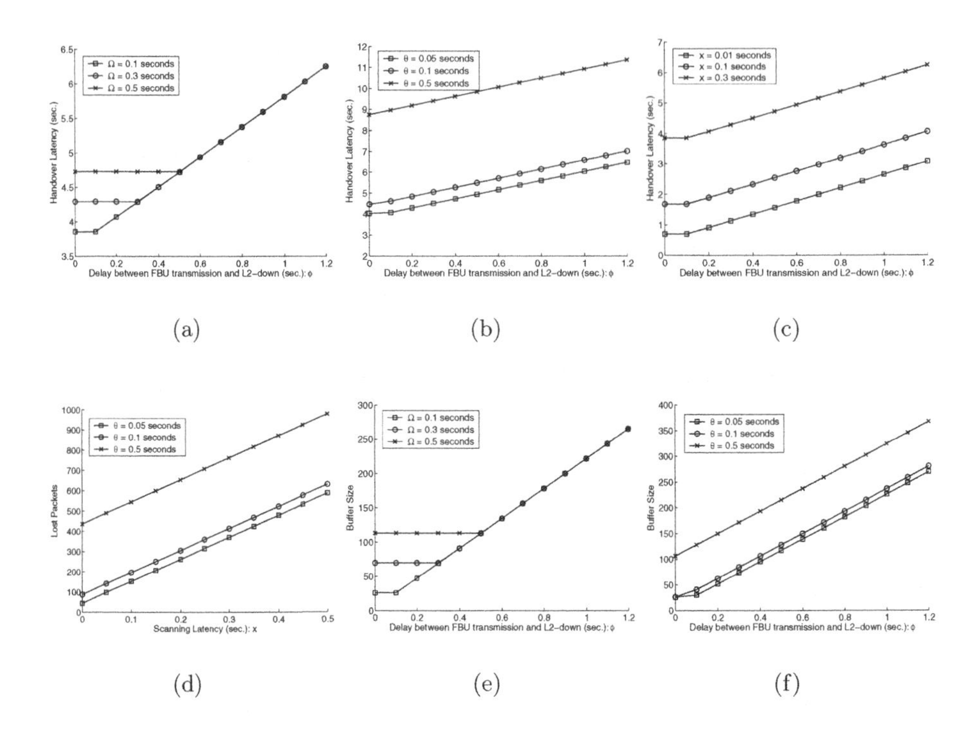

Fig. 4. Numerical Results I

Irrespective of whether MN receives FBAck on the previous subnet or not, the handover procedure of FMIPv6 is divided into two independent procedures; P_I, the procedure to be executed by MN itself, and P_{II}, the procedure to be executed by only both PAR and NAR in order to establish the bidirectional tunnel and send the tunneled packets. The two separated procedures will start when MN sends FBU to PAR, and combine into one when NAR receives FNA from MN. We assume that NAR has already received at least HI from PAR, when it receives FNA from MN. Processing FNA is also assumed to be executed after the completion of tunnel establishment between NAR and PAR. Until the two procedures P_I and P_{II} combine into one, the completion times of each procedure are defined as follows:

- $C_{P_I} = \phi + \theta + t_I$.
- $C_{P_{II}} = t_I + (2 + \tau)t_F + \Omega$.

If $C_{P_I} > C_{P_{II}}$, NAR has buffered the packets tunneled from PAR and forwards them to MN when it receives FNA. Otherwise, NAR waits the packets which will be tunneled from PAR (or runs its NCoA confirmation procedure and sends FBU encapsulated in FNA to PAR) when it receives FNA.

After announcing its attachment to NAR and receiving the tunneled packets, MN registers its new CoA to HA, and to CNs sequentially. In FMIPv6, the handover latency HL_F in a session time is define as follows (see Fig.2):

$$HL_F = \sum_{K=0}^{\infty} K\Pi_p(K)(x+\theta) + \sum_{K=0}^{\infty} K\Pi_s(K)(MAX\{C_{P_I}, C_{P_{II}}\} - t_I - \theta + \tau t_I). \quad (5)$$

Since FMIPv6 supports packet buffer function, packet losses does not occur during MN's subnet movement. Therefore, the number of lost packets PL_F in a session time is defined as follows.

$$PL_F = \lambda \sum_{K=0}^{\infty} K\Pi_p(K)(x+\theta) - \lambda \sum_{K=0}^{\infty} K\Pi_s(K)\theta. \quad (6)$$

The required buffer size is represented by the sum of the buffer sizes required at both PAR and NAR. The required buffer size BS_F in a session time is represented as follows:

$$BS_F = \lambda \sum_{K=0}^{\infty} K\Pi_s(K)(\Omega + 2t_F) + \lambda \sum_{K=0}^{\infty} K\Pi_s(K) \cdot MAX\{C_{P_I} - C_{P_{II}}, 0\}. \quad (7)$$

where λ is the average packet arrival rate.

3.4 Numerical Results

For examinations, the following fixed parameters are used: $\eta = 0.01 sec.$, $\epsilon = 0.005 sec.$, $\tau = 1.2$, $a = 1$, $b = 2$, $n = 3$ (subnet layer is 3), $\lambda_p = 0.033$ (that is, the mean of AP area residence time is 30 seconds), $\nu = 1$, $\lambda = 100$ *packets/sec.*, $y = 0.01 sec.$. As the target of investigation, we select the following changeable parameters and their default values:$x = 0.3 sec.$, $\theta = 0.03 sec.$, $\Omega = 0.1 sec.$, and $\phi = 0.02 sec.$. While we select one parameter and change its value, the remaining parameters values are set to their default values during the following investigation.

Fig.4 explains the handover latency, the number of lost packets, and the required buffer size in FMIPv6. As the latency ϕ between FBU transmission and L2-down trigger increases, the handover latency and required buffer size also increase. Therefore, ϕ should be as small as possible. In addition, when Ω, θ and x are low, the performance of FMIPv6 is high. The number of lost packets depends on only Layer 2 handover latency, that is θ and x.

From Fig.4 (a), we can find that the handover latency in FMIPv6 is unchanged as lowest when ϕ is so low (e.g. for $\phi = 0 \sim 0.3 sec.$ and $\Omega = 0.3 sec.$). It is resulted from that the handover procedure of FMIPv6 is divided into two independent procedures; P_I and P_{II}. If $C_{P_I} > C_{P_{II}}$, the performance depends largely on ϕ. Otherwise, ϕ does not affect the performance and Ω plays a important role of changing the performance.

4 Impacts of Signal Time on FMIPv6

Lots of research has already shown that the FMIPv6 has less handover latency (and less packet loss) than MIPv6 [4,5,9]. In this paper, therefore, we focus on the effect of signaling time on FMIPv6 performance rather than the performance comparisons between MIPv6 and FMIPv6.

As we discussed in the above section, the handover latency and the required buffer size depend largely on ϕ (the latency between FBU transmission and L2 down event) and Ω (address confirmation latency). For the exchange of FBU and FBAck to be completely done in the previous link, ϕ should be high and thereby C_{P_I} will be larger than $C_{P_{II}}$. It means the performance of FMIPv6 will depend largely on ϕ. MN can send FBU early in the overall handover duration to receive FBAck. In this case, however, the more buffering is required at PAR and NAR. If the packet arrival rate λ is very high, the buffers at PAR and NAR will be overflowed and packets will be lost. Therefore, the duration between FBU transmission and FBAck reception (and thereby ϕ) should not be too long.

On the other hand, thinking about the case that MN does not receive the FBAck, there are two reasons: one reason is that FBU is delivered to PAR, but MN moves to a new area before receiving FBAck, and the other is that FBU itself is lost or that FBU is delivered to PAR, yet sequential HI is not delivered to NAR. In any case, if MN does not receive FBAck in previous link, MN should re-send FBU, which is encapsulated in FNA, via new link since MN does not know whether or not FBU was delivered to PAR. If FBU is lost, then the anticipated fast handovers are not feasible. So we assume in this paper if FBU is not lost, then NAR receives at least HI when NAR receives the FNA. When NAR receives FNA encapsulating FBU, if NAR is still confirming NCoA presented by HI, it reserves the FNA encapsulating FBU in its local memory. If the confirmation procedure is a success, then NAR sends HAck to PAR and waits for the tunneled packets from PAR. After then NAR processes the FNA in temporal storage, and forwards the tunneled packets to MN's NCoA. If the confirmation of NCoA is failed, then NAR assigns new CoA and include it HAck. However, if FBU is lost, so NAR does not receive HI from PAR and NAR receives FNA with FBU, then NAR forwards FBU to PAR to setup tunnel between PAR and NAR. From the observation, we can separate the handover processing duration of FMIPv6 as three parts as follows: Let assume T the duration between FBU transmission and FNA delivery, it is same with C_{P_I}, T' the duration between FBU transmission and HI delivery, and T'' the duration between FBU transmission and NAR's reception of the first packet tunneled from PAR, it is same with $C_{P_{II}}$.

Theorem: When $T' < T < T''$, the handover latency in FMIPv6 has the minimal value.

Proof: From the Fig.2, each duration T, T', and T'' is represented as follows:

$$
\begin{aligned}
T &= C_{P_I} = \phi + \theta + t_I \\
T' &= t_I + t_F \\
T'' &= C_{P_{II}} = t_I + (2+\tau)t_F + \Omega.
\end{aligned}
$$

For $S = \{\phi, \theta, \Omega, t_F, t_I, \tau\}$, we define a function $C(S)$ indicating the duration between MN's FBU transmission and MN's reception of the first tunneled packet in the new link.

$$C(S)=\begin{cases} T' + (4+\tau)t_F + \Omega + \tau t_I, & for\ T \leq T' \quad (a) \\ T'' + \tau t_I, & for\ T' < T < T'' \quad (b) \\ T + \tau t_I, & for\ T \geq T'' \quad (c) \end{cases} \tag{8}$$

The handover latency is determined by this function $C(S)$ and L2 scanning duration x. From Eq.8, the handover latency HL_F should be re-defined as follows:

$$HL_F=\begin{cases} \sum_{K=0}^{\infty} K\Pi_p(K)(x+\theta) \\ +\sum_{K=0}^{\infty} K\Pi_s(K)\{T' + (4+\tau)t_F + \Omega - t_I - \theta)\}, \\ \qquad for\ T \leq T' \\ \sum_{K=0}^{\infty} K\Pi_p(K)(x+\theta) \\ +\sum_{K=0}^{\infty} K\Pi_s(K)\{T'' - t_I - \theta\}, \\ \qquad for\ T' < T < T'' \\ \sum_{K=0}^{\infty} K\Pi_p(K)(x+\theta) \\ +\sum_{K=0}^{\infty} K\Pi_s(K)\{T - t_I - \theta\}, \\ \qquad for\ T \geq T'' \end{cases} \tag{9}$$

By using the above parameters definitions, we can infer the following inequalities for ϕ, which make the FMIPv6 handover latency distinct.

i) $T < T'$
$\equiv \phi + \theta + t_I < t_I + t_F$
$\equiv \phi \leq t_F - \theta$

ii) $T' \leq T \leq T''$
$\equiv t_I + t_F < \phi + \theta + t_I < t_I + (2+\tau)t_F + \Omega$
$\equiv t_F - \theta < \phi < (2+\tau)t_F + \Omega - \theta$

iii) $T \geq T''$
$\equiv \phi + \theta + t_I \geq t_I + (2+\tau)t_F + \Omega$
$\equiv \phi \geq (2+\tau)t_F + \Omega - \theta$

Let the minimal value of $C(S)$ be $C_{min}(S)$. The minimal $C_{min}(S)$ is found from the above equation as the followings.

$$\begin{aligned} & Eq.8(a) - Eq.8(b) \\ = & \{T' + (4+\tau)t_F + \Omega + \tau t_I\} - \{T'' + \tau t_I\} \\ = & 3t_F > 0. \\ \therefore & Eq.8(a) > Eq.8(b). \end{aligned} \tag{10}$$

$$\begin{aligned} & Eq.8(c) - Eq.8(b) \\ = & T + \tau t_I - T'' - \tau t_I \\ = & \phi + \theta - \Omega - (2+\tau)t_F \end{aligned}$$

$$= \phi + \theta - \Omega - (2+\tau)t_F > 0$$
$$(\because \ \phi > (2+\tau)t_F + \Omega - \theta \ from \ iii))$$
$$\therefore \ Eq.8(c) > Eq.8(b). \tag{11}$$

Therefore, Eq.8 (b) is the lowest value of $C(S)$. So $C_{min}(S)$ is derived as followings:

$$C_{min}(S) = T'' + \tau t_I, \qquad for \ T' < T < T''. \tag{12}$$

Therefore, when $T' < T < T''$, the $C(S)$ is lowest as shown in Fig.5(a) and consequently, the handover latency will be lowest.

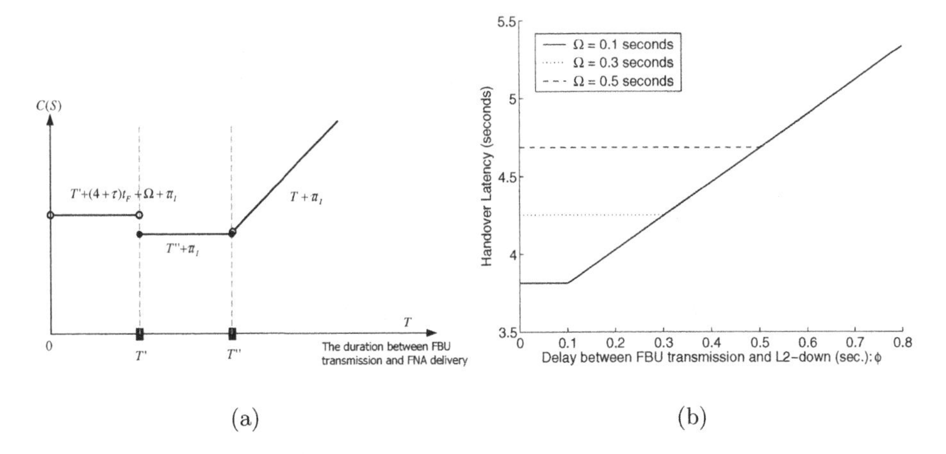

Fig. 5. The time to send FBU and Ω to minimize the handover latency

Corollary 1) For lowest handover latency in FMIPv6, FBAck should not always be delivered in previous link.

From Theorem, $C_{min}(S)$ is lowest for $T' < T < T''$. $T' < T < T''$ means that FBU should be delivered in previous link and NAR should receive HI until FNA is delivered to NAR. Therefore FBack reception on previous link is not necessary condition for effective handover in FMIPv6.

Corollary 2) If the delivery latency between PAR and NAR t_F is lower than link switching latency θ, when $0 < \phi < (2+\tau)t_F + \Omega - \theta$, the C(S) has the minimal value $C_{min}(S) = T'' + \tau t_I$ and consequently the handover latency is the smallest.

In Theorem, $T' < T < T''$ is represented as follows

$$T' < T < T''$$
$$\equiv t_I + t_F < \phi + \theta + t_I < t_I + (2+\tau)t_F + \Omega$$
$$\equiv t_F - \theta < \phi < (2+\tau)t_F + \Omega - \theta$$
$$\equiv 0 < \phi < (2+\tau)t_F + \Omega - \theta$$
$$(\because assuming \ \ t_F < \theta \ \ in \ most \ cases). \tag{13}$$

Since $t_F < \theta$, when $0 < \phi < (2+\tau)t_F + \Omega - \theta$, the handover latency has the minimal value as shown Fig.5 (b). This implies that even if HI is not delivered and FNA is delivered to NAR, the handover latency in FMIPv6 shows lowest value.

5 Conclusions

Fast Handovers for IPv6 (FMIPv6) protocol is proposed to reduce the handover latency in Mobile IPv6 standard protocol. In this paper, we inspected the mechanism of FMIPv6 protocol over IEEE 802.11 wireless network in detail, and analyzed numerically the performance of FMIPv6 in terms of handover latency, packet loss, and required buffer size using proposed models. From the numerical results, we found that the performance is very different according to the signals delivery time of FMIPv6, especially FBU. To make FMIPv6 more effective, we calculated the optimal time for FBU delivery, such as $0 < \phi < (2+\tau)t_F + \Omega - \theta$ under condition that FBU is not lost. In addition, there needs not any buffer in NAR.

References

1. D. Johnson, C. Perkins, and J. Arkko, "Mobility Support in IPv6," draft-ietf-mobileip-ipv6-24.txt, Internet draft (work in progress), June 2003.
2. S. Thomson and T. Narten, "IPv6 Stateless Address Autoconfiguration," IETF RFC 2462, December 1998.
3. R. Koodli, "Fast Handovers for Mobile IPv6," draft-ietf-mobileip-fast-mipv6-08.txt, Internet draft (work in progress), October 2003.
4. S. Pack and Y. Choi, "Performance Analysis of Fast Handover in Mobile IPv6 Networks" accepted for Lecture Notes in Computer Science (LNCS), Springer-Verlag, 2003.
5. X.P. Costa, R. Schmitz, H. HArtenstein and M. Liebsch, "A MIPv6, FMIPv6 and HMIPv6 Handover Latency Study: Analytical Approach" Proc. of IST Mobile and Wireless Telecommunications Submit, June 2002.
6. T. Janevski, *Traffic analysis and design of wireless IP networks*, Arttech House, 2003.
7. A. Mishra, M.H. Shin, and W. Arbaugh, "An empirical analysis of the IEEE 802.11 MAC layer handoff process", ACM SIGCOMM Computer Communication Review, Vol. 33, Issue 2, Pages: 93 - 102, 2003.
8. T. Narten, E. Nordmark and W. Simpson, "Neighbour Discovery for IP version 6", IETF RFC 2461, December 1998.
9. M. Torrent-Moreno, X. Perez-Costa, and S. Sallent-Ribes, "A Performance Study of Fast Handovers for Mobile IPv6", in Proc. of the 28th Annual IEEE International Conference on Local Computer Networks (LCN'03),
10. Y.H.Han, "Hierarchical Location Caching Scheme for Mobility Management", Ph.D. thesis, Dept. of computer science and engineering, Korea University, December, 2001.
11. S.H. Hwang, Y.H. Han, S.G. Min, and C.S. Hwang, " An Address Configuration and Confirmation Scheme for Seamless Mobility Support in IPv6 Network" Lecture Notes in Computer Science, Vol. 2957, pp. 74-86, Feb. 2004.

An Efficient Binding Update Scheme Applied to the Server Mobile Node and Multicast in the MIPv6

Hye-Young Kim[1] and Chong-Sun Hwang[1]

Department of Computer Science and Engineering,
Korea University, Seoul, Korea
{khy, hwang}@disys.korea.ac.kr

Abstract. Mobile nodes are changing their point of attachment dynamically such as a network deployed in an aircraft, a boat, a train, or a car. These mobile nodes move together and share the same mobility properties. Therefore, this paper addresses an efficient location management scheme that utilizing for mobile nodes to move as a group. Our proposed method adds Server Mobile Node(SMN) to the existing Mobile IPv6 protocol, which maintains the functioning of mobile communication and stores the necessary information there. By implementing this method, it is expected that the lifetime of the mobile nodes re-setting efficiently. By combining the multicast routing with Mobile IPv6 method, in addition, the number of the binding upgrade(BU) messages produced in the home agent and CNs will not increase almost, although the number of the mobile nodes increases. We address the key functions for our proposed scheme including system configuration, location registration, and packet delivery. Our analytical model shows the usefulness of our proposed mechanism using analytical models and compares the results with the previous researches.

1 Introduction

In recent years, we have a rapid growth in the need to support mobile nodes(MNs) over global Internet-based mobile computing system. In addition, third-generation systems such as International Mobile Telecommunication System 2000(IMT-2000) and the Universal Mobile Telecommunications System(UMTS) seeks to unify existing cellular, Mobile IP cordless, and paging networks for universal use, the next generation will have the additional goal of offering heterogeneous services to mobile hosts that may roam across various geographical and network boundaries. In the system, MNs require special support to maintain connectivity as they change their point of attachment. When MN moves to another subnet out of its home network for smooth communication on the International Engineering Task Force(IETF)[1] Mobile Internet Protocol version 6(MIPv6) protocol, MN performs location registration to the home agent(HA) and the correspondent nodes(CNs) of MN even during its mobility

I. Niemegeers and S. Heemstra de Groot (Eds.): PWC 2004, LNCS 3260, pp. 14–28, 2004.

by using the binding update(BU) messages in order to inform them of its current location after getting Care-Of-Address(COA)[2].

To support user's mobility, MN frequently creates the BU messages which in turn causes network overload because the additional signaling unnecessarily consumes the frequency bandwidth. The increase in the BU messages, therefore, has emerged as one of serious barriers to efficient location management. Therefore to deploy this mobile IP service widely, the Hierarchical Mobile Internet Protocol Version6(HMIPv6)[3] is also being researched in IETF. By adding a Mobility Anchor Point(MAP) in a visited network to manage local mobility there, we can limit HAs to providing only global or inter-MAP mobility management. HMIPv6 lets us to avoid frequent locational registration of MNs with HAs, which may be a long way from the MNs, and to reduce the time required for handovers. In the above approach to location management, the MNs are considered to move independently and so register their location with the network individually.

The evolution of information technology has enabled a mobile object such as person, car, bus, train, airplane, or ship to carry a plethora of information devices. Consider the situation in which a large number of MNs are riding on the same train or bus or aircraft. As it may contain a lot of mobile nodes, each communicating with several peers, the questions of locating, optimal routing and signaling overload are significantly more important. Therefore we consider an efficient location management scheme that mobile nodes move as a group. However, some CNs might even receive duplicate BU message carrying the same address in case they are corresponding with several MNs residing in the subnet. We therefore consider a solution based on multicast routing protocols for delivering network scope binding updates. In this paper, we propose a method to reduce the inefficient flood of BU messages by adding the Server Mobile Node(SMN) to the IETF MIPv6 protocol and by applying multicast to the transmission of the BU messages. As it provides a mobile networking function, an information keeping function, and a buffering function for the transmission packets, SMN performs the efficient setting of lifetime on the binding update messages and minimizes the delay or loss of transmitted packets. This can be applied to the transmission of the BU messages so that despite the increase of MN or CNs, the number of the BU messages will not increase almost. Our proposed method minimizes the signaling due to the minimized addition of the BU messages and due to the expansion of MN.

The organization of this paper is as follows. This introduction section is followed by the literature review of related work. The section 3 is the main body of the paper, where we explain a new model for efficient location management, including a location registration scheme and its algorithm. The section 4 shows the mathematical modeling, its simulation, and the comparison of performance between the existing IETF Mobile IPv6 and the proposed new model. The final section suggests a future study direction.

2 Related Work

IETF which is in charge of global standardization on the internet protocol has the Mobile IP Working Group to support connection and mobility through the internet of MNs. The recently provided MIPv6 protocol is registered as an internet draft and sometimes it is also adopted as the standard of IETF. However, signaling and required bandwidth created for mobility management on the MIPv6 protocol causes network overload, leading to traffic bottleneck. Particularly, it has been a concern that a registration process may cause too much traffic between the visited networks and home networks.

In order to resolve this problem, [3] and [4] have proposed a hierarchical mobility scheme based on MIPv6. Because of MIPv6's flexibility, their proposal can be deployed easily, and can interact and work well with MIPv6. In this scheme, a site can be an Internet Service Provider(ISP) network. A mobility network of a site is a LAN(Local Area Network) that defines an address space for the MNs roaming within a site. As routers for the MN, Mobility Servers(MSs) maintain a binding for every mobile node that currently visited the site. Hierarchical Mobile Internet Protocol version 6(HMIPv6), location registration is performed by sending the BU messages in accordance with changes in MN location, from the MS where they are currently encompassed to the top-ranked parent nodes of those MS, after MSs are tree-structured. In [5], by contract, the border router is placed for a communication with a separate mobility agent, and an external network for location management, based on MIPv6. In this case, location registration is performed by sending the BU messages only to the mobility agent in line with changes in MN location. In some studies [3, 4, 5], signaling is curtailed by reducing the number of the BU messages through the separation of the micro areas from the macro areas, considering the mobility of users on the MIPv6 and MIPv4 protocol. Each MN considers its location using location information from nearest MS or MAP, and then binding updates its location information individually if it detects that it has entered a new location registration area. Consider the very situation in which a large number of MNs within the MS or MAP are moving on the same mobility. All MNs try to access the network individually and simultaneously to update their location information. Therefore, they have inefficient flood of binding updates.

Concatenated Location Management(CLM) is used in [6] so that signaling created by the updated binding messages is minimized. Its basic idea is to treat MNs that share the same movement characteristics, as a single entity for location management. In order to update the location information of MN, it is to establish Intermediate Radio Stations(IRS) on each vehicle and CLM update the location information of all MNs with one action. So CLM, greatly enhanced the usage efficiency of the network since each group of users that moves together is treated as a single entity for location update. But [6] is considering only movement of vehicle or train which is a group of MNs without the respect to cases about individual MN's mobility.

Therefore, this paper proposes a method to efficiently create the BU messages by managing the local mobility of MNs similar to stratified MIPv6 by considering cases of mobility of MNs by adding SMN.

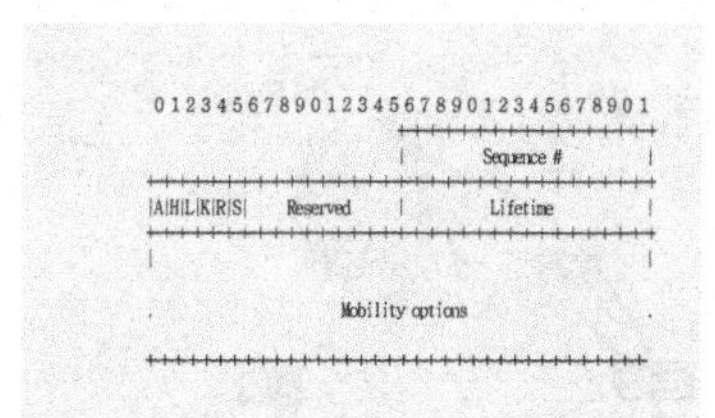

Fig. 1. BU Message Formats for Proposed System Configuration

Table 1. The Elements Consisting of SMN

SMN_{IP}	COA_{SMN}	SMN_{PREFIX}	T_s	MN_1	MN_2	...	MN_n	CN_1	...	CN_n
...	...	...	...	...	...	...	...	...	...	...
...	...	...	...	...	...	...	...	...	...	...

3 System Configuration

3.1 Proposed System Configuration

Our proposed scheme manages network mobility and maintains binding information on MN by adding SMN to the IETF MIPv6 structure. One SMN has an area for location registration by the subnet, connected with many MNs. SMN is connected to wireless networks due to the mobility, for example, routers used on the aircraft, bus or train, it has a Server Gateway(SG) that manages network connection of SMN. HA that is in the home link of SMN will have the home address of SMN acquired upon location registration, and will have COA_{smn} acquired when it moves to another subnet. The SMN_{prefix} is a bit string that consists of some number of initial bits of the home address SMN_{IP} which identifies the home link within the internet topology. All MNs share the same IP prefix[7]. The SMN performs location management through the BU messages received like MN does. As shown in Table 1, SMN keeps information about each of MNs for location management of MNs.

We add a new flag(S) which is included in the BU message to indicate to the HA if the BU message is coming from a SMN and not from a MN. SMN provides connectivity to nodes in the mobile network, it indicates this to the HA by setting a flag(S) in BU message as same scheme of Network Mobility(NEMO)[8]. The Figure 1 shows the BU message format in this paper. It sends the BU messages to HA and CNs of MN through SMN's SG right before the expiration of lifetime set in the BU messages and when it moves to another subnet. It is performed by the SMN, the COA would then be sent periodically to each CN corresponding with MNs. Therefore, the BU messages of SMN sent to CNs are exploded. Each CN would receive exactly the same COA_{smn}. However some CNs might even receive duplicate BU carrying the same COA_{smn} in case they are corresponding with several MNs residing in the mobile network.

Therefore in our proposed system configuration, if IPv6 multicast is supported on network, SMN uses the method of sending the binding update mes-

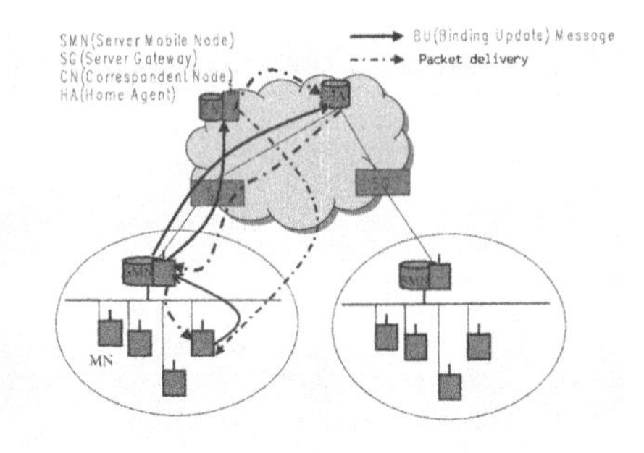

Fig. 2. Proposed System Configuration

sages to a Multicast Router(MR) that has a multicast group and multicast information to solve this problem. The subnetwork has a permanent multicast address which the SMN registers in the DNS[9, 10]. The SMN sends periodic BU s containing a binding between its SMN_{prefix} and its SMN_{COA} to the multicast address. All CNs of MN use the IPv6 multicast method to join the multicast group. Many studies on the multicast method for MNs have been conducted[11, 12], but this subject is beyond the scope of the current study. Figure 2 shows the system configuration proposed in this paper.

3.2 Location Registration

SMN advertise an address on its subnet, and binds its current location with its own regional COA and list of the address prefixes of the links for the MNs. When MN moves to a new and different subnet, it gets two COAs. One is the SMN's COA(COA_{SMN}) and the other is the MN's COA(COA_{MN}) with a prefix equal to COA_{SMN}. The MN registers the binding between its HA and the COAs, COA(COA_{SMN}) and COA_{MN}, to its HA and CNs through SMN. When the HA receives this BU message it create a binding cache entry binding the SMN's Home Address to it COA at the current point of attachment. The HA acknowledged the BU by a Binding Acknowledgement(BA) to the SMN through SG. A positive acknowledgement means that HA has set up forwarding for the network. Once the binding process completes, a bi-directional tunnel is established between the HA and SMN. The tunnel end points are SMN's Address and the HA's address. Then, when the HA receives packets for the MN, it can directly send them to the SMN by using an IP-in-IP tunnel. At that time SMN broadcasts the BU to MNs connected with it.

If MN moves within a domain, it only needs to change its (COA_{MN}), and the (COA_{SMN}) remains the same. Notice that it does not register any binding to its HA. MN has its own lifetime T_m and lifetime T_s of SMN. It is should be noted, however, that the determining of lifetime values and the frequency of binding refresh will also affect the signaling load of a network. If lifetimes have small values and the frequency of binding refresh is high, the signaling bandwidth for updating the COAs of an MN is also high. On the other hand, if the lifetimes have large values and the frequency of binding refresh is low, the signaling bandwidth for updating the COAs is also low. This reduction effect will increase as the number of MN increases. However, if long lifetime is set, MN stays in shorter time than lifetime set in a subnet. If MN moves to another subnet, this may cause the loss of packets because binding information kept by MN and CNs and that of HA sometimes have no information about MN location. MN takes steps to identify and authenticate the communication path and the communicating agent through lifetime set in its location registration[2]. If only lifetime of MN is simply set too long, therefore, increased are the chances of having a security problem between MNs and CNs. But in our proposed system configuration, MN's lifetime, T_m, becomes possible to set long in spite of no movement of MN, because BU procedure is happened by lifetime of SMN. These lifetimes are initialized from the lifetime field continued in the BU and are decreased until it reaches zero. At this time, the entry must be deleted from the binding cache of CNs and the BU list of MNs. Because the MN requires the binding service from this node to go beyond this period, it must send a new BU to the HA, the SMN, and CNs before the expiration of the period, in order to extend the lifetime. Lifetime of SMN has to be the same as or longer than lifetime of MN. (That is, $T_m \leq T_s$).

MN's mobility in this proposed scheme can be classified into three cases and the proposed scheme for its location management is as follows:

Case 1. When MN moves within SMN
1) MN sends BU_a, BU messages, to SMN. BU_a contains MN's home address, COA, COA with a prefix equal to SMN_{prefix} and lifetime in the form of $\{MN_{IP}, COA_{MN}, COA_{SMN}, T_m\}$
2) SMN updates or creates the information list of MN kept as contents of BU_a, $\{MN_{IP}, COA_{MN}, COA_{SMN}, T_m\}$ and sends BA responding to MN.
3) SMN sends the BU_a of the CNs through the SG. The CNs create or update the MN's information.

Case 2. When MN moves to another SMN
1) MN sends BU_a, BU messages, to new SMN in the area it moved to. The MN obtains COA_{SMN} on the foreign link and BU_a contains of its home address, COA, COA with a prefix equal to SMN_{prefix} and lifetime in the form of $\{MN_{IP}, COA_{MN}, COA_{SMN}, T_m\}$ as information on MN.
2) SMN creates MN's information by using the contents of BU_a, and sends BU_b, BU messages, to HA of MN through SG. A BU_b contains $\{MN_{IP}, SMN_{IP}, COA_{SMN}, COA_{MN}, T_s, T_m\}$ as information on SMN.
3) HA updates or creates information as content of BU_b,$\{MN_{IP}, SMN_{IP}, COA_{SMN}, COA_{MN}, T_s, T_m\}$, and sends BA, response to BU_b, to the SMN through the SG.
4) SMN sends a BA, response to BU_b, to the MN.
5) SMN sends a BU_b to the CNs of MN through the SG. The CNs create or update the MN's information as content of BU_b, $\{MN_{IP}, SMN_{IP}, COA_{SMN}, COA_{MN}, T_s, T_m\}$.

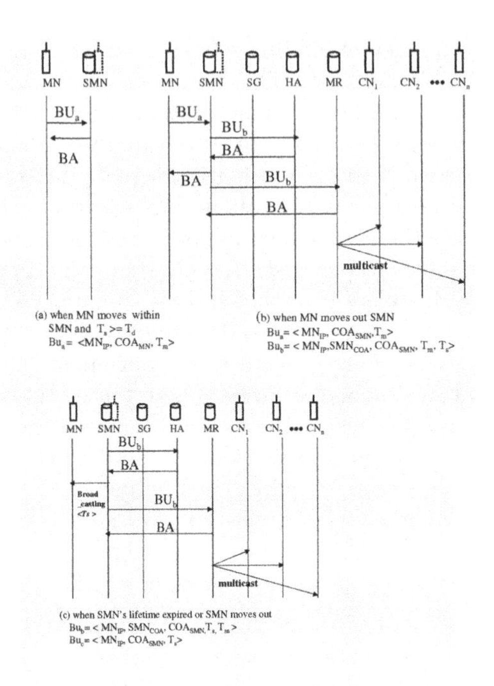

Fig. 3. BU Procedures of Proposed Scheme

Case 3. The expiration just before lifetime T_s of SMN or when SMN moves
1) SMN sends BU_b, BU messages, to the HA of SMN through SG. As described in section 3.2, BU message is included a flag, S.
2) HA updates or creates information as content of BU_b, $\{MN_{IP}$, SMN_{IP}, COA_{SMN}, COA_{MN}, T_s, $T_m\}$, and sends BA, response to BU_b, to the SMN through the SG.
3) SMN broadcasts information as content of BU_c,$\{SMN_{IP}$, COA_{SMN}, $T_s\}$ to all connected MNs.
4) MNs of the SMN updates information kept as contents of BU_c.
5) SMN sends BU_b to a CNs of SMN through the SG. The CNs create or update the MN's information as content of BU_b, $\{MN_{IP}$, SMN_{IP}, COA_{SMN}, COA_{MN}, T_s, $T_m\}$.

Figure 3 is simple representation of the above explanation as a picture. Similar to [16], we define the following parameters for location update in the rest of this paper:
C_{ms} : The transmission cost of location between the MN and the SMN.
C_{ng} :The transmission cost of location between the SMN and the SG.
C_{sh} :The transmission cost of location between the SG and the HA.
C_{hr} : The transmission cost of location between the HA and the MR.
C_{rc} : The transmission cost of location between the MR and the CN.
U_s : The processing cost of location update at the SMN.
U_g : The processing cost of location update at the SG.

U_h : The processing cost of location update at the HA.
U_r : The processing cost of location update at the MR.

Here, If the MN moves in the SMN, it is referred to $C_{mn\_in\_move}$, and if the MN moves out of the current SMN to another SMN, it is referred to $C_{mm\_out\_move}$, and if the SMN moves, it is referred to C_{smn}. The following computations can apply to each case for one time BU cost. Since usually the costs for a wireless communication network is higher than a wire communication network, the ratio is considered by using ρ to distinguish the difference between wire and wireless communication networks(ρ is dependent on the network, which is $\rho > 1$) The communication cost for a wire communication network is referred to δ, and the transmission cost for a wireless communication is referred to $\delta\rho$.

The multicast Listener Discovery(MLD) protocol[14] is used by IPv6 multicast routers to learn the existence of multicast group members on their connected links. The collected multicast group informantion is then provided to the multicast routing protocol. MLD is derived from version 2 of the Internet Group Management Protocol in IPv6. Therefore, we are considered by using γ the cost of multicast in our proposed scheme. Because the SMN can be mobile, a communication network between the SMN and the SG is assumed to be a wireless one. Therefore, C_{ms}, C_{sc}, and C_{sh} are the wireless communication networks, which incur most expensive communication costs. Because the number of hops or the distance between routers does not have much influence on the number of the BU messages or on costs, they are not considered in this paper. The following computations can apply to each case for one time BU cost.

$$C_{mn\_in\_move} = 2\delta\rho + U_s \quad (1)$$

$$C_{mn\_out\_move} = 8\delta\rho + U_s + U_g + U_h + U_r + \gamma \cdot \delta\rho \quad (2)$$

$$C_{smn} = 7\delta\rho + U_g + U_h + U_r + \gamma \cdot \delta\rho \quad (3)$$

3.3 Packet Delivery

When the HA receives pakets for the MN, it can directly send them to the SMN by using an IP-in-IP tunnel. The SMN forward them to the MN by using another IP-in-IP tunnel, with the routing header option reffering to the binding entry in the SMN. If a packet with a source address belonging to the SMN's prefix is received from the network, the SMN reverse-tunnels the packet to the HA through this tunnel. This reverse-tunneling is done by using IP-in-IP encapsulation[15]. The HA decapsulates this packet and forwards it to the CN. For traffic or originated by itself, the SMN can use either reverse tunneling or route optimization as specified in [2].

When a data packet is first sent by a CN to a MN, it gets routed to the HA which currently has the binding for the SMN. When the HA receives a data packet meant for a MN, it tunnels the packet to SMN's current COA. The SMN decapsulates the packet and forwarding to MN. Then, the CN can send packets directly to the MN.

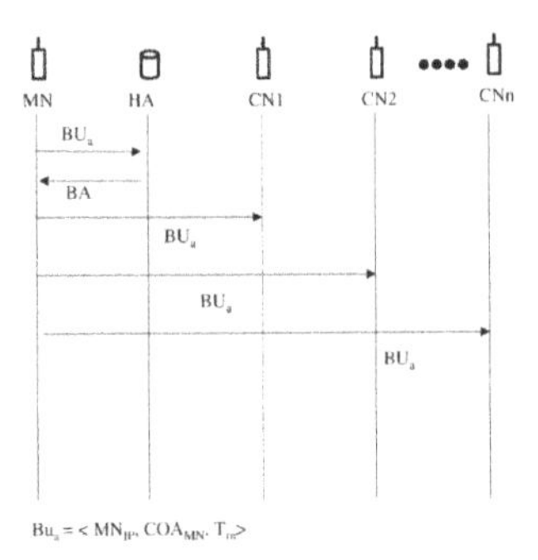

Fig. 4. BU Procedures of IETF Protocol

4 Performance Analysis

In this section, we make a performance analysis the proposed scheme in section 3. We compare the performance of Mobile IP and our proposed scheme.

4.1 An Analytical Model of the Binding Update Costs in IETF MIPv6

Costs for the binding update of mobile nodes incur when they move to another subnet and right before their lifetime expires. If the number of the CNs is referred to c, the binding update process by means of the MIPv6 protocol of IETF can be described as in Figure 4. As in the section 3.2, the following variables are used for computing costs for binding update.

C_{mh} : The transmission cost of location between the MN and the HA.

C_{mc} :The transmission cost of location between the MN and the CN.

U_h : The processing cost of location update at the HA.

U_c : The processing cost of location update at the CN.

Binding update costs for location registration when MN leaves its current location or right before expiration of its lifetime can be computed as follows:

$$C_{mn\_in\_move} = 2\delta\rho + 2 \cdot c \cdot \delta\rho + U_h + c \cdot U_c \tag{4}$$

In this paper, we assume that MN that has lifetime T_m moves on the path R to compute total cost C_{tot_mip} for binding update of MN on the MIPv6 protocol of IETF. The duration of time that MN stays in a location registration area is referred to a random variable, T.

Therefore, if an average of total costs for location update by the binding update messages that incur while MN stays in a location registration area is referred to C_{tot_mip}, it is shown as follows:

$$C_{tot\_mip} = (\lceil \frac{\frac{1}{\lambda}}{T_m} \rceil + 1) \cdot 2\delta\rho + 2 \cdot c \cdot \delta\rho + U_h + c \cdot U_c \tag{5}$$

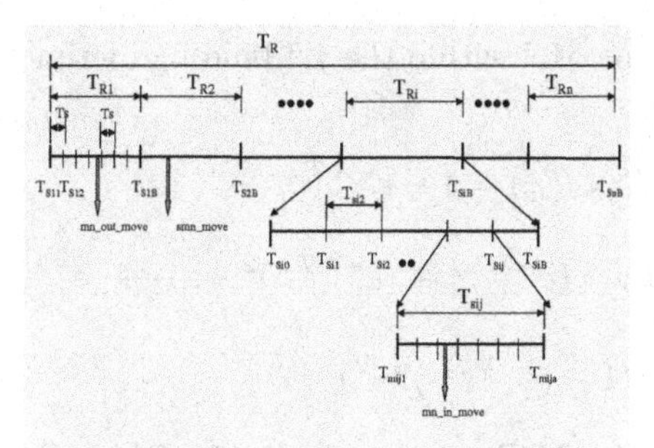

Fig. 5. The Relationship Between T_m,T_s and T_R

4.2 An Analytical Model of the Binding Update Costs in the Proposed Scheme

To analyze the binding update costs of the proposed method in this paper, it is assumed that MN moves on the path R that is divided into as many as n of location registration areas. If the staying time of MN on the ith path, R_i is shown as, T_R, staying time of MN, T_{R_i}, follows the exponential distribution that has the average $\frac{1}{\lambda}$. T_{R_i} is assumed. T_s represents the binding lifetime of SMN in a location registration area. T_m represents MN's binding lifetime. The relation with T_{R_i}, T_s, T_m is $T_{R_i} \geq T_s \geq T_m (1 \leq i \leq n)$, and in the case of $T_{R_i} < T_s < T_m$, it is excluded from the analytical model because no binding update occurs. $T_s = \alpha \cdot T_m$(provided that$\alpha \geq 1$), $T_{R_i} = \beta_i \cdot T_s$ (provided that $\beta_i \geq 1$) is assumed. $T_{s_{ij}}$ represents the jth T_s section of SMN in the path R_i. (provided that$j \geq 0$) That leads to $\sum_{i=0}^{n-1} T_{R_i} + T_s \leq T_{s_{ij}} < \sum_{i=0}^{n-1} T_{R_i} + j \cdot T_s$ represents the kth T_m section within $T_{s_{ij}}$, SMN's binding lifetime. That leads to $(k-1) \cdot T_m \leq T_{m_{ijk}} < k \cdot T_m$ within $T_{S_{ij}}$. The formula for time required for MN to move on the path can be shown as Figure 5, it becomes, $T_R = \sum_{i=1}^{n} T_{R_i}$, $T_{R_i} = \sum_{j=1}^{\beta_i} T_{s_{ij}}$, $T_{s_{ij}} = \sum_{k=1}^{\alpha} T_{m_{ijk}}$ assume that the initial state refers to time when MN in possession of binding lifetime T_s stays within a SMN on the path R_i for cost analysis. If MN's staying time, T_{R_i} on R_i is shown as random variable X, as assumed above, if X follows the exponential distribution that has the average $\frac{1}{\lambda}$, the Cumulative Density Function $F_i(t)$ for the probability that MN stays for the time less than within its moving area R_i is shown as follows:

$$F(x) = P(X \leq x) = \int_{x=0}^{x=T_{Ri}} \lambda \cdot e^{-\lambda x} dx \qquad (6)$$

An analysis model of binding update cost for MN's movement within the given SMN. If the probability that MN within the jth binding lifetime section of SMN on the path R_i stays in the kth binding lifetime section $(k \geq 1)$ is $P(k-1, k)$. Before MN that stays in the jth binding lifetime section of SMN exceeds the kth binding lifetime section on the path R_i, the frequency of the BU messages caused by MN becomes $k-1$. The average frequency of location update that is caused by MN within the jth binding lifetime section of SMN can be expressed as follows:

$$P(0,1) = P_r[0 \leq X < T_m] = 1 - e^{-\lambda \cdot T_m}$$

$$\sum_{k=1}^{\alpha-1} k \cdot P(k-1,k) = (e^{-\lambda \cdot T_m} + (e^{-\lambda \cdot T_m})^2 + \cdots + (e^{-\lambda \cdot T_m})^{\alpha-1}) \cdot P(0,1)$$
$$= e^{-\lambda \cdot T_m} \cdot (1 - (e^{-\lambda \cdot T_m})^{\alpha-2})$$

The BU costs of the MN's movement only within SMN can be shown as in Formula(1) of 3.2. Therefore, $E_{[mn\_in\_move]}$, the average total costs for MN's location update within only SMN are as follows:

$$E_{[mn\_in\_move]} = \sum_{k=1}^{\alpha-1} k \cdot P(k-1,k) \cdot BU\ cost = e^{-\lambda \cdot T_m} \cdot (1 - (e^{-\lambda \cdot T_m})^{\alpha-2}) \cdot C_{mn\_in\_move} \quad (7)$$

An analytical model of the binding update costs for moving to another SMN after staying within the given SMN. MN is assumed to be within the jth binding lifetime of SMN on the path R_i. Before MN within the jth binding lifetime section of SMN on the path R_i moves to another SMN that exceeds the kth binding lifetime, the frequency of BU messages caused due to its moving to another SMN becomes $k-1$. Therefore, the average frequency of location update caused by the movement of MN to another SMN within the kth binding lifetime section of the SMN can be expressed as follows:

$$\sum_{k=1}^{\alpha-1} k \cdot Q(j,k-1,k) = \sum_{k=1}^{\alpha-1} k \cdot e^{-\lambda \cdot (k-1) \cdot T_m} \cdot Q(j,k-1,k)$$
$$= e^{-\lambda \cdot T_s} \cdot (1 - (e^{-\lambda \cdot T_m})^{\alpha-2}) \quad (8)$$

When MN in possession of binding lifetime $T_m$within the jth binding lifetime section of SMN on the path R_i moves to another SMN on the path R_i , if the frequency of its movement is w, the frequency for $T_m$results in$r = \frac{w}{\frac{T_{R_i}}{T_m}}$ and because the assumption based on 4.2 leads to $T_{R_i} = \alpha \cdot beta_i \cdot T_m$, $r = \frac{w}{\alpha \cdot \beta_i}$, results. The binding update costs for MN caused by the movement of MN to another SMN within the kth binding lifetime section of SMN can be shown as in formula(2) of the section 3.2. The average total costs of location update $E_{[mn\_out\_move]}$ that are incurred because MN within the jth lifetime section of SMN moves to another SMN are as follows:

$$E_{[mn\_out\_move]} = \sum_{k=1}^{\alpha-1} Q(j,k-1,k) \cdot k \cdot frequency \cdot BU\ cost$$
$$= e^{-\lambda \cdot T_s} \cdot (1 - (e^{-\lambda \cdot T_m})^{\alpha-2}) \cdot \frac{w}{\alpha \cdot \beta_i} C_{mn\_out\_move} \quad (9)$$

An analytical model of the binding update costs for moving to another SMN after staying within the given SMN. When SMN moves from $j-1$ to j of binding lifetime T_s on the path R_i, the frequency of location update by the binding update messages of MN created in the movement becomes $j-1$. Therefore, the number of location update by the MN's binding update messages are as follows:

$$\sum_{j=0}^{\beta_i-1} j \cdot R(j-1,i) = e^{-\lambda \cdot T_s} \cdot (1-(e^{-\lambda \cdot T_s})^{\beta_i-2})$$

When SMN moves from $j-1$ to j of binding lifetimeT_s on the path R_i, the MN's binding costs incurred in this movement can be shown as in Formula(3) of the section 3.2. Therefore, the average total costs for location update by the MN's binding update messages created, $E_{[mn\_out\_move]}$ are as follows:

$$\begin{aligned} E_{[smn\_move]} &= \sum_{k=1}^{\beta_i-1} j \cdot R(j-1,j) \cdot BUcost \\ &= e^{-\lambda \cdot T_s} \cdot (1-(e^{-\lambda \cdot T_s})^{\beta_i-2}) \cdot C_{smn\_move} \end{aligned} \quad (10)$$

Accordingly, if the total average costs for location registration by the MN's binding update messages in the proposed method are referred to $E_{[TOT]}$

$$\begin{aligned} E_{[TOT]} &= e^{-\lambda \cdot T_m} \cdot (1-(e^{-\lambda \cdot T_m})^{\alpha-2}) \cdot C_{mn\_in\_move} \\ &+ e^{-\lambda \cdot T_s} \cdot (1-(e^{-\lambda \cdot T_m})^{\alpha-2}) \cdot \frac{\omega}{\alpha \cdot \beta_i} C_{mn\_out\_move} \\ &+ e^{-\lambda \cdot T_s} \cdot (1-(e^{-\lambda \cdot T_s})^{\beta_i-2}) \cdot C_{smn\_move} \end{aligned} \quad (11)$$

4.3 Comparison and Analysis for Binding Update Costs

We have analyzed the binding update costs of MN depending on the number of CNs, the number of MNs, the mobility of MN, and the progression of time, and compared them with IETF MIPv6. The the results were computed by MATLAB, and Table 2 lists some of the parameters used in our performance analysis[16].

Table 2. The Elements consisting of SMN

U_s	U_h	U_c	U_r	U_g	N	ρ	γ	α	β	M	C
15	25	10	10	10	10	10	260	2	1..10	15	5

In the Table 2, m represents the number of MNs, n represents the maximum number of paths on which MN moves, and c represents the maximum number of CNs, α and β represent the update frequency of MN and the sever moving nodes while MN stays within a SMN on a given path. Besides, ρ is a variable that distinguishes the difference in communication costs between wire and wireless communication networks.

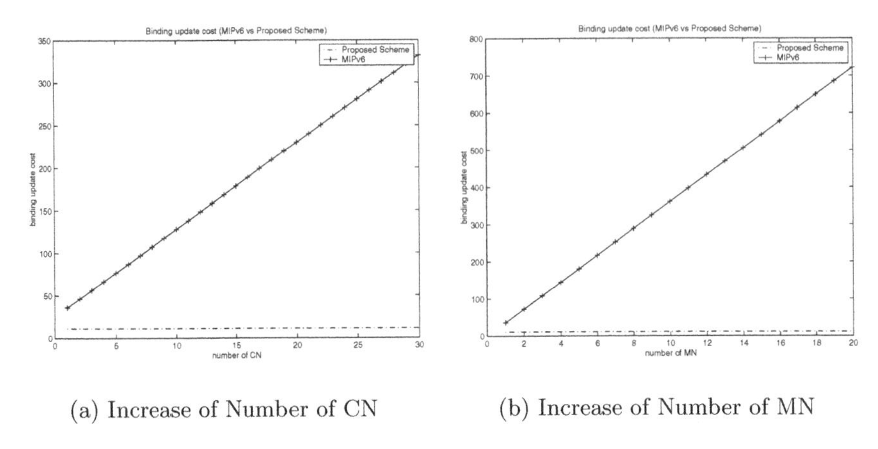

(a) Increase of Number of CN (b) Increase of Number of MN

Fig. 6. Binding Update Cost

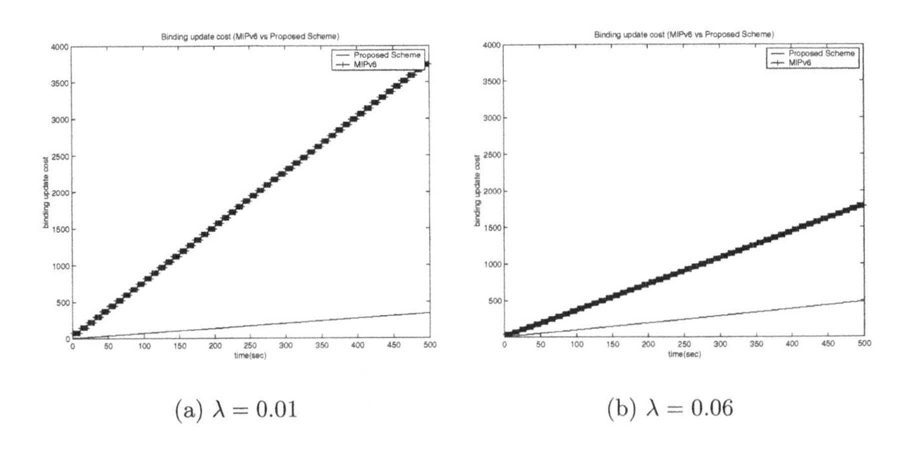

(a) $\lambda = 0.01$ (b) $\lambda = 0.06$

Fig. 7. Comparision of Binding Update costs According to Mobility

γ is a variable for multicast costs in network. On each link, one multicast router is elected to act as a querier. It periodically sends out a query onto this link. The default value for the query interval is 125 second. And the default value of the maximum response delay is 10 second. A group membership timer is using the default values for, $2 \cdot$ query + response delay, results in a default value 260 second[17].

U_s, U_h, U_r, U_g and U_r mean costs for location update performance from SMN, HA, and the multicast router, respectively. First, MN's binding update costs have been computed by increasing the number of CNs from 1 to 30.

If the multicast router is supported as in Figure 6 (a), the BU coast in our proposed scheme is not increase almost by the increased number of CNs. It shows a growing difference in the BU costs as the number of CNs increases, compared with the protocol of IETF MIPv6. The MN's binding update costs are also computed, as the number of MN connected to HA on the IETF MIPv6 protocol and the number of MN connected to SMN are increased up to 20. The BU coast in our proposed scheme is not increase almost by the increased number of MN as in Figure 6 (b). To analyze the MN's BU costs according to MN's mobility, λ of $\frac{1}{\lambda}$the average staying time of MN is changed for the MN's BU costs in the cases of 0.01 and 0.06. As is smaller, MN's mobility becomes frequent. When compared with IETF MIPv6 as in Figure 7 (a) and (b), it is known from frequent MN's movement and the section of $\lambda = 0.01$ that our proposed method is more efficient. Besides, when time for MN is increased, it is shown that the BU costs are not increase almost.

5 Conclusion and Future Work

This paper adds SMN that has mobility on the IETF MIPv6 protocol, and considers two lifetimes.(i.e., MN's lifetime and SMN's lifetime)

Presented in this paper is a technique that can minimize costs for MN's binding update by applying the multicast method with the multicast router for CNs. Also, we propose the location update methods about each case which BU of MN happens, we calculate location update costs to be required in each cases. Presented in this paper is a technique that can minimize costs for MN's BU by applying the multicast method with the multicast router for CN and adding the SMN. And we calculate location update costs each technique of IETF MIPv6 and our proposed method and do the comparison analysis. Our proposed method is maintained without taking the influence almost by the increased number of MNs. If the staying time of MN follows the exponential distribution that has the average $\frac{1}{\lambda}$, as λ is smaller, MN's mobility becomes frequent. When compared with IETF MIPv6, it is know from frequent MN's movement that our proposed method is more efficient.

Consequently our proposed method, when the number of MN increases, and when MN has frequent mobility, is efficient.

References

1. The Internet Engineering Task Force
2. D. B. Johnson and C. Perkins, "Mobility Support in Pv6", IETF(International Engineering Task Force) Internet Draft, draft-ietf-mobileip-ipv6-24.txt, June 30, 2003
3. H.soliman, C.Castelluccia, K. El-malki, and L. Beller, "Hierarchical MIPv6 mobility management(HMIPv6)", dfraft-ietf-mobileip-hmipv6-06.txt, July 2002
4. C.Castelluccia, "HMIPv6: A hierachical mobile ipv6 proposal", ACM Mobile Computing and Communication Review, vol. 4, no. 1, pp.48-59, Jan. 2000
5. C. E. Perkins, "Mobile-IP Local Registration with Hierarchical Foreign Agents", Internet Draft-perkins-mobileip-hierfa-00.txt, Feb. 1996.

6. Hideaki YUMIBA, Koji SASDA, and MAsami YABUSAKI, "Concatenated Location MAnagement", IEICE Transaction Communication Vol. E85-B, No. 10, pp. 2083-2089, 2002
7. Thierry Ernst, Alexis Olivereau, Ludovic Bellier, Claude Castelluccia, Hong-Yon Latch, "Mobile Networks Support in Mobile IPv6(Prefix Scope Binding Updates)", Internet-Draft draft-ernst-mobileip-v6-network-03.txt, Internet Engineering Task Force(IETF), March 2002
8. Vijay Devarapalli, Ryuji Wakikawa, Alexandru Petrescu and Pascal Thubert, "Network Mobility(Nemo) BAsic Support Protocol", IETF Internet draft, dfraft-ietf-nemo-basic-support-02.txt, Dec. 2003
9. Levon Esibov, Bernard Aboda, Dave Thaler, "Linklocal Multicast Name Resolution(LLMNR)", Internet-Draft draft-ietf-dnsext-mdns-30.txt, DNSEXT Working Group Internet Engineering Task Force(IETF), 17 March 2004
10. B.Haberman, J.Martin,"Multicast Router Discovery", MAGMA Working Group IETF Internet draft, dfraft-ietf-magma-mrdisc-00.tat, Feb. 2004
11. R. Vida, L.Costa, "Multicast Listener Discovery(MLDv2) for IPv6", Internet-Draft draft-vida-mld-v2-08.txt, Internet Engineering Task Force(IETF), June 2004
12. George Xylomenos, George C.Polyzos, "IP Multicast for Mpbile Hosts", IEEE Communication Magazine, Vol 35, Issue 1, pp. 54-58, Jan. 1997
13. RalphDroms, C. Perkins, Jim Bound, M. Carney, "Dynamaic Host Configuration Protocol for IPv6", INTERNET-DRAFT,¡draft-ietf-dhc-dhcpv6-20.txt¿, September 2001.
14. S.Deering, W.Fenner, and B.Haberman, "Multicast Listener Discovery(MLD) for IPv6", RFC 2710. Oct. 1999
15. A.Conta and S. Deering, "Generic Packet Tunneling in IPv6 Specification", RFC 2473, IETF December 1998
16. Jiang Xie, Lan F. Akyildiz, "A Novel Distributed Dynamic Location Management for Minimizing Signaling Coasts in Mobile IP", IEEE Transaction on Mobile Computing vol, 1, No. 3, pp. 163-175, 2002
17. Christian Bettstetter, Anton Riedl, Gerhard GeBler, "interoperation of Mobile IPv6 and Protocol Independent Multicast Dense Mode"

A Unified Route Optimization Scheme for Network Mobility*

Jongkeun Na[1], Jaehyuk Choi[1], Seongho Cho[1], Chongkwon Kim[1], Sungjin Lee[2], Hyunjeong Kang[2], and Changhoi Koo[2]

[1] School of Electrical Engineering and Computer Science, Seoul National University, Seoul, Republic of Korea
{jkna,jhchoi,shcho,ckim}@popeye.snu.ac.kr

[2] Telecomunication R&D Center, Samsung Electronics, Suwon, Republic of Korea
{steve.lee,hyunjeong.kang,chkoo}@samsung.com

Abstract. Recently Network Mobility(NEMO) is being concerned as new mobility issue. Lots of NEMO issues are already being touched in IETF NEMO WG but the solution is still premature especially to the Route Optimization (RO). NEMO has several problem spaces that need RO such as nested tunnels problem. Unfortunately, there is no solution that can be universally applied as one for all that results in supporting the coherent network mobility. In this paper, we propose a unified route optimization scheme that can solve several types of RO problem by using Path Control Header (PCH). In our scheme, Home Agent (HA) does piggyback the PCH on the packet which is reversely forwarded from Mobile Router (MR). That enables any PCH-aware routing facility on the route to make a RO tunnel with MR using the Care-of address of MR contained in the PCH. By applying to some already known NEMO RO problems, we show that our scheme can incrementally optimize the routes via default HA-MR tunnel through the simple PCH interpretation.

1 Introduction

Along with the proliferation of mobile communication networks such as wireless LAN (Local Area Network), PAN (Personal Area Network) and CAN (Car Area Network), most of public transportation systems (e.g. bus, train, airplane) are envisioned to have a permanent connectivity to the Internet even while moving around. In these communication environments, the new mobility problem occurs due to the mobile networks in which the network itself is moving entirely. By now, the literature has only considered how to support host mobility. However, we have to also consider about network mobility because lots of small or medium sized networks require the mobile behavior as same as the current mobile node has in the coming future.

---

* This work was supported by Samsung Electronics and the Brain Korea 21 Project.

I. Niemegeers and S. Heemstra de Groot (Eds.): PWC 2004, LNCS 3260, pp. 29–38, 2004.

The existing Mobile IP [2] solution cannot provide the network mobility because it has different characteristics in comparison with the problem of supporting host mobility. With the need of new protocol for network mobility, a basic protocol was proposed in [1]. It supports transparent mobility to every node in mobile network by using a bi-directional tunnel between the Mobile Router (MR) and the Home Agent (HA). As a protocol extending Mobile IPv6, the MR registers its network prefix as well as its Care-Of Address (CoA) through the extended binding update (a.k.a Prefix Scoped Binding Update [6]) so that HA can properly intercept and tunnel the packets whose destination address belongs to the mobile network prefix to MR's CoA. Basically, this basic protocol can be accepted as a complete network mobility support protocol if we can ignore the routing inefficiency inherently inherited by the bi-directional tunneling. However, the routing efficiency could be more important metric in supporting network mobility than in host mobility. A mobile network which consists of several nodes can consume more link bandwidth and routing resource along the path via HA in IP routing infrastructure than a single mobile node. Therefore, the route optimization for the efficient IP routing must be considered in this literature along with NEtwork MObility (NEMO) basic protocol [7].

There are some efforts for Route Optimization (RO). For RO in IP routing infrastructure, Some approaches such as [3–5] require a special router or the extension of the existing router which can handle the packet redirection to gain RO effect. The RO schemes belong to this category can be applied to both Mobile IP and NEMO in IP routing infrastructure. On the other hand, There are other kinds of the NEMO-specific RO problem. [8] well defines RO problem spaces of NEMO and briefly analyzes the proposed interim solutions such as [9, 10]. Typically, one of NEMO-specific RO problem is a nested tunnels problem that can be formed due to the network mobility. Most of proposed solutions are for solving that problem. As of now, it's not easy to say how RO problems in NEMO can be best solved in the reasonable manner. However, the sure thing is that current proposed solutions can be applied only to one problem space of RO. That is an uncomfortable and unnatural facet in supporting coherent network mobility. We need a simple and effective, unified route optimization solution for network mobility.

In this paper, we propose a route optimization scheme based on Path Control Header (PCH) piggybacking by HA. The scheme is a unified solution that can solve several types of route optimization problem with applying the same principle to the routing facilities such as HA, MR and Correspondent Router (CR), e.g. ORC router in [5]. In the proposed scheme, HA does piggyback the PCH on the packet which is reversely forwarded from MR through the bi-directional MR-HA tunnel. PCH is a hop-by-hop option header so that it can be processed by all of the routing facilities on the path that is from HA to Correspondent Node (CN). HA forwards the PCH piggybacked packets toward CN for the route optimization. CR on the path can make a RO tunnel with MR using the information like the CoA of MR contained in the PCH.

Our proposed RO scheme, PCH piggybacking by HA, is a simple and effective one in solving the problems of route optimization without any incompatibility

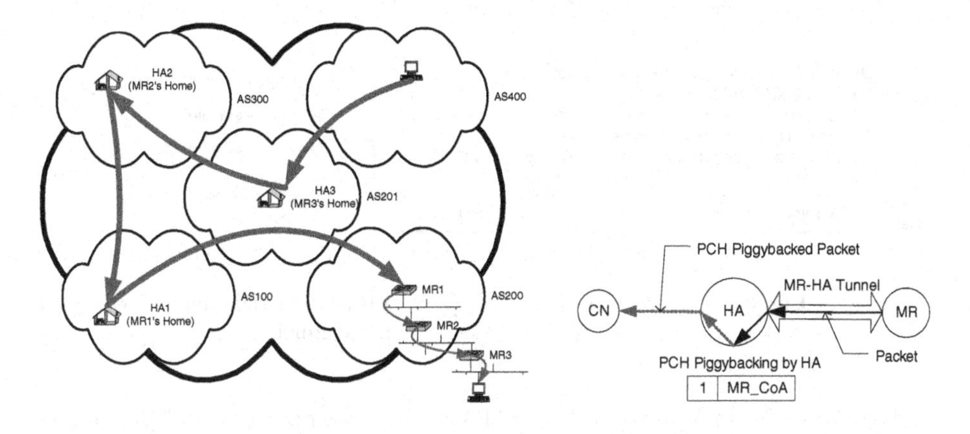

Fig. 1. Nested tunnels optimization in nested mobile networks

Fig. 2. PCH piggybacking by HA

with the basic NEMO protocol [7]. By taking the functional extension of routing facilities such as HA, MR and CR, we can incrementally optimize the routes over CN-HA-MR without the loss of transparency to CN. And also, we expect that the basic concept of this scheme can be used to support other mobility-related route optimizations as a unified solution, not limited to network mobility.

The rest of this paper is organized as follows: the related works is mentioned in Section 2. In Section 3, we describes the basic operation of PCH based RO scheme. Then, we show how to apply our scheme on RO problem spaces in Section 4. Finally, we conclude in Section 5.

2 Related Work

There are some types of route optimization problems. In particular, we can summarize two problems related to NEMO. One is to the route optimization in IP routing infrastructure. The other is to nested tunnels optimization in nested mobile networks. In the former case, CR based approach was introduced in [5]. In there, CR provides the same service to all of CN behind it as if MR supports the transparent mobility service to nodes behind it. This approach can reduce the overhead of applying Mobile IPv6 route optimization to each CN because all of CN behind CR can share the optimized tunnel, i.e. RO tunnel, between MR and CR. Also, for the packets reversely forwarded from MR to any CN behind CR, they can be passed to the optimized tunnel, not to the default tunnel between MR and HA.

In the latter case, it is another type of route optimization problem in NEMO. If multiple mobile networks are nested as Fig.1, that brings a routing overhead to us which is well known as "pinball" or "dog-leg" routing (for the details, see [8]). The packets sent from CN to LFN (Local Fixed Node) get follow the routing path like CN→HA3→HA2→HA1→MR1→MR2→MR3→LFN by IP routing

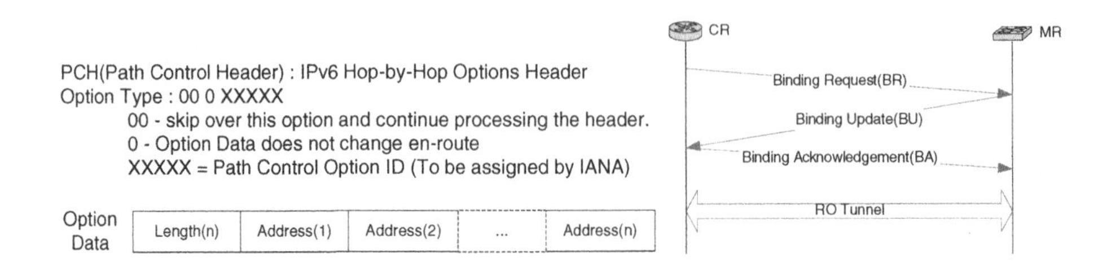

Fig. 3. PCH option format

Fig. 4. The signaling procedure for RO Tunnel

and the basic NEMO protocol. In NEMO context, we need to avoid the nested tunneling like this because it incurs very inefficient routing depending on the relative location of HAs. Currently several interim solutions are being proposed in [1] that conceptually convert nested tunnels into a flat tunnel. For the details of NEMO, problem statements and related work can be referred in [1, 7–10].

As a result of looking into the route optimization for NEMO, first, we need to devise a solution to allow the route optimization without the limitations such as CN-aware, not scalable, load imbalanced, insecure. Second, we need a unified route optimization scheme that can solve most of problem types related to NEMO.

3 PCH (Path Control Header) Scheme

In this section, we introduce the basic concept and operation of our proposed RO scheme in NEMO context.

3.1 PCH Piggybacking by HA

To route optimization, HA does piggyback PCH on the packet which is reversely forwarded from MR through a bi-directional MR-HA tunnel. PCH is a hop-by-hop option header so that it can be processed by all of the routing facilities on the path that is from HA to CN. The mentioned routing facility means an entity which can play a role of the transparent routing agent that can support the packet redirection service like HA. The router in the Internet that implements such an agent function provides the packet redirection service to the nodes behind it by intercepting the packets sent from them and redirecting to the RO tunnel. We call it a CR in here if it is functioning for CNs. The RO tunnel between CR and MR can be established when CR gets know the existence of HA by processing the packet with PCH.

In Fig.2, HA de-capsulates the encapsulated packet forwarded from MR via MR-HA tunnel and then forwards the PCH piggybacked packet to CN for the route optimization. Any existing CR on the path from HA to CN can catch the path control information as examining PCH in the packet. Therefore, the

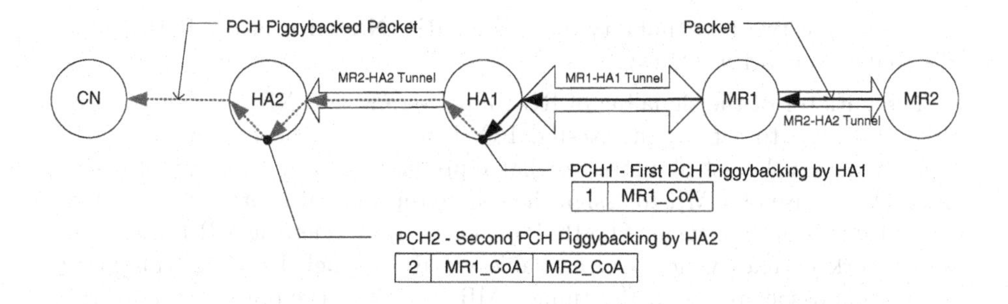

Fig. 5. Nested PCH piggybacking by HAs

CR can initiate the procedure of making a RO tunnel between itself and MR using MR's CoA which is contained in PCH. After setting up the RO tunnel, the packets of CN will be redirected to the RO tunnel at CR.

This scheme is simple and effective in respect of RO. It only requires a little effort of HA to provide the RO tunnel between CR and MR. HA does PCH piggybacking on the packet which is following a non-optimized path of MR-HA tunnel. In here, we can say that CR may be an access router that providing the routing service for a few of subnets or a border router that runs BGP routing protocol in one AS [8].

Fig.3 shows the structure of PCH. PCH includes an address information as an option data. In here, the address information represents the list of IPv6 addresses. The address contained in PCH indicates the CoA of MR in MR-HA relationship. Through PCH, CR gets know the CoA of MR so that CR can initiate the signaling for RO tunnel.

In Fig.5 that shows the case of forming nested tunnels, PCH gets contain two CoAs, each of MR1 and MR2. HA2 gets to know the fact that its MR2-HA2 tunnel is nested under the outer MR1-HA1 tunnel after taking a look at the packet with PCH1. The nested HA just adds the CoA of its MR on the received PCH to make its PCH. Then, HA2 does piggyback PCH which includes the CoA of MR1 (i.e. the exit point of the outer tunnel) and the CoA of MR2 (i.e. the exit point of its tunnel). In this case, one CR on the path between HA2 and CN will be able to make RO tunnel with MR2 by using the nested address information carried in PCH.

3.2 Making a Route Optimization (RO) Tunnel

The CR can make a RO tunnel after getting the piggybacked PCH from HA. The signaling to construct a RO tunnel between CR and MR is done with 3-way handshake as in Fig.4. The messages defined in here are carried by Mobility Header defined in [2]. We define new message called BR (Binding Request) to notify MR of the need of RO tunnel. BU (Binding Update) and BA (Binding Acknowledgement) are used for the same purpose as defined in [2] and [7]. And also, we define two new mobility options : NRP (Nested Routing Path), RNP (Reachable Network Prefixes).

The initiator of the signaling of RO tunnel should add NRP mobility option in BR message to set up the Nested RO Tunnel with the nested MR. NRP option contains the list of addresses that represents the tree topology of nested MRs. That is used for MR to assign the source routing path that is necessary to nested tunnels optimization. The RNP option is used to let the MR know about the network prefixes which are reachable via RO tunnel. By using this prefix information associated with RO tunnel, MR can select the optimized path (i.e. RO tunnel) for the out-going packets. This option should be contained in BA message.

If the 3-way handshaking RO signaling between MR and CR is done with success, the routing table of both includes new entry for directly reachable prefixes via RO tunnel. By referring that entry, MR can forward the packets to the established RO tunnel because they are destined to the network that is reachable via it. The CR can do the same thing for the prefix of mobile network that is bound through BU from MR. CR intercepts the packets destined to the prefix and redirects them to the RO tunnel.

3.3 Extensions

For route optimization, MR should understand BR message sent from routing facilities such as CR. According to [2], MR must maintain Binding Update List (BU List). In managing BU List, the following information must be maintained additionally to use RO tunnel defined in this proposed solution. The successful establishment of RO tunnel allows the ready of RO-enabled tunnel interface that would be associated with the correspondent entry of BU List. That tunnel interface should be setup to add IPv6 RH0 (Routing Header Type 0) optional header at the encapsulation of tunneled packets if the NROT (Nested Route Optimization Tunnel) flag is set. The reason why it should do will be explained in Section 4.3. And, MR should maintain the RO tunnels in its own context. In other words, MR can tear down less necessary RO tunnels according to its own criterion such as Least Recently Used (LRU) in the case of resource shortage.

For route optimization, HA should maintain the state of PCH piggybacking for per traffic flow. The traffic flow can be classified by the destination address of the packets. HA does piggyback PCH on one packet per the traffic flow. The piggybacking state should be managed by the soft-state. The piggybacking state of per traffic flow comes to be set when the first packet is piggybacked and reset when the state timer is expired. HA doesn't need to piggyback PCH on the packets belong the traffic flow while the correspondent piggybacking state is set. The overhead of managing the piggybacking state can be minimized by the careful implementation.

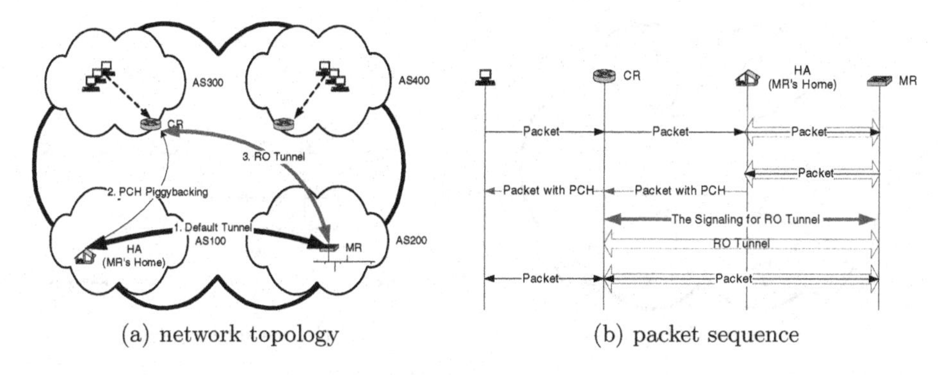

(a) network topology (b) packet sequence

Fig. 6. Route Optimization by single CR

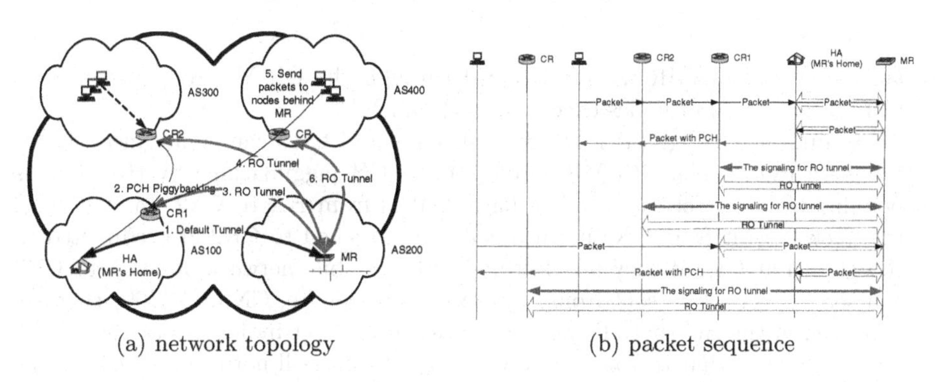

(a) network topology (b) packet sequence

Fig. 7. Route Optimization by multiple CRs

4 Route Optimization Using PCH

In this section, we illustrate how to apply our RO scheme on several types of route optimization in NEMO context. For easy understanding, each of route optimization procedures is described together with network configuration and message flow diagram.

4.1 Route Optimization by CR

Depending on the location of CR and how many CRs on the path between CN and HA, the various types of route optimization can be reflected. In here, we typically show two cases that expose the effect of PCH based RO. First, we assume the network configuration like Fig.6(a). There is one CR on HA-CN path in the border side of AS in which CN exists. In this case, all of other CNs belong the AS can get the gains of route optimization through CR-MR RO tunnel that is pre-established by the PCH piggybacked packet forwarded from HA to CN. Fig.6(b) is showing the procedure of RO tunnel establishment between CR and MR. Once established, the real communication between CNs behind CR

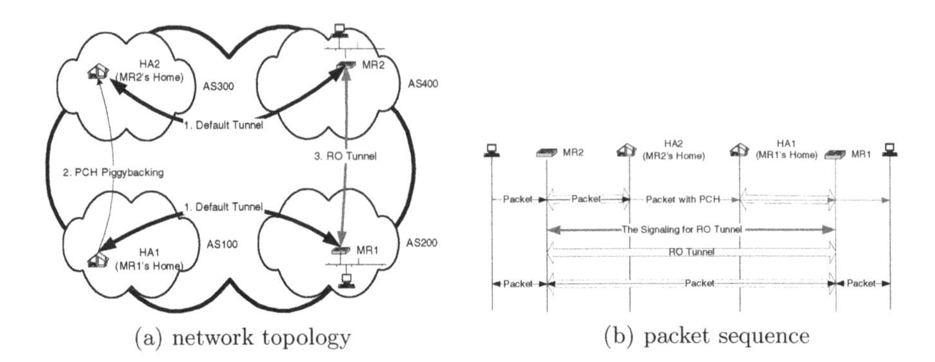

(a) network topology (b) packet sequence

Fig. 8. Route Optimization over MR-to-MR

and nodes behind MR will be realized through the CR-MR RO Tunnel. It is transparent to all of nodes except for CR and MR.

Second, as in Fig.7(a) and Fig.7(b), CR1 and CR2 can simultaneously establish a RO tunnel with MR through one PCH piggybacking by HA. This is possible because both are on the path that is from HA to CN2. In that case, the packets sent from CNs in all of subnets attached to CR2 are redirected to RO tunnel at CR2 if they are destined to the mobile network of the MR. CR1 can serve the packets sent from any CNs (in the figure, CN in AS400) that are scattered in the Internet. The packets reached on CR1 indicate that there is no CR in the path that is from CN to CR1, or CR but still not received PCH. The packets from CN are redirected at CR1 and, reversely the packets from MR are forwarded via HA. At the next time, the CR on the CR1-CN path can make a RO tunnel by picking up on PCH in the reversely forwarded packet from HA. As a result of PCH piggybacking by HA, we can serve the incremental route optimization to all of CNs.

4.2 Route Optimization over MR-to-MR

As in Fig.8(a) and Fig.8(b), we can get the RO tunnel over MR-to-MR by using PCH piggybacking. MR per se interprets PCH piggybacked from the HA of the other MR and initiates the signaling for RO tunnel with the other MR. As a result of that, the nodes behind one MR can directly communicate with the nodes behind the other MR without any routing overhead.

4.3 Nested Tunnels Optimization (NTO)

Our scheme can also be applied to solve the nested tunnels problem without the loss of generality. We assume the 3-level nested network configuration as Fig.9(a) to show NTO using PCH based scheme. In nested mobile networks, the RO tunnel is called NROT (Nested RO Tunnel) because we introduce the source routing concept in handling the nested tunnels optimization. To the correct

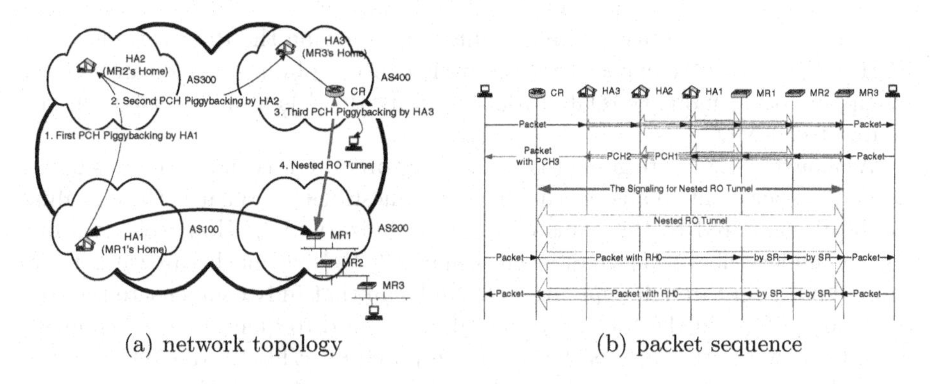

(a) network topology (b) packet sequence

Fig. 9. Nested Tunnels Optimization by CR

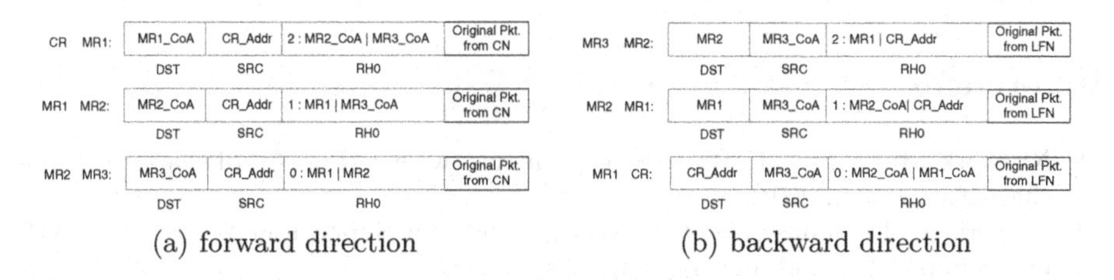

(a) forward direction (b) backward direction

Fig. 10. Packet Delivery via Nested Route Optimization Tunnel

routing in the nested network configuration, we take advantage of IPv6 Routing Header Type 0 (RH0) in NROT.

In Fig.9(a), CR gets to know the existence of nested tunnels through PCH information (MR1's CoA and MR2's CoA, MR3's CoA) and then initiate the signaling for NROT to MR3 via nested tunnels. At this time, the Binding Request (BR) message contains the NRP Option. The NRP Option is used to inform MR3 of the nested path information. If MR3 receives the BR message having the NRP option, MR3 also gets know that it is nested. Therefore, the tunnel between CR and MR3 becomes a NROT.

In a NROT, the entry point of tunnel adds RH0 at encapsulation. Reversely, the exit point of tunnel deletes RH0 at decapsulation. For the packets tunneled from CR to MR3, the packet forwarding is done with source routing of RH0 (MR1→MR2→MR3). For the packets tunneled from MR3 to CR, the reverse source routing (MR2→MR1→CR) occurs. Fig.9(b) shows message flow for NTO by CR. Fig.10(a) and Fig.10(b) show the content of RH0 packet at the packet delivery via NROT.

5 Conclusion and Future Work

As a unified solution for NEMO RO, we introduced the concept and basic operations of our proposed scheme which implemented by PCH Piggybacking in the HA. We proved that the proposed scheme can be used to solve most of the RO problems defined in [8] as a unified solution by showing the RO cases based on PCH in Section 4. We expect that the basic concept of our scheme can be used to support other mobility-related route optimizations as a unified solution, not limited to NEMO.

As a next step, we are going to evaluate our proposed scheme through simulation and experiment to answer the following questions. 1) What is the signaling overhead compared to the gains of route optimization? 2) How well does our scheme under the various handover scenarios? 3) How much can we get the throughput gain through applying RO? And, can that fully compensate the cost of installing CRs in the Internet? And also, we need to quantitatively compare our scheme with other approaches being proposed in IETF NEMO WG or other research area. Lastly, we also leave the detailed security consideration into the future work.

References

1. NEtwork MObility: NEMO web page, http://www.ietf.org/html.charters/nemo-charter.html, Jan. 2004.
2. C. Perkins, D. Johnson and J. Arkko, "Mobility Support in IPv6," draft-ietf-mobileip-pv6-18 (work in progress), July 2002.
3. Fumio Teraoka, Keisuke Uehara, Hideki Sunahara, and Jun Murai, "VIP : A Protocol Providing Host Mobility," Aug. 1994
4. Weidong Chen, Eric Lin, "Route Optimization and Location Updates for Mobile Hosts," International Conference on Distributed Computing Systems,1996
5. Ryuji Wakikawa, Susumu Koshiba, Keisuke Uehara, Jun Murai, "ORC: Optimized Route Cache Management Protocol for Network Mobility," Proc. of ICT2003, Nov 2003.
6. Ernst, T., Castelluccia, C., Bellier, L., Lach, H. and A. Olivereau, "Mobile Networks Support in Mobile IPv6 (Prefix Scope Binding Updates)," IETF draft-ernst-mobileip-v6-network-03 (work in progress), March 2002.
7. Vijay, D., Ryuji, W., Alexandru, P., Pascal, T.,"NEMO Basic Support Protocol," IETF draft-ietf-nemo-basic-support-02(work in process), December 2003.
8. Thubert, P., and Molteni, M., "Taxonomy of Route Optimization Models in the NEMO Context," IETF draft-thubert-nemo-ro-taxonomy-02 (work in progress), February 2004.
9. Thubert, P., and Molteni, M., "IPv6 Reverse Routing Header and Its Application to Mobile Networks," IETF draft-thubert-nemo-reverse-routing-header-04 (work in progress), February 2004.
10. Chan-Wah Ng, and Takeshi Tanaka, "Securing Nested Tunnels Optimization with Access Router Option," IETF draft-ng-nemo-access-router-option-00 (work in progress), Oct. 2002.

Paging Load Congestion Control in Cellular Mobile Networks

Ioannis Z. Koukoutsidis

School of Electrical and Computer Engineering,
National Technical University of Athens,
Zografou 157 73, Greece
koukou@telecom.ntua.gr

Abstract. This paper addresses the requirement for a congestion control mechanism, in order to efficiently handle paging traffic over cellular networks. The standard simultaneous paging approach can easily result in congestion, for relatively small values of the offered load. Instead of differentiating paging capacity based on the incoming load, this approach is more oriented towards a limited-resource system. Considering that a paging channel usually experiences high and low utilization periods, it is proposed to differentiate the paging mechanism according to the processed load. Given medium or even mild overload conditions, several forms of sequential paging substantially decrease blocking probabilities, while presenting good delay behavior. Based on a queueing analysis, we are able to quantitatively estimate potential improvements, and point out the basic points of such an adaptive scheme.

1 Introduction

In mobile communication networks, the need for expanded system capacity grows with the number of users and the amount of information required for a given service. However, the increase in demand always seems to be one step ahead of our capability to satisfy it. In view of this, congestion problems are often inevitable, especially in the wireless channels where capacity is even more scarce. Today, the increased user population densities and emerging bandwidth-consuming technologies make the balance even more unfavorable. Since bandwidth scarceness is a reality, new methods are essential to handle congestion problems.

Apart from the data traffic case, signaling congestion is a significant part of the overall problem. As communication technology becomes more complex, signaling plays a continuously augmenting role in establishing and maintaining connectivity. A large part of the signaling traffic in wireless networks is due to the location management operations [1]. These involve the *location update* and *paging* procedures, which are necessary to track user location in the presence of mobility. Location update is essentially a reporting mechanism, by which a moving subscriber informs network databases of its approximate position within a fixed or dynamic *location area* [2]. On the other hand, paging is a search procedure by which the exact cell where a user currently resides is retrieved. The

I. Niemegeers and S. Heemstra de Groot (Eds.): PWC 2004, LNCS 3260, pp. 39–53, 2004.

paging mechanism involves sending a *page request* (PR) message over a forward channel in all cells where the user is likely to be present, as indicated approximately by the location update procedure. Albeit location update and paging are antagonizing procedures, they work complimentarily to each other to provide a location management service. In absolute terms, location update is a more prolonged and costly procedure, however it is possible to view paging as the fundamental operation for call establishment [3], since, simply stated, the principal and foremost goal of location management is to retrieve the serving base station (BS) of the moving subscriber. Paging can also be considered more fundamental with respect to congestion problems since its largest part is associated with the wireless interface, which suffers from lack of resources.

Paging load control involves the critical issue of handling *mass* paging requests for different users. Large paging loads can cause severe congestion problems in system queues (switches, controllers, transceivers, etc.), which may lead to large delays or blocking of incoming call requests. In this paper, we study the congestion problem in paging channels within a cellular network, in presence of mobility. We focus mainly on the so-called 'control center'[1] of the network which distributes page requests, and the congestion problem that exists subsequently in BS channels. After studying the problem, we proceed to formulate a new paging load control method which dynamically adjusts the paging mechanism depending on the offered load in the system.

The rest of the paper is succinctly organized in two parts. The first part addresses the congestion problem and its associated parameters. The two basic mechanisms, *blanket* and *sequential* paging are discussed and a queueing analysis is presented with finite buffer capacity. Subsequent numerical results clarify the problem characteristics and lay the ground for the formulation of a paging load control method. The second part of the paper is devoted to devise the new method and its implementation characteristics. Finally, the paper ends with a discussion of the most important issues and a recapitulation of the major contribution.

2 Blanket and Sequential Paging: A Congestion Perspective

Simultaneous polling of cells in a location area is currently used to track user position. This procedure has been eloquently named *blanket paging* (BP), since it covers every possible cell where the user might be located at once. However, BP is also associated with high signaling cost and responsible for the majority of congestion problems. The flooding of downlink broadcast channels with paging messages and the apparent redundancy that inheres can lead to blocking of new calls and long delays at queues, especially at peak traffic periods. In all cases, it is possible to calculate the paging channel bandwidth in order to achieve certain quality constraints [4]. Still, it is equally challenging to reduce congestion problems in limited-capacity systems.

[1] In current systems, this is the Mobile Switching Center (MSC).

Almost all research on improved paging techniques has focused on some form of *sequential paging* (SP) [5]. Sequential paging attempts to reduce the existing redundancy by polling locations separately, instead of simultaneously. To minimize the associated paging cost, cells should be queried in order of decreasing location probability [6]. Sequential paging also alleviates the problem of congestion by reducing the mean number of page requests to be handled by a base station. On the downside, the major drawback is the introduced delay in establishing a call connection by performing successive steps, at each step waiting for a certain period until the system perceives if the user resides in the selected cell or not. Bounds on delay lead to a constrained optimization problem, which has been solved efficiently in many essays [5],[6],[7],[8]. The essence of tackling delays lies in forming sets of cells which are queried simultaneously at each step, otherwise called *sequential group paging* (SGP). It can be understood that SGP is an intermediate approach between BP and SP, yielding mean values of total cost C and delay D, so that

$$C_{SP} < C_{SGP} < C_{BP},$$
$$D_{SP} > D_{SGP} > D_{BP}.$$

Efforts to view and understand the queueing aspects of paging were made in [9],[10]. In similar approaches there, base station channels were modelled as $M/M/1$ queues in a system where PRs are distributed by a central control to the appropriate base stations. It was shown that sequential paging distributes the load more evenly to BS queues, which under very high loads can also reduce the overall delay in call establishment. The reduction in delay is due to the fact that PRs experience less time waiting to be served, which benefit outweighs the increase in delay caused by sequential paging. In other words, paging is performed *simultaneously* for *different* subscribers in the *same* time slot in *different* base stations, where of course there is a probability of locating a user. The authors conclude that when paging channels are not heavily loaded, flooding results in the shortest delay; however, when paging channels face congestion, flooding places a high volume of messages on these channels, which results in very high queueing delays. From our perspective, since congestion also entails the notion of delay, there is no clear distinction between the loss of performance attributed to a specific paging mechanism. The objective should be to increase the effective rate of incoming messages, while eliminating symptoms of congestion that lead to a degradation in system performance.

The analytical approaches in [9],[10] are similar and focus on upbringing the issue of delay, assuming infinite capacity queues. The reduction in congestion is only suggested indirectly by the reduction in the effective paging load to be accommodated in each queue. In an effort to quantitatively show the relationship, we adopt in the following a more realistic model with finite capacity queues and blocking. The acquired results are valuable in designing an efficient paging load control method.

3 Queueing Analysis

We assume that page requests are distributed to BSs by a central control, as shown in Fig. 1. Base stations have finite buffers in which to store messages. Without significant loss of generality, a single broadcast channel can be dedicated to each BS. Thus, an $M/M/1/K$ queue is proposed as a representation of the real-life system.

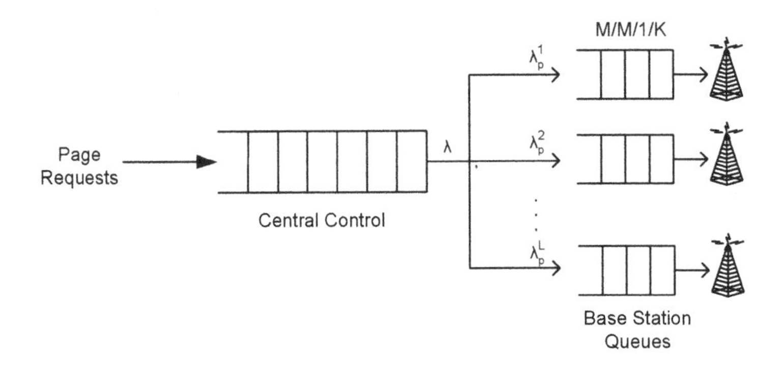

Fig. 1. Queueing system representation of page requests

The following basic notation is used in the analysis:
λ: aggregate PR arrival rate
λ_p: individual PR arrival rate at each BS
μ: service rate at each BS
K: total storage capacity (queue+server)
L: number of BSs in system
S: mean number of searches to locate a mobile
N: mean number of PRs in a system queue
T: mean waiting time in system (average response time)

We progressively complicate the analysis by studying each paging mechanism separately (BP, SP, SGP) and initially admitting a uniform distribution of user location. We aim to derive analytical results for the blocking probability and average response time of a PR. Incoming message requests are assumed to be served in chronological order (FIFO queues).

3.1 Blanket Paging

This is the simplest case; since all BS cells in a system area are polled simultaneously, the aggregate mean call arrival rate is transferred to all BS queues. As the behavior of all queues is identical, it suffices to study the single queue where the user will be found. The load in a single queue is given by $\rho = \lambda/\mu$. From

$M/M/1/K$ queueing analysis [11], the equilibrium distribution for the number of paging messages in the system is

$$\pi_n = \begin{cases} \frac{1-\rho}{1-\rho^{K+1}} \cdot \rho^n, & \text{if } \rho \neq 1 \\ \frac{1}{K+1}, & \text{if } \rho = 1 \end{cases} \tag{1}$$

where $n = 0, 1, \ldots, K$. Substituting $n = K$, we have the blocking probability of a page request, $P_B = \pi_K$. The mean number of PRs in the system is

$$N = \sum_{n=0}^{K} n \cdot \pi_n = \sum_{n=0}^{K} n \cdot \frac{1-\rho}{1-\rho^{K+1}} \cdot \rho^n = \frac{1-\rho}{1-\rho^{K+1}} \cdot \sum_{n=0}^{K} n \cdot \rho^n = \\ = \frac{\rho[1-(K+1)\rho^K + K \cdot \rho^{K+1}]}{(1-\rho)\cdot(1-\rho^{K+1})} \tag{2}$$

if $\rho \neq 1$ and

$$N = \sum_{n=0}^{K} n \cdot \frac{1}{K+1} = \cdots = \frac{K}{2} \tag{3}$$

if $\rho = 1$. To find the mean waiting time in system, which is also the average response time of a PR, we may apply Little's law, considering the actual rate of PRs that are admitted into the system. The fraction of arrivals who are served is

$$\lambda_s = \lambda(1 - \pi_K) \tag{4}$$

Hence, the mean response time for each served PR is

$$T = \frac{N}{\lambda(1-\pi_K)} \tag{5}$$

3.2 Sequential Paging

Uniform Case. We first consider the uniform case where a user has an equal probability to be located at each cell at the time of a call arrival, denoted as $p_i = 1/L\ (i = 1, 2, \ldots, L)$. A uniform distribution produces the highest average number of attempts and hence provides a lower bound in the algorithm performance [6]. Assuming sequential steps are selected at random by the central control, each BS will initially receive exactly $1/L_{th}$ of the total load. This must be multiplied by the mean number of searches to locate a mobile; hence, the individual arrival rate at each BS channel is

$$\lambda_p = \lambda \cdot \frac{S}{L} \tag{6}$$

Given that $S < L$, we have that $\lambda_p < \lambda$. Hence as anticipated, the individual arrival rate of incoming PRs is always less in the SP case. For a uniform distribution, the mean number of paging attempts equals $S = \frac{L+1}{2}$.

Substituting $\rho = \frac{\lambda_p}{\mu}$ in Eqs. (1) and (2), we get the blocking probability and mean number of PRs at a single queue. However, considering that the system will make on average S attempts until the requested user is found, the total blocking probability is

$$P_B = S \cdot \pi_K \tag{7}$$

The average response time of a PR is the sum of the individual times at each queue. Hence, in an analogous manner, we have that

$$T = S \cdot \frac{N}{\lambda_p(1 - \pi_K)} \tag{8}$$

Conditioning on the event that a mobile user will eventually be found, the mean number of searches is unaffected and equals $\frac{L+1}{2}$. However, due to blocking, the mean number of searches may be reduced to:

$$S = (1 - \pi_K) + (1 - p_1)(1 - \pi_K)^2 + \cdots + (1 - p_1 - \cdots - p_{L-1})(1 - \pi_K)^L \tag{9}$$

where p_i, $i = 1, \ldots, L$ is the location distribution and each of the summands corresponds to the probability of making each successive paging step. In the case of a uniform distribution, we have

$$S = \sum_{i=1}^{L} \frac{L - (i - 1)}{L} (1 - \pi_K)^i \tag{10}$$

Non-Uniform Case. Assume now that each BS has a different load, as a result of a non-uniform location distribution. Then the arrival rate at each paging channel can be represented as

$$\lambda_p^i = \phi_i \cdot \lambda$$

($i = 1, \ldots, L$), where ϕ_i is the fraction of the load distributed at each queue, based on a sequential polling mechanism and the underlying location distribution, $\{p_i\}$. Then we should calculate the blocking probability and delay separately at each queue. Mean values for the parameters discussed can then be produced by taking the average, weighted by the probability of a PR being processed at each queue. The mean number of paging attempts is in general given by $S = \sum_{i=1}^{L} i \cdot p_i$ and the mean number of searches w.r.t. blocking by (9).

The detailed analysis of how the underlying location distribution affects the individual incoming rates λ_p^i is an arduous task and is bypassed here. With much less complication, the behavior of the system in the non-uniform case can be shown by admitting an identical mean number of searches for all users, for which holds $S < \frac{L+1}{2}$. We also assume that polled locations are selected at random, so that the exact same load is delivered at each queue. In so doing, the previous sequential paging analysis can be applied, with parameters modified by S, $\{p_i\}$.

3.3 Sequential Group Paging

Uniform Case. Let us consider N_G polling groups, each group containing $\frac{L}{N_G} = \theta$ cells. Assume for simplicity that θ is an integer number[2] and each group is chosen randomly, so that every BS receives the exact same load. Similarly to the previous analysis, the arrival rate at each paging channel is

$$\lambda_p = \lambda \cdot \frac{S}{N_G} \tag{11}$$

where $S = \frac{N_G+1}{2}$. To find the mean response time of a PR, we have to consider the ensemble of all BS queues. Let $N_1, N_2, \ldots, N_\theta$ denote the number of PRs at the cell BSs of any given group. For $\rho \neq 1$, we have that

$$Pr\{N_i = n\} = \frac{1-\rho}{1-\rho^{K+1}} \rho^n$$

where $i = 1, 2, \ldots, \theta,\ n = 0, 1, \ldots, K$. Hence the cumulative distribution function (cdf) is

$$Pr\{N_i \leq n\} = \frac{1-\rho^{n+1}}{1-\rho^{K+1}} \tag{12}$$

Similarly, for $\rho = 1$ the cdf becomes

$$Pr\{N_i \leq n\} = \frac{n+1}{K+1} \tag{13}$$

At each paging stage j ($j = 1, 2, \ldots, N_G$), all BSs in the j_{th} group are paged to find the mobile. Therefore, the delay in the j_{th} stage is directly related to the maximum queue length amongst the θ BSs that page in that group. Let N_m denote the maximum number of messages in any of the θ queues at any time instance. The cdf of N_m occurs as follows

$$Pr\{N_m \leq n\} = [Pr\{N_i \leq n\}]^\theta = \begin{cases} \left[\frac{1-\rho^{n+1}}{1-\rho^{K+1}}\right]^\theta, & \text{if } \rho \neq 1 \\ \left[\frac{n+1}{K+1}\right]^\theta, & \text{if } \rho = 1 \end{cases} \tag{14}$$

To find the mean value of N_m, we have that

$$E[N_m] = \sum_{n=0}^{K} Pr\{N_m > n\} = \begin{cases} \sum_{n=0}^{K} \left[1 - \left(\frac{1-\rho^{n+1}}{1-\rho^{K+1}}\right)^\theta\right], & \text{if } \rho \neq 1 \\ \sum_{n=0}^{K} \left[1 - \left(\frac{n+1}{K+1}\right)^\theta\right], & \text{if } \rho = 1 \end{cases} \tag{15}$$

The total blocking probability is calculated by $P_B = S \cdot \pi_K$, as the blocking within other BSs in the same group is irrelevant when the mobile is not found

[2] In the case where L is not an integer multiple of N_G, $\theta_0 = \lfloor \frac{L}{N_G} \rfloor$ cells should be assigned to the first $N_G - (L - \theta_0 \cdot N_G)$ groups and $(\theta_0 + 1)$ to the remaining $(L - \theta_0 \cdot N_G)$, in an optimal partition [8]. The analysis must be differentiated, as in the non-uniform case below.

there. A problem is encountered when calculating the mean response time; here, the average delay in an unsuccessful step is generally different from a successful one, since in the first case we have to consider the longest delay of all θ queues. If the user is eventually found after S steps, the total average delay can be calculated as

$$T = \frac{N}{\mu} + (S-1)(\frac{E[N_m]}{\mu}) \tag{16}$$

If not, the mean response time should be given by

$$T = S \cdot \frac{E[N_m]}{\mu} \tag{17}$$

The mean number of searches w.r.t blocking can be calculated similar to (9):

$$S = (1-\pi_K^\theta) + (1-p_1)(1-\pi_K^\theta)^2 + \cdots + (1-p_1-\cdots-p_{N_G-1})(1-\pi_K^\theta)^{N_G} \tag{18}$$

where π_K^θ is the combined probability of blocking in the whole group and $\{p_j\}$, $j = 1, \ldots, N_G$ is the total location probability in a group of cells. However, after S searches, there is no way of knowing whether a user has been found or not. Due to the non-independence of successive paging steps, we note that there generally exists no closed-form solution for T.

Non-Uniform Case. This case is very complicated and outside the limited scope of this paper. We outline the solution as following. The delay and blocking probability are computed based on the BSs in each group. Here, the number and identity of BSs in a group are of primary importance. These should be calculated based on a system partition, as close to the optimal as possible. The total parameters must be calculated as weighted averages on all groups. The random selection hypothesis could also apply here, in order to simplify the solution.

Remark. It should be added that in real-life systems, there exists a timeout interval W, during which the system awaits the mobile unit's response [9]. For simplicity, we have incorporated the timeout into the service time, assuming the time for the system to perceive user presence is the same, whether he responds or not. For more accurateness the extra period $(W - \frac{1}{\mu})$ should be added to the SP and SGP schemes, for all $(S-1)$ unsuccessful attempts.

4 Numerical Results

In this section, numerical results are presented that show the comparative behavior of blanket and various forms of sequential paging. It is appropriate to consider a 'system scale', whereupon a larger system also has a higher storage capacity. In the case of a small system we have $L = 20$ locations and buffer size $K = 30$, whereas in the larger system we specify that $L = 40$ and $K = 100$.

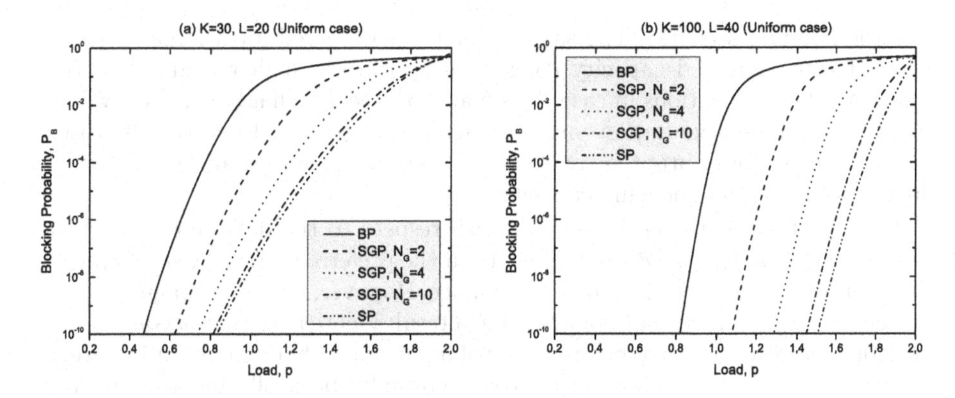

Fig. 2. Blocking probability vs. aggregate load for blanket and sequential paging cases (a) $K = 30$, $L = 20$, (b) $K = 100$, $L = 40$

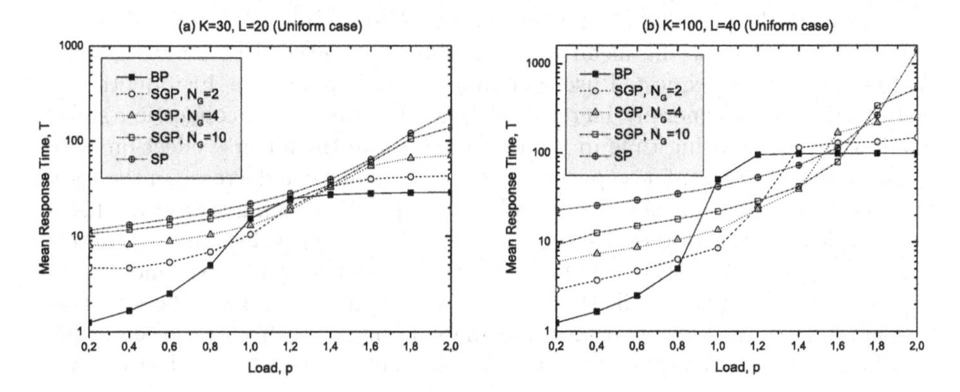

Fig. 3. Mean response time of a served PR vs. aggregate load for blanket and sequential paging cases (a) $K = 30$, $L = 20$, (b) $K = 100$, $L = 40$

Apart from the load parameter ρ, the aggregate incoming call rate determines the waiting time in system. For simplification, the service service rate at BS channels is taken equal to unity, so that $\lambda = \rho$ in all test cases.

Figures 2(a),(b) present the blocking probabilities under a uniform location distribution, for various values of the load parameter ρ. The y-axis' are drawn on a logarithmic scale to efficiently represent blocking behavior. We note that generally in communications systems, the fraction of blocked calls should be kept less than 10^{-2}. The blocking probability is always less in the sequential cases, despite the fact that the system might make numerous attempts to locate the mobile. This implies much less congestion in system queues, as a result of the distribution of the aggregate load. Indeed, blocking decreases when increasing the number of sequential steps, with the exception of very high loads, where the combination of increased congestion and multiple paging attempts can increase the percentage of rejected calls. Hence, for normal and increased loads, sequential paging in its various forms can achieve a notable decrease in blocking probability over the flooding scenario. The same behavior is observed for a larger system, where due to increased capacity congestion appears at higher loads. Also the higher number of locations in case (b) expands the search in a larger area, which yields a better relative improvement for sequential paging. Finally, it is worth adding that a more abrupt increase of blocking probabilities always occurs in the case of increased queueing capacity.

Clearly, if we had been indifferent with respect to the delay introduced by successive paging steps, SP would have been the preferred strategy, since almost always it achieves better blocking performance. However, numerous paging stages in such a transparent system would lead to a result identical to that of congestion, since delay is also excessive before establishing a call. What's more, in the event of excessive delay the calling party would normally back off, and thus paging should be cancelled, resulting –similarly to blocking– in uncompleted calls.

Fig. 3 shows respective results in terms of the mean response time of a PR. The y-axis' are again logarithmic to better discern the growth of the curves. In order to avoid the pitfall of producing smaller response times for more congested channels, results show the mean response time of a *served* PR, i.e. we assume that eventually the requested user is found. As anticipated, the introduction of more paging steps generally increases delays. However, when congestion starts building up, the waiting time in a queue might have the adverse effect on total response times. In fact, for a specific range of load values, a decrease in the total response time can be achieved if we adopt a sequential paging strategy. This is more evident in the case of a larger system, where for ρ values approximately in the range $(1 \leq \rho \leq 1.2)$, BP performs even worse than the extreme case of sequentially polling each cell. However, for very high load values, the situation is 'back to normal', with sequential paging suffering both from congestion and very large delays. This characteristic behavior is encountered in all test cases.

It is worth noting that a different number of paging groups will be optimal for different load values. For example, in Fig. 3(b) choosing $N_G = 2$ is optimal for $\rho = 1$, yet it performs worst when $\rho = 1.4$. Finally, it is noted that if we had a higher queueing capacity, together with a small number of locations, the delay performance of SP would be improved accordingly to the confinement of the search space.

The effects of a non-uniform location distribution are depicted in Fig. 4. Here, no partitioning is assumed and cells are polled sequentially. System parameters are set to $K = 30$, $L = 20$. Non-uniform cases (SP-NU) are portrayed by reducing the mean number of paging attempts to locate a mobile, according to an underlying location distribution. Results are also compared against the BP scheme.

For illustrative purposes, let us define a parameter $a = \frac{S}{L}$, where $\frac{1}{L} \leq a \leq \frac{L+1}{2L}$, depending on the mean number of searches. This can be called the 'reciprocal of search concentration', as for small values of a there exists a large concentration of location probabilities and vice-versa. Then in general, SP can sustain loads $\frac{1}{a}$ times greater than the BP case, which is extremely important in congestion situations. The effects of this are transferred to the blocking probability curves of Fig. 4(a). For more concentrated distributions the delay performance

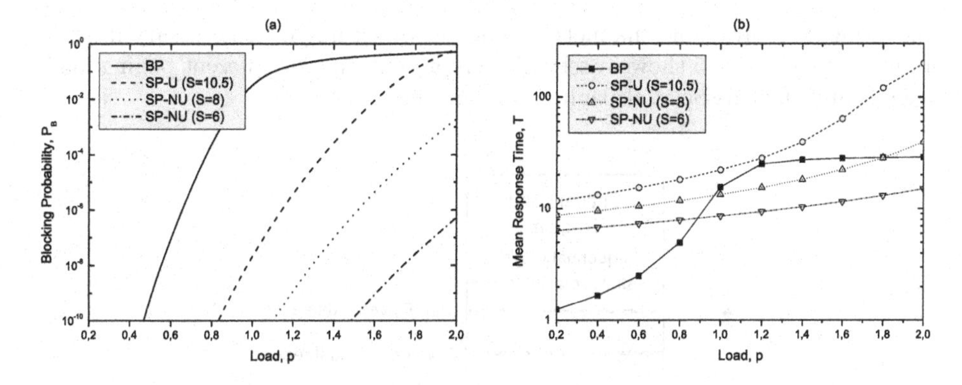

Fig. 4. Blocking probabilities (a) and mean response times (b) vs. increasing values of the aggregate load, in a general case with uniform and non-uniform location distributions. Results refer to the BP and SP cases in a system where $K = 30$, $L = 20$

is improved as well, as shown in Fig. 4(b). Despite the fact that all cells were polled sequentially in this example, the concentration of search can give smaller response times than the simultaneous paging approach, for a specific range of load values.

5 Paging Load Control

Having seen the basic dimensions of the problem, it is concluded that neither form of paging can efficiently handle the flow of messages in a wireless network. Each strategy performs better at a specific range of load values. Therefore a control mechanism should be envisaged, especially in view of a limited resource system. The goal is to increase the rate of *served* paging messages, while withholding the delay beneath acceptable levels. From an analytical view, this is a difficult optimization problem, since blocking and delay may be contradictory constraints. Also, the behavior of the system can change with unpredictable load variances.

Here, an efficient control method is proposed for handling the overall problem. Our proposal is outlined as follows. In a paging system queue, there are usually alternating intervals of low and high activity periods. In order to reduce congestion, the system (i.e. the MSC) can monitor, or receive as feedback, the state of its BS paging channels. The state depends on the offered paging load and can be represented in multiple scales. Depending on the queue load state, the system differentiates its paging policy: under low load conditions, the system does not face congestion problems; so it can increase throughput by simultaneously polling all cells. However, in high load situations, it is best that polling is done sequentially, so as to distribute PRs among BSs and reduce the average load. The case of very high loads is an extreme situation, since neither sequential paging performs well. Therefore, it should be treated differently. Essential methods are discussed in [12], for example by applying admission control policies or allocating more available channels for paging, which might be necessary in such an utmost case. Fig. 5 shows a schematic representation of different paging load regions and their associated paging mechanism.

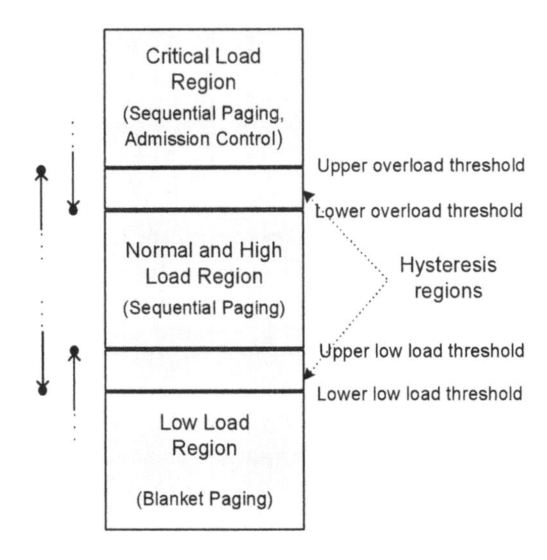

Fig. 5. Paging congestion control regions with upper and lower load thresholds

Control triggers must be generated periodically by the system in order to check congestion conditions. This could be done by sending polling control messages and waiting for the system's response, or by the BSs automatically reporting congestion conditions, in a feedback fashion [13]. Alternative ways of load sampling are counting the number of messages in the queue, the number of idle periods, or estimating blocking probabilities. The frequency of load sampling can also be programmed in advance. Generally, congestion detection should be performed at a smaller time scale for increasing paging traffic; besides this being the most important region, system behavior can change more abruptly for such values. In addition, overloading usually occurs in specific periods of a day, special events or occasions (e.g. New Year's eve, big athletic events, natural disasters, etc.). At those times and places, triggering should be performed more frequently to efficiently control congestion levels.

The choice of load thresholds is the most critical part of a congestion control algorithm. Systems should be designed so that even at peak throughput, the queueing lengths and delays are kept within predefined bounds. Attention should be drawn to the fact that we're dealing with a real-time system, with often rapid and unpredictable changes in behavior. The choice of the threshold level and the associated paging mechanism should depend on the number of channels and buffer capacities, as well as the location distribution. In the case of $M/M/1$ queues, the authors in [10] have defined a *crossover load*, above which sequential paging delivers better delay performance. However, in the case of finite capacities two such roots of a higher-degree polynomial equation are required, as indicated by the numerical results presented in Section 4. This is very difficult to solve analytically or numerically. But even if we can end up in approximate solutions, delay should not be the sole criterion to differentiate the paging mechanism, as we would ignore the large benefits acquired by reducing blocking probabilities, even with a sacrifice in delay. Hence, the choice of threshold values should be a compromise between blocking and delay. Based on the evaluated test cases, a rough estimate is as follows: BP should be applied when the offered load is $\rho < 0.8$. In the range $0.8 < \rho < 1.5$, some form of sequential group paging should be applied; it is noted that SP should generally be avoided, except in the case of very small number of locations or very concentrated distributions. Above $\rho = 1.5$ is generally a critical overload region.

In order to better cope with the real time environment, hysteresis regions are proposed, as shown in Fig. 5. Marked-end arrows show the points of changing polling behavior upon entering a new load region, depending on the previous queue state. In general, hysteresis is essential to improve stability and robustness of the system. A hysteresis region leaves an error margin and prevents frequent changes in the paging mechanism. This should also be designed to give the congestion control software some time to initiate action and bring the resource occupancy down. If the congestion trigger is delayed, the system might reach pretty close to 100% occupancy, thus leading to terrible service to all users in the system.

Based on the results in Section 4, the benefits of the outlined control algorithm are straightforward. A large scale simulation with realistic traffic conditions should be further conducted in order to study the statistics of load variance. This would enable us to explicitly define upper and lower thresholds for a specific system implementation.

6 Ending Notes

In the last section of the paper, several important issues are left for discussion. First, despite the fact that given specific parameters of the problem, sequential paging can decrease the mean response time of a PR, using SP or SGP as a means to tackle the delay is not a sensible approach. Sequential paging should be used primarily to relieve blocking, even with a small increase in delay. The definition of a *utility function* which compromises blocking and delay and would be an appropriate index of performance remains under investigation.

Also, sequential paging might not always solve congestion hardships. Instead, it might just transpose the problem; essentially, the execution of the paging procedure in steps postpones the query of certain cells at later time epochs. So we might just transfer the problem, if at later time epochs arrives a significant number of page requests for other users at the same base station. This is worsened by the fact that PRs for different subscribers can arrive irregularly, or in the case of large inhomogeneities in paging traffic load amongst different BSs. Such characteristics can lead to hysteresis-type congestion problems, as outlined in

[14]. In view of this, a realistic simulation study becomes more crucial to view the actual benefits of such a scheme.

In addition, a detailed view of the system provides more insight. As it has been noted, the reduction in delay produced by sequential paging is due to paging different subscribers at the same time slot, in different BSs. Therefore, in a microscopic look, the problem becomes one of concurrently serving different page requests in a slotted system. As it was shown in [15], the optimal solution to the concurrent search problem, when searching for n terminals, is to solve n independent sequential paging problems. However, practically this is not possible, if the system does not have multiple (theoretically $O(n)$) paging channels or if it does not have the capability to send more than one message in a single slot. If we were to increase the effective throughput with no concern for delays, the PR with the highest probability of locating a mobile should be processed at each slot of the paging channel. However, in order to combat delays, a priority metric should be introduced, both in terms of oldness in the system and location probability [15]. The issues of conflicts and synchronization are more critical at slot level here. On the other hand, our proposal aims at a higher level control which can be more applicable in a real-life system.

It is equally remarked that the sequence in which jobs are served once they are in the queue does not generally affect mean performance parameters [11]. Nevertheless, it does affect the mean response time of a *specific* page request. Here we have considered a FCFS strategy, which is also the most realistic case. It is not considered that other service disciplines are widely applicable in the context of study. However, the analysis can be extended to cover such issues, especially in the direction of prioritized page requests.

Finally, the analysis assumed an error-free environment. In practice, paging errors are a cause of re-propagation of messages and thus a cause of congestion. Paging strategies in the presence of transmission errors have been previously presented in [16],[17]. Adjusting the polling mechanism to an erroneous environment is necessary, however the essentials of the higher-level control algorithm remain unaffected.

Returning to the context of this work, a more precise statement regarding the expected gain of the proposed adaptive method is closely tied with the detailed association of load ranges with paging mechanisms and must be further researched. Of equivalent interest are further analytical performance evaluation results for inhomogeneous user location distributions and different service or storage capacities at each cell. In this respect, a more powerful modeling approach by means of a network of queues is deemed necessary.

In conclusion, despite the variety of open issues for analysis, as well as implementation, it is evident that if the control mechanism –as outlined above– is properly applied, an improved system performance can easily be attained.

References

1. Pollini, G.P., Meier-Hellstern, K.S., Goodman, D.J.: Signaling traffic volume generated by mobile and personal communications. IEEE Comm. Mag. **33** (1995) 60–65
2. Akyildiz, I.F., McNair, J., Ho, J.S.M., Uzunaliŏglu, H., Wang, W.: Mobility management in next-generation wireless systems. Proc. of the IEEE **87** (1999) 1347–1384
3. Bhattacharya, A., Das, S.K.: Lezi-update: An information-theoretic approach to track mobile users in PCS networks. In Proc. ACM/IEEE Mobicom '99 (1999) 1–12
4. Saraydar, C.U., Rose, C.: Minimizing the paging channel bandwidth for cellular traffic. In Proc. ICUPC '96 (1996)
5. Krishnamachari, B., Gau, R.-H., Wicker, S.B., Haas, Z.J.: Optimal sequential paging in cellular networks. Wireless Networks **10** (2004) 121–131
6. Rose, C., Yates, R.: Minimizing the average cost of paging under delay constraints. Wireless Networks **1** (1995) 211–219
7. Abutaleb, A., Li, V.O.K.: Paging strategy optimization in personal communication systems. Wireless Networks **3** (1997) 195–204
8. Wang, W., Akyildiz, I.F., Stüber, G.L.: An optimal paging scheme for minimizing signaling costs under delay bounds. IEEE Comm. Letters **5** (2001) 43–45
9. Goodman, D.J., Krishnan, P., Sugla, B.: Minimizing queueing delays and number of messages in mobile phone location. Mobile Networks and Applications **1** (1996) 39–48
10. Rose, C., Yates, R.: Ensemble polling strategies for increased paging capacity in mobile communication networks. Wireless Networks **3** (1997) 159–167
11. Bose, S.K.: An Introduction to Queueing Systems. Kluwer Academic/Plenum Publishers (2000)
12. Keshav, S.: Congestion Control in Computer Networks. PhD Thesis, UC Berkeley TR-654 (1991)
13. Shenker, S.: A theoretical analysis of feedback flow control. In Proc. ACM Sigcomm '90 (1990)
14. Ackerley, R.G.: Hysteresis-type behavior in networks with extensive overflow. BT Tech. Journal **5** (1987) 42–50
15. Gau, R.-H., Haas, Z.J.: Concurrent search of mobile users in cellular networks. IEEE/ACM Trans. on Networking **12** (2004) 117–130
16. Awduche, D.O., Ganz, A., Gaylord, A.: An optimal search strategy for mobile stations in wireless networks. In Proc. ICUPC '96 (1996)
17. Verkama, M.: Optimal Paging—A search theory approach. In Proc. ICUPC '96 (1996)

Opportunistic Scheduling for QoS Guarantees in 3G Wireless Links

Jeong Geun Kim and Een-Kee Hong

Department of Electronics and Information,
Kyung Hee University, Suwon, Republic of Korea
{jg_kim,ekhong}@khu.ac.kr

Abstract. Proportional Fair (PF) share policy has been adopted as a downlink scheduling scheme in CDMA2000 1xEV-DO standard. Although it offers optimal performance in aggregate throughput conditioned on equal time share among users, it cannot provide a bandwidth guarantee and a strict delay bound, which is essential requirements of real-time (RT) applications. In this study, we propose a new scheduling policy that provides quality-of-service (QoS) guarantees to a variety of traffic types demanding diverse service requirements. In our policy data traffic is categorized into three classes, depending on sensitivity of its performance to delay or throughput. And the primary components of our policy, namely, Proportional Fair (PF), Weighted Fair Queuing (WFQ), and delay-based prioritized scheme are intelligently combined to satisfy QoS requirements of each traffic type. In our policy all the traffic categories run on the PF policy as a basis. However the level of emphasis on each of those ingredient policies is changed in an adaptive manner by taking into account the channel conditions and QoS requirements. Such flexibility of our proposed policy leads to offering QoS guarantees effectively and, at the same time, maximizing the throughput. Simulations are used to verify the performance of the proposed scheduling policy. Experimental results show that our proposal can provide guaranteed throughput and maximum delay bound more efficiently compared to other policies.

1 Introduction

With the explosive growth of the Internet and rapid proliferation of personal communication services, the demand for data services in wireless networks has been ever-increasing. Responding to this demand, wireless networks have been evolving toward packet-switched architectures that are more flexible and efficient in providing packet data services. The CDMA2000 1xEV-DO standard (abbreviated hereafter as 1xEV-DO) is one of such architectures, that is designed to support data services in the third generation wireless network [1]. Compared to previous wireless networks often characterized by circuit-switched and voice-oriented architecture, the 1xEV-DO system has several unique features. Notably among them is the "opportunistic" scheduling used to schedule downlink (or equivalently forward link) transmission.

I. Niemegeers and S. Heemstra de Groot (Eds.): PWC 2004, LNCS 3260, pp. 54–68, 2004.

The basic idea behind the opportunistic scheduling is that temporal channel variation of multiple users is taken into account in scheduling [2]. Thus at a given time the scheduling decision favors a mobile user currently seeing a better channel. If all the traffic is "elastic" (i.e., having flexible service requirements), such opportunistic scheduling mechanisms then greatly improve throughput performance and yield higher bandwidth utilization.

The 1xEV-DO standard adopted an opportunistic scheduling called Proportional Fair (PF) share policy as a downlink scheduler. Although the PF policy offers optimal performance in aggregate throughput conditioned on equal time share among users [3], it cannot provide bandwidth guarantee and strict delay bound, which is essential requirements of real-time applications. As the 1xEV-DO system expands its service realm from non-real-time (NRT) to real-time (RT) applications like video streaming, such limitation triggers a lot of research efforts toward expanding its capability for supporting various quality-of-service (QoS) requirements of RT traffic.

The authors in [4] proposed a scheduling discipline called the Exponential Rule in which the queue with larger weighted delay of head-of-line (HOL) packet gets higher priority in transmission. This rule behaves like the PF policy when the weighted delay difference is not relatively large. As the difference becomes significant, this policy gracefully adapts from the PF policy to the Exponential rule. However, this policy suppresses the chance the HOL packet with relative lower weighted delay gets selected for transmission even when it currently sees a better channel. This may lead to lower throughput performance. In [5], Liu et al. proposed an utility-based scheduling algorithm in which utility is a decreasing function of packet delay. By scheduling the transmission at each time in a way the total utility rate is maximized, this scheme can provide a strict delay bound. However, it is questionable to how the utility function can be devised for throughput-sensitive applications. Furthermore existence of the utility functions and their derivatives is not obvious for applications with diverse requirements in reality. To address the issues of fairness and throughput guarantee in the PF policy, a dynamic rate control algorithm was proposed in [6]. In this scheme, target throughput is set along with the associated minimum and maximum bounds, and the scheduler attempts to maintain user's perceived throughput within those bounds. However this scheme can only work with throughput-sensitive applications, but may not suit other traffic categories.

The main contribution of this paper is that we propose a novel "opportunistic" service rule that can flexibly support a wide range of traffic categories. Most of previous works in [5]-[6] focuses on applications sensitive to either delay or throughput. In contrast, our scheme can deal with QoS needs from both traffic categories using a simple metric. In addition, throughput guarantee can be supported, depending on urgency, in either a strict or relaxed manner via configurable parameters. Such capability is useful when throughput guarantee is not necessarily required over microscopic time-scale, but rather needed on long-term basis. Moreover in order to maximize the link capacity, the RT traffic runs on the PF policy as long as delay or throughput performance does not deviated much from its prescribed target value. However, as its QoS target is more likely to be missed, the corresponding traffic gains more weight in scheduling decision and increases the chance for transmission. This feature of graceful adaptation from the PF policy to the prioritized mode can provide maximized throughput and, at the same time, offers an accurate tool for exercising QoS. Our numerical results indicate that the proposed rule can deliver QoS guarantees in multiservice wireless environments that are often characterized by heterogeneous channel conditions and mixed QoS requirements. Moreover, throughput gains are observed against other schemes and the flexible throughput guarantees are proven to work effectively as designed and give a capacity improvement.

The rest of the paper is organized as follows. In Section 2 we give overview of CDMA2000 1xEV-DO standard and describe the proposed scheduling algorithm. Numerical results and simulations are reported in Section 3, followed by concluding remarks in Section 4.

2 Scheduling Algorithm

2.1 Background

To utilize temporal channel variation, the PF share policy in the 1xEV-DO requires the mobile terminals (MTs) to report continuously their channel state information to the scheduler in the access point (AP). Accordingly, the MTs report back the channel condition to the network every 1.667 msec. The pilot bursts from the AP enable the MT to accurately estimate the channel conditions. The channel state information is sent back to the AP in the form of data rate request (see Table 1) through data rate request channel (DRC) in the reverse link [1]. Once data on DRC from each MT is gathered, the PF share scheduler selects the MT i^* which satisfies

$$i^* = \arg\max_i \frac{\mathrm{DRC}_i(t)}{\hat{R}_i(t)} \tag{1}$$

where $\mathrm{DRC}_i(t)$ is the data rate request of MT i at time t, $\hat{R}_i(t)$ is average transmission rate of MT i by time t.

In 1xEV-DO, downlink bandwidth is shared among multiple MTs in time-division multiplexing (TDM) basis, where fixed-size time-slots are dedicated to each MT based on scheduling policy. In particular, different modulation schemes including QPSK, 8PSK, and 16QAM are employed in an adaptive manner, depending on the channel condition of the target MT. The set of available data rates for the corresponding channel conditions is listed in Table 1. To provide fairness in channel usage among the MTs, the average rate $\hat{R}_i(t)$ of the ith MT is updated every time-slot as follows:

$$\hat{R}_i(t+1) = \left(1 - \frac{1}{t_c}\right)\hat{R}_i(t) + \frac{1}{t_c}R_i(t)$$

where t_c is the time constant set to 1000 slots [2], and $R_i(t)$ is the current rate of the ith MT. At the end of current transmission the scheduler determines the next MT to transmit based on the criterion in (1) and this procedure repeats.

Table 1. Data rate options in CDMA2000 1xEV-DO.

Nominal Data Rate (Kbps)	Nominal Slots per PHY Packet	Total Bits per PHY Packet	E_c/N_t (dB) thresholds for DRC selection
38.4	16	1024	-13.5
76.8	8	1024	-10.5
153.6	4	1024	-7.4
307.2	2	1024	-4.3
614.4	1	1024	-1.0
921.6	2	3072	1.5
1228.8	1	2048	3.7
1843.2	1	3072	7.1
2457.6	1	4096	9.1

2.2 Algorithm

In a majority of previous works, prioritized services and associated QoS capabilities have been implemented by incorporating a weighting function $\mathbf{F}_w(\cdot)$ into the PF metric in (1). The modified metrics in general have the following form:

$$i^* = \arg\max_i \frac{\mathrm{DRC}_i(t)}{\hat{R}_i(t)} \mathbf{F}_w(\cdot).$$

Different weight functions $\mathbf{F}_w(\cdot)$'s have been chosen depending on which performance factor (e.g., bandwidth, delay) is prioritized. The Exponential rule in [4] has a weight function given by

$$\mathbf{F}_w(\cdot) = \exp\left(\frac{a_i W_i(t) - \overline{aW}}{1 + \sqrt{\overline{aW}}}\right) \tag{2}$$

where a_i is the weight of the ith flow, $W_i(t)$ is the waiting time of the packet, and $\overline{aW}$ is mean weighted delay. In [6], $\mathbf{F}_w(\cdot)$ is defined as a group of functions whose values are proportional to deviation from the target rate. See [6] for details.

As pointed earlier, these rules lack the flexibility of handling diverse QoS requirements. To overcome this limitation, our rule is designed to support three primary types of traffic into which most network applications can be categorized:

- Class I: Delay-sensitive
- Class II: Throughput-sensitive
- Class III: Best-effort.

Such capability of providing differentiation in service is essential in operating an anticipating multi-service wireless networks where a wide array of traffic types coexist demanding diverse requirements. In our rule, the metric is simple to calculate and the control parameters are flexibly configured depending on the traffic category.

We propose the rule named the "Adaptive policy" whose metric is given by:

$$i^* = \arg\max_i \frac{\mathrm{DRC}_i(t)}{\hat{R}_i(t)} \left(C_{wd}(t)(\overline{D_i(t)} + 1)^\alpha + 1 \right)^{w_i(t)} \tag{3}$$

In the above equation $C_{wd}(t)$ is a coefficient that determines the impact of the weight function and is set to the maximum ratio of DRC over the average rate at each scheduling epoch, that is,

$$C_{wd}(t) = \max_i \frac{\mathrm{DRC}_i(t)}{\hat{R}_i(t)}.$$

α is a configurable parameter for prioritizing traffic sensitive to either delay or throughput, and $w_i(t)$ is the weighting factor controlling the level of emphasis on the weighting function in (3). Configuration of these parameters will be discussed later in the next section.

$\overline{D_i(t)}$ is the normalized waiting time given by

$$\overline{D_i(t)} = \frac{D_i(t) - D_{i,max}}{D_{i,max}} \tag{4}$$

where $D_i(t)$ is the waiting time of the HOL packet in the ith flow (destined to MT i) and $D_{i,max}$ is the maximum tolerable delay of the HOL packet. $D_{i,max}$ is specified as a QoS parameter for each class and is given by:

$$D_{i,max} = \begin{cases} D_{i,max}, & \text{for Class I} \\ \max(t_{a,i}, F_{i,-1}) + \frac{L_i}{R_i}, & \text{for Class II} \\ \infty, & \text{for Class III.} \end{cases} \tag{5}$$

For delay-sensitive traffic (Class I), the maximum delay is specified by $D_{i,max}$ as input QoS parameter. For best-effort traffic (Class III), the maximum delay is set to infinity. Thus the weighting function in (3) becomes the unity and the bandwidth is shared by the PF policy among the MTs. For Class II traffic which is sensitive to throughput, $D_{i,max}$ is set to the finish time of the HOL packet. The notion of finish time is central to service discipline called weighted fair queueing (WFQ) and finish time for the ith flow is given by [7]:

$$F_i(t) = \max(t_{a,i}, F_{i,-1}) + \frac{L_i}{R_i} \tag{6}$$

where $t_{a,i}$ is the arrival time of the HOL packet, $F_{i,-1}$ is the finish time of the previous packet, L_i is the packet length, and R_i is the promised rate (or bandwidth). WFQ scheme attempts to emulate packet flow in ideal fluid model by calculating the departure time of a packet (i.e., finish time) in a corresponding fluid model and using this virtual time stamp to schedule packets. The advantage in offering throughput guarantees via WFQ-like policy against the periodic counter in [4] is more accurate and fair in distribution of bandwidth among the competing flows.

Since there often exists mismatch between physical packet size and data size from the higher layer, the expression of finish time in (6) needs to be modified to take into account the case in which multiple data packets are transmitted over a single packet. In this case, the representative arrival time $t_{a,i}$ for a group of packets including the HOL and its subsequent ones is set to that of the HOL packet. Once such a flow is selected for transmission, the packets behind the HOL enjoy a free ride and the flow gets more bandwidth than necessary. To correct this, the actual finish time is calculated after transmission. The actual finish time $F_{i,k+n-1}$ for n packets conditioned that the kth packet of the ith flow is at HOL, is

$$\begin{aligned} F_{i,k+n-1}(t) \quad & = \max(t_{a,(i,k)}, F_{i,k-1}(t)) + \textstyle\sum_{j=k}^{k+n-1} \frac{L_{i,j}}{R_i} \\ & + \textstyle\sum_{j=k}^{k+n-2} \left(t_{a,(i,j+1)} - F_{i,j}(t)\right) \cdot \mathbf{1}_{\{t_{a,(i,j+1)} > F_{i,j}(t)\}} \end{aligned} \quad (7)$$

where $\mathbf{1}_x$ is the indicator function whose value is 1 if the condition x is satisfied and 0 otherwise. Previous finish time $F_{i,-1}$ in (6) becomes equivalent to (8).

2.3 Configuration of Parameters

As delay of the HOL packet approaches the prescribed target, how the weighting function $(C_{wd}(t)(\overline{D_i(t)} + 1)^\alpha + 1)$ varies is depicted in Fig. 1. Here, $C_{wd}(t)$ is set to 63 for illustration. As the figure indicates, the rule is designed so that the flows are scheduled on the PF policy as long as delay of the HOL packet has a sufficient margin from the target, but the delay-based priority part in the metric gradually overrides the PF policy as the delay approaches to the target. By scheduling this way our proposed rule can utilize temporal channel variation and consequently maximize the channel capacity. In contrast, the Exponential rule (2) suppresses selection of the HOL packets whose weighted delays are less than mean value, regardless of the channel conditions.

The weighting factor $w_i(t)$ is another parameter configured at the connection setup phase, according to the service requirements of traffic. The values of $w_i(t)$ for each traffic class is set as follows:

$$w_i(t) = \begin{cases} 1, & \text{for Class I} \\ \text{variable (0-1)}, & \text{for Class II} \\ 0, & \text{for Class III.} \end{cases} \quad (8)$$

The $w_i(t)$ is fixed at 1 for Class I. In this case the resulting rule behaves like a delay-based priority scheme for lagging flows ($\overline{D_i(t)} > 0$), whereas leading or in-sync flows ($\overline{D_i(t)} \leq 0$) still run on the PF policy. For Class III, $w_i(t)$ is fixed

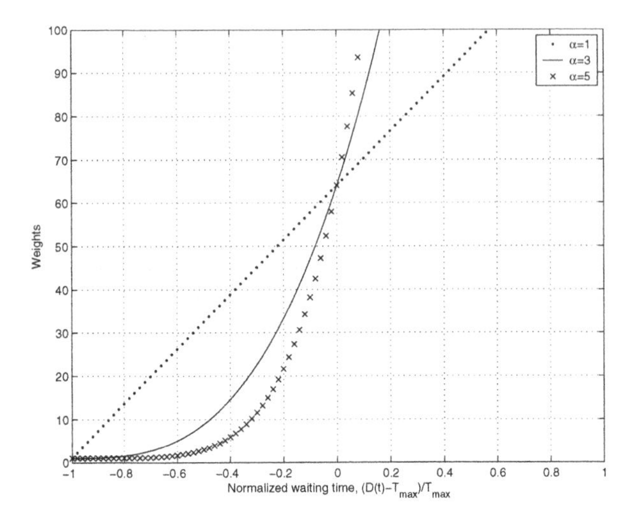

Fig. 1. Weight function $(C_{wd}(t)(\overline{D_i(t)}+1)^{\alpha}+1)$ versus $\overline{D_i(t)}$.

at 0, suppressing the weighting function and equalizing the rule to the PF policy. For Class II, $w_i(t)$ varies depending on how strictly bandwidth guarantee is required. Together with $w_i(t)$ an associated parameter denoted by Δ is used to indicate tolerable deviation from the target throughput. The rationale behind introducing this parameter is to control the way of allocating bandwidth over time. Depending on the type of applications, MTs may have different level of expectation on bandwidth guarantees. For applications that require a constant bandwidth over time, the rule needs to behave like WFQ rule. For the opposite case, some applications, e.g., Web browsing, may care about just the total amount of bandwidth allocated over the entire session, but not about strict guarantees over a microscopic time scale. For Class II, $w_i(t)$ is initially set to 0, and is reset to w_0 as the average rate $R_i(t)$ deviates from the promised rate $\widetilde{R_i}$ by Δ amount. From then on $w_i(t)$ is multiplied with the constant β until it reaches 1 or the average rate exceeds the promised rate. The following summarizes the algorithm:

1. Initialize $w_i(t)$,

$$w_i(t) = 0$$

2. If the average rate deviates from the promised rate by larger than the tolerance Δ, reset $w_i(t)$ to a non-zero value w_0.

$$w_i(t) = w_0, \qquad \text{if } \frac{R_i(t)-\widetilde{R_i}}{\widetilde{R_i}} < -\Delta$$

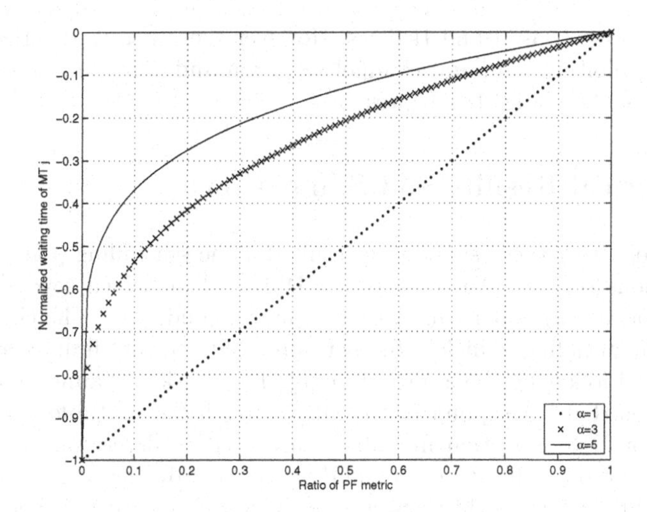

Fig. 2. Prioritization of leading flow over in-sync flow.

3. Update $w_i(t)$ at each scheduling epoch, i.e.,

$$w_i(t) = \begin{cases} \beta \times w_i(t), & w_i(t) < 1, \\ 1, & w_i(t) \geq 1 \end{cases}$$

4. Repeat Step 3 until $R_i(t) > \widetilde{R_i}$
5. If $R_i(t) > \widetilde{R_i}$, then $w_i(t) = 0$

Whereas the parameter $C_{wd}(t)$ is designed to prioritize Class I and II traffic over Class III, the parameter α is introduced to prioritize among the flows belonging to Class I or II. To analyze the impact of this parameter, we derive the condition in which a leading flow j gets scheduled before an in-sync flow i, i.e., $\overline{D_i(t)} = 0$. Assuming both the flow i and j belong to Class I for brevity, the metric of the flow j must satisfy:

$$\frac{\mathrm{DRC}_i(t)}{\hat{R}_i(t)}(C_{wd}(t)+1) < \frac{\mathrm{DRC}_j(t)}{\hat{R}_j(t)}\left(C_{wd}(t)(\overline{D_j(t)}+1)^{\alpha}+1\right).$$

Since $C_{wd}(t) \gg 1$, $1/C_{wd}(t) \approx 0$ and $\overline{D_j(t)}$ must satisfy

$$\overline{D_j(t)} > \left(\frac{\mathrm{DRC}_i(t)}{\hat{R}_i(t)} \bigg/ \frac{\mathrm{DRC}_j(t)}{\hat{R}_j(t)}\right)^{1/\alpha} - 1.$$

Figure 2 shows the conditions in which a leading flow j gets prioritized over an in-sync flow i for $\alpha = 1, 3, 5$. As α becomes larger, the leading flow must see much better channel and the waiting time of the HOL packet must further approach the target delay bound. Larger α leads to much stricter delay-based scheduling among Class I and II flows. However there is a tradeoff in choosing α between exploiting temporal channel variation and delivering more accurate QoS requirements as Fig. 1 indicates.

3 Numerical Results and Simulations

In this section, we present simulation results for the scheduling policy described in the previous section. We consider a CDMA cell in which several MTs run different applications under time-varying channel conditions. The channel models are Rayleigh fading, which is further approximated into finite-state Markov channel model following the approach in [8]. The Markov channel model has 10 channel states that correspond to the DRC rates listed in Table 1. The E_c/N_t thresholds for DRC selection in Table 1 are used to determine the ranges of E_c/N_t mapping onto the states of the discrete Markov channel model [9]. The speed of a MT determines how fast the channel varies, which will be characterized by the transition probability of the Markov channel model.

We simulate a scenario of mixed traffic in which MTs with RT traffic or with NRT traffic are uniformly positioned in a cell, competing for downlink bandwidth. We assume that NRT traffic generates packets following a Poisson process, whereas RT source traffic is modelled by the real-time video streaming model in [10]. In this model, a video streaming session consists of a sequence of frames with the period of 0.1 seconds. The number of packets generated in each frame is fixed at 8, and the packet inter-arrival time and packet size is distributed by truncated Pareto distribution:

$$F(x) = \begin{cases} 1 - \frac{K^a}{x^a}, & x < m, \\ 1, & x \geq m, \end{cases}$$

with $a = 1.2$, $K = 0.0025$ seconds, $m = 0.00125$ seconds for the inter-arrival time, and $a = 1.2$, $K = 20$ bytes, $m = 125$ bytes for the packet size, respectively. With these parameters the mean rate of a single video traffic is 32 Kbps. For the sake of simplicity, no packet error is assumed. Since DRC rate is selected to target at 1 % packet error rate [1], we believe the impacts of packet error on system performance can be safely ignored. Unless noted otherwise, the results are obtained with $\alpha = 3$. All the simulation results are reported based on 1000 seconds run, equivalent to 1,670,000 slots. Table 2 summarizes the values of the various parameters in simulation.

Figure 3 depicts throughput for three different scheduling policies: PF, EXP (Exponential rule in [4]), and Adaptive rule proposed in this paper. All the MTs are under homogeneous channel conditions, i.e., mean $E_c/N_t = 0$ dB and $v = 3$ Km/h. The number of MTs with RT application is fixed at 5. In terms of throughput, the PF policy shows the best performance due to its property of placing absolute priority on the MT seeing the best channel. However, throughput of RT traffic decreases proportionally as the shared link becomes loaded with more traffic. Obviously it is because RT traffic in the PF policy is equally

Table 2. Parameter values used in the simulations.

Parameters	Symbol	Value
Number of MTs with RT application	N_{rt}	3 - 5
Number of MTs with NRT application	N_{nrt}	10 - 50
Maximum inter-arrival time of NRT traffic	-	0.01 seconds
Mean packet size of NRT traffic	-	640 bits
Video frame period	-	0.1 seconds
Number of packets in a video frame	-	8
Mean rate of video traffic	-	32 Kbps
Mean inter-arrival time of video packets	-	0.006 seconds
Maximum inter-arrival time of video packets	-	0.0125 seconds
Mean packet size of video traffic	-	50 bytes
Maximum packet size of video traffic	-	125 bytes
Speed of MT (Km per hour)	v	3 - 120 Km/h

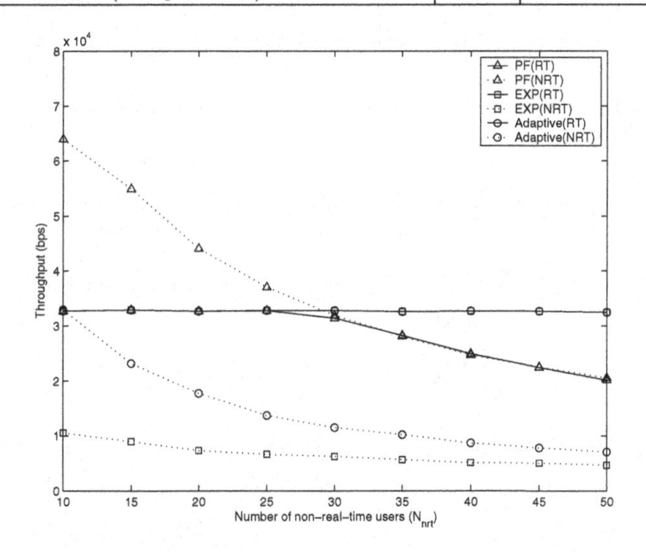

Fig. 3. Throughput of MTs with RT and NRT traffic ($v = 3$ Km/h).

treated with other NRT traffic. In contrast, the EXP and Adaptive policies guarantee a constant throughput (i.e., 32 Kbps) irrespective of traffic conditions. In particular, the Adaptive policy yields higher throughput in NRT traffic than the EXP policy. For $N_{nrt} = 10$, the Adaptive policy achieves twice the throughput of NRT traffic that the EXP policy does.

Figure 4 and 5 shows throughput and delay CDF of RT traffic, respectively, under heterogeneous channel conditions in which mean E_c/N_t of 0 dB, -3 dB, and 3 dB is given to three MTs. For the case of $v = 3$ Km/h in Fig 4, throughput decreases proportionally with worse channel conditions under the PF policy, whereas the Adaptive policy offers a steady throughput irrespective of channel conditions. In delay CDF, both the EXP and the Adaptive policies miss 0.2

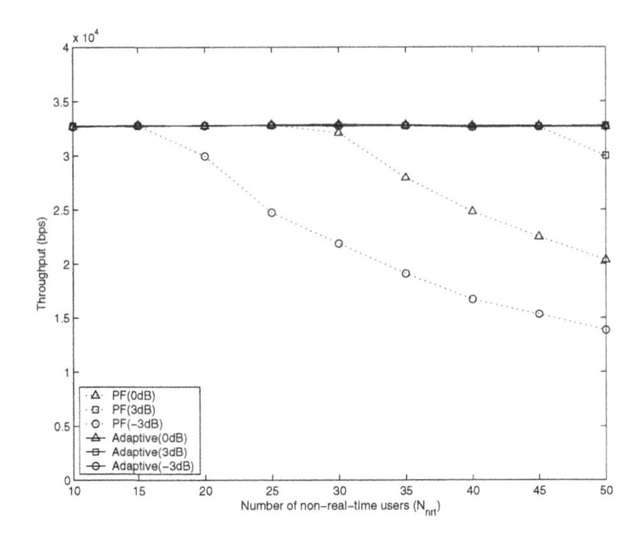

Fig. 4. Throughput under heterogeneous channel environments ($v = 3$ Km/h, $N_{nrt} = 30$).

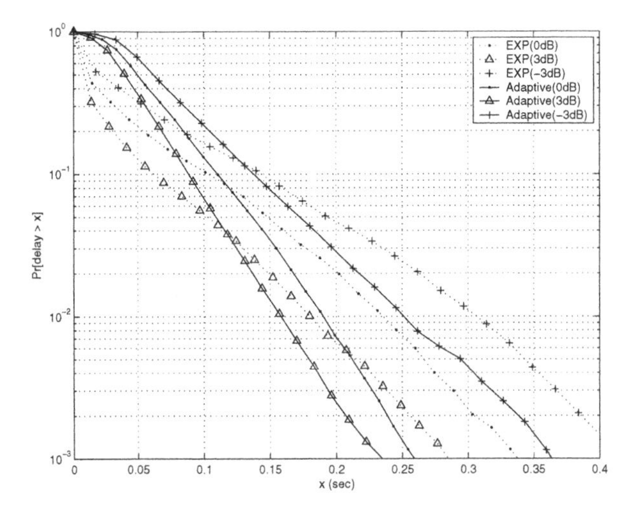

Fig. 5. Delay CDF under heterogeneous channel environments ($v = 3$ Km/h, $N_{nrt} = 30$).

seconds delay bound. However, the EXP policy fails at 0 and -3 dB, but the Adaptive policy does only at -3 dB.

Delay performance under heterogeneous QoS requirements is shown in Fig. 6 and 7. Five MTs having RT traffic set maximum delay bound at 0.1, 0.2, 0.3, 0.4, and 0.5 seconds, respectively. In the Adaptive policy, those bound are set to the parameter of $D_{i,max}$. In the EXP policy, the weight factor is set following

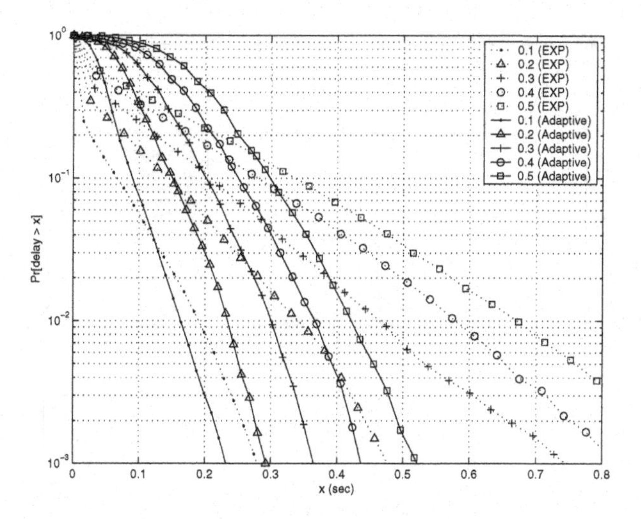

Fig. 6. Delay CDF with heterogeneous QoS requirements ($v = 3$ Km/h, $N_{nrt} = 50$).

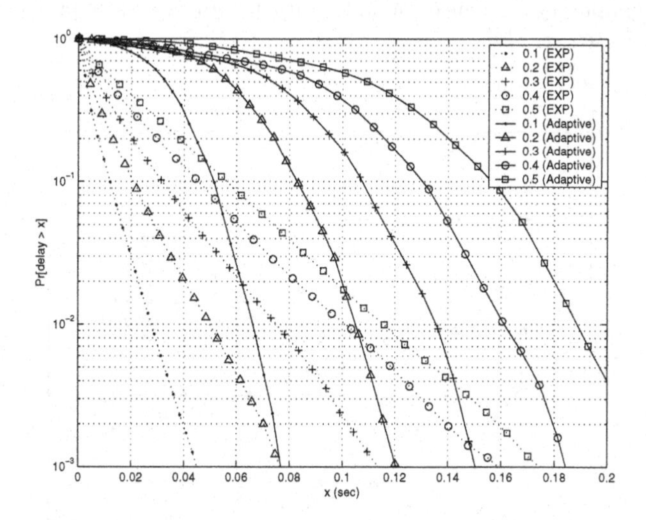

Fig. 7. Delay CDF with heterogeneous QoS requirements ($v = 50$ Km/h, $N_{nrt} = 50$).

the formula $a_i = -\log(\delta_i)/T_i$ suggested in [4]. Here δ_i and T_i is derived from the following delay requirement:

$$P[\text{Delay} > T_i] \leq \delta_i.$$

Thus, δ_i is set to 0.01 and T_i is set to from 0.1 to 0.5 as above. For the case of $v = 3$ Km/h, both policies did not deliver successfully the specified QoS requirements. However when it comes to the level of deviation, the EXP policy far exceeds the desired delay bound, whereas the Adaptive policy offers a delay bound relatively close to the specified values. For $v = 50$ Km/h, both policies

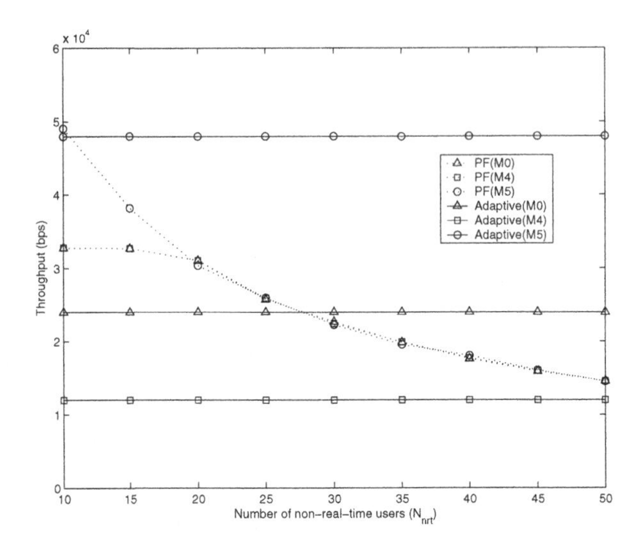

Fig. 8. Throughput performance for MTs with heterogeneous requirements (v = 3 Km/h, Mean $E_c/N_t = -3$dB).

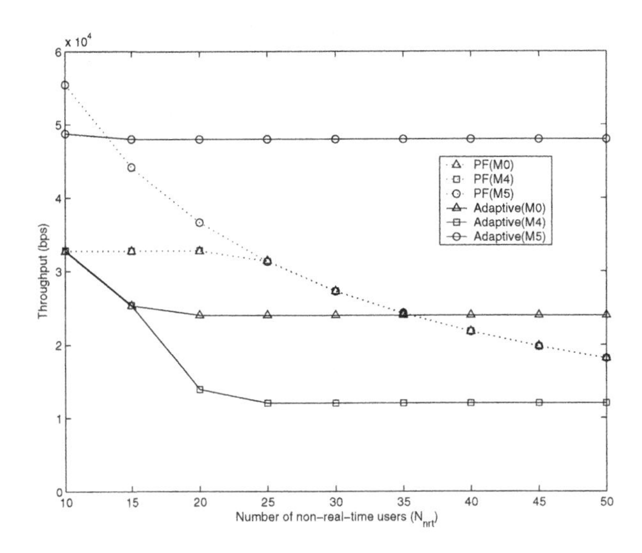

Fig. 9. Throughput performance for MTs with heterogeneous requirements (v = 50 Km/h, Mean $E_c/N_t = -3$dB).

meet the maximum delay bound as specified, but in a somehow conservative manner. In particular, the EXP policy exercises too much bandwidth as shown in Fig. 7. From those results, the Adaptive policy can offers a maximum delay bound more accurately relative to the EXP policy.

Figure 8 and 9 shows throughput performance of the Adaptive policy in a scenario where MTs have traffic requiring a strict throughput guarantee. Two MTs, M0 and M4, have mean data rate of 32 Kbps for both and target throughput of 12 Kbps and 24 Kbps, respectively. And the MT M5 has the mean data of 64 Kbps and target throughput of 48 Kbps. Under the PF policy all MTs yield basically equivalent throughput given the same channel conditions. Due to this property of fair sharing, throughput per a MT decreases as the number of MTs increases for $v = 3, 50$ Km/h. Thus, it is not possible to guarantee the requested bandwidth with the PF policy. In contrast, the Adaptive policy, as shown in the figures, provides a steady bandwidth without being affected by traffic conditions. Notably two MTs, M0 and M4, requiring the target of 12 and 24 Kbps, respectively, get some extra bandwidth at $N_{nrt} = 10, 20$ (See Fig. 8 and 9) since overall link capacity increases with faster speed and the system is not highly loaded. This indicates the Adaptive policy basically runs on the PF policy, and switches to WFQ-like mode when QoS cannot be satisfied with the PF policy. Such capability of adaptation is significant in that it can exploit better throughput performance of the PF policy while offering QoS guarantees when needed.

4 Conclusions

In this paper, we proposed a new scheduling policy that provides QoS guarantees to a variety of traffic types demanding diverse service requirements. In our proposal, data traffic is categorized into three classes, depending on sensitivity of its performance to delay or throughput. And the primary components of our policy, namely, the PF, WFQ, and delay-based prioritized scheme are intelligently combined to satisfy QoS requirements of each traffic type. Our policy changes the level of emphasis on each of those ingredient policies in an adaptive manner, taking into account the channel conditions and QoS requirements. Such flexibility, as shown in numerical results, leads to offering QoS guarantees effectively and, at the same time, maximizing the throughput.

References

1. Bender, P., Black, P., Grob, M., Padovani, R., Sindhushayana, N., Viterbi, A.: CDMA/HDR: a bandwidth-efficient high-speed wireless data service for nomadic users. IEEE Commun. Mag. **38** (2000) 70–77
2. Jalali, A., Padovani, R., Pankaj, R.: Data throughput of CDMA-HDR: a high efficiency, high data rate personal communication wireless system. Proc. IEEE VTC '00, **3** (2000) 1854–1858
3. Kelly, F.: Charging and rate control for elastic traffic. European Trasactions on Telecommunications **8** (1997) 33–37
4. Shakkottai, S., Stolyar, A. Scheduling algorithms for a mixture of real-time and non-real-time data in HDR. Proc. International Teletraffic Congress '2001 (2001)
5. Liu, P., Berry, R., Honig, M.: Delay-sensitive packet scheduling in wireless networks. Proc. IEEE WCNC '03, **3** (2003) 1627–1632

6. Huang, C., Su, H., Vitebsky, S., Chen, P.: Schedulers for 1xEV-DO: third generation wireless high-speed data systems. Proc. IEEE VTC '03 **3** (2003) 1710–1714
7. Zhang, H., Service discipline for guaranteed performance service in packet-switching networks. Proc. IEEE **83** (1995) 1374–1396
8. Wang, H., Moayeri, N.: Finite-state Markov channel–a useful model for radio communication channels. IEEE Trans. Veh. Technol. **44** (1995) 163–171
9. Paranchych, D., Yavuz, M.: Analytical expression for 1xEV-DO forward link throughput. Proc. IEEE WCNC '03 **1** (2003) 508–513
10. 3GPP2: 3GPP2 1xEV-DV Evaluation Methodology (2001)

Erlang Capacity Analysis of Multi-access Systems Supporting Voice and Data Services

Insoo Koo[1], Anders Furuskar[2], Jens Zander[2], and Kiseon Kim[1]

[1] Dept. of Infor. and Comm., Gwang-Ju Institute of Science and Technology, 1 Oryong-dong, Puk-gu, Gwangju, 500-712, Korea
{kiss and kskim}@kjist.ac.kr

[2] Radio Communication Systems, S3
Isafjordsgatan 30B, KISTA, Stockholm, Sweden.
anders.furuskar@ericsson.com, and jens.zander@radio.kth.se

Abstract. In this paper, we analyze and compare the Erlang capacity of multi-access systems supporting several different radio access technologies according to two different operation methods: separate and common operation methods. In the common operation, any terminal can connect to any sub-system while each terminal in the separate operation only can connect to its designated sub-system. In a numerical example with GSM/EDGE-like and WCDMA-like sub-systems, it is shown that we can get up to 60% Erlang capacity improvement through the common operation method when using a near optimum so-called service-based user assignment scheme. Even with the worst-case assignment scheme, we can still get about 15% capacity improvement over the separate operation method.

1 Introduction

Future mobile networks will consist of several distinct radio access technologies, such as WCDMA or GSM/EDGE, where each radio access technology is denoted as "sub-system." Such future wireless networks demanding utilizing the cooperative use of a multitude of sub-systems are named multi-access systems. In the first phase of such multi-access systems, the radio resource management of sub-systems may be performed in a separate way to improve the performance of individual systems independently, mainly due to the fact that the sub-systems have no information of the situation in other sub-systems, or that the terminals do not have multi-mode capabilities. Under such a separate operation method, an access attempt is only accepted by its designated sub-system if possible, and otherwise rejected.

Intuitively, improvement of multiple-access systems is expected in a form of common resource management where the transceiver equipment of the mobile stations supports multi-mode operations such that any terminal can connect to any sub-system. This may be accomplished either through parallel tranceivers in hardware, or using software radio [1]. The common radio resource management functions may be implemented in existing system nodes, but inter-radio access

I. Niemegeers and S. Heemstra de Groot (Eds.): PWC 2004, LNCS 3260, pp. 69–78, 2004.

technology signaling mechanisms need to be introduced. In order to estimate the benefit of such common resource management of the multi-access systems, some studies are necessary especially in aspects of quantifying the associated Erlang capacity.

As an example of improving the performances of common resource management, for single-service scenarios, the "trunking gain" of multi-access system capacity enabled by the larger resource pool from common resource management has previously been evaluated in [2] through simulation, and multi-service allocation is not considered. In multi-service scenarios, it is expected that the capacity of multi-access systems also depends on how users of different services are assigned on to sub-systems. The gain that can be obtained through the employed assignment scheme can be named as "assignment gain", and further the capacity gain achievable with different user assignment principles has been estimated in [3,8]. These studies however disregard trunking gains. In this paper, we combine both trunking and assignment gains, and quantify the Erlang capacity of the multi-access system. Further, we provide the upper and lower bounds of the Erlang capacity of the multi-access system by considering two extreme cases; one is that all terminals can not support multi-mode function, and the other one is that all terminals can do, which corresponds to the separate and common operations of multi-access systems. In the case of the common operation, we also consider two kinds of user assignment schemes; service-based user assignment [3] as the best case, which roughly speaking assigns users to the sub-system where their service is most efficiently handled, and the rule opposite to the service-based assignment as the worst case reference. With the consideration of these extreme cases, the Erlang capacity of multi-access systems would be a useful guideline for operators of multi-access systems.

2 Erlang Capacity Analysis of the Multi-access Systems

In this section, we consider a multi-access system consisting of two sub-systems supporting two services, voice and data, and analyze the Erlang capacity of the multi-access systems according to two different operation methods; separate and common operations. However, it is noteworthy that generalization of this analysis to a case with arbitrary number of sub-systems and services is straightforward, but for lack of space reasons left out.

2.1 Separate Operation of the Multi-access Systems

In the separate operation method of the multi-access systems, an access attempt is accepted by its designated sub-system if possible, and otherwise rejected. In order to evaluate the performance of the separate operation in the aspects of traffic analysis, at first we need to identify the admissible region of voice and data service groups in each sub-system. Let Q_v^l and Q_d^l be the link qualities such as frame error rate that individual voice and data users experience in the sub-system l ($l = 1, 2$) respectively, and $Q_{v,min}$ and $Q_{d,min}$ be a set of minimum link quality level of each service. Then, for a certain set of system parameters such as service quality requirements, link propagation model and system assumption, the admissible region of the sub-system l with respect to the simultaneous number of users satisfying service quality requirements in the sense of statistic, $S_{sub,l}$ can be defined as

$$
\begin{aligned}
&S_{sub,l} \\
&=\{(n_{(v,l)}, n_{(d,l)}) | P_r(Q_v^l(n_{(v,l)}, n_{(d,l)}) \geq Q_{v,min} \ and \\
&\qquad\qquad\qquad\qquad Q_d^l(n_{(v,l)}, n_{(d,l)}) \geq Q_{d,min}) \geq \beta\%\} \\
&=\{(n_{(v,l)}, n_{(d,l)}) | 0 \leq f_l(n_{(v,l)}, n_{(d,l)}) \leq 1 \ and \ n_{(v,l)}, n_{(d,l)} \in Z_+\} \ for \ l = 1, 2
\end{aligned} \tag{1}
$$

where $n_{(v,l)}$ and $n_{(d,l)}$ are the admitted number of calls of voice and data service groups in the sub-system l respectively, $\beta\%$ is the system reliability defined as minimum requirement on the probability that the link quality of the current users in the sub-system l is larger than the minimum link quality level, which is usually given between 95% and 99%, and $f_l(n_{(v,l)}, n_{(d,l)})$ is the normalized capacity equation of the sub-system l. In the case of linear capacity region, for example $f_l(n_{(v,l)}, n_{(d,l)})$ can be given as $f_l(n_{(v,l)}, n_{(d,l)}) = a_{lv} \cdot n_{(v,l)} + a_{ld} \cdot n_{(d,l)}$ for l=1,2. Such linear bounds on the total number of users of each class, which can be supported simultaneously while maintaining adequate QoS requirements, are commonly found in the literature for CDMA systems supporting multi-class services [4,5]. Further provided the network state lies within the admissible region, then the QoS requirement of each user will be satisfied with $\beta\%$ reliability.

In the aspects of network operation, it is of vital importance to set up a suitable policy for the acceptance of an incoming call in order to guarantee a certain quality of service. In general, call admission control (CAC) policies can be divided into two categories: number-based CAC (NCAC) and interference-based CAC (ICAC) [6]. In the case of ICAC, a BS determines whether a new call is acceptable or not by monitoring the interference on a call-by-call basis while the NCAC utilizes a pre-determined CAC threshold. In this paper, we adopt a NCAC-type CAC since its simplicity with which we can apply general loss network model to the system being considered for the performance analysis, even though the NCAC generally suffers a slight performance degradation over the ICAC [6] and the CAC thresholds for the NCAC should be specifically redesigned in the case of changes in propagation parameters or traffic distributions. That is, a call request is blocked and cleared from the system if its acceptance would move the next state out of the admissible region, delimited by Eqn.(1), otherwise it will be accepted.

In order to focus on the traffic analysis of sub-systems under the CAC policy of our interest, we also consider the standard assumptions on the user arrival and departure processes. That is, we assume that call arrivals from users of class j in the sub-system l are generated as a Poisson process with rate $\lambda_{(j,l)}$ $(j = v, d)$. If a call is accepted then it remains in the cell and sub-system of its origin for an exponentially distributed holding time with mean $1/\mu_{(j,l)}$ which is independent of other holding times and of the arrival processes. Then, the offered traffic load of the j-th service group in the sub-system l is defined as $\rho_{(j,l)} = \lambda_{(j,l)}/\mu_{(j,l)}$.

With these assumptions, it is well known from M/M/m queue analysis that for given traffic loads, the equilibrium probability for an admissible state $N_l(\equiv (n_{(v,l)}, n_{(d,l)}))$ in the sub-system l, $\pi(N_l)$ can have a product form on the the truncated state space defined by the call admission strategy such that it is given by [7]:

$$\pi(N_l) = \begin{cases} \frac{\rho_{(v,l)}^{n_{(v,l)}} \rho_{(d,l)}^{n_{(d,l)}}}{n_{(v,l)}! n_{(d,l)}!} / \sum_{N_l \in S_{sub,l}} \frac{\rho_{(v,l)}^{n_{(v,l)}} \rho_{(d,l)}^{n_{(d,l)}}}{n_{(v,l)}! n_{(d,l)}!}, & N_l \in S_{sub,l}, \\ 0 & \text{otherwise.} \end{cases} \tag{2}$$

Subsequently, the blocking probability for a user of class j in the sub-system l is also simply expressed as

$$B_{(j,l)} = \sum_{N_l \in S^j_{blk,l}} \pi(N_l) \tag{3}$$

where $S^j_{blk,l}$ is the subset of states in $S_{sub,l}$ which states must move out of $S_{sub,l}$ with the addition of one user of class j. Here, it is noteworthy that $\pi(N_l)$ and $B_{(j,l)}$ are dependent on the admission region $S_{sub,l}$, and the traffic loads $\rho_{(j,l)}$. Then, the Erlang capacity of the sub-system l, defined as the set of supportable offered traffic load of each service, can be calculated as a function of supportable offered traffic loads of all service groups such that

$$C_{(Erlang,l)} = \{(\rho_{(v,l)}, \rho_{(d,l)}) | \ B_{(v,l)} \leq P_{(B,v)_{req}} \ and \ B_{(d,l)} \leq P_{(B,d)_{req}}\} \tag{4}$$

where $P_{(B,j)_{req}}$ is the required call blocking probability of service j and can be considered as GoS requirement.

Finally, the combined Erlang capacity, C_{Erlang}, of the multi-access systems under the separate operation is the sum of those of sub-systems such that

$$C_{Erlang} = \{(\rho_v, \rho_d) | (\rho_v, \rho_d) \equiv \sum_{l=1}^{2} (\rho_{(v,l)}, \rho_{(d,l)}), \ (\rho_{(v,l)}, \rho_{(d,l)}) \in C_{(Erlang,l)} \ for \ l = 1, 2\} \tag{5}$$

2.2 Common Operation of the Multi-access Systems

In the common operation of the multi-access systems, the admissible region of the considered multi-access systems depends on how users of different services are assigned onto the sub-systems. That is, according to the employed user assignment scheme in the common operation, the admissible region of multi-access systems can be one of subset of the following set:

$$S_{system} = \{(n_v, n_d)|\ (n_v, n_d) \equiv \sum_{l=1}^{2}(n_{(v,l)}, n_{(d,l)}), and\ 0 \leq f_l(n_{(v,l)}, n_{(d,l)}) \leq 1\ for\ l = 1, 2\} \tag{6}$$

where n_v and n_d are the admissible number of users of voice and data in the multi-access system.

In the reference [3], Furuskar discussed principles for allocating multiple services onto different sub-systems in multi-access wireless systems, and further derived the favorable near-optimum sub-system service allocation scheme through simple optimization procedures that maximizes the combined multi-service capacity, which here is named as "service-based assignment algorithm." In order to focus on investigating the Erlang capacity improvement through the common operation of multi-access systems by simultaneously considering the trunking gain and the assignment gain, in this paper we only consider two kinds of user assignment schemes for the common operation of multi-access systems, that is, a service-based assignment algorithm, which was proposed in [3] as a near-optimum user assignment method, and a rule opposite to the service-based assignment algorithm as a worst case assignment method. It is noteworthy that the rule opposite to service-based assignment is not likely to be sued in reality but here we adopt it as a interesting reference for the worst case of the common operation method.

In the service-based assignment algorithm, we assign users into the sub-system where their expected relative resource cost for the bearer service type in question is the smallest. That is, when a user with service type j is coming in the multi-access system ($j = v$ or d), then we assign the user to the sub-system $\hat{l}$ that meets the following.

$$\hat{l} = arg\left\{\min_l\left(\frac{\partial f_l(n_{(v,l)}, n_{(d,l)})}{\partial n_{(j,l)}} / \frac{\partial f_l(n_{(v,l)}, n_{(d,l)})}{\partial n_{(\sim j,l)}}\right)\right\} \tag{7}$$

where $\sim j$ is the 'other service' that is, if j=v then, $\sim j$ is d. For the case that each sub-system has a linear capacity region, then the assignment rule can be simply expressed as $\hat{l} = arg\left\{\min_l\left(\frac{a_{lj}}{a_{l\sim j}}\right)\right\}$.

On the other hand, in the rule opposite to the service-based assignment algorithm, we assign the user having service type j to the sub-system $\hat{l}$ that meets the following.

$$\hat{l} = arg\left\{\max_l\left(\frac{\partial f_l(n_{(v,l)}, n_{(d,l)})}{\partial n_{(j,l)}} / \frac{\partial f_l(n_{(v,l)}, n_{(d,l)})}{\partial n_{(\sim j,l)}}\right)\right\} \tag{8}$$

According to the employed user assignment scheme, we can obtain corresponding admissible region of the multi-access systems under the common operation. If we denote $S_{s-based}$ as the admissible region of the multi-access systems with the service-based assignment scheme, and $S_{opp-s-based}$ as one with the rule

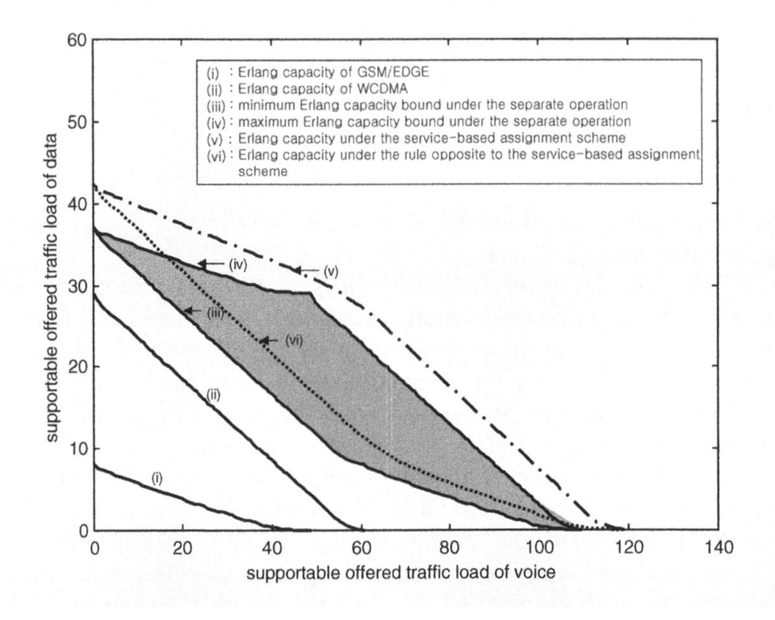

Fig. 1. Erlang capacity of a GSM/EDGE-like and WCDMA-like multi-access system.

opposite to the service-based assignment scheme, respectively. Then, we can calculate the corresponding Erlang capacities for above two assignment scheme using the similar method presented in previous section, by using Eqns(2-4) after replacing $S_{sub,l}$ with $S_{s-based}$ and $S_{opp-s-based}$, respectively.

3 Numerical Results

So far we have presented analytical procedures to investigate the Erlang capacity of multi-access systems. In this section, we will consider two set of scenarios with different bearer capabilities and quality requirements of sub-systems to visualize the capacity gains of multi-access system according to the two operation methods. First scenario is a case with coexisting GSM/EDGE-like and WCDMA-like sub-systems as a practical example. The other is more artificial case to consider the effect of sub-system capacities on the Erlang capacity of multi-access systems.

As a practical example, let's firstly consider a case, $P = 2$ with coexisting GSM/EDGE-like and WCDMA-like sub-systems. When a spectrum allocation of 5 MHz is assumed for both systems, admissible capacity regions of both systems supporting mixed voice and data traffics are modeled as a linear region such that $f_l(n_{(v,l)}, n_{(d,l)})$ is given as $a_{lv} \cdot n_{(v,l)} + a_{ld} \cdot n_{(d,l)}$ for l=1,2 where GSM/EDGE-like system is denoted as sub-system 1 while WCDMA-like system as sub-system 2. Further, $(a_{1v}\ a_{1d})$ and $(a_{2v}\ a_{2d})$ are given as (1/62 1/15) and (1/75 1/40) respectively, for standard WCDMA and EDGE data bearers and a circuit switched equivalent (CSE) bitrate requirement of 150Kbps [8]. Fig. 1 shows the resulting Erlang capacity regions when the required call blocking probability is set to 1%. Fig. 1(i) and (ii) show the Erlang capacity of GSM/EDGE and WCDMA, respectively. Then, the Erlang capacity of multi-access system under the separate operation can be given as the vector sum of those of sub-systems as like Fig. 1. It is noteworthy that the Erlang capacity line, stipulating the Erlang capacity region of multi-access system, depends on the service mix in the sub-systems, and lies between the minimum bound line (Fig. 1(iii)) and the maximum bound line (Fig. 1(iv)). This means that the shadowed traffic area, delimited by Fig. 1(iii) and Fig. 1(iv), is not always supported by the multi-access system under the separate operation. For example, the traffic load of (46, 29) can be supported only when the GSM/EDGE supports the voice traffic of 46, and the WCDMA supports the data traffic of 29, but this occasion is very rare. Subsequently, we should operate the system with the Erlang capacity region stipulated by Fig. 1(iii) for the sake of stable system operation. On the other hand, Fig. 1(v) shows the Erlang capacity region of multi-access system under the service-based assignment algorithm. In this case, with the service-based assignment scheme we assign voice users to GSM/EDGE as far as possible, and data users to WCDMA since GSM/EDGE is relatively better at handling voice users than WCDMA, and vice-versa for data users. As a result, it is observed that we can get about 95% capacity improvement through the service-based assignment algorithm over the separate operation where we utilize the supportable area of Erlang capacity for the performance comparison. Fig. 1(vi) also shows the Erlang capacity region of multi-access system when assigning users according to the rule opposite to the service-based assignment algorithm. In this case, the voice users are as far as possible assigned to WCDMA and as many data users as possible to GSM/EDGE, which corresponds to the worst-case scenario in the common operation. The resulting Erlang capacity is dramatically lower than that of the service-based assignment algorithm (roughly 54% capacity reduction). However, ever in the worst case, we know that the Erlang capacity of multi-access system under the common operation still can provide about 27% capacity improvement over the separate operation.

Secondly, we consider an artificial case to consider the effect of air-link capacities of sub-systems on the Erlang capacity of multi-access systems where the admissible regions of each sub-system are also delimited by the linear bound, and $(a_{1v}\ a_{1d})$ and $(a_{2v}\ a_{2d})$ are given as (1/10 1/10) and (1/20 1/10), respectively. Fig. 2 shows the resulting Erlang capacity regions for the two operation methods. With the service-based assignment scheme, in this case we assign voice users to sub-system 2 as far as possible, and data users to sub-system 1 since sub-system 2 is relatively better at handling voice users than sub-system 1, and vice-versa for data users. As a result, we can achieve a gain of up to 37% over the rule opposite to the service-based assignment through the service-based user assignment, and the gain of up to 88.5% over the separate operation method. When comparing these results with those of the previous example, we also know that the Erlang capacity gains of multi-access systems, which can be achieved

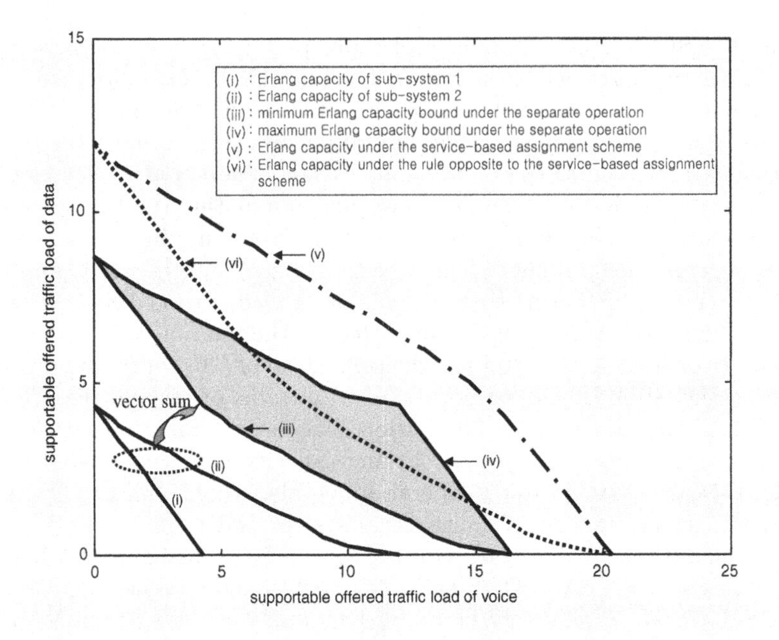

Fig. 2. Erlang capacity of a multi-access system for the two operation methods.

by the operation methods, are very sensitive to the sub-system capacities such as the shape and the area of the capacity.

Finally, Fig. 3 shows the Erlang capacity gain of a multi-access system according to the traffic mix-ratio between voice and data for the previous two numerical examples. Here, we define the traffic mix-ratio as $\rho_v/(\rho_v + \rho_d)$, and use the total supportable traffic load of the system for the performance comparison, i.e the sum of the maximum supportable voice and data traffic for a certain service-mix ratio. Noting that the Erlang improvement of common mode operation over the separate operation converges into a trunking gain as the traffic mix-ratio between voice and data goes to 0 or 1, we know that the Erlang improvement of common mode operation is mainly due to the trunking efficiency gain when the rule opposite to the service-based assignment scheme is used, and that the gain is less sensitive to the traffic mix-ratio between voice and data while it is sensitive to the sub-system capacities. On the other hand, Fig. 1 shows that the Erlang capacity improvement in the case of the service-based assignment scheme varies according to the traffic mix-ratio between voice and data. This means that in this case we can get both a trunking efficiency gain and a service-based assignment gain simultaneously. It is noteworthy that the trunking efficiency gain is rather insensitive to the service mix, whereas the service-based assignment gain depends significantly on the service mix. The service-based assignment scheme is thus more beneficial in mixed service scenarios.

This paper can be further extended, by considering different service control mechanisms, and the capabilities that each terminal can handle multi-mode

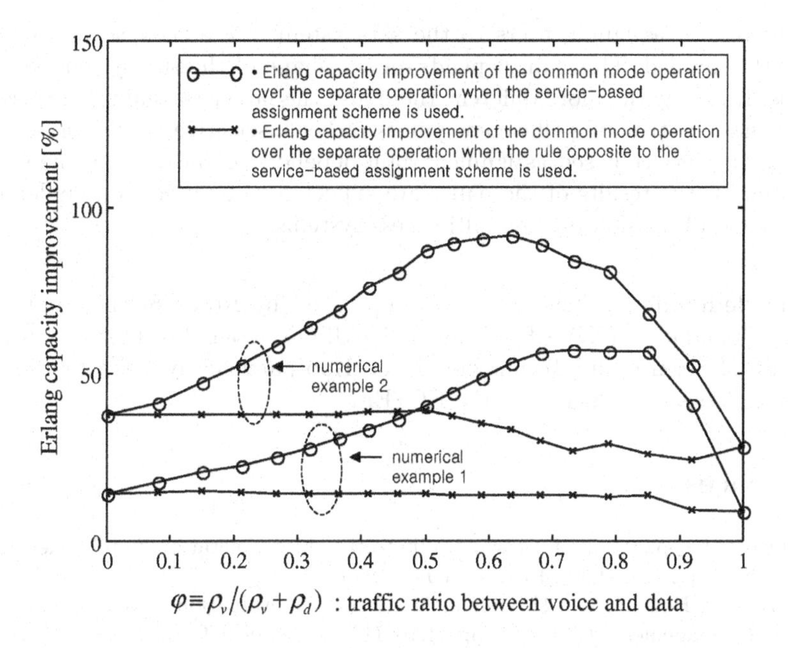

Fig. 3. Erlang capacity improvement of a multi-access system according to the traffic mix-ratio between voice and data, φ

function. For the lack of space, in this paper we have only dealt with two extreme cases; separate and common operations of multi-access systems. For the case that non-multi-mode and multi-mode terminals coexist, however we can expect that the corresponding Erlang capacity will fall into the regions between the Erlang capacity under the separate operation and that under the common operation, even though we can not exactly mention how and how much the fraction of non-multi-mode terminals affects overall performance of multi-access system. The corresponding Erlang capacities according to different service control mechanisms will also exist in the regions between the Erlang capacity under the service-based user assignment and that under the rule opposite to the service-based assignment, since the service-based user assignment is proved to be the optimum sub-system service allocation scheme [3].

4 Conclusion

In this paper, we investigate the Erlang capacity of multi-access systems according to two different operation methods: separate and common operation methods. Through numerical examples, we observe that the Erlang capacity improvement that can be obtained through common operation method is two-fold. First, a trunking efficiency gain is achieved due to the combining of resource pools. This gain depends on the sub-system capacities, for small sub-system capacities the gain is significant. Secondly, a service-based assignment gain can be achieved by assigning users to the sub-system where their service is most efficiently handled. This gain depends on the shape of the sub-system capacity regions. Roughly, the more different these are, the larger the gain. It is also observed that the trunking efficiency gain is rather insensitive to the service mix, whereas the service-based assignment gain depends significantly on the service mix. Finally, the results of the paper are expected to be used as a guideline to operate and dimension future multi-access systems.

Acknowledgement. This work was supported by Korea Science and Engineering Foundation (KOSEF) through the UFON research center at Gwangju Institute of Science and Technology. Insoo Koo particularly was supported by the Basic Research Program of the KOSEF.

References

1. S. Ogose, "Application of software radio to the third generation mobile telecommunications," pp.1212-1216, Proc. of VTC, 1999.
2. A. Tolli, P. Hakalin, and H. Holma, "Performance evaluation of common radio resource management (CRRM)," pp.3429- 3433, Proc. of ICC, 2002.
3. A. Furuskar, "Allocation of multiple services in multi-access wireless systems," pp.261-265, Proc. of MWCN, 2002.
4. J. Yang, Y. Choi, J. Ahn and K. Kim, "Capacity plane of CDMA system for multimedia traffic," IEE Electronics Letters, vol.33, no.17, pp.1432-1433, Aug. 1997.
5. A. Sampath, P. Kumar and J. Holtzman, "Power control and resource management for a multimedia CDMA wireless system," pp.21-25, Proc. of PIMRC, 1995.
6. Y. Ishikawa and N. Umeda, "Capacity design and performance of call admission cotnrol in cellular CDMA systems," IEEE JSAC, vol. 15, pp.1627-1635, Oct. 1997.
7. F. Kelly, "Loss networks," *The Annuals of Applied Probability*, Vol.1, pp.319-378, 1999.
8. A. Furuskar, "Radio resource sharing and bearer service allocation for multi-bear service, multi-access wireless networks," Ph.D thesis, available at http://www.s3.kth.se/radio/Publication/Pub2003/af_phd_thesis_A.pdf

Mobility Management in Cellular Communication Systems Using Fuzzy Systems

J.J. Astrain[1], J. Villadangos[2], M. Castillo[1], J.R. Garitagoitia[1], and F. Fariña[1]

[1] Dpt. Matemática e Informática
[2] Dpt. Automática y Computación
Universidad Pública de Navarra
Campus de Arrosadía
31006 Pamplona (Spain)
josej.astrain@unavarra.es

Abstract. Mobility management in cellular communication systems is needed to guarantee quality of service, and to offer advanced services based on the user location. High mobility of terminals determines a high effort to predict next movement in order to grant a correct transition to the next phone cell. Then a fuzzy method dealing with the problem of determining the propagation path of a mobile terminal is introduced in this paper. Since multi-path fading and attenuation make difficult to determine the position of a terminal, the use of fuzzy symbols to model this situation allows to work better with this imprecise (fuzzy) information. Finally, the use of a fuzzy automaton allows to improve significatively the final recognition rate of the path followed by a mobile terminal.

1 Introduction

The scenario of this work is a mobile cellular communication system, where mobility of terminals (mobile stations, MS), users and services must be granted and managed.

Mobile communication systems are limited in terms of bandwidth, and they must deal with non-uniform traffic patterns. So, the development of a mobility management model is justified in order to inter-operate different mobile communication networks. The mobility management model considers three tasks: *location*, mobile stations should be located in order to find them as fast as possible when a new incoming call occurs; *directioning*, the knowledge about the direction followed by a mobile station during its movement across the cell allows to offer some intelligent services; and *handover* control.

Handover processes are the transitions between two contiguous cells performed by a mobile station when travelling across a cellular system. Resources allocated by a particular mobile station when it is placed in a cell must be renegotiated with the next arrival cell when travelling across the network, and disposed after the cell transition. In homogeneous networks, the handover process occurs when signal-noise ratio (or another parameter) is below a given level.

I. Niemegeers and S. Heemstra de Groot (Eds.): PWC 2004, LNCS 3260, pp. 79–91, 2004.

In heterogeneous ones, the handover takes specific parameters of the networks, aspects about quality of service, and user's preferences into account.

In the literature a great variety of algorithms to manage the mobility of users or terminals has been described, the main part of them considering predictive algorithms because in the new communication environments, both resources and services present mobility. A lot of concepts of mobility were first introduced in cellular mobile networks for voice, now data and multimedia is also included. Mobility models classification (used in computing and communication wireless networks) is provided in [3]. Mobility models (either macroscopic or microscopic) are fundamental to design strategies to update location and paging, to manage resources and to plan the whole network. The mobility model influence will increase as the number of service subscribers grows and the size of the network units (cells) decreases.

The progress observed on wireless communications makes it possible to combine a predictive mobility management with auxiliary storage (dependent on the location, caching, and prefetching methods). Mobile computing in wireless communication systems opens an interesting number of applications including quality of service management (QoS) and final user services. Some examples of predictive applications are: a new architecture to browse web pages [6], where the predictive algorithm is based on a learning automaton that attributes the percentage of cache to the adjacent cells to minimize the connection time between servers and stations; the adaptation of the transport protocol TCP to wireless links in handover situations [7], that uses artificial intelligence techniques (learning automaton) to carry out a trajectory prediction algorithm which can relocate datagrams and store them in the caches of the adjacent cells; an estimation and prediction method for the ATM architecture in the wireless domain [10], where a method improving the reliability on the connection and the efficient use of the bandwidth is introduced using matching of patterns with Kalman filters (stability problems); and so on.

The solution here presented is based on the concept of Shadow Cluster that defines dynamically the influence area of the mobile element, getting an interesting prediction of the trajectory of the mobile and therefore of its influence area. This solution follows a dynamic strategy in contrast to a simple periodic updating of the position of the mobile stations. The proposed system includes some intelligence level needed to interact with the quasi-deterministic behaviour of the user profile movements, and the prediction of the random movements.

Sensitivity to deviations from the real path followed by a MS is controlled by means of editing operations [8]. Paths are modelled as strings of symbols that are classified performing an imperfect string matching between the path followed and the dictionary containing the pattern paths existing in the current cell. The number of patterns that must be saved in the user's profile (user dictionary) is reduced, because only the most frequent paths followed by the user and the main paths followed by the rest of users are considered. The degree of resemblance between two patterns is measured in terms of a fuzzy similarity [5] by way of a fuzzy automaton. Fuzzy techniques perform better than statistical ones for this problem because in cellular systems, the signal power used to determine the location of a MS are affected by fading, attenuation and multi-path propagation problems. So, location uncertainties due to the signal triangulation are better described and managed using fuzzy techniques. An advantage of this predictive fuzzy method is that it deals with a non limited number of signal propagation errors and its independence from the system architecture.

Future cellular systems will use microcells [4] to answer the high density of calls. The direct consequence of this will be an increment of the number of handoffs/handovers and so algorithms that complete as fast as possible these processes will be required. The fuzzy system here presented represents an alternative to the growth of the offered services and the effective use of the bandwidth compared with natural solutions such as changes in modulation techniques or special codifications. The research referred here shows the open investigation lines in wireless networks and its integration in the current technology.

The rest of the paper is organized as follows: section 2 is devoted to present the mobility management system model used; section 3 introduces the predictive capabilities of the mobility management system model previously presented; section 4 evaluates the results obtained for this technique; finally conclusions and references end the paper.

2 Mobility Management System Model

As introduced in [9], mobility modelling, location tracking and trajectory prediction are suitable in order to improve the QoS of those systems. By predicting future locations and speeds of MSs, it is possible to reduce the number of communication dropped and as a consequence, a higher QoS degree. Since it is very difficult to know exactly the mobile location, an approximate pattern matching is suitable.

In [9], movement is split into the regular part of the movement and the random part. The latter part is modelled according to a Markov chain and is described with a state transition matrix of probabilities, but the random character is not fixed completely in any movement. The regular part of movement lets high prediction velocities and easy calculations while the random part associates with a big process power and a low prediction speed. This prediction algorithm works with regular movements, cycles and simple itineraries, and with new patterns. Besides, the algorithm uses information of the previous stage and probabilistic data and physical restrictions to predict the movement. Mobility prediction is easier when user movements are regulars. Because of that, the first step in the predictive mobility management algorithm consists of detecting roaming patterns that must be stored in a database. Once this information is obtained, which allows to build a cell path dictionary containing the most frequent paths followed by the users of a cell when travelling. After that, user's dictionary is built taking into account the regular movements followed by a user across a specific cell. Both, user and cell dictionaries, constitute a hybrid dictionary that is used by an imperfect string matching system to analyze, by terms of edit operations, the similarity between the path followed by a MS and all the paths included in that dictionary.

When the MS follows a non regular movement, the path followed is not included into its user dictionary, but the hybrid dictionary also contents the most frequent paths followed by all the users of this cell. So, the only case whenever the system fails is when the movement followed by the MS is not usually followed neither the user, neither the rest of users. At this point, no prediction must be performed, but in order to avoid this circumstance later, the string corresponding to that movement is added to the user dictionary. Such as, next time the movement could be predicted.

MSs can measure seven different pilot signals powers, since each cell of a cellular system is enclosed by six cells. So, a fuzzy symbol including the ownership degree for each cell is built every time that the MS measures that power. During its movement across the network, a MS builds a string of fuzzy symbols (the known locations of the MS) by concatenating those symbols. The strings obtained are matched with the strings contained in the hybrid dictionary, obtaining a similarity measure calculated by a fuzzy automaton [1,5]. Then, a macroscopic mobility model is obtained, but a microscopic one is also desired. Then, the same concept is applied to a single cell, but now, the cell is split into different divisions. Three different codification schemes to describe the paths are considered. First one considers cell division in non-overlapped rectangular zones, each one uniquely identified by an alphabetic symbol. Second scheme considers sectorial division of a cell, and third one considers a k-connectivity scheme, where eight alphabetic characters represent all the possible movements that the MS can perform (see figure 1). Now, the ownership degrees managed correspond to the different divisions performed for a cell. In [2], codifications are analyzed.

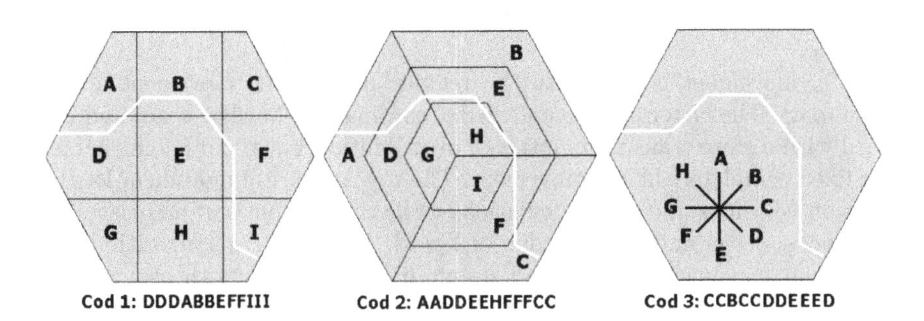

Fig. 1. Path codification schemes.

3 Trajectory Prediction

Trajectory prediction can be performed by the MS or by the base station (one BS for each cell of the system). If the prediction is performed by the BS, the storage of the dictionaries containing the user and cell more frequently followed paths takes place at the BS side. Furthermore, the BS must predict the movement followed for each user situated on its cell. So, the MS only transmits each fuzzy symbol (pilot signal power measured for each one of the adjacent cells) to its BS associated whenever this measure takes place (each 480 ms by standard). If the prediction is performed by the MS, dictionaries resides in the own terminal. When the MS is going to access a new cell, the BS of the incoming cell transmits the cell profile (cell dictionary) to the MS. Since the user dictionary is stored in the MS, and the cell dictionary has already been recovered, the MS can perform the calculus of its trajectory itself. So, the MS stores the dictionaries (user and cell profiles), measures the signal power in order to obtain the fuzzy symbol, and calculates the similarity between the string containing the path followed by the MS and the possible paths contained in the hybrid dictionary (more frequent paths of the user and cell dictionaries). In this paper, we have selected this last option, because nowadays mobile phone facilities and resources have considerably grown. Multimedia services require storage and computational resources in the mobile terminals that can be used to implement the fuzzy system here presented. In terms of bytes used, dictionary storage for a cell only requires 10 Kbytes for 50 cell paths.

Each time that the MS calculates the fuzzy symbol $\tilde{\alpha} = \{S_1\ S_2\ S_3\ S_4\ S_5\ S_6\ S_7\}$ by measuring the pilot signal power for each one of the adjacent cells, it obtains the existing similarity between the string of fuzzy symbols built by concatenation of the observed symbols and the pattern strings contained in the dictionary. As it can be seen in figure 2 (left), a MS located in cell 1 has 6 adjacent cells, so the given fuzzy symbol has seven components. Signal triangulation (see figure 2(right)) allows the MS to estimate the proximity degree for each cell (S_i, $\forall\ i = 1, 2 \ldots 7$). Since the proximity degree is certainly a fuzzy concept, a fuzzy tool is needed to work with the fuzzy symbol obtained $\tilde{\alpha}$ (composed by these seven proximity degrees). Then, a fuzzy automaton [5] is proposed to deal with the required imperfect string of fuzzy symbols comparison.

3.1 Formal Trajectory Prediction System

We propose the use of finite fuzzy automata with empty string transitions AFF_ε $(Q, \Sigma, \mu, \mu_\varepsilon, \sigma, \eta)$ where:

Q : is a finite and non-empty set of states.

Σ : is a finite and non-empty set of symbols.

μ : is a ternary fuzzy relationship over $(Q \times Q \times \Sigma)$; $\mu : Q \times Q \times \Sigma \to [0, 1]$. The value $\mu(q, p, x) \in [0, 1]$ determines the transition degree from state q to state p by the symbol x.

μ_ε : is a ternary fuzzy relationship over $Q \times Q$; $\mu_\varepsilon : Q \times Q \to [0, 1]$. The value $\mu_\varepsilon(q, p) \in [0, 1]$ determines the transition degree from state q to state p without spending any input symbol.

σ : is the initial set of fuzzy states; $\sigma \in \mathcal{F}(Q)$.

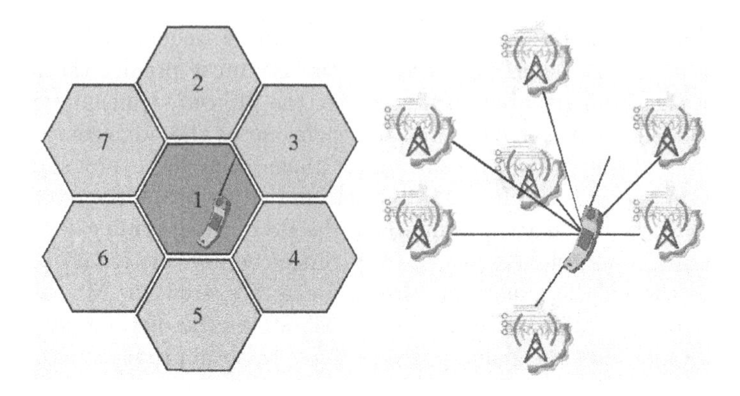

Fig. 2. (left) Cellular communication system. (right) Signal triangulation.

η : is the final set of fuzzy states; $\eta \in \mathcal{F}(Q)$; $\eta : Q \rightarrow [0,1]$.

The AFF_ε operation over a string $\alpha \in \Sigma^*$, is defined by $(AFF_\varepsilon, T, \hat{\mu}, \hat{\mu}_\varepsilon, \mu^*)$, where:

(i) $AFF_\varepsilon \equiv (Q, \Sigma, \mu, \mu_\varepsilon, \sigma, \eta)$ is finite fuzzy automaton with transitions by empty string.
(ii) T is a t-norm $T : [0,1]^2 \rightarrow [0,1]$.
(iii) $\hat{\mu} : \mathcal{F}(Q) \times \Sigma \rightarrow \mathcal{F}(Q)$ is the fuzzy state transition function. Given a fuzzy state $\tilde{Q} \in \mathcal{F}(Q)$ and a symbol $x \in \Sigma$, $\hat{\mu}(\tilde{Q}, x)$ represents the next reachable fuzzy state. $\hat{\mu}(\tilde{Q}, x) = \tilde{Q} \circ_T \mu[x]$ where $\mu[x]$ is the *fuzzy* binary relation over Q obtained from μ by the projection over the value $x \in \Sigma$. Then, $\forall\, p \in Q : \hat{\mu}(\tilde{Q}, x)(p) = max_{\forall q \in Q}\{\mu_{\tilde{Q}}(q) \otimes^T \mu(q,p,x)\}$.
(iv) $\hat{\mu}_\varepsilon : \mathcal{F}(Q) \rightarrow \mathcal{F}(Q)$, is the fuzzy states transition by empty string. Given a fuzzy state $\tilde{Q} \in \mathcal{F}(Q)$, $\hat{\mu}_\varepsilon(\tilde{Q})$ represents next reached fuzzy state without consuming an input symbol. $\hat{\mu}_\varepsilon(\tilde{Q}) = \tilde{Q} \circ_T \hat{\mu}_\varepsilon^T$ where $\hat{\mu}_\varepsilon^T$ is the T-transitive closure of the fuzzy binary relationship $\mu_\varepsilon \circ_T \hat{\mu}_\varepsilon^T = \mu_\varepsilon^{(n-1)T}$ if Q has cardinality n. So, $\forall\, p \in Q : \hat{\mu}_\varepsilon(\tilde{Q})(p) = max_{\forall q \in Q}\{\mu_{\tilde{Q}}(q) \otimes^T \hat{\mu}_\varepsilon^T(q,p)\}$.
(v) $\mu^* : \mathcal{F}(Q) \times \Sigma^* \rightarrow \mathcal{F}(Q)$, is the main transition function for a given string $\alpha \in \Sigma^*$ and it is defined by:
   a) $\mu^*(\tilde{Q}, \varepsilon) = \hat{\mu}_\varepsilon(\tilde{Q}) = \tilde{Q} \circ_T \hat{\mu}_\varepsilon^T$, $\forall\, \tilde{Q} \in \mathcal{F}(\mathcal{Q})$.
   b) $\mu^*(\tilde{Q}, \alpha x) = \hat{\mu}_\varepsilon(\hat{\mu}(\mu^*(\tilde{Q}, \alpha), x)) = (\mu^*(\tilde{Q}, \alpha) \circ_T \mu[x]) \circ_T \hat{\mu}_\varepsilon^T$, $\forall\, \alpha \in \Sigma^*$, $\forall\, x \in \Sigma$, $\forall\, \tilde{Q} \in \mathcal{F}(\mathcal{Q})$.

As it shows figure 3, the fuzzy automaton is used to compare each string of fuzzy symbols with all the known paths for the current cell. Those paths, stored in terms of symbol strings into a dictionary, can be the paths most frequently followed by a certain user or/and by all the users of this cell. The automaton provides as output the ownership degree of the string of fuzzy symbols to the paths contained in the dictionary. $\frac{D_1}{D_{2 \ldots k}} > 10^{dd}$ where k is the number of strings (paths) contained in the dictionary, and dd is the decision degree. dd can be

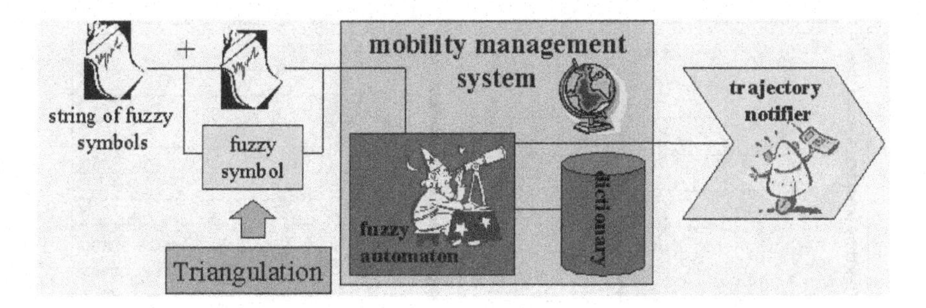

Fig. 3. Mobility management system architecture.

selected in order to increase or reduce the recognition rate, taken into account that a low value for *dd* can increase the number of false recognitions. A false recognition is the fact to decide that a MS follows a certain path when it does not follow this path. Section 4 explains better the way to select the value of *dd*.

As soon as the MS identifies the movement pattern followed, it notifies the predicted trajectory to the BS. Then, the local BS can contact the BS placed in the destination (incoming) cell. Such as, BS can manage adequately available resources in order to grant quality of service. In the same way, once known the path followed, the MS can receive information about some interesting places or shops placed in its trajectory. It can be seen as a way of sending/receiving selective and dynamic publicity or/and advertisements.

The detection of the trajectory followed does not stop the continuous calculus performed by the MS, because the user can decide to change its movement, and then, prediction must be reformulated. However, the MS only contacts the BS when prediction is performed. Absence of information makes the BS suppose that no prediction has obtained; or if a previous one was formulated, that it has not been modified. Then, the number of signaling messages is considerably reduced, the computation is performed in the MS (client side) and traffic decisions can be taken in advance to the *hand-off* process.

4 Robustness of the Mobility Management System Model

Figure 4 (left) shows the influence of the selected codification in terms of path discrimination. Cuadricular codification needs less (average) time to determine the path followed than other codifications (sectorial and 8-connectivity). Figure 4 (right) presents the results obtained when increasing the number of location zones (not available for 8-connectivity codification). The cell is divided in 1, 4, 9, 16 and 25 zones, and we can observe that as the number of zones considered grows, the average time needed to determine the path followed by a Ms inside a cell decreases.

Multiple-classification methods allows to combine the similarity degrees obtained for the three different proposed codifications in order to increase the final recognition rate. Since we are always working with fuzzy symbols and similarity

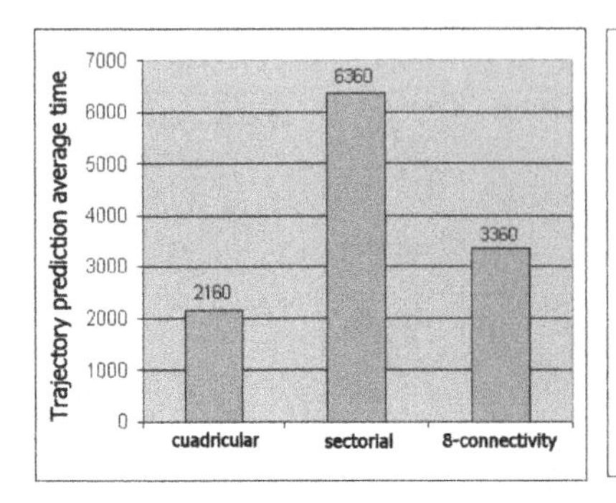

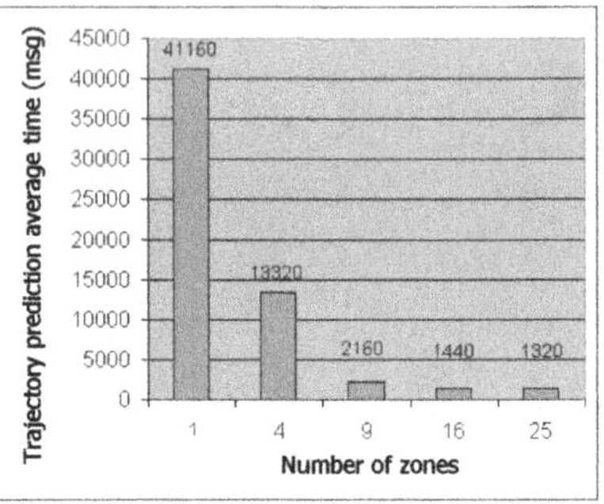

Fig. 4. Trajectory detection average time for different codifications (9 zones) (left) and different number of zones (right).

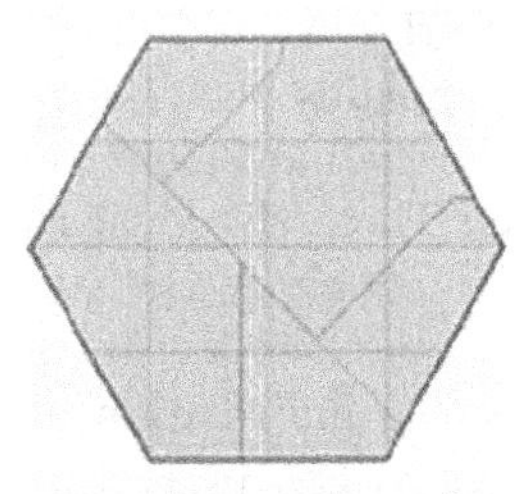

Fig. 5. Most frequent followed trajectories of a cell.

measures, fuzzy inference methods can also be considered to increase the final recognition rate in future works.

In order to evaluate the robustness of the mobility management system model proposed, we have chosen the example described in figure 5 and table 1.

Table 1. Simulation parameters.

Cell length (m)	1000
Location notification time (ms)	480
Terminal speed (m/s)	15
Codification	cuadricular
Number of zones	16
Number of paths included in the dictionary	4
Average path length (♯ symbols)	86
Standard deviation path length (♯ symbols)	20,4

We have selected a fuzzy automaton using parametric Hamacher t-norms and t-conorms [5] to predict the trajectories followed by a MS in the related scenario in absence (figure 6) and in presence (figure 7) of signal triangulation errors. The presence of errors is due to the attenuation of the signal and to the multi-path propagation. Attenuation can produces the loss of a symbol, multi-path propagation can introduce a new symbol, and both of them can produce a change of two symbols.

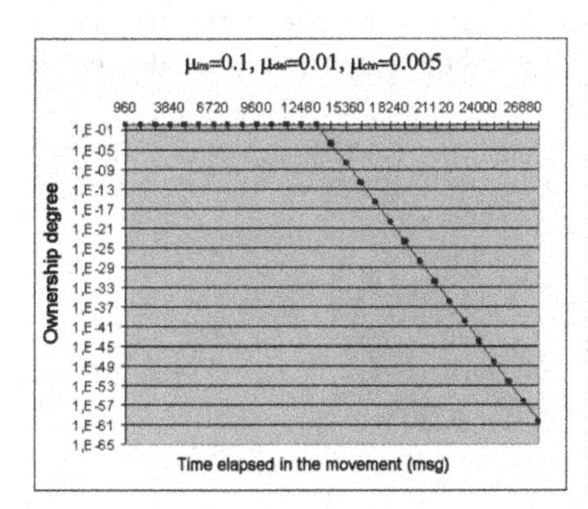

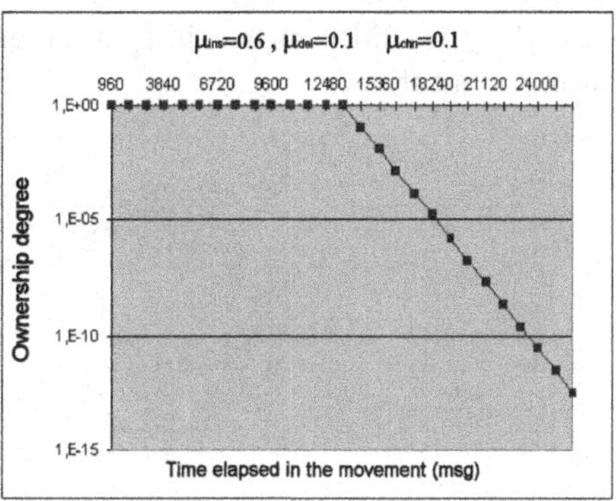

Fig. 6. Ownership degree evolution for the rest of the MS trajectories for different automata configurations (in error absence).

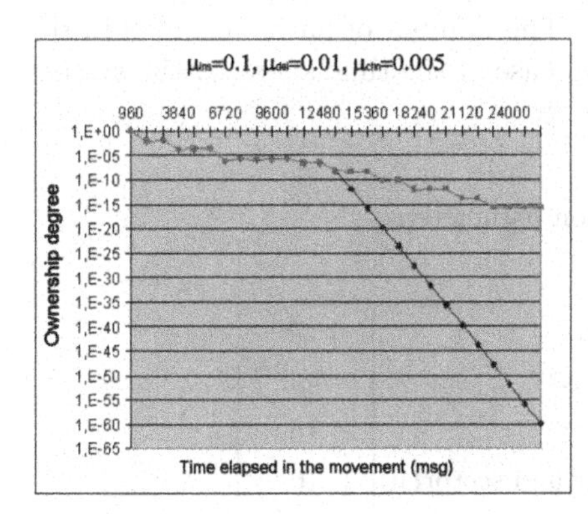

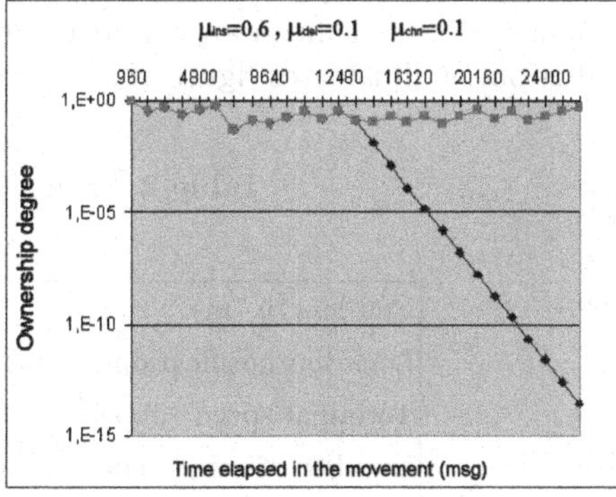

Fig. 7. Ownership degree evolution for the rest of the MS trajectories for different automata configurations (in error presence).

The value of the decision degree dd can be fixed to 5 when the triangulation is error free (non errors are considered) and to 10 when errors in the triangulation are considered. Increasing the value of dd we have a higher certainty in the path estimation, but we need more time to obtain the path estimation. Other values can be selected according to the difficulty of the estimation (number of similar/different paths in the cell), the certainty degree required in the estimation, the time required to establish the resources allocation or many other parameters. So, an interesting tuning mechanism to improve the robustness of the mobility management system is introduced.

As figure 8 shows, the mobility of an MS can be represented with a low number of symbols (figure 8 a)) or with a higher number (figure 8 b)). Due to the behaviour of the fuzzy automata, the system can correctly estimate the trajectory followed by the MS in both cases. The automaton deals with insertions, deletion and changes of symbols, so the fact to represent the trajectory followed by the MS with a high number of symbols (high precision degree) is non representative.

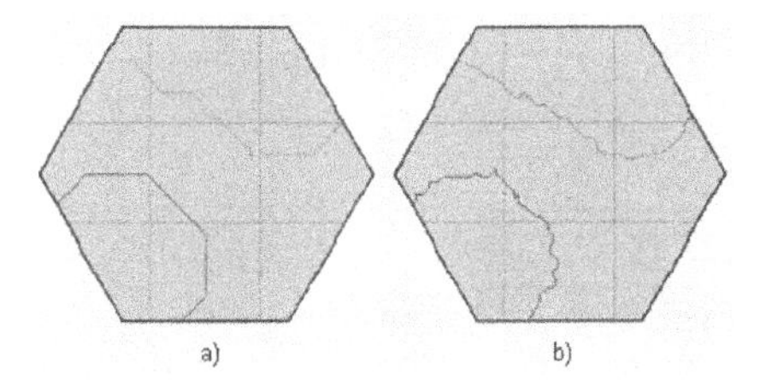

Fig. 8. Mobility representation, different lengths in the string representation.

In order to evaluate the influence of the number of paths considered in a cell, we have defined six different situations illustrated in figure 9. The parameters of the experiment are described in table 2. The number of paths included in the dictionary does not mean an important increase of the time spent by the system prediction, as illustrates figure 10.

Table 2. Simulation parameters.

Cell length (m)	1,000
Location notification time (ms)	480
Terminal speed (m/s)	15
Number of zones (cuadricular and sectorial)	16
Number of paths included in the dictionary	2-20
Average path length (♯ symbols)	84.37
Standard deviation path length (♯ symbols)	30,08
Average path length (ms)	40,497.6
Standard deviation path length (ms)	14,438.35

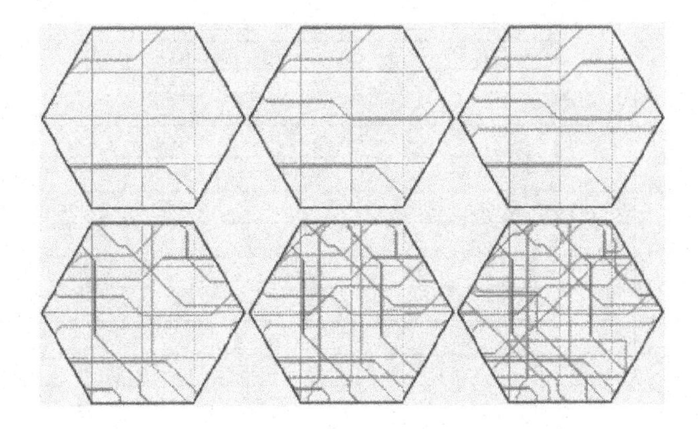

Fig. 9. Different cellscenarios.

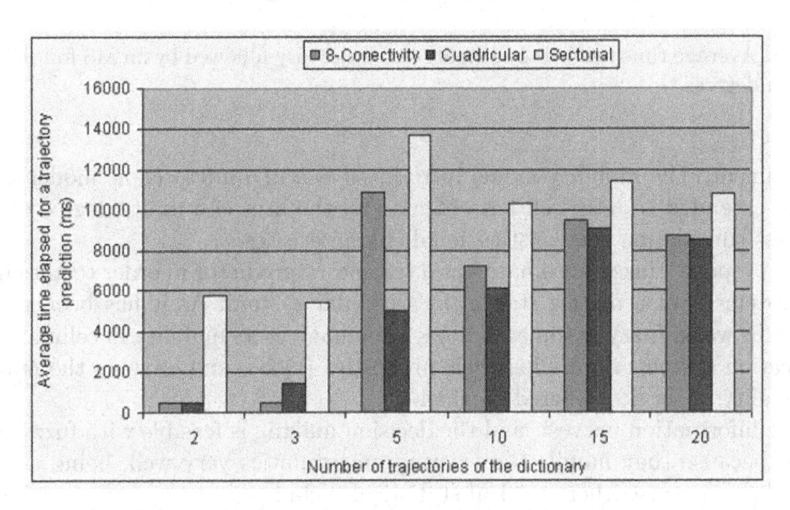

Fig. 10. Influence of the number of paths considered.

Finally, figure 11 presents the average time (measured in milliseconds) needed to estimate the trajectory followed by an MS for the worst case considered in the previous scenario (20 paths). The error rate considered are: 0, representing that non errors are introduced in the location estimation; 1/5, representing an edition error (insertion, deletion or change) of a symbol each five symbols of the string; and 1/10, representing an edition error each ten symbols of the string. Ten different decision degrees *dd* are considered in order to evaluate the time needed to ensure a certainty degree in the estimation.

5 Conclusions

The new application that is exposed in this article deals with a general problem related to the mobility. Mobility management increases benefits of any mobile

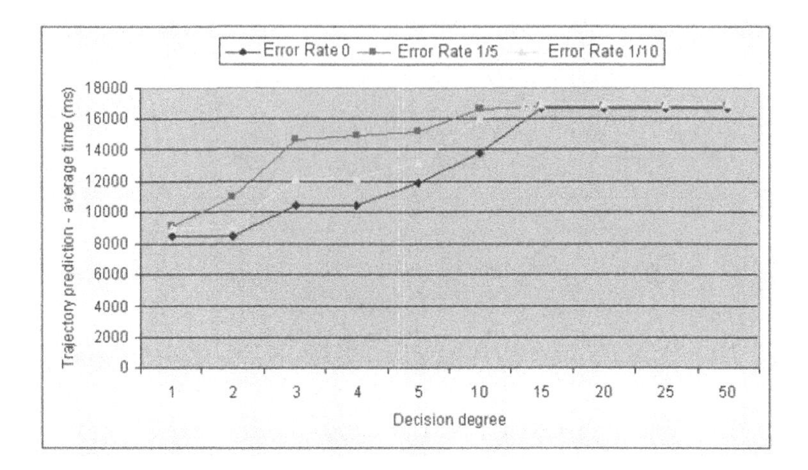

Fig. 11. Average time needed to estimate the trajectory followed by an MS for different decision degrees.

environment. The mobility model introduced is a 2D, microscopic model which can be extended to inter-cell movements and the user can fix its degree of randomness considering the constitutional characteristics.

We propose a fuzzy automaton as a trajectory predictor in order to anticipate the movement of a mobile station in a cellular system. As it has been proved along this work, fuzzy automata allows to manage users mobility in cellular communication systems improving their properties (QoS), and growing the number and quality of services offered by them.

The information harvest, and the decision making is feasible with fuzzy techniques because they handle the existing uncertainties very well, being a really good alternative to other classical statistical techniques.

Three different path codification schemes are considered. Automaton performance for them is evaluated showing that the average time to discriminate the path decrease as the number of zones increase being the rectangular logical division the most interesting codification scheme.

The effect of different fuzzy parameter values for the automaton is analyzed, showing the relevance of them in order to minimize the time needed to discriminate the path followed by the MS among the set of paths contained in the dictionary. In the same way, the automaton works with imperfect strings of symbols due to the vagueness of the location estimation realized by the MS.

Different scenarios and decision degrees are studied in order to show the robustness of the mobility management technique proposed.

References

1. J. J. Astrain, J. R. Garitagoitia, J. Villadangos, F. Fariña, A. Córdoba and J. R. González de Mendívil, *An Imperfect String matching Experience Using Deformed Fuzzy Automata*, Frontiers in Artificial Intelligence and Applications, Soft Computing Systems, vol. 87, pp. 115-123, IOS Press, The Nederlands, Sept. 2002.
2. J. J. Astrain, J. Villadangos, P. J. Menéndez, J. Domínguez, "Improving mobile communication systems QoS by using deformed fuzzy automata", *Proceedings of the International Conference in Fuzzy Logic and Technology EUSFLAT 2003*, Zittau (Germany), 2003.
3. C. Bettsetter, "Mobility modeling in wireless networks: categorization, smooth movement, and border effects", *ACM Mobile Computing and Communications Review*, vol. 5, no. 3, pp. 55-67, July 2001.
4. G. Edwards, A. Kandel and R. Sankar, "Fuzzy Handoff Algorithms for Wireless Communication", *Fuzzy Sets and Systems*, vol. 110, no. 3, pp. 379-388, March 2000.
5. J. R. Garitagoitia, J. R. González de Mendívil, J. Echanobe, J. J. Astrain, and F. Fariña, "Deformed Fuzzy Automata for Correcting Imperfect Strings of Fuzzy Symbols", *IEEE Transactions on Fuzzy Systems*, vol. 11, no. 3, pp. 299–310, 2003.
6. S. Hadjefthymiades and L. Merakos, "ESW4: Enhanced Scheme for WWW computing in Wireless communication environments", *ACM SIGCOMM Computer Communication Review*, vol. 29, no. 4, 1999.
7. S. Hadjefthymiades, S. Papayiannis and L. Merakos, "Using path prediction to improve TCP performace in wireless/mobile communications", *IEEE Communications Magazine*, vol. 40, no. 8, pp. 54-61, Aug. 2002.
8. V. I. Levenshtein, *Binary codes capable of correcting deletions, insertions, and reversals*, Sov. Phys. Dokl., vol. 10, no. 8, pp. 707–710, 1966.
9. G. Liu and G. Q. Maguire Jr., "A Predictive Mobility Management Scheme for Supporting Wireless Mobile Computing", *Technical Report*, ITR 95-04, Royal Institute of Technology, Sweden, Jan. 1995.
10. T. Liu, P. Bahl and I. Chlamtac, "Mobility Modeling, Location Tracking, and Trajectory Prediction in Wireless Networks", *IEEE JSAC*, vol. 16, no. 6, pp. 922-936, Aug. 1998.

Admission Control and Route Discovery for QoS Traffic in Ad Hoc Networks

Zhong Fan and Siva Subramani

Toshiba Research Europe Ltd.
Telecommunications Research Laboratory
32, Queen Square, Bristol BS1 4ND, UK
{zhong.fan,siva.subramani}@toshiba-trel.com

Abstract. As wireless networks become more widely used, there is a growing need to support advanced services, such as real-time multimedia streaming and voice over IP. Real-time traffic in wireless ad hoc networks often have stringent bandwidth or delay requirements. For example, a delay of over 150 ms in voice transmissions is felt as disturbing by most users. To support quality of service (QoS) requirements of real-time applications, various traffic control mechanisms such as rate control and admission control are needed. In this paper, an admission control and route discovery scheme for QoS traffic is proposed. It is based on on-demand routing protocols and performs admission control implicitly during the route discovery process. Local bandwidth measurements are used in admission control decisions. Simulation results have shown that this admission control scheme can greatly improve network performance such as packet delivery ratio and delay.

1 Introduction

A wireless mobile ad hoc network (Manet) is an autonomous system where all nodes are capable of movement and can be connected dynamically in an arbitrary manner. Without a network infrastructure, network nodes function as routers which discover and maintain routes to other nodes in the network. As wireless networks become more widely used, there is a growing need to support advanced services, such as real-time multimedia streaming and voice over IP. Real-time traffic in wireless ad hoc networks often have stringent bandwidth or delay requirements. For example, a delay of over 150 ms in voice transmissions is felt as disturbing by most users. QoS support for real-time applications in wireless ad hoc networks is a challenge. It has been recognized that traditional mechanisms for providing guaranteed or hard QoS (such as ATM, or IP integrated services) are not suitable for wireless networks where network conditions constantly change due to node mobility and shared medium access.

Since shared wireless resources are easily over-utilized, traffic load in the network must be controlled properly so that acceptable quality of service for real-time applications can be maintained. In other words, the wireless channel must be kept from reaching the congestion point where loss and delay increase

I. Niemegeers and S. Heemstra de Groot (Eds.): PWC 2004, LNCS 3260, pp. 92–106, 2004.

rapidly with the traffic load. For example, Figure 1 (obtained from an analytical model of the IEEE 802.11 MAC [1]) shows that for a typical 802.11 network (basic access, without RTS/CTS), both throughput efficiency and packet delay deteriorate considerably as the number of traffic sources increases. Therefore admission control is necessary to maintain the QoS performance within an acceptable region.

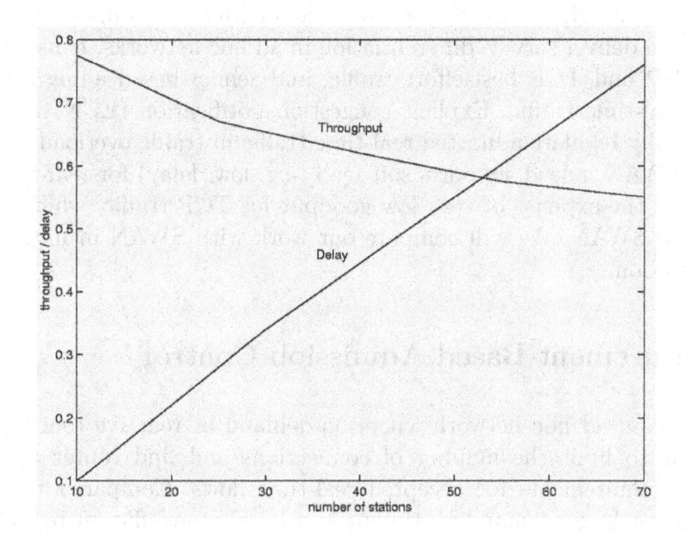

Fig. 1. Throughput efficiency and packet delay vs. number of sources

In this paper, an admission control and route discovery scheme for QoS traffic is proposed. It is based on on-demand routing protocols and performs admission control implicitly during the route discovery process. Local bandwidth measurements are used in admission control decisions. Simulation results have shown that this admission control scheme can greatly improve network performance such as packet delivery ratio and delay.

2 Related Work

QoS has attracted a lot of attention recently in the ad hoc research community. At the MAC layer, the dynamic nature of ad hoc networks makes it difficult to assign a central controller to maintain connection states and reservations. Therefore best-effort distributed MAC controllers such as the IEEE 802.11 Distributed Coordination Function (DCF) are widely used in existing ad hoc networks. Recently a number of distributed control schemes have been proposed to support service differentiation at the MAC layer, e.g. [2].

At the network layer, most of the ad hoc routing schemes proposed so far are also best-effort, i.e. no QoS support. INSIGNIA [3] is an in-band signalling system that supports adaptive reservation-based services in ad hoc networks. It represents a general-purpose approach to delivering QoS (mainly the signalling aspect) and does not address the issue of admission control. In [4] a distributed call admission controller is introduced, which is based on service curve provisioning. The drawback of this scheme is that it is difficult to accurately measure the service curve of the network. Further, their admission criterion is the deterministic universal service curve which can be very conservative. In [5], Ahn *et al.* proposed SWAN, a stateless network model which uses distributed control algorithms to deliver service differentiation in ad hoc networks. It uses rate control for UDP and TCP best-effort traffic, and sender-based admission control for UDP real-time traffic. Explicit congestion notification (ECN) is employed to dynamically regulate admitted real-time traffic in traffic overload conditions. Although SWAN indeed supports soft QoS (e.g. low delay) for real-time traffic, it does so at the expense of very low goodput for TCP traffic, which is a main drawback of SWAN. We will compare our work with SWAN in more detail in the next section.

3 Measurement-Based Admission Control

We consider an ad hoc network where on-demand or reactive routing is used. The goal is to limit the number of connections and find routes that satisfy bandwidth requirements for accepted real-time flows. Compared to proactive routing protocols, on-demand routing protocols are more efficient by minimizing control overhead and power consumption since routes are only established when required. AODV (ad hoc on-demand distance-vector) [6] and DSR (dynamic source routing) [7] are two of the on-demand protocols currently under active development in the IETF Manet working group.

In an on-demand protocol, when a source is in need of a route to a destination, it broadcasts a Route Request (RREQ) message to its neighbors, which then forward the request to their neighbors, and so on, until the destination is located. The RREQ packet contains the source node's address and current sequence number, the destination node's address and last known sequence number, as well as a broadcast ID. During the process of forwarding the RREQ, intermediate nodes record in their route tables the address of the neighbor from which the first copy of the broadcast packet is received, thereby establishing a reverse path. Here we assume intermediate nodes do not reply to RREQs. Once the RREQ reaches the destination, the destination responds by unicasting a Route Reply (RREP) back to the source node. Information obtained through RREQ and RREP messages is kept with other routing information in the route table of each node. To provide quality of service, extensions (extra fields) can be added to these messages during the route discovery process. Here we consider bandwidth as the QoS parameter and add an extra field in the RREQ packets for bandwidth requirement of the real-time flow. Only bandwidth is considered here, because it has been regarded as the most important QoS parameter for most real-time applications. In addition, our simulation results show that admission control based on bandwidth requirement not only improves traffic throughput, but also reduces average delay. In [5] admission control is also used to provide differentiated QoS for real-time flows. However our approach is different from the SWAN mechanism of [5] in that, unlike SWAN that uses separate probing packets after route establishment, our method piggybacks QoS (bandwidth) information with route request packets, which incurs minimum control overhead. In this sense route discovery messages in our scheme also act as probes for distributed admission control. Moreover, in our method intermediate nodes make admission decisions based on local measurements, while in SWAN it is the source that accepts/rejects incoming connections based on the probe response.

In the proposed framework, each node needs to periodically measure the bandwidth of its outgoing wireless links. Bandwidth measurements are used here for admission control and route discovery for traffic flows with certain amount of bandwidth requirements. More specifically, each node measures the number of packet transmissions in its carrier sensing range over a specific time interval (T) and calculates the occupied average bandwidth of existing real-time traffic using the following weighted moving average:

$$B_{avg}(j) = B_{avg}(j-1) * \alpha + (1-\alpha) * Rate(j), \quad (1)$$

where α is the weight (or smoothing factor), $B_{avg}(j)$ is the average bandwidth at the measurement window j and $Rate(j)$ is the measured instantaneous bit rate at j. Our simulations indicate that the above measurement method is able to provide bandwidth estimates accurate enough for the purpose of admission control, and is easy to implement and cost effective.

A requested bandwidth extension is added to RREQ messages, which indicates the minimum amount of bandwidth that must be made available along an acceptable path from the source to the destination. When any node receives such a RREQ, it will perform the following operation: if its available bandwidth cannot satisfy the QoS (bandwidth) requirement, the node drops the RREQ and does not process it any further. Otherwise, the node continues processing the RREQ as specified in the best-effort on-demand routing protocol. For admission control purposes, let B_{th} be the admission threshold rate. Then to admit a new QoS flow the available bandwidth (B_{avail}) must be greater than the required bandwidth (B_{req}) of the new flow:

$$B_{avail} = B_{th} - B_{avg} > B_{req} \quad (2)$$

Here B_{th} is a critical parameter. Due to the MAC overhead and the fact that interferences from signals transmitted by nodes located further away than one hop (e.g. nodes within the carrier sensing range) may have an impact on the channel medium [8], the admission threshold can be very conservative, say, a small percentage of the total channel capacity. We will elaborate more on this later in Sections 4 and 5 of this paper.

After a new flow is admitted, it immediately starts transmission and consumes network bandwidth. Since the available bandwidth is measured continuously at each node, the newly admitted flow will be taken into account for any future admission control decisions. Similarly, when a flow stops, the available bandwidth will increase, creating space for other flows to be admitted.

4 Performance Evaluation

Extensive simulations have been conducted using the network simulator ns-2 [9]. In our simulations, we use the two-ray ground propagation model and omni-directional antennas. The wireless channel capacity is 11 Mbps. Ten nodes are randomly distributed in an area of 670 m by 670 m. Each node has a transmission range of 250 m and carrier sensing range of 550 m. The routing protocol we use is the AODV routing protocol [6] and the MAC protocol is IEEE 802.11. The bandwidth measurement interval T is 1 second. The traffic used in these simulations is generated by a number of video connections. Each video connection is modelled as a 200 kbps constant bit rate (CBR) source sending 512-byte packets at the rate of 50 packets per second. So B_{req} is 200 kbps. CBR connections are established between any two nodes selected randomly. We increase the number of connections one by one every 5 seconds until the number of connections reaches eight, hence after 40 seconds of simulation time all connections are active.

Figure 2 shows the measured traffic rate at a single node. It can be seen that at this particular node the traffic fluctuates over time and the maximum throughput achieved is around 1.4 Mbps. This may be a discouraging result at first sight given the total traffic load of 1.6 Mbps and the channel capacity of 11 Mbps. However, both theoretical analyses and previous experiments have shown that the usable bandwidth (or throughput) in a wireless network is often far less than the nominal channel capacity. For example, the authors of [10] showed that, the theoretical maximum throughput (TMT) of the IEEE 802.11 MAC is given by:

$$TMT(x) = \frac{8x}{ax + b} \quad \text{Mbps} \tag{3}$$

where x is the MSDU (MAC service data unit) size in bytes, a and b are parameters specific to different MAC schemes and spread spectrum technologies. For example, when the channel data rate of a high-rate direct sequence spread spectrum MAC is 11 Mbps, MSDU is 512 bytes and the RTS/CTS scheme is used, the theoretical maximum throughput is only 2.11 Mbps ($a = 0.72727, b = 1566.73$). Further, this TMT is defined under a number of idealized assumptions such as zero bit error rate, no losses due to collisions, etc. In an ad hoc environment we also have to consider the effect of interference of traffic from neighboring nodes that are possibly more than one hop away (e.g. hidden nodes, exposed nodes [11])[1]. Therefore it is not surprising that effective throughput is even less than the theoretical maximum throughput. Figure 2 indicates the need of admission control for real-time traffic, because a lot of packets would have been lost due to medium contention and congestion if no admission control is implemented. It

[1] Generally, how to identify interfering nodes intelligently and precisely is still an open question.

also gives us valuable insight into the optimal range of selecting the admission threshold rate.

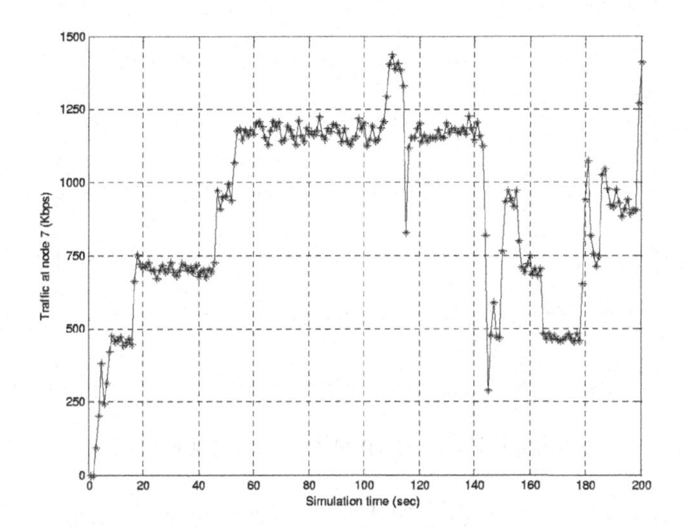

Fig. 2. The measured traffic at a particular node

Figure 3 compares the packet drop ratios of the network with admission control and that without control. In ad hoc networks both unavailability of routes and congestion (queue overflow) result in packet drops. It is clear that admission control with a threshold of 1 Mbps is very effective in reducing packet loss (drop ratio close to zero). Figure 4 presents the packet delivery ratios under different traffic loads. Admission control with threshold of 1 Mbps achieves nearly 100% packet delivery ratio, while the performance without admission control degrades significantly as traffic load increases. Admission control with a threshold of 1.2 Mbps has some improvements compared to the case of no control, but not as pronounced as the case with an admission threshold of 1 Mbps. Clearly a trade-off exists here: smaller admission threshold guarantees better QoS at the expense of denying more connection requests and potentially wasting available bandwidth, whereas larger threshold accepts more connections at the cost of degraded QoS. From another perspective, Figures 5 and 6 present the number of packets sent and successfully received during the simulation for both cases of control and no-control. It is evident that lack of an admission control protocol results in significant packet loss, particularly when the traffic load is high. With admission control, most of the packets are delivered to their destinations, which ultimately gives better packet delivery ratio.

Figure 7 plots the total number of RREQ packets processed over a time period of 100 seconds. It is clear that with admission control quite a few RREQ packets are dropped due to insufficient bandwidth at high loads, which results in an effect of call blocking similar to that in connection-oriented networks. This

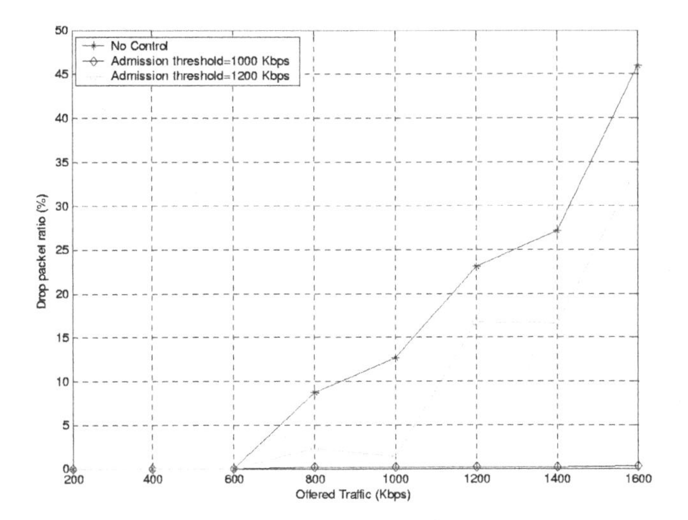

Fig. 3. Packet drop ratio vs. offered load

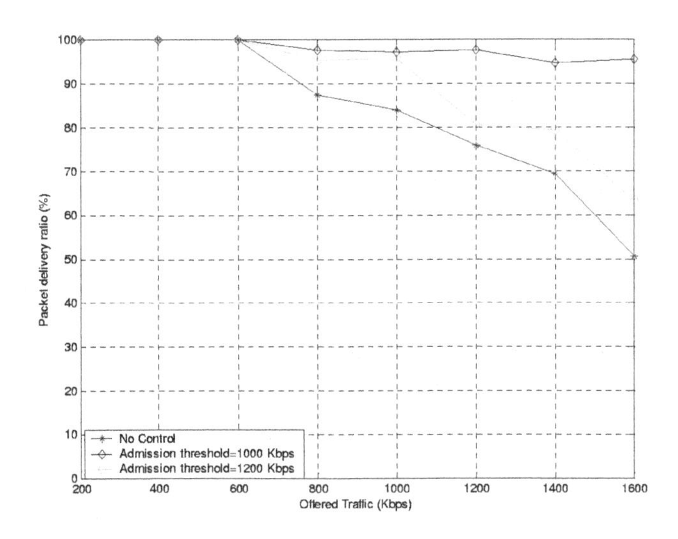

Fig. 4. Packet delivery ratio vs. offered load

effect is also evident from Figure 5 and Figure 6, where we can see that with admission control the total number of packets sent at high load is less than that in a network without control.

The overall average end-to-end delay is shown in Figure 8 and the delay for a particular node is shown in Figure 9. The offered traffic load is 1000 kbps. With admission control the delay is not only smaller, but also nearly constant (small jitter), which is an important requirement of real-time multimedia applications. From Figure 9, it can be noted that without control the delay can peak up

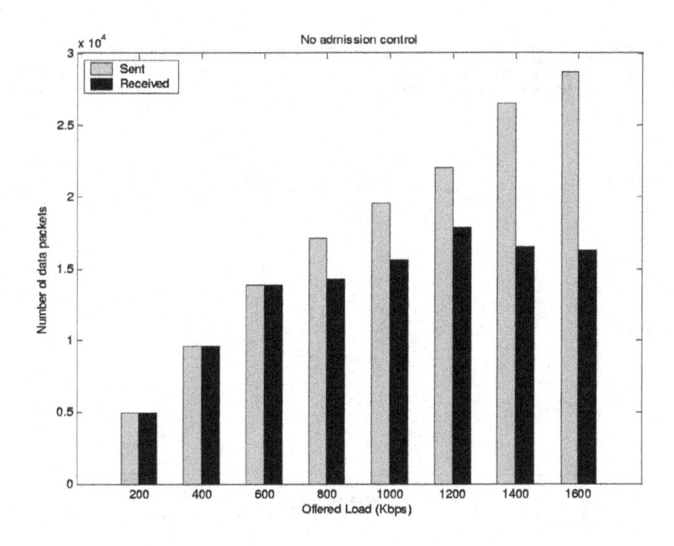

Fig. 5. Number of packets sent/received vs. offered load (without control)

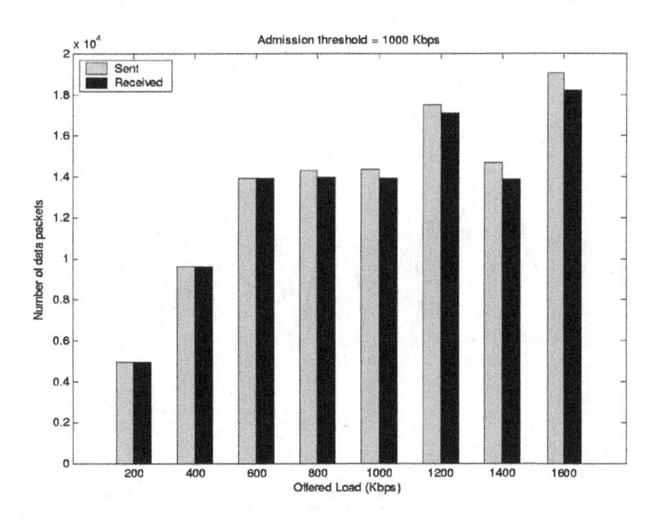

Fig. 6. Number of packets sent/received vs. offered load (with admission control)

into the range of 15 to 20 msec. This is because, when the channel becomes congested packets are queued heavily resulting in prolonged delays and large delay variations. This is avoided using admission control. Figure 10 plots the overall delay against traffic load, which shows admission control improves packet delay as offered load increases.

The above simulation results have been obtained for the scenario of stationary nodes. Next we introduce mobility into the model. We use the random waypoint model where the pause time is 25 seconds, and each node moves with a maximum

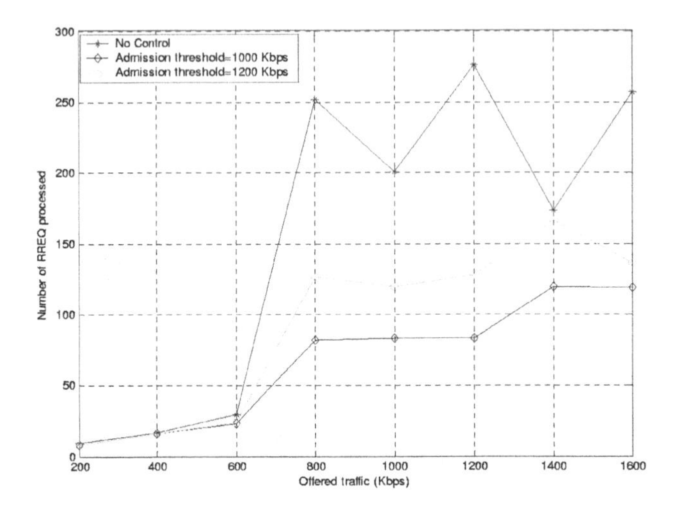

Fig. 7. Number of RREQs processed

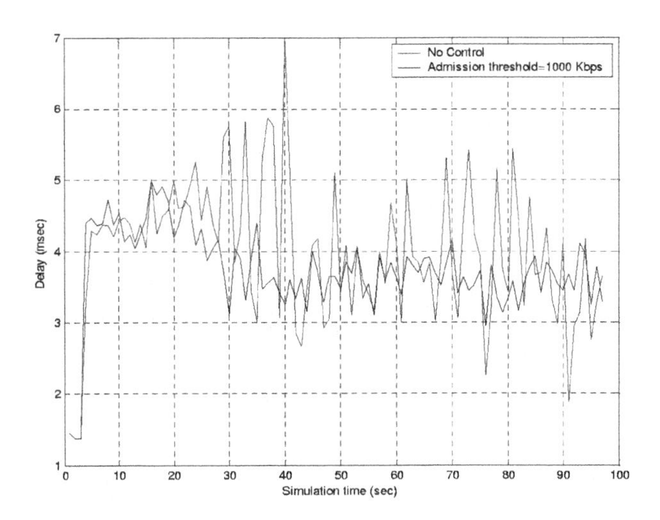

Fig. 8. Overall average delay for all the packets

speed of 5 m/sec. Figure 11 clearly shows the improvement of packet delivery ratio when admission control is implemented. On the other hand, mobility indeed degrades the performance due to topology change and re-routing, as shown in Figure 12.

To test our algorithm further, we add two TCP connections as background traffic. Figure 13 demonstrates the superior performance of admission control.

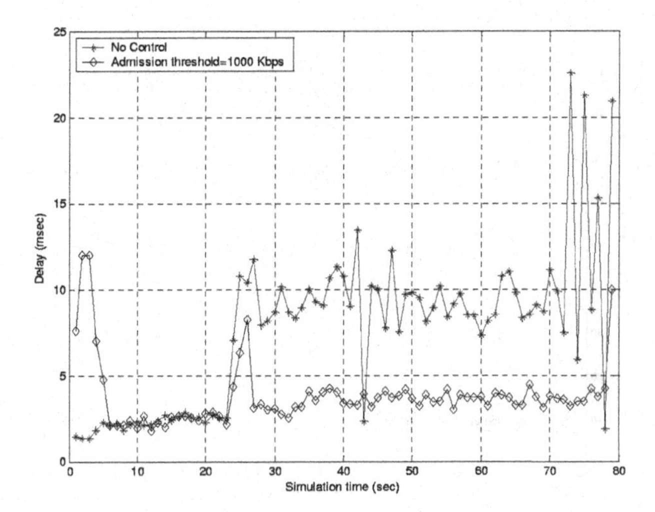

Fig. 9. Average delay of a single node

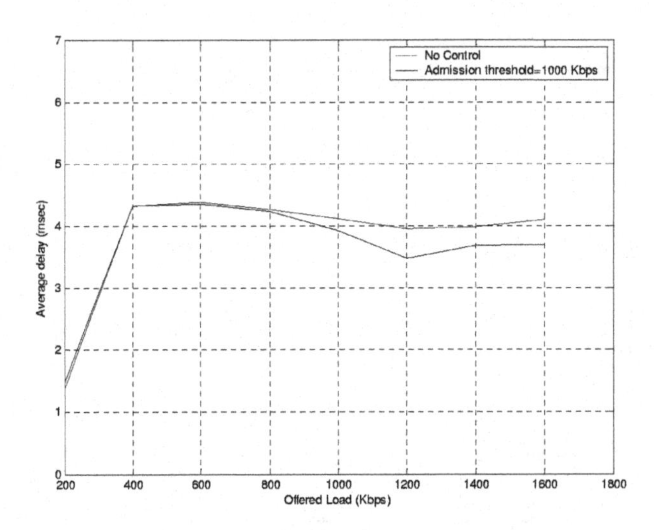

Fig. 10. Average delay vs. offered load

In summary, our admission control mechanism is able to achieve high packet delivery ratio while keeping constant low delay. It therefore can be used effectively to support real-time video or voice applications with QoS requirements.

5 Adaptive Admission Control

As mentioned previously the usable bandwidth (or throughput) in a wireless ad hoc network is often far less than the nominal channel capacity. Therefore to be

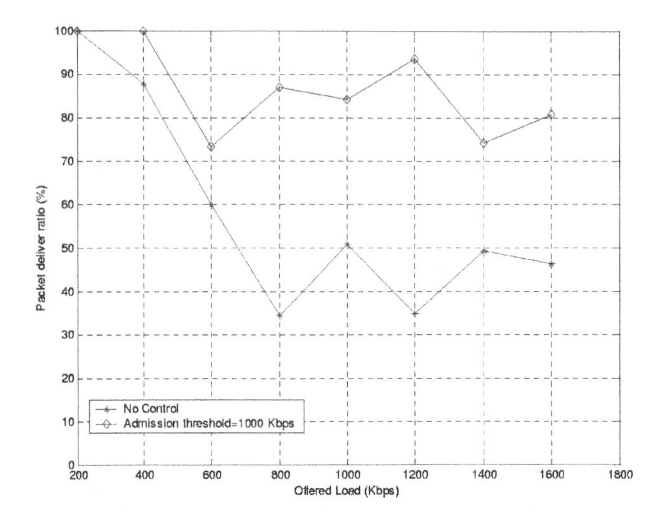

Fig. 11. Packet delivery ratio vs. offered load (mobility scenario, with control and without control)

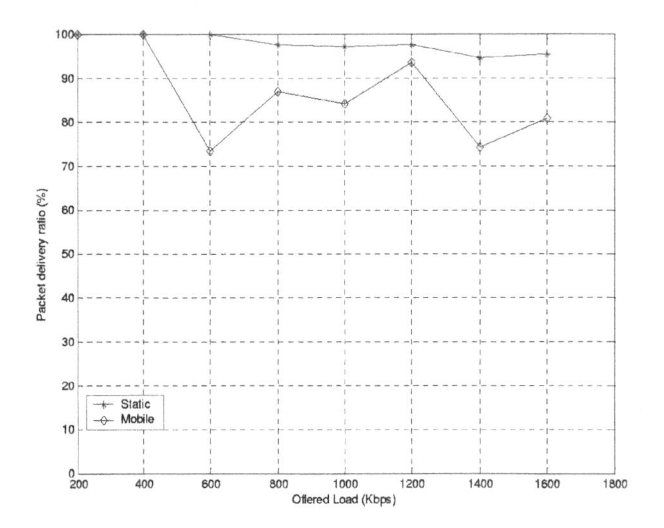

Fig. 12. Packet delivery ratio vs. traffic load (stationary and mobile nodes, both with admission control)

on the safe side (avoiding the violation of QoS), admission policies often have to be quite conservative. For example, in a network of 11 Mbps channel capacity, the admission threshold rate is often set as merely 1 Mbps.

To have a more intelligent way of determining this admission threshold with the aim of maintaining QoS while achieving high network utilization, we propose to adapt the threshold according to channel conditions (e.g. the number of nodes

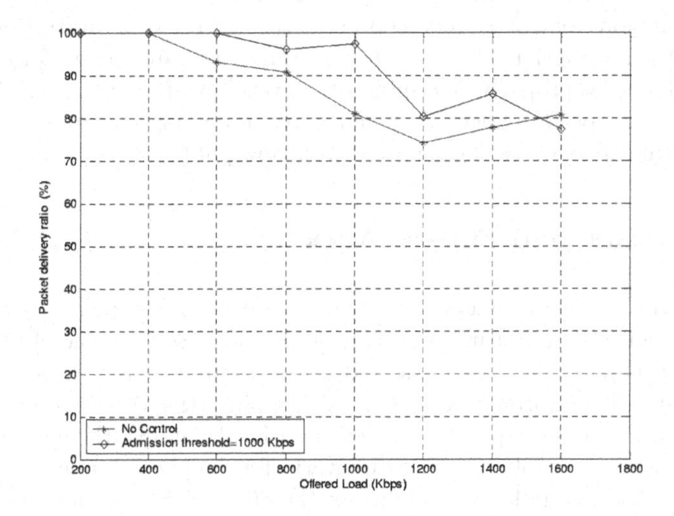

Fig. 13. Packet delivery ratio vs. offered load (with background TCP traffic)

contending for the channel). Here we can use MAC delay as an indication of the channel condition. MAC delay (in a RTS-CTS-DATA-ACK cycle) is a very useful metric to identify congestion hotspots and measure link interference in an ad hoc network [12], and it continuously fluctuates throughout the time. The MAC delay of a packet represents the time it takes to send the packet between the transmitter and receiver including the total deferred time (including possible collision resolution) as well as the time to fully acknowledge the packet. It is easy to measure: at the source node subtracting the time that a packet is passed to the MAC layer (ts) from the time an ACK packet is received from the receiver (tr): $D = tr - ts$.

Assume that we have two thresholds A_1 and A_2, where $A_1 > A_2$. For instance, in an 11 Mbps network, $A_1 = 1.3$ Mbps, $A_2 = 1$ Mbps. Then switching of admission threshold is dependent on two key parameters [12]: (i) MAC delay threshold D, and (ii) N, the number of times that the measured MAC delay measurements exceed a predetermined threshold D consecutively. Specifically, at the beginning the admission threshold is A_1. When the measured MAC delay measurements exceed a predetermined threshold (i.e., D) for more than N times consecutively, the threshold is changed to a more conservative one, A_2. Hence when the wireless medium is busier (MAC delay is higher) the admission control policy becomes more strict, i.e. accept fewer connections. When the admission threshold is changed to A_2, it remains so for at least 5 seconds. After that if the MAC delay is smaller than D, the threshold is reverted back to A_1 automatically.

The two parameters D and N have an impact on network performance. When D and N are configured as large values, admission control is too loose and accepts too many connections rendering the protocol to be less effective against moderate congestion. In contrast, when N and D are configured with small values, admission control is very conservative and network utilization is low. Therefore, the appropriate choice of these parameters is important for admission control to function properly. We intend to investigate the choices of N and D and evaluate the performance of this adaptive admission control scheme using simulations and results will be reported in future publications.

6 Conclusion and Future Work

Real-time applications such as video streaming often have stringent QoS requirements. Current ad hoc routing protocols do not address the issue of QoS provisioning. In this paper QoS is considered in the route discovery process and admission control is performed implicitly by dropping RREQ packets. Bandwidth is considered as the QoS parameter here and local bandwidth measurement is used in admission control and route discovery. The bandwidth monitoring mechanism makes use of the built-in ability of the 802.11 wireless channel and only incurs moderate CPU overhead (simple bandwidth calculation). The admission control process does not incur excessive overhead either, since QoS information is carried together (piggybacked) with route discovery packets. It has been shown that the proposed admission control scheme is very effective in keeping packet loss and delay very low for real-time flows.

It is worth pointing out that our method does not rely on a QoS-capable MAC, so it can be readily applied to current 802.11-based ad hoc networks. On the other hand, it would be interesting to investigate the performance gain of admission control when it is combined with some of the recently proposed QoS MAC protocols, e.g. AEDCF [13]. These MAC protocols can provide priorities to real-time traffic over best-effort traffic and admission control is only applied to real-time services. Wireless bandwidth measurement is another important research issue. Other measurement techniques such as those proposed in [14] [15] and their applicability in admission control in ad hoc networks are topics of further study.

The proposed QoS method does not depend on any particular routing protocol: it can generally be applied to any on-demand routing protocol. Admission control policies based on other QoS metrics can also be incorporated into the framework, e.g. MAC delay, buffer occupancy and packet loss. Recent experimental studies have suggested that more attention be paid to link quality when choosing ad hoc routes [16]. To this end, route discovery to find routes with better signal quality and stability can be done using the approach proposed in this paper for on-demand protocols. The main challenges, however, involve practical estimates of link quality (e.g. signal-to-noise ratio) and techniques to find feasible paths based on these link metrics.

As an example, we consider the problem of finding bandwidth-constrained least delay routes for real-time flows. In our admission control and route discovery framework, when the destination node generates a RREP in response to a RREQ, it includes in the RREP a delay field whose initial value is zero. As the RREP propagates along the reverse path, each intermediate node forwarding the

RREP adds its own estimated MAC delay time to its one hop upstream node to the delay field. Thus when the source node receives the final RREP message, it would obtain an estimate of the path delay. The above strategy will result in multiple paths, which can provide a more robust packet delivery. The source node then can choose the minimum delay route that (automatically) satisfies the bandwidth constraint (not necessarily the minimum hop-count route). This idea is deemed as a subject of our future work.

Acknowledgements. This work was conducted at the Telecommunications Research Laboratory of Toshiba Research Europe Limited. We would like to thank Gary Clemo for his continued support and useful comments. The constructive comments from the anonymous reviewers are also gratefully acknowledged.

References

1. H. Wu, Y. Peng, K. Long, S. Cheng, and J. Ma. Performance of reliable transport protocol over IEEE 802.11 wireless LAN: Analysis and enhancement. In *IEEE INFOCOM*, 2002.
2. I. Aad and C. Castelluccia. Differentiation mechanisms for IEEE 802.11. In *IEEE INFOCOM*, 2001.
3. S. Lee, G. Ahn, and A. Campbell. Improving UDP and TCP performance in mobile ad hoc networks with INSIGNIA. *IEEE Communications Magazine*, June 2001.
4. S. Valaee and B. Li. Distributed call admission control for ad hoc networks. In *IEEE VTC*, 2002.
5. G. Ahn, A. Campbell, A. Veres, and L. Sun. SWAN: Service differentiation in stateless wireless ad hoc networks. In *IEEE INFOCOM*, 2002.
6. C. Perkins, E. M. Belding-Royer, and S. Das. Ad hoc on-demand distance vector (AODV) routing. *RFC 3561*, 2003.
7. D. Johnson, D. A. Maltz, and Y. Hu. The dynamic source routing protocol for mobile ad hoc networks (DSR). *Internet Draft*, 2003.
8. K. Bertet, C. Chaudet, I. Guerin Lassous, and L. Viennot. Impact of interferences on bandwidth reservation for ad hoc networks: a first theoretical study. In *IEEE GLOBECOM*, 2001.
9. K. Fall and K. Varadhan. ns notes and documentation. Technical report, The VINT Project, 2000.
10. J. Jun, P. Peddabachagari, and M. Sichitiu. Theoretical maximum throughput of IEEE 802.11 and its applications. In *IEEE International Symposium on Network Computing and Applications*, 2003.
11. S. Xu and T. Saadawi. Does the IEEE 802.11 MAC protocol work well in multihop wireless ad hoc networks. *IEEE Communications Magazine*, pages 130–137, June 2001.
12. S. Lee and A. Campbell. HMP: hotspot mitigation protocol for mobile ad hoc networks. In *IEEE/IFIP IWQOS*, 2003.
13. L. Romdhani, Q. Ni, and T. Turletti. Adaptive EDCF: enhanced service differentiation for IEEE 802.11 wireless ad-hoc networks. In *IEEE WCNC*, 2003.
14. M. Kazantzidis and M. Gerla. End-to-end versus explicit feedback measurement in 802.11 networks. In *IEEE ISCC*, 2002.

15. K. Chen, K. Nahrstedt, and N. Vaidya. The utility of explicit rate-based flow control in mobile ad hoc networks. In *IEEE WCNC*, 2004.
16. K. Chin, J. Judge, A. Williams, and R. Kermode. Implementation experience with MANET routing protocols. *ACM Computer Communication Review*, 32(5), 2002.

Adaptive Multi-mode Routing in Mobile Ad Hoc Networks

Jeroen Hoebeke, Ingrid Moerman, Bart Dhoedt, and Piet Demeester

Department of Information Technology (INTEC), Ghent University - IMEC vzw
Sint-Pietersnieuwstraat 41, B-9000 Ghent, Belgium
Tel: +32-(0)92649970, Fax: +32-(0)92649960
{jeroen.hoebeke, ingrid.moerman, bart.dhoedt, piet.demeester}@intec.UGent.be

Abstract. Mobile ad hoc networks are wireless multi-hop networks with a completely distributed organization. The dynamic nature of these networks imposes many challenges on mobile ad hoc routing protocols. Current routing protocols do not take into the account the network context and therefore their performance is only optimal under certain network conditions. This paper proposes a novel concept for routing in mobile ad hoc networks, adaptive multi-mode routing, and demonstrates its feasibility and effectiveness.

1 Introduction

A mobile ad hoc network (MANET) is an autonomous system consisting of mobile nodes that communicate with each other over wireless links [1]. The network does not rely on any fixed infrastructure for its operation. Therefore nodes need to cooperate in a distributed manner in order to provide the necessary network functionality. One of the primary functions each node has to perform is routing in order to enable connections between nodes that are not directly within each others send range. Developing efficient routing protocols is a non trivial task because of the specific characteristics of a MANET environment [2]: the network topology may change rapidly and unpredictably because of node movements, the available bandwidth is limited and can vary due to fading or noise, nodes can suddenly join or leave the network... All this must be handled by the routing protocol in a distributed manner without central coordination. Consequently, routing in ad hoc networks is a challenging task and much research has already been done in this field, resulting in various routing protocols [3]. However, most current routing protocols are general purpose routing protocols that do not take into account the specific network conditions they operate under. As a consequence and as shown by various performance evaluation studies, their performance is only optimal under certain network conditions and no overall winner can be designated. In section 2, we discuss two commonly known routing techniques, proactive and reactive routing, and show that their performance strongly depends on the network conditions. This justifies the need to use different routing techniques depending

I. Niemegeers and S. Heemstra de Groot (Eds.): PWC 2004, LNCS 3260, pp. 107–117, 2004.

on the network conditions. Therefore, in section 3 we propose a solution to this problem through the development of an adaptive multi-mode routing protocol and discuss the advantages and implementation issues of this new approach. In section 4, the feasibility and possible performance gain of our approach is demonstrated. Finally, conclusions are made in section 5.

2 Performance of Existing Routing Protocols

2.1 Classification of Routing Protocols for Mobile Ad Hoc Networks

Over the last few years, numerous routing protocols have been developed for ad hoc networks. Basically, these protocols can be categorized in the following two classes depending on the way they find routes: *proactive* routing protocols and *reactive* routing protocols.

Proactive routing protocols or table-driven routing protocols attempt to have at all times an up-to-date route from each node to every possible destination. This requires the continuous propagation of control information throughout the entire network in order to keep the routing tables up-to-date and to maintain a consistent view of the network topology. These protocols are typically modified versions of traditional link state or distance vector routing protocols encountered in wired networks, adapted to the specific requirements of the dynamic mobile ad hoc network environment.

Reactive protocols or on-demand routing protocols only set up routes when needed. When a node needs a route to a destination, a route discovery procedure is started. This procedure involves the broadcasting of a route request within the network. Once a route is established by the route discovery phase, a route maintenance procedure is responsible for keeping the route up-to-date as long as it is used.

Most other types of routing protocols [4] can be seen as variants of proactive and reactive techniques. Hybrid routing protocols try to combine proactive and reactive techniques in order to reduce protocol overhead. Nearby routes are kept up-to-date proactively, while far away routes are set up reactively. Position-based routing protocols use geographical information to optimize the routing process. Finally, hierarchical protocols, such as clustering protocols, introduce a hierarchy in the network in order to reduce the overhead and to improve the scalability. In the remainder of the paper we focus on the fundamental proactive and reactive techniques.

2.2 Performance Evaluation of Proactive and Reactive Routing Protocols

In the literature, many simulation studies have been performed in order to evaluate the performance of proactive and reactive routing protocols [5]. They all come to the conclusion that each technique has its advantages and disadvantages and can outperform the other depending on the network conditions. To illustrate this observation, we extensively simulated the performance of WRP (Wireless Routing Protocol) and

AODV (Ad Hoc On-Demand Distance Vector Routing) in the network simulator Glomosim [6].

WRP [7] is a proactive distance vector protocol in which nodes communicate the distance and the second-to-last hop for each destination through update messages sent at periodic times and on link changes. On receiving an update message, the node modifies its distance table and looks for better paths using this new information. The extra second-to-last hop information helps remove the counting-to-infinity problem most distance vector routing algorithms suffer from. Also, route convergence is speeded up when a link failure occurs.

AODV [8] builds routes using a route request - route reply query cycle. A source node that needs a new route, broadcasts a route request packet across the network. Nodes receiving this packet set up backwards pointers to the source node. If a node is either the destination or has a valid route to the destination, it unicasts a route reply back to the source, otherwise the request is rebroadcasted. As the reply propagates back to the source, nodes set up forward pointers to the destination. Once the source node receives the reply, data can be forwarded to the destination. On a link break in an active route, the node upstream of the link break propagates a route error message to the source node after which it can reinitiate route discovery.

In order to illustrate the dependence of protocol performance on the network conditions, we present the simulation results for both protocols in a 50 node static network, with nodes randomly distributed in a rectangular region of size 600m by 600m. Packets of size 512 bytes are sent at a rate of 10 packets per second by 5 and 20 sources respectively. The transmission range of all nodes is approximately 200 meter and the MAC layer model used is 802.11b direct sequence spread spectrum at 2Mbit/s. The performance metrics considered are the packet delivery ratio, the end–to-end delay and the number of control packets per data packet delivered.

Figure 1 presents the simulation results. When there are few traffic sources present in the network, both protocols succeed in delivering almost all data packets. However, the number of control packets AODV needs in order to deliver the data is significantly lower than WRP. In a static network, AODV only uses control packets to set up routes when they are needed, whereas WRP periodically exchanges routing update messages in order to keep all routes up to date. The end-to-end delay of both protocols is comparable. In these network conditions, the deployment of a reactive protocol is preferred.

When the number of sources increases to 20, WRP does not need additional control messages for keeping its routing tables up to date, which results in a lower number of control packets per data packet delivered and is opposite to the behavior of AODV. For these high network loads, the results become completely different. Both protocols suffer from a lower packet delivery ratio, because more transmitting nodes contend for the wireless medium causing congestion and packet loss. However, the effect of high network loads is more distinct for AODV. When packets are dropped due to congestion, the MAC layer protocol notifies AODV of the loss. AODV will assume a link break has occurred and reacts by sending a route error message back to the source and reinitiating a new route discovery, during which additional packets in the saturated buffers are dropped. This effect, together with the higher number of control packets results in a significant increase in number of control packets per data packet delivered and in end-to-end delay. This means that for high network loads, the use of proactive routing is advised.

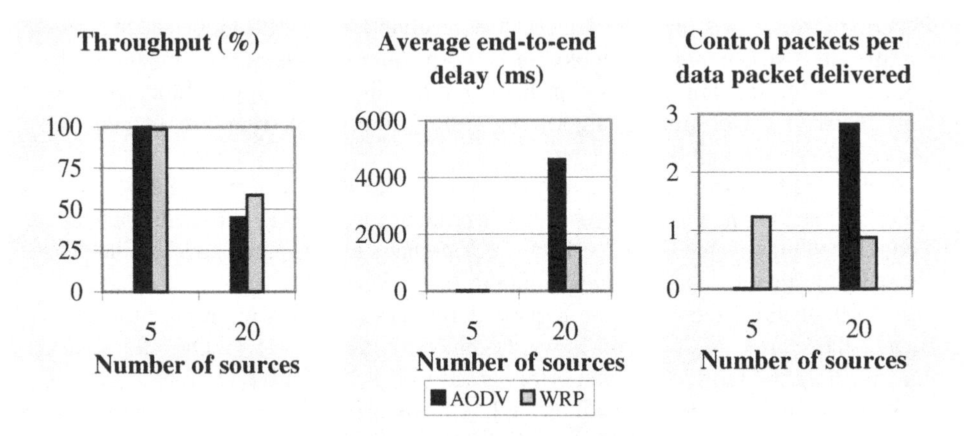

Fig. 1. Performance evaluation of WRP (proactive) and AODV (reactive) in a 50 node static network with nodes randomly distributed in a rectangular region of size 600m by 600m. Traffic is sent at a rate of 10 packets per second by 5 and 20 nodes respectively

This simulation result proves that the performance is strongly dependent on the network conditions. Apart from the number of traffic sources, also the network size, node density, send rate and mobility have influence on the performance.

So, basically both approaches rely on the propagation of control messages throughout the entire network in order to establish routes, but the way in which the broadcasting of control messages is applied differs completely. As a consequence, their performance will be different, with one technique outperforming the other depending on the network conditions.

3 Adaptive Multi-mode Ad Hoc Routing Framework

3.1 The Need for Adaptive Routing

The example presented in section 2.2 clearly shows the strong dependency of protocol performance on the network conditions. Existing routing protocols are unable to adapt to the networking context, which can result in a severe performance degradation. Ideally, devices should choose the optimal routing technique depending on the type of ad hoc network they participate in and the current network conditions in this network. For overhead and compatibility reasons it is currently not feasible having devices implementing different protocols and switching protocol according to the network conditions. Hybrid routing protocols such as the Zone Routing Protocol, Fisheye State Routing and SHARP [9] are already a first step into the development of routing protocols that combine multiple routing techniques, but they do not obtain the degree of adaptation we envision.

Therefore, we propose the development of an adaptive multi-mode routing protocol that has multiple compatible modes of operation (e.g. proactive, reactive, flooding or variants), where each mode is designed to operate as efficiently as possible in a given networking context. Simulation studies or analytical studies can be used to determine the optimal network conditions of the different modes. The main issue in the development of such a framework is that nodes need to be capable of monitoring and estimating the network conditions in their environment with as little overhead as possible. Based on these predictions nodes can adapt their mode of operation to the networking context and perform the best possible routing. In the following sections we present the framework of our novel adaptive multi-mode routing protocol.

3.2 The Adaptive Multi-mode Ad Hoc Routing Framework

Our framework (see figure 2) consists of two main components, a monitoring agent and the actual routing protocol, which we will now discuss in more detail.

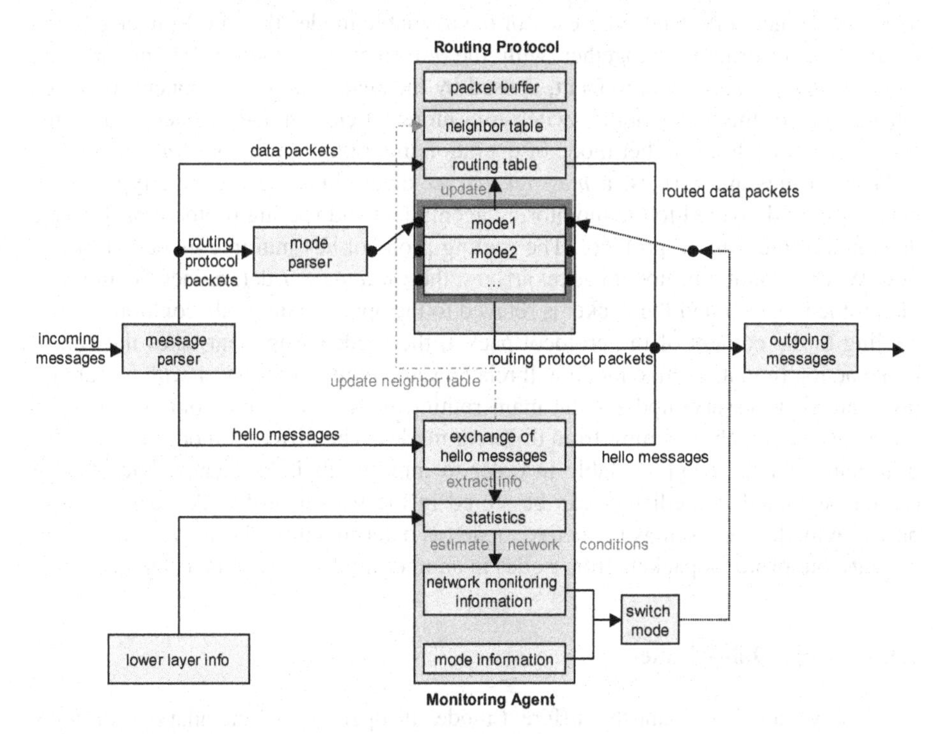

Fig. 2. Framework of the proposed adaptive multi-mode routing protocol

The monitoring agent is responsible for collecting information about the network conditions in the environment of the node. This is done in two ways. First of all, *local statistics* from the network layer or other layers are collected in the *statistics* component of the monitoring agent. These statistics can include, but are not limited to: the number of data packets routed, signal strength of the received packets, number of packets dropped due to congestion... Secondly, *non-local statistics* are collected through the periodic broadcasting of hello messages to the neighboring nodes. These hello messages provide two types of information. By receiving or not receiving hello messages the connectivity to other nodes is determined and link breaks are detected. In addition, the hello messages contain statistics and network monitoring information collected by the sender of the hello message, such as the observed network load and the mode of operation the node is currently in. In this way, by receiving hello messages, nodes are provided with information from their immediate environment.

When a node receives a hello message, the information in the message is extracted and stored in the statistics component. In addition, the connectivity and mode information is used to update the *neighbor table* in the routing protocol.

Periodically, the network monitoring information component processes the collected statistics. This component is responsible for extracting useful information about the networking context such as the network load or mobility. Based on simulation studies or analytical models, the *mode information* component has knowledge under which network conditions each of the available modes their performance is optimal. This information, together with the information provided by the *network monitoring information* component, is used by the *switch mode* component to decide whether or not the node should switch to a more efficient mode of operation. If the node has to switch to another mode of operation the routing protocol is informed.

When a message arrives, a *message parser* determines the message type. Hello messages are delivered to the monitoring agent; data and routing protocol packets are delivered to the routing protocol. The routing protocol has multiple modes of operation. When a routing protocol packet arrives, the *mode parser* determines the mode of the protocol packet and the packet is relayed to the appropriate mode component. According to the content of the protocol packet, the mode component takes the appropriate action (e.g. a reactive mode will relay a route request or answer with a route reply) and, if necessary, updates the main routing table. This table contains all valid route entries, possible coming from different modes. The different modes can use the information in the neighbor table in order to improve their efficiency. Packets that cannot be routed immediately can be stored in the packet buffer. At each moment only one mode is chosen as the active mode (as determined by the *switch mode* component), but protocol packets from nodes in another mode can also be received.

3.3 Compatibility Issues

As already stated, we want the different modes of operation of the adaptive protocol to be compatible. In this way, different modes in different parts of the network can coexist and each node can decide in a distributed manner when to switch to another mode. For instance, consider a large ad hoc network with a number of heavily loaded clusters of nodes. In these clusters it will be more efficient to proactively set up routes, whereas in the other parts of the network reactive routing is advised (assuming we only have a proactive and reactive mode). Another example is a static ad hoc network with a lot of traffic and a few highly mobile nodes. In this case, it would be more efficient that the highly mobile nodes set up their routes reactively and do not take part in the proactive routing process in the remainder of the network.

The development of compatible modes requires some compatibility issues to be resolved, which we will now illustrate for two cases, assuming the protocol has a proactive and reactive mode.

Case 1: Node *n* does not have a route entry for a data packet with destination *d* that has to be routed

- and *n* is currently in a reactive mode: if the route request was broadcasted throughout the entire network and no reply was received, node *d* is unreachable and the data packet is dropped
- and *n* is currently in a proactive mode: destination *d* can be located outside a proactive part of the network. Therefore, node *n* can use the functionality of the reactive mode to find a route by broadcasting a route request.

Case 2: Node *n* its mode is proactive

- and a neighbor *m* changes its mode from reactive to proactive: node *n* should send its current proactive tables to node *m*
- and a neighbor *m* changes its mode from proactive to reactive: node *n* removes all information related to node *m* from its proactive tables, as node *m* does not participate anymore in the proactive routing process. However, by simply removing this information, active connections that use node *m* as relay will now have a sub-optimal route or no route at all. Therefore, before cleaning up the proactive tables, this information will be used to create reactive entries for the active connections that use node *m* as relay node. As a consequence, running connections will not be influenced by the change in protocol mode. Finally, node *n* will send an update packet to inform neighboring nodes in proactive mode of this change.

The above examples illustrate that during the implementation of new modes care should be taken to sustain compatibility with the existing modes. Also, when writing modes, generic functions need to be provided, in order to easily integrate new modes.

3.4 Advantages of Adaptive Multi-mode Routing

Adaptive multi-mode routing has numerous advantages:

- Improved efficiency by adaptation: by its capability to adapt to the network, the routing protocol can provide better routing in networks with varying conditions. As mobile ad hoc networks are intrinsically characterized by a very dynamic nature, this is certainly a big advantage opposed to existing routing protocols that are not aware of the network context.
- Compatibility: when the modes are developed with built-in compatibility in mind, different modes in different parts of the network can coexist.

- User friendliness: devices can participate seamlessly in different types of ad hoc networks without the need to manually switch to another protocol, because the protocol will adapt itself to the current network conditions. This user friendliness can certainly be an advantage.
- Future proof: the use of different modes eases the future development of the protocol. Existing modes can be extended or enhanced or new modes can be added without the need to completely change or rethink the protocol design.

4 Performance Evaluation

Based on the framework described in section 3, we developed a proof of concept version of the proposed adaptive multi-mode ad hoc routing protocol with two compatible modes of operation, one proactive mode and one reactive mode. These modes are based on WRP and AODV respectively. The functionality of the network monitoring agent is currently limited to determining the network load based on the number of packets to route and the number of neighbors, which are affected by the packet transmissions. This information is exchanged with the neighboring nodes by broadcasting hello messages. When the observed network load exceeds a certain threshold, which was now manually determined, nodes change their mode of operation from reactive to proactive. Once the load falls below this threshold, the mode is set back to reactive.

We simulated the performance of the initial implementation of our adaptive multi-mode ad hoc routing protocol (AMAHR) in a 50 node static network, with nodes randomly distributed in a rectangular region of size 600m by 600m. Packets of size 512 bytes are sent at a rate of 10 packets per second. The number of sources is initially set to 5. After 1300 seconds the number of sources is increased to 20 and after 2500 seconds the number of sources is set back to 5. The transmission range of all nodes is approximately 200 meter and the MAC layer model used is 802.11b direct sequence spread spectrum at 2Mbit/s. The hello interval is 1 second in the proactive mode and 5 seconds in the reactive mode. Again, the performance metrics considered are the packet delivery ratio, the end–to-end delay and the number of control packets per data packet delivered. Figure 3 shows the evolution of these three performance metrics over time.

The results clearly show that AMAHR combines the advantages of both proactive and reactive routing by its capability to adapt to the network context. Initially, the observed traffic load is low and nodes set up routes reactively. Once the number of traffic sources increases to 20, the network monitoring component detects the increase in network load. The observed network load then exceeds the defined threshold and nodes switch to proactive routing.

As a consequence, under low network loads our adaptive protocol has the high packet delivery ratio and low control overhead of AODV. The number of control packets per data packet delivered is slightly higher than the reactive routing protocol, due to the periodic exchange of hello messages needed for monitoring the network environment. However, this is not necessarily a drawback, as the neighborhood information provided by the hello messages could be used for implementing a more efficient broadcasting scheme, thereby reducing the control overhead.

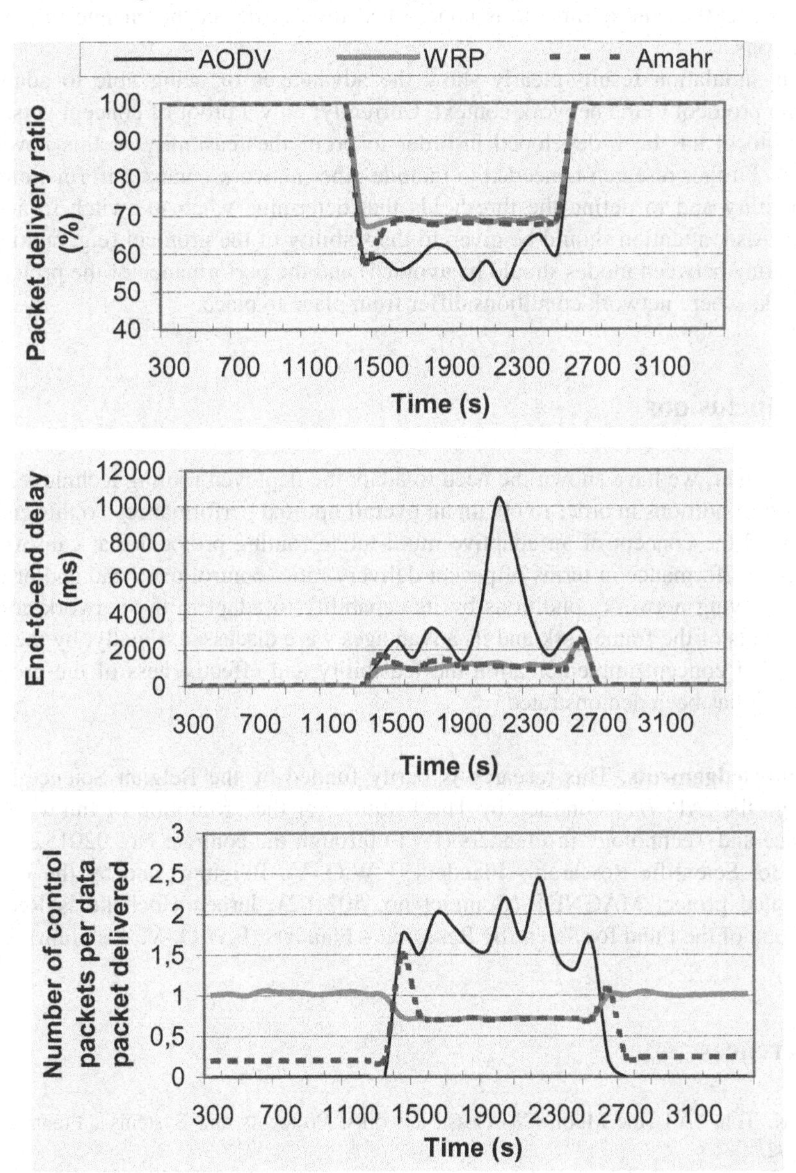

Fig. 3. Performance evaluation of WRP (proactive), AODV (reactive) and AMAHR in a 50 node static network with nodes randomly distributed in a rectangular region of size 600m by 600m. Traffic is sent at a rate of 10 packets per second and the number of sources is increased from 5 to 20 after 1300 seconds and back decreased to 5 sources after 2500 seconds

By changing its mode from reactive to proactive when the traffic load increases, AMAHR achieves the high packet delivery ratio and low control overhead and end-to-end delay of WRP. Only at the time nodes switch to another mode, the performance is less than the optimum, as nodes need time to detect the change in network conditions.

Our simulation results clearly show the advantages of being able to adapt the routing protocol to the network context. Currently, only a proof of concept version of the protocol has been developed in order to proof the feasibility of this novel approach. Further research is needed to include other network context information such as mobility and to define the thresholds that determine when to switch to another mode. Also, attention should be given to the stability of the protocol (e.g. continuous alternating between modes should be avoided) and the performance of the protocol in networks where network conditions differ from place to place.

5 Conclusions

In this paper, we have shown the need to adapt the deployed routing technique to the network conditions in order to obtain an overall optimal performance. To this end, we presented the concept of an adaptive multi-mode routing protocol that can offer an optimal performance in terms of packet delivery ratio, control overhead and/or delay under varying network conditions by its capability to adapt to the network context. The details of the framework and its advantages were discussed. Finally, by means of a proof of concept implementation the feasibility and effectiveness of the proposed approach has been demonstrated.

Acknowledgements. This research is partly funded by the Belgian Science Policy through the IAP V/11 contract, by The Institute for the Promotion of Innovation by Science and Technology in Flanders (IWT) through the contract No. 020152, by the Fund for Scientific Research - Flanders (F.W.O.-V., Belgium) and by the EC IST integrated project MAGNET (Contract no. 507102). Jeroen Hoebeke is Research Assistant of the Fund for Scientific Research – Flanders (F.W.O.-V., Belgium)

References

1. C-K. Toh, "Ad Hoc Mobile Wireless Networks: Protocols and Systems", Prentice Hall, 2002.
2. S. Corson and J. Macker, "Mobile Ad hoc Networking (MANET): Routing Protocol Performance Issues and Evaluation Considerations," RFC 2501, Jan. 1999.
3. E. Royer and C. Toh, "A Review of Current Routing Protocols for Ad Hoc Mobile Wireless Networks", *IEEE Personal Communications*, vol. 6, no. 2, Apr. 1999, pp. 46-55.
4. X. Hong, K. Xu and M. Gerla, "Scalable Routing Protocols for Mobile Ad Hoc Networks", *IEEE Network*, vol. 16, no. 4, July-Aug. 2002, pp. 11-21.

5. S. R. Das, R. Castaneda and J. Yan, "Comparative Performance Evaluation of Routing Protocols for Mobile Ad Hoc Networks", *Proceedings of 7th Int. Conf. on Computer Communications and Networks (IC3N)*, Lafayette, LA, Oct., 1998, pp. 153-161.
6. X. Zeng, R. Bagrodia and M. Gerla, "GloMoSim: a Library for Parallel Simulation of Large-scale Wireless Networks", *Proceedings of the 12th Workshop on Parallel and Distributed Simulations (PADS'98)*, Banff, Alberta, Canada, May 1998.
7. S. Murthy and J.J. Garcia-Luna-Aceves, "An Efficient Routing Protocol for Wireless Networks", *Mobile Networks and Applications*, vol.1, no.2, Oct. 1996, pp. 183-197.
8. C.E. Perkins and E.M. Royer, "Ad-hoc On-Demand Distance Vector", *Proceedings 2nd IEEE Workshop on Mobile Computing Systems and Applications*, New Orleans, LA, U.S.A, Feb. 1999, pp. 90-100.
9. V. Ramasubramanian, Z. Haas and E.G. Sirer, "SHARP: A Hybrid Adaptive Routing Protocol for Mobile Ad Hoc Networks", *Proceedings of the 4th ACM International Symposium on Mobile Ad Hoc Networking and Computing*, Annapolis, Maryland, U.S.A, June 2003.

Service Location and Multiparty Peering for Mobile Ad Hoc Communication

Dirk Kutscher and Jörg Ott

Technologiezentrum Informatik (TZI), Universität Bremen,
Postfach 330440, 28334 Bremen, Germany,
{dku|jo}@tzi.uni-bremen.de

Abstract. Flexible personal communications may require dynamically discovering, using, and combining a number of services to support the activities of a mobile user. However, many service discovery and service control protocol frameworks are not designed with requirements for ad-hoc and group communication in a changing environment in mind. In this paper, we motivate the case for personalized group communications based upon a (static) office application scenario featuring simple remote device control and then enhance the scope towards service location and dynamic establishment of group communications for mobile users: ad-hoc multiparty peering. We particularly explore the issues relating to group communication setup and robustness in the presence of changing connectivity and present a framework for mobile multiparty ad-hoc cooperation.

1 Introduction

A variety of mobile personal computing and communication equipment is available today providing a wide range of functions, e.g., to enable rich interactions with other persons or to gain access to a broad spectrum of information resources. Such personal devices are usually optimized for a certain class of tasks and, mostly due to their form factors, none of them is likely to meet all the conflicting user requirements at the same time: e.g., devices are not light and small *and* offer a comfortable keyboard and a large display. Given the diversity of specializations, it appears sensible to combine the strengths of the individual components dynamically as needed to carry out a certain task – rather than attempting to manufacture them into a single device that would be subject to the aforementioned tradeoffs.

Component-based architectures have been used for many years e.g., for flexibly composing sophisticated communication endpoints from independent application entities or devices with well-defined interfaces. Each of these entities focuses on dedicated services; they are combined to form a coherent system by means of a component infrastructure, and different (wireless) personal (e.g., Bluetooth) and local area (e.g., WLAN) networking technologies may provide the basis for communications.

A particular set of issues arises when such modular systems comprise more than two entities and operate in mobile and highly dynamic ad-hoc environments and are required to cooperate with an arbitrary number of peers to fulfill a certain task: 1) the individual entities cannot rely on well-identifiable service brokers to locate peers providing desired services; 2) different components may be located in different networks without direct

I. Niemegeers and S. Heemstra de Groot (Eds.): PWC 2004, LNCS 3260, pp. 118–131, 2004.

communication paths between all of them; 3) network attachments and addresses may change; 4) (group) communications between all involved entities needs to be provided; and 5) secure communications without pre-shared configurations is needed.

In this paper, we address these issues for the deployment of group communication technologies in mobile ad-hoc communication environments. In section 2, we describe a sample scenario for multiparty peering, followed by a brief discussion of related work in section 3. Section 4 outlines our approaches to group communication (Mbus) and service location and association (DDA) and reviews their present limitations. Section 5 introduces extensions of the DDA approach for accommodating multiparty peering scenarios, and section 6 enhances the Mbus protocol to provide group communication services in dynamically changing network topologies. Finally, section 7 concludes this paper with a summary.

2 Modular Mobile Multimedia Conferencing

Our (non-ad-hoc, non-mobile) starting point for such a scenario is a user's multimedia conferencing endpoint that is made up of numerous devices and software components available around the user's desk: an IP telephone for placing and receiving regular phone calls, a workstation with camera, video communication tool, and application sharing software; a laptop with the user's mail folder and all current documents; and a PDA, containing an address book and a calendar. When the user establishes a call, she may combine any of these entities according to her needs for a particular conversation. Application state (such as the input focus in a distributed editor, the current speaker, etc.) is shared locally so that the application entities can act coherently.

This scenario can be extended to include mobile (nomadic) users who do not have an associated desk area environment – at a certain time or permanently. Examples include employees visiting co-workers in other offices or subsidiaries who want to stay connected as well as environments with non-territorial offices. Such users need to dynamically locate the devices or services they want to use in an ad-hoc fashion: in the simplest case they may just require access a single device in a foreign environment, e.g. to place a phone call controlled from a portable device, so that only point-to-point communication needs to be established dynamically, e.g. from a PDA to an IP phone[KO03]. If more application components need to be involved as for multimedia conferencing, group communication sessions need to be set up among a number of components in an ad-hoc fashion.

What is needed is an ad-hoc communication environment that allows components to dynamically locate each other based upon the services they offer (*service discovery and selection*), securely establish bi-lateral or group communication relationships among selected entities (*service association*), cooperate to carry out the respective task(s) (*service control*), and, finally, disband the coupling again (*service dissociation*).

3 Related Work

Ad-hoc communication for the coordination of application components is essentially related to two research domains: 1) coordination-based protocols and architectures for ad-hoc communication and 2) component-based application development. In addition to ad-hoc communication aspects, there is also the issue of (auto-)configuration, e.g., through mechanisms such as DHCP, IPv6 stateless auto-configuration and zeroconf protocols.

Frameworks for ad-hoc communications typically provide a set of functions, i.e., service/peer location, association procedures and the actual communication mechanisms that may comprise different communication patterns such as request/response communication and event notifications. In the following, we briefly describe two prominent solutions in that area: the *Service Location Protocol* and *Universal Plug and Play*. Jini and Salutation are additional ad-hoc communication frameworks. We have provided a more detailed discussion of ad-hoc communication solutions in [KO03].

- The *Service Location Protocol* (SLP, [GPVD99]) is a framework for service discovery and selection, intended for automatically configuring applications that want to use services such as network printers or remote file systems in an administrative domain. SLP was designed to only solve the problem of locating services, not associating to the services themselves. Service association (i.e., allocating a service resource, exchanging confidential access credentials) needs to be done in a second step, with a different protocol so that SLP alone is rather incomplete for our purposes [KO03].
- *Universal Plug and Play* (UPnP, [Cor00]) is an architecture for device control and provides protocols for discovery, device description, transmission of control commands and event notification. UPnP is intended to provide peer-to-peer communication between different types of devices without manual configuration, e.g., in home networks where devices from different vendors are connected spontaneously. To allow these devices to interwork they must dynamically learn their capabilities and exchange information without knowing each other in advance.

For most environments, such as enterprise application development, component-based software development can now be considered an acknowledged design principle. We can distinguish between frameworks for the composition of program binaries such as COM and JavaBeans and distributed component architectures such as DCOM [BK98] and Corba [OMG]. The latter support spreading system components across various devices (e.g. PCs, laptops, stand-alone networked appliances, PDAs, etc.) allowing to employ dedicated pieces of equipment where appropriate or just spread the work load as needed.

Many existing protocols rely on point-to-point communication between components and employ classical RPC interactions. Group communication is only used as a rendezvous mechanism for service discovery because multicast connectivity is usually restricted to single links and thus cannot be relied upon. Furthermore, many protocols are designed for static scenarios where service availability and reachability do not change after the discovery process. Such protocols are difficult to deploy in ad-hoc communication scenarios, where services can appear and disappear dynamically and where both service providers and clients may be mobile.

However, we argue that group communication is an important element for service coordination because it is a natural solution for many coordination scenarios and may simplify several kinds of interactions. This is especially true for the development of systems intended for dynamic environments where network addresses are not necessarily persistent for the duration of a session. Group communication in conjunction with, e.g., *soft-state* protocols help to make these systems more robust against intermittent connectivity and changing service availability.

Our approach is significantly simpler than a generic multicast routing protocol for ad hoc networks, e.g., as described in [OT98] and [GLAM99], because we do not have to take large scale ad hoc networks with unconstrained mobility into account. In particular, our approach is driven by application scenarios where a user device coordinates distributed application components in a dynamic, heterogeneous network environment – which typically comprises infrastructure-based components to many of which mobile ad hoc networking (MANET) protocols are not applicable. For example in an office environment, a user's laptop and wireless IP telephone would be connected using the fixed WLAN infrastructure and would thus rely on the available IP connectivity. In these scenarios, mobility is an issue when users and/or devices enter or leave networks, but it is not issue with respect to node mobility and changing network topologies. In our scenario, where a central user device is connected to a set of personal devices in the environment using different link layer technologies, it would not be helpful to implement a MANET protocol implementation on each device, because the device would not have to participate in MANET routing and packet forwarding anyway. Instead, it is sufficient (and more practical) to have the central user device become aware of the connected network links and the available communication peers.

4 Mbus/DDA-Based Ad Hoc Peer-to-Peer Cooperation

In [KO03], we have presented a local coordination environment for mobile users that addresses service discovery and remote control of multimedia communication equipment (such as IP phones) in mobile ad-hoc networks. The *Dynamic Device Association (DDA)* concept is used to locate, select, get hold of, and release devices and services as described in section 4.1. The actual device interaction takes place using the Mbus we have developed for group communications component-based systems with a particular focus on multimedia conferencing (section 4.2).

4.1 Dynamic Device Association

The DDA framework addresses the discovery of and the secure association with services. It comprises five phases:

1. *Service and device discovery*
   Fixed and mobile devices announce their availability in regular intervals; in addition a query mechanism is supported. The announcements include service descriptions and rendezvous URIs to allow other entities to contact them as needed. The Session Announcement Protocol (SAP) [HPW00] is used for the announcements, augmented by a dedicated query message. Device and service descriptions are represented using the Session Description Protocol (SDP) [HJ98].

2. *Device selection*
   User devices searching for peers with certain functionality scan the announcements for the service sought. Once a set of suitable devices has been found, one or more of them are selected as required by the user or some automated algorithm.
3. *Service and device association.*
   The purpose of the association step is to initiate an application protocol session between two entities. For each device and service chosen, an association process is invoked by the user device. The selected device is contacted and, if necessary, an authentication procedure is carried out. Finally, the intended application protocol is bootstrapped. To support all kinds of application protocols (including SIP [RSC+02], Mbus [OPK02], HTTP [FGM+99], and SOAP [BEK+02]), the DDA session description language may contain arbitrary key-value-pairs for protocol-specific information.
4. *Application protocol operation*
   The application protocol runs in the context of the established association. This may involve all kinds of interactions between the mobile and the associated device. When no longer needed, the application protocol session is terminated (which may but need not lead to a device dissociation). In our scenario, we invoke Mbus [OPK02] as application protocol running the call and conference control profile [OKM01].
5. *Service and device dissociation*
   Eventually, when the associated device is no longer needed, the user's mobile device dissociates from the device and potentially makes the device fully available again to the public.

In figure 1, a service client (the PDA) receives service announcements from two entities (IP phones), selects one phone and initiates the association process. After authenticating the PDA's user, the phone answers the association request and provides the application session configuration parameters in an association process. Both parties join the corresponding application session and communicate over the specific session protocol.

4.2 Group Communication with the Mbus

We have developed a group communication environment, the *Message Bus* (Mbus, [OPK02]) for component-based systems in multicast-capable environments. Mbus is a message-oriented coordination protocol that provides group and point-to-point communication services, employing a network-layer addressing scheme. Each Mbus session member provides a single, unique Mbus address chosen by the application. A fully qualified Mbus address is used for Mbus unicasting, partly qualified and empty addresses for multicasting and broadcasting, respectively.

The payload of an Mbus message is a list of *Mbus commands* with application-defined semantics. Commands may represent status updates, event notifications, RPCs and other interaction types. Each may include a list of parameters, such as strings, numbers and lists. Mbus also provides a set of control messages, particularly for dynamic group membership tracking based upon regular announcements (`mbus.hello()` messages).

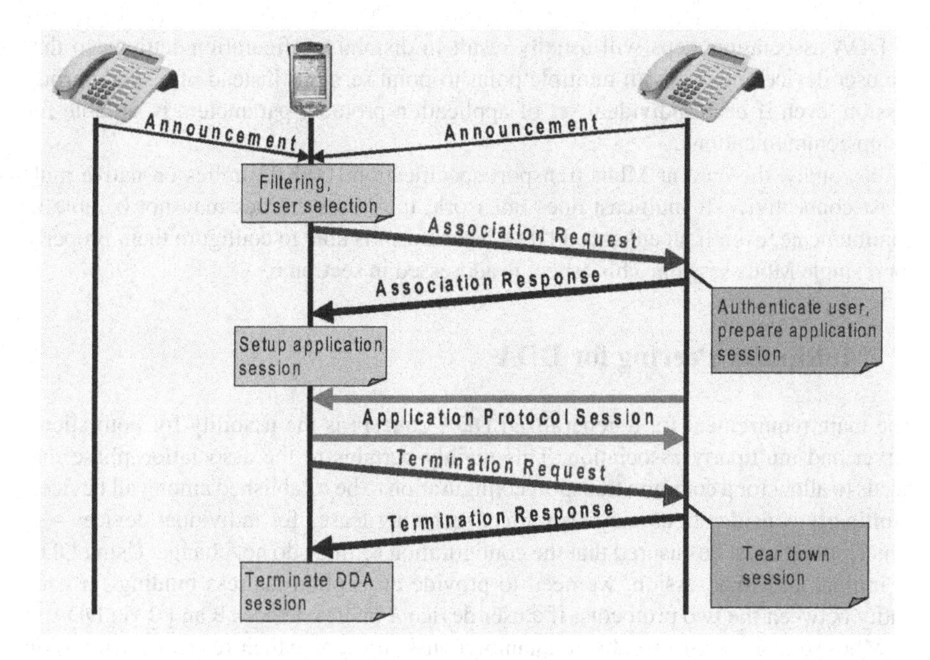

Fig. 1. The DDA process

Mbus messages may be transmitted as UDP multicast and unicast datagrams. A receiver-driven filtering process delivers those messages to applications that carry a destination address matching the own fully-qualified entity address. Messages that are directed to a single entity only (i.e., providing a sufficiently qualified destination address) can optionally be sent via unicast. The necessary Mbus-to-UDP address mapping can be inferred from the regular membership announcements as each Mbus messages carries the sender's fully qualified Mbus address.

The Mbus provides security services such as authentication and messages integrity based on hashed message authentication codes (HMACs) and confidentiality based on encryption (relying on symmetric cryptography and shared keys). In order to join an Mbus session, an application has to know the Mbus transport address (i.e., multicast address and port number) and security parameters, which are distributed out-of-band, i.e., configured manually or communicated using other protocols. In this particular case, we use the DDA protocol.

While DDA and Mbus address the need for secure establishment of ad-hoc cooperation scenarios, the present protocol mechanisms are still not capable of supporting group communications – although Mbus in principle does. This is mainly because of two reasons:

Firstly, similar to other approaches, the DDA model relies on the concept of a one-to-one relationship of service providers and users. A service user contacts exactly one DDA service provider and requests the necessary configuration data to establish a point-to-point communication session. For multiparty peering, the same user device needs to contact multiple service providers (aiming at setting up a group). But applying a series of DDA association steps will usually result in disjoint configuration settings so that the user device ends up with multiple point-to-point sessions instead of a single group session, even if each individual set of application protocol parameters is suitable for group communications.

Secondly, the current Mbus transport specification [OPK02] relies on native multicast connectivity. If multicast does not work, the Mbus entities may not be able to communicate, even if an enhanced DDA mechanism is able to configure them properly for a single Mbus session. This aspect is addressed in section 6.

5 Multiparty Peering for DDA

The main requirement for a generalized DDA concept is the usability for both client-server and multiparty association. This mainly pertains to the association phase that needs to allow for a common transport configuration to be established among all devices. Multiparty considerations may also affect renewing leases for individual devices – in which case it must be ensured that the configuration settings do not change. Using DDA to initiate an Mbus session, we need to provide an explicit address binding for each entity between the two protocols: If a user device `A` invites devices `B` and `C` via DDA to an Mbus session, it must be able to unambiguously recognize their respective entities in order to exclude address clashes.

Finally, if a user device has multiple network interfaces to receive service announcements and to communicate with its peers, it needs to coordinate the use of multicast addresses and choose the same for Mbus communications across all interfaces. This is needed to avoid reconfiguration in case of network topology changes, e.g., if peers move from one network to another. A user device may need to allocate multicast addresses and ensure that the same address is available on all locally connected links prior to initiating the first association.

The basic extension to the DDA process is to generalize the association and to enable DDA clients to not only request a configuration from a DDA service but to optionally *invite* a service into a session by providing it with the required session parameters. Where applicable, service entities should allow for both forms of association, i.e., be able to offer a session configuration and to be invited into a session. Service entities that are restricted to either mode indicate their preferred association mode in their service announcement to avoid unnecessary requests/response cycles.

For the HTTP-based DDA protocol we have implemented the "invitation" mode with an HTTP POST [FGM+99] request. Note that the authentication requirements do not change for association invitations: for digest authentication, the DDA client would still provide the credentials in the request message. For DDA for Mbus sessions, we have defined additional attributes for the session description that allow both parties to express their Mbus and corresponding UDP/IP endpoint addresses. Because both parties have to know each other's addresses in advance, we allow for both the request and the response in every DDA HTTP request (GET and POST) to contain a message body. For example, when a DDA service is invited and has received a corresponding association invitation, it will send a session description fragment responding to the request and thus provide the required address information.

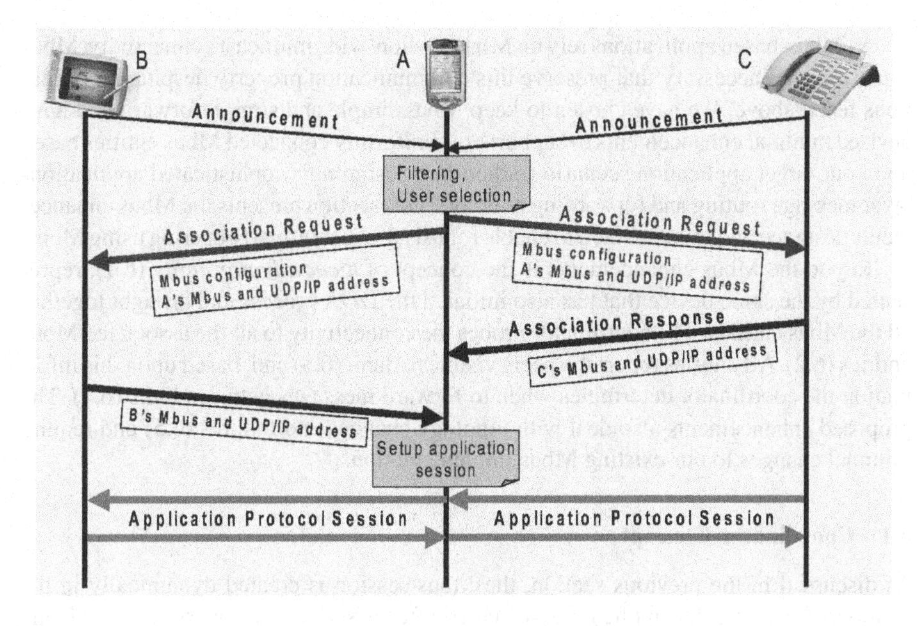

Fig. 2. DDA in invitation mode

Figure 2 depicts the message exchange for the DDA invitation mode. A chooses an application session configuration and conveys this information to both B and C. After the two DDA associations have been completed, all three entities can join the Mbus session and check for the availability of the other sides, using the Mbus address information that has been exchanged in the DDA process. For scenarios where IP multicast connectivity is not available we have defined a *probing* process that helps to determine the optimal communication mode (e.g., multicast or direct unicast) between the initiator (the user device) and the invited service entity. The probing process is described in section 6.2.

6 Multiparty-Peering for Mbus

The initial scenario described in section 2 has implicitly assumed full multicast connectivity between all application entities – a safe assumption for an environment built around a single link of a local area network with today's operating systems. However, this assumption is unlikely to hold as soon as mobile devices (such as PDAs or laptop computers) become involved making use of WLAN infrastructure or even engage into ad-hoc communications with other devices using dedicated Bluetooth or infrared links, in addition to their connection to the (W)LAN. For security reasons, WLANs are usually connected via access routers or firewalls which may prevent multicast forwarding (deliberately or accidentally), WLAN access points may have multicast forwarding disabled to protect the bandwidth constrained wireless network. If different subnetworks are involved, any regular router (even if multicast-enabled) prevents Mbus messages from propagating as they are constrained to link-local communications.

As Mbus-based applications rely on Mbus session-wide multicast connectivity, Mbus extensions are necessary that preserve this communication property despite the limitations listed above. We have chosen to keep Mbus simple and straightforward and have devised minimal enhancements to support non-uniformly connected Mbus entities based upon our target application scenario (rather than designing a sophisticated application-layer message routing and forwarding overlay). This section presents the Mbus enhancements designed and implemented to enable robust ad-hoc multiparty peering using Mbus.

Key to the Mbus enhancements is the concept of a *coordinator entity* (6.1), represented by the same device that has also initiated the DDA process and brought together all the Mbus entities. The coordinator probes the connectivity to all the associated Mbus entities (6.2). All entities report the peers visible to them (6.3) and, based upon this information the coordinator determines when to forward messages between links (6.4). The proposed enhancements also deal with topology changes and failures (6.5) and require minimal changes to our existing Mbus implementation.

6.1 Coordinator Concept

As discussed in the previous section, the Mbus session is created dynamically in the context of a particular application scenario: after a service discovery phase, all the services are contacted by the initiating entity. Obviously, this very entity – the *coordinator* – has a complete overview of the components it has sought to carry out the intended task. Furthermore, the coordinator may have used different link layer technologies to contact the various peers so that it is the only one in a position to take up the responsibility of initially establishing reachability between all involved parties and also act as a hub if native multicast connectivity is not available. Finally, the coordinator is the only entity capable of re-invoking the DDA procedures, e.g., to update Mbus configuration parameters or to prolong service leases.

Usually centralized architectures are considered risky as they introduce a potential bottleneck and a single point of failure. With the DDA scenario in mind, however, there may be no other way to establish connectivity between the entities in the first place. And, as the Mbus is used to communicate control messages only (rather than large data volumes), processing power and communication bandwidth are not considered to be problematic. While the coordinator could obviously be a single point of failure, we achieve fate-sharing with the intended application as the coordinator is also in control of the other devices; hence, its failure will likely cause the application to fail anyway. Nevertheless, we consider enhanced robustness here an important subject of further work.

Figure 3 shows three conceivable settings with a coordinator `A` and three devices `B`, `C`, and `D`. In setting a), full multicast connectivity is available so that there is no need for the coordinator to perform any kind of message forwarding. Setting b) shows device `B` being on a separate link: `A` and `B` can communicate via unicast and multicast, and so can `A`, `C`, and `D`. But `B` has no way to talk to `C` and `D` and vice versa, neither with unicast nor with multicast. Setting c) depicts a scenario with `B` on a separate link again, with `C` and `D` sharing the same link while `A` is connected via a router (or some other entity blocking link-local multicast). As a result, `A` and `B` can talk via unicast and multicast and so can `C` and `D`. `A` and `C` as well as `A` and `D` can only communicate via unicast.

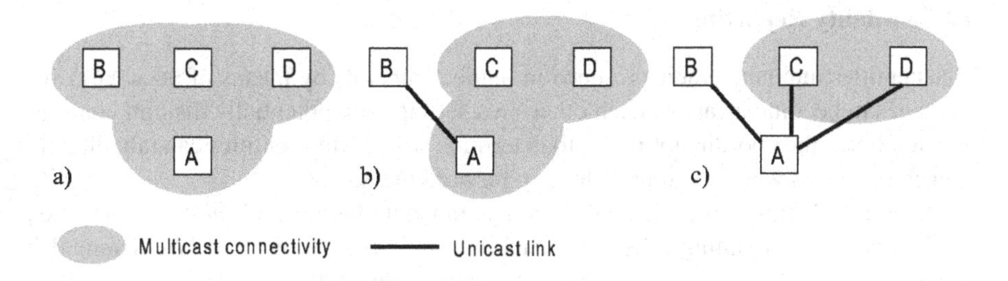

Fig. 3. Three multiparty peering scenarios

Those three settings can be taken representatively for most connectivity variants that one may experience in local ad-hoc communications. Note also that even though we do not explicitly discuss asymmetric connectivity in this paper, we have verified that the algorithms presented below will work in those cases, too.

6.2 Connectivity Discovery

As outlined above, the coordinator needs to determine what kind of connectivity is available to its peers. As it was able to initially contact them and create an association, plain IP connectivity is obviously available. The next step is to determine whether the respective entities are also reachable via multicast or only via unicast.

For this purpose, we introduce `mbus.probe(m|u seq-no)` messages that are parameterized with a flag indicating whether this message is sent, at the IP layer, via unicast ("u") or multicast ("m") and with a sequence number (for matching probes and their responses). The coordinator starts sending `mbus.probe` messages to each of the newly associated entities using their Mbus unicast addresses (learned from the DDA association messages). These messages are sent once via IP unicast (using the "u" flag) and once via IP multicast. (using the "m" flag). For each message sent, regardless whether unicast or multicast, the sequence number is incremented by one. The coordinator retransmits the messages up to three times to deal with possible packet loss.

A receiver of such a message responds to each of the messages, again once by unicast and once by multicast – so that up to six messages are exchanged in total. Each response message – `mbus.probe.ack (m|u seq-no*)` – again contains a flag indicating whether the message was sent via unicast or multicast and contains a list of `mbus.probe` sequence numbers received from the coordinator for the last few seconds.

If the coordinator receives `mbus.probe` responses via unicast *and* multicast, acknowledging both unicast and multicast probes, full unicast and multicast connectivity is available. Otherwise, the combination of response messages received (via unicast and/or multicast) and their acknowledged sequence numbers reveal in which direction multicast connectivity is available, if at all. For simplicity, in all cases but the first, the communication between the coordinator and the probed Mbus entity will only use unicast communication. The result of this process is used to configure the message routing for both the coordinator and the Mbus entity. Connectivity probing may be repeated when topology changes are suspected, e.g., when an entity has become invisible on a link.

6.3 Visibility Reporting

Mbus entities announce their presence in regular intervals by means of `mbus.hello()` messages to become aware of each other. In a setting with potentially disjoint communication links, the coordinator needs to determine which Mbus entities can talk directly to each other and which require its help to forward messages.

To establish this view, each Mbus entity transmits periodic *visibility reports*, i.e., Mbus messages containing a list of other Mbus entities it is aware of. We distinguish two kinds of visibility: *native visibility* refers to Mbus entities whose `mbus.hello()` messages were received directly, i.e., without the help of the forwarding coordinator; *effective visibility* refers to all Mbus entities from which `mbus.hello()` messages have been received recently. An Mbus visibility report is defined as `mbus.visible ((<Mbus address> [native])*)`, i.e., it provides a list of peers, each indicating the Mbus address being reported and a "`native`" flag showing whether the message has been received directly from the respective entity. If both native and relayed messages are received from another entity, the native reporting takes precedence.

When all Mbus entities start communicating, only native visibility reports are possible. If the coordinator observes that all entities can see all others natively no further actions are necessary on its part. Otherwise, the coordinator can determine from the visibility reports how the Mbus session is partitioned and start forwarding messages between those partitions. All Mbus messages except for `mbus.hello()` are forwarded unchanged. `mbus.hello()` needs to receive special processing to allow distingushing native messages from relayed ones: a single parameter peer `(via <Mbus coordinator-address>)` is inserted into the message yielding `mbus.hello ((via <coordinator-address>))`.

As soon as the coordinator starts forwarding messages, Mbus entities will add also those peers to their visibility reports whose messages have been forwarded. The coordinator uses the effective visibility to determine when full connectivity of the Mbus session has been achieved. It continues to use the native visibility to constantly monitor the overall connectivity and adapt its forwarding behavior when necessary.

6.4 Message Transmission and Forwarding

Mbus message transmission is conceptually extended to support multiple *interfaces* per Mbus entity. A regular Mbus entity (i.e., not the coordinator) provides a multicast interface and may provide one or more unicast interfaces and uses only a single link. Each interface is basically similar to a link layer interface with routing table entries (based on Mbus addresses) pointing to this interface. The original Mbus design has a default route for all traffic pointing to the multicast interface and may have one unicast interface per known Mbus entity for the unicast optimization.

For Mbus sessions with partial multicast connectivity and a coordinator acting as a "hub", the transmission behavior of Mbus entities needs to be adapted only slightly. Mbus entities that have full multicast connectivity with their coordinator do not need to change; the above rules just work. Mbus entities that have only unicast connectivity to their coordinator and no multicast connectivity to other entities (i.e., do not see any native visibility reports except from the coordinator) use their unicast interface to the coordinator as default interface. Mbus entities that have directly reachable multicast peers but only a unicast interface to the coordinator, create two default routes and thus duplicate their outgoing Mbus messages (except for those using the unicast optimization) transmitting them via multicast and sending them to the coordinator.

The coordinator is responsible for relaying Mbus messages and modifying `mbus.hello` messages in transit. Its forwarding functions are configured based on the visibility reporting and the connectivity discovery. It may have any number of unicast and multicast interfaces on different links. Each incoming Mbus message is examined with respect to its target Mbus address. If this is a multicast address, the coordinator forwards the message to all interfaces (except for the one it has been received on) and forwards a local copy to its own application. Otherwise, the coordinator examines – based upon native visibility reports – to which interface the message needs to be forwarded or hands the message to its local application.

6.5 Change and Failure Handling

The coordinator permanently monitors effective and native visibility as reported from each endpoint. In case multicast connectivity improves, the coordinator will notice further entities reporting native visibility of each other and so the coordinator can reduce forwarding. If multicast connectivity is lost, incomplete effective visibility reports indicates that additional forwarding needs to be installed. If the coordinator looses contact to an entity, (e.g. by missing `mbus.hello()` messages), it may need to re-enter the *connectivity discovery* again. If this does not reveal ways to re-establish connectivity (e.g. because the entity is not longer reachable), the coordinator may attempt a DDA re-association or go through the entire service location procedure again to look for a different device offering the same services.

7 Conclusions

We have described scenarios and solutions for multiparty peering of service entities, considering the aspects *service discovery* and *group communication* and the special issues for ad-hoc communication scenarios, such as changing network topologies and peer mobility. The discussion of these scenarios has shown that group communication is a desirable feature for many component-based services in local networks. However, it has also been evident that its implementation is not always trivial, because general multicast connectivity cannot be assumed and because dynamic communication scenarios require concepts that address potential changes in network topology while maintaining a continuous group communication session at the application layer. The service discovery approach that we have presented addresses the requirements for ad-hoc communication by employing a service announcement scheme based on soft-state communication that has been designed with respect to scalability and efficiency. We have extended the DDA service association protocol to support multiparty peering and have discussed the use of these extensions for establishing Mbus sessions.

Using the Mbus as a basis, we have developed a group communication model that provides the concept of a group communication session that can encompass multiple underlying multicast and unicast sessions. One key aspect of this model is a central coordinating entity that manages the individual sessions, monitors entity visibility and provide message relay functions where appropriate. By adding some minimal changes to the Mbus protocol (adding new membership information messages) and by extending the Mbus implementation requirements slightly, Mbus entities can accommodate changing multicast connectivity, dynamic changes to the group membership and mixed multicast/unicast environments – characteristics that are typical for in dynamic communication scenarios. The central role of the coordinator is not considered an issue for most cases because the fate of the coordinator is coupled to the user's application anyway.

The link-state monitoring and forwarding functions that we have described are not to be misinterpreted as elements of a general layer 3 ad-hoc networking routing protocol such as AODV [PBRD03]. While routing protocols for ad-hoc networks provide multihop routing between potentially mobile hosts in order to establish and maintain an ad-hoc IP network, our approach is much simpler and highly efficient: The main goal is to provide group communication in scenarios where no comprehensive multicast connectivity between the intended group member can be established, and the forwarding is restricted to specific messages of a selected application protocol (Mbus). Moreover, we rely on a special case, where there is always a central entity (the coordinator) that has a direct link to each of the session members.

In summary, the DDA framework peered with Mbus provides a lightweight yet powerful infrastructure for ad-hoc group cooperation in dynamic mobile environments. Our approach largely builds upon existing and well-established protocols simplifying integration with all kinds of personal devices.

References

[BEK⁺ 02] Don Box, David Ehnebuske, Gopal Kakivaya, Andrew Layman, Noah Mendelsohn, Henrik Frystyk Nielsen, Satish Thatte, and Dave Winer. Simple Object Access Protocol (SOAP) 1.1. W3C Note 08, May 2002.

[BK98] N. Brown and C. Kindel. Distributed Component Object Model Protocol – DCOM/1.0, January 1998. draft-brown-dcom-v1-spec-03.txt.

[Cor00] Microsoft Corporation. Universal Plug and Play Device Architecture. available online at http://www.upnp.org/, June 2000.

[FGM⁺ 99] Roy T. Fielding, Jim Gettys, Jeffrey C. Mogul, Henry Frystyk Nielsen, Larry Masinter, Paul Leach, and Tim Berners-Lee. Hypertext Transfer Protocol – HTTP/1.1. RFC 2616, June 1999.

[GLAM99] J. J. Garcia-Luna-Aceves and Ewerton L. Madruga. A multicast routing protocol for ad-hoc networks. In *INFOCOM*, pages 784–792, 1999.

[GPVD99] Erik Guttmann, Charles Perkins, John Veizades, and Michael Day. Service Location Protocol, Version 2. RFC 2608, June 1999.

[HJ98] Mark Handley and Van Jacobsen. Session Description Protocol. RFC 2327, April 1998.

[HPW00] Mark Handley, Colin Perkins, and Edmung Whelan. Session Announcement Protocol. RFC 2974, October 2000.

[KO03] Dirk Kutscher and J rg Ott. Dynamic Device Access for Mobile Users. In *Proceedings of the 8th Conference on Personal Wireless Communications*, 2003.

[OKM01] Jörg Ott, Dirk Kutscher, and Dirk Meyer. An Mbus Profile for Call Control. available online at http://www.mbus.org/, February 2001. Internet Draft draft-ietf-mmusic-mbus-call-control-00.txt, Work in Progress.

[OMG] OMG. CORBA/IIOP Specification. available online at http://www.omg.org/docs/formal/02-12-06.pdf.

[OPK02] Jörg Ott, Colin Perkins, and Dirk Kutscher. A Message Bus for Local Coordination. RFC 3259, April 2002.

[OT98] K. Obraczka and G. Tsudik. Multicast routing issues in ad hoc networks, 1998.

[PBRD03] Charles E. Perkins, Elizabeth M. Belding-Royer, and Samir R. Das. Ad hoc On-Demand Distance Vector (AODV) Routing, February 2003. Internet Draft draft-ietf-manet-aodv-13.txt, Work in Progress.

[RSC+ 02] Jonathan Rosenberg, Henning Schulzrinne, Gonzalo Camarillo, Alan Johnston, Jon Peterson, Robert Sparks, Mark Handley, and Eve Schooler. SIP: Session Initiation Protocol. RFC 3261, June 2002.

Ad Hoc Routing with Early Unidirectionality Detection and Avoidance*

Young-Bae Ko[1], Sung-Ju Lee[2], and Jun-Beom Lee[1]

[1] College of Information & Communication, Ajou University, Suwon, Korea
[2] Mobile & Media Systems Lab, Hewlett-Packard Laboratories, Palo Alto, CA 94304

Abstract. This paper is motivated by the observation that current research in ad hoc networks mostly assumes a physically flat network architecture with the nodes having homogeneous characteristics, roles, and networking- and processing-capabilities (e.g., network resources, computing power, and transmission power). In real-world ad hoc networks however, node heterogeneity is inherent. New mechanisms at the network layer are required for effective and efficient utilization of such heterogeneous networks. We discuss the issues and challenges for routing protocol design in heterogeneous ad hoc networks, and focus on the problem of quickly detecting and avoiding unidirectional links. We propose a routing framework called Early Unidirectionality Detection and Avoidance (EUDA) that utilizes geographical distance and path loss between the nodes for fast detection of asymmetric and unidirectional routes. We evaluate our scheme through ns-2 simulation and compare it with existing approaches. Our results demonstrate that our techniques work well in these realistic, heterogeneous ad hoc networking environments with unidirectional links.

1 Introduction

Starting from the days of the packet radio networks (PRNET) [10, 13] in the 1970s and survivable adaptive networks (SURAN) [11] in the 1980s to the global mobile (GloMo) networks [14] in the 1990s and the current mobile ad hoc networks (MANET) [6], the multi-hop ad hoc network has received great amount of research attention. The ease of deployment without any existing infrastructure makes ad hoc networks an attractive choice for applications such as military operations, disaster recovery, search-and-rescue, and so forth. With the advance of IEEE 802.11 technology and the wide availability of mobile wireless devices, civilians can also form an instantaneous ad hoc network in conferences or in class rooms.

Recent research in ad hoc networks has focused on medium access control and routing protocols. Because of shared wireless broadcast medium, contention and hidden terminals are common in ad hoc networks and hence MAC is an important problem. Routing is also an interesting issue as routes are typically multi-hop. When the end-to-end source and destination are not within each other's transmission range, routes are multi-hop and they rely on intermediate nodes to forward the packets. The construction and maintenance of the routes are especially challenging when nodes are mobile.

* This work was in part supported by grant No. R05-2003-000-10607-02004 from Korea Science & Engineering Foundation, and University IT Research Center project.

I. Niemegeers and S. Heemstra de Groot (Eds.): PWC 2004, LNCS 3260, pp. 132–146, 2004.

Although there has been a great amount of work in these areas, most of the research assumes the nodes are homogeneous. All nodes are assumed to have the same or similar radio propagation range, processing capability, battery power, storage, and so forth. Even the schemes that utilize the hierarchy of the nodes [15, 25] assume a flat physical network structure and the hierarchies are merely logical. In reality however, nodes in ad hoc networks have heterogeneity. In the military scenarios for instance, the troop leader is usually equipped with more powerful networking devices than the private soldiers of the troop. Radios installed in the vehicles such as tanks and jeeps have more capabilities than radios the soldiers carry, as vehicles do not have the same size- or power-constraints as the mobile soldiers have. Another reason could be the financial cost. The state-of-the-art equipments are very expensive and hence only a small number of nodes could be supplied with such high-end devices. Similarly, civilians possess different types of mobile devices ranging from small palm-pilots and PDAs to laptops.

The heterogeneity of the ad hoc network nodes creates challenges to current MAC and routing protocols. Many MAC protocols use the request-to-send/clear-to-send (RTS/CTS) handshake to resolve channel contention for unicast packets. The assumption here is that when node A can deliver RTS to node B, node A will also be able to receive CTS from node B. Routing protocols in ad hoc networks typically assume bidirectional, symmetric routes, which do not always hold true when node heterogeneity is introduced. The performance of these protocols may degrade in networks with heterogeneous nodes [23].

One of the major challenges in ad hoc networks with heterogeneous nodes is the existence of "unidirectional links." Along with the medium access control, routing performance can be suffered from the existence of unidirectional links and routes. Unidirectional links may exist for various reasons. Different radios may have different propagation range, and hence unidirectional links may exist between two nodes with different type of equipments. IEEE 802.11b uses different transmission rates for broadcast and unicast packets. That creates gray zones [17] where nodes within that zone receive broadcast packets from a certain source but not unicast packets. The hidden terminal problem [28] can also result in unidirectional links. Moreover, interference, fading, and other wireless channel problems can affect the communication reachability of the nodes. Some recent proposals have nodes adjust the radio transmission range for the purpose of energy-aware routing [9] and topology control [26]. The nodes in these schemes transmit packets with the radio power just strong enough to reach their neighbors. When nodes move out of that range, the link turns into unidirectional, when in fact it could be bidirectional when each node sends packets with the maximum transmission range. The unidirectional links (and routes) are therefore, quite common in ad hoc networks.

In this paper, we focus on the issues and challenges for routing protocol design in heterogeneous ad hoc networks. Specifically, we focus on the heterogeneity of node transmission power and unidirectional links resulting from it. We propose a routing technique EUDA (Early Unidirectionality Detection and Avoidance) that proactively detects unidirectional links and avoids constructing routes that include such links. We introduce two approaches: (i) a network-layer solution that utilizes node location information and (ii) a cross-layer solution based on a path-loss model.

The rest of the paper is organized as follows. Related work on heterogeneous ad hoc networks is covered in Section 2. We then study how existing ad hoc routing protocols handle unidirectional links in Section 3. Section 4 introduces EUDA, followed by ns-2 simulation results in Section 5. We conclude in Section 6.

2 Related Work

There has been recent research interest in heterogeneous ad hoc networks. DEAR (Device and Energy Aware Routing) protocol [1] considers the heterogeneity of the nodes in terms of the power source. Nodes that have continuous energy supply from external power forward more packets than nodes that are running on battery. As the main goal of DEAR is energy-awareness and it only addresses power source heterogeneity, it does not investigate other issues such as unidirectional links that may result from the node heterogeneity. ISAIAH (Infra-Structure AODV for Infrastructured Ad Hoc networks) [16] introduces "pseudo base stations (PBS)" that are immobile and have infinite amount of power supply. Its routing protocol selects paths that include such PBS nodes instead of regular mobile nodes. Similar to DEAR, it does not address the problem of unidirectional routes. The notion of reliable nodes that are secure and robust to failure is used in [31]. This work focuses on the optimal placement of such reliable nodes and subsequent route construction. Heterogeneity of ad hoc network node is also studied in [30]. Initially, the nodes are grouped into clusters. Each cluster elects a backbone node based on node capabilities. The backbone nodes themselves form a network called Mobile Backbone Network (MBN) for efficient, scalable communication. The spirit is similar to existing hierarchical, clustering work, but it uses the "physical" hierarchy of the nodes. The optimal number of backbone nodes is obtained analytically, and the clustering and routing schemes are introduced. By using simulations, MAC performance in ad hoc networks with heterogeneous node transmission power is analyzed in [23]. This study illustrates the negative impact of unidirectional links on handshake-driven MAC protocols. Ad hoc network heterogeneity is also investigated in [3], but it only considers heterogeneous network interfaces.

There has been recent attention on routing in ad hoc networks with unidirectional links [2, 24, 27]. These schemes however, rely on proactive routing mechanisms where each node periodically exchanges link information for route maintenance. Various performance studies [4, 7, 12] report that proactive table-driven routing protocols do not perform well in ad hoc networks, especially in highly mobile, dynamic situations. Although there have been numerous on-demand ad hoc routing protocols proposed, very few give attention to unidirectional links. Dynamic Source Routing (DSR) [8] could operate with unidirectional links, but it comes at the cost of excessive messaging overhead, as two network-wide flooding is required for each route construction; one from the source to the destination and the other from the destination to the source. Ad hoc On-Demand Distance Vector (AODV) [22] does not work well in the presence of unidirectional links. There have been proposals to solve this problem, which is the topic of the next section.

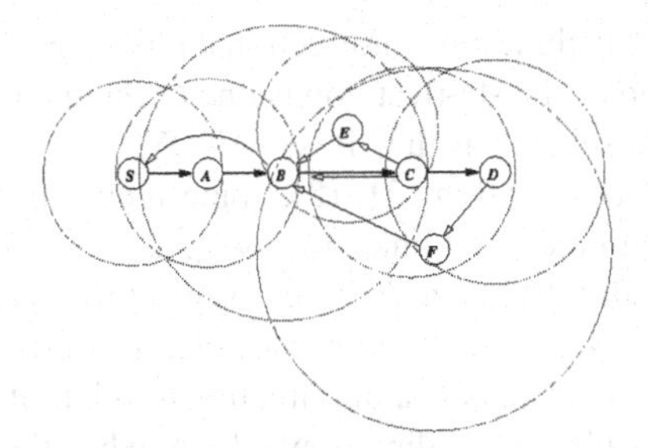

Fig. 1. Routing in networks with unidirectional links. The link between B and C is a unidirectional link. The packets transmitted by node B reach node C, but not vice versa.

3 Unidirectional Links in AODV

On-demand routing is generally shown to give good performance than proactive schemes when node mobility is high and network connectivity changes frequently. Although our idea to be presented in Section 4 can be applied to proactive routing protocols, we focus here on on-demand routing. Here we illustrate how unidirectional routes are formed in existing on-demand routing approaches.

Nodes in on-demand routing protocols do not maintain routes to all nodes in the network. They build routes to only nodes they need to communicated with, only when they have data packets to send to. Routes are usually constructed by flooding a ROUTE REQUEST (RREQ) packet to the entire network. When the destination receives multiple ROUTE REQUEST packets, it selects the best route based on the route selection algorithm (e.g., minimum delay, shortest hop, minimum energy, etc.). The destination then sends a ROUTE REPLY (RREP) message to the source via the reverse of the chosen path. This ROUTE REPLY will reach the source of the route through the selected path *only if* the route is bidirectional. In addition to ROUTE REPLY packets, ROUTE ERROR (RERR) messages, which are used to inform the source node of the route disconnection, are transmitted through the reverse route. When there is a unidirectional link in the route, these on-demand routing protocols cannot operate correctly.

Let us investigate how existing, popular ad hoc on-demand routing protocols function in the presence of unidirectional links. AODV (Ad hoc On-Demand Distance Vector) routing algorithm [22] for example assumes that all links between neighboring nodes are symmetric (i.e., bi-directional links). Therefore, if there are unidirectional links in the network and these unidirectional links are included on a reverse path, AODV may not be successful in a route search. As an example in Figure 1, a RREQ packet generated by a source node S traverses the path $< S - A - B - C - D >$ until it arrives at a destination node D. Each circle in the figure represents the node transmission range. When node D receives the RREQ, it sends a RREP back to node S via the reverse path, $< D - C - B - A - S >$. Note that however, the RREP is not able to reach from node C to node B because node B is not located within node C's transmission range. As a result of a RREP delivery failure by node C, the source S cannot receive the corresponding RREP packet and hence it experiences a route discovery failure in its first trial. Such a failure will repeatedly cause route discovery processes with no benefit. Although there exists a route that does not include any unidirectional link, $< S - A - B - E - C - D >$, this route cannot be found as the shortest hop is one of the main route selection criteria in AODV. We refer to this scheme as the "Basic-AODV."

In the latest AODV specification [21, 18], some mechanisms are newly added to handle the problem of unidirectional links. One way to detect unidirectional links is to have each node periodically exchange *hello* messages that include neighboring information. This scheme however, requires large messaging overhead. Another solution is *blacklisting*. Whenever a node detects a unidirectional link to its neighbor, it blacklists that neighbor from which a link is unidirectional. Later when the node receives a RREQ from one of the nodes in its blacklist set, it discards the RREQ to avoid forming a reverse path with a unidirectional link. Each node maintains a blacklist and the entries in the blacklist are not source-specific. In order to detect a unidirectional link, a node sets the "Acknowledgment Required" bit in a RREP when it transmits the RREP to its next hop. On receiving this RREP with the set flag, the next hop neighbor returns an acknowledgment (also known as RREP-ACK) to the sending node to inform that the RREP was received successfully. In the case when RREP-ACK is not returned, the node puts its next hop node on its *blacklist* so that future RREQ packets received from those suspected nodes are discarded. We refer to this version of AODV as the "AODV-BL (Blacklist)" scheme.

Again, let's use Figure 1 to illustrate the AODV-BL scheme. Here, node C cannot be acknowledged by node B when delivering RREP and therefore it will put node B in its blacklist set. Later when node S re-broadcasts a new RREQ packet, node C will ignore this RREQ received from B but forward another copy of the RREQ from node E. Finally, a destination node D will receive the RREQ through a longer route at this time. Node D returns a RREP back to the source S via a reverse path; in this case, the reverse path is $< D - C - E - B - A - S >$, having no unidirectional links.

The AODV-BL scheme may be efficient when there are few unidirectional links. However, as the number of asymmetric links increase, its routing overhead is likely to become larger since a source node will always suffer from a failure in its first trial of route discovery and need to flood RREQ messages more than once to find a route with all bidirectional links. It also results in an increase of route acquisition delay.

4 Ad Hoc Routing with EUDA

4.1 Basic Mechanism

In EUDA (Early Unidirectionality Detection and Avoidance), a node detects a unidirectional link *immediately* when it receives a RREQ packet. Remember that in AODV-BL, such a detection will be done much later with its RREP-ACK and blacklisting mechanisms. Our goal is to detect a unidirectional link immediately in RREQ forwarding process. The basic idea is that, when node X receives a RREQ from node Y, node X compares its transmission range using the highest power level to an estimated distance between them. If the value of estimated distance from node X to Y is larger than the transmission range of node X, node X considers its link to Y as a unidirectional link, resulting in RREQ packet drop without any further forwarding. Only when a transmission range of node X is equal to or larger than its estimated distance towards node Y,

RREQs from Y will be processed. In EUDA, all nodes receiving RREQ packet are required to decide whether to forward it or not. This decision is based on the comparison of transmission range with distance between the nodes. Note that duplicate detection is still enforced. If a node has already forwarded a RREQ from a source with a specific sequence number, it will drop the rest of the RREQs with the same <source, sequence number> entry. When a node has received RREQs but has not forwarded any of them (i.e., they are from nodes connected through a unidirectional link), it will process subsequent RREQs. Consequently, a node will forward at most one RREQ from a source with a specific sequence number.

Again, see Figure 1 as an example. When node C receives a RREQ from node B, it detects a unidirectional link immediately using the proposed scheme and discards the packet. Later, node C may receive another RREQ from node E. Node C this time will forward this RREQ as it came through a bi-directional link. Finally, the destination D receives the first RREQ via nodes E and C, forming the reverse path $< D - C - E - B - A - S >$ and replying its RREP through this path. Eventually, node S obtains a path having only the bidirectional links. Note that in EUDA, this successful route discovery is achieved from the first attempt by the source, as long as there is at least one bidirectional route from a source to a destination. When no such route exists, the source will time-out without receiving any RREP. It will re-try few times before giving up when no routes can be found.

4.2 Distance Estimation

The next question is how to calculate an estimated distance between two communicating nodes, so that it can be compared with the radio transmission range to determine whether the link between two nodes are bidirectional or unidirectional. This can be done in two different ways as described below.

Network Layer Solution with Location Information. Suppose that the nodes know their geographical position.[1] A transmitter node X includes its own location information in RREQ to be broadcasted. When a node receives a RREQ from X, it calculates the estimated distance (*d*) based on its own physical location. This method can be useful and is easy to implement, but requires information of physical location of the participating nodes.

Cross Layer Solution. As an alternative, we can utilize a wireless channel propagation model, i.e., the two-ray ground path loss model that is designed to predict the mean signal strength for an arbitrary transmitter-receiver separation distance. In wireless networks, if we know the transmitted signal power (P_t) at the transmitter and a separation distance (*d*) of the receiver, the received power (P_r) of each packet is given by the following equation:

[1] Commonly, each node determines its own location by using GPS or some other techniques for positioning service [5]. Recent research has shown that location information can be utilized for ad hoc routing protocols to improve their performance [19, 29].

Table 1. Variables and their default values in ns-2.

Parameters	Definition	Default Value
G_t	Transmitter antenna gain	1.0
G_r	Receiver antenna gain	1.0
h_t	Transmitter antenna height	1.5 m
h_r	Receiver antenna height	1.5 m
L	System loss factor	1.0 (i.e., no loss)

$$P_r = \frac{P_t * G_t * G_r * (h_t^2 * h_r^2)}{d^4 * L} \tag{1}$$

Table 1 lists all the variables used in the equation above and their default values commonly set in *ns-2* simulations. Now we can derive the following equation from Eq. (1) to compute the distance:

$$d = \sqrt[4]{\frac{P_t * G_t * G_r * (h_t^2 * h_r^2)}{P_r * L}} \tag{2}$$

The above Eq. (2) states that the distance between two communicating nodes can be estimated at a receiver side, if the transmitted power level P_t of the packet transmitter and the power received at the receiver P_r are known.

To implement this method, the transmitter should make the transmitted power information available to the receiver, by putting the power information either on a RREQ or a MAC frame. However, the latter approach may cause a compatibility problem with IEEE 802.11, as it has to modify the format of the MAC header in order to specify P_t. Therefore, we use the former approach of modifying a RREQ packet format.

4.3 Discussion on the Distance Estimation Scheme

One assumption behind the unidirectionality detection methods presented above is that any node's communication range is constant when it transmits with its maximum power P_{max}. Remind that this theoretically maximum transmission range is compared with the estimated distance between the two nodes to detect unidirectional links. In reality however, there may be some attenuation of the transmitter power over distance. Therefore, one would argue that such an assumption is unrealistic because transmission range of the RREQ receiver (thus, a potential RREP forwarder) can vary due to several negative environmental factors such as obstacles, reflections, fading, etc.

To take this argument into account, we modify the proposed distance estimation based comparison method so that unidirectional link detection is made with more realistic parameters of the channel gain, the receiver sensitivity, and the receiver's signal-to-noise ratio.

Let us assume that there are two nodes *i* and *j*. When node *j* receives a RREQ from node *i*, it measures the received signal power, $P_r(j)$. The channel gain, G_{ij}, is computed as the received power ($P_r(j)$) at node *j* over the transmitted power ($P_t(i)$) at node *i* (see Eq. (3) below):

$$G_{ij} = \frac{P_r(j)}{P_t(i)} \tag{3}$$

We assume that the sender power $P_t(i)$ is advertised in the RREQ packet. We also assume that the channel gain G_{ij} between two nodes i and j is approximately the same in both directions— note that the same assumption was also made in [20]. Given the transmitter/receiver power information and the channel propagation characteristics, the received power at node i must at least be equal to its minimum receiving threshold RX_Thresh_i, in order for node i to receive any packet successfully from node j with the transmission power $P_t(j)$.

$$P_r(i) = G_{ij} * P_t(j) \geq RX\_Thresh_i \tag{4}$$

This implies that if node j transmits at the maximum power $P_t(max)$ (i.e., replacing $P_t(j)$) satisfies Eq. (4), it can successfully deliver packets to node i. Observe that the value of RX_Thresh_i is related to the receiving sensitivity at node i.

We now define one additional equation such that the observed signal-to-noise ratio SNR_i for the transmission at node i must at least be equal to its minimum SNR_Thresh_i (representing the channel status observed at node i):

$$SNR_i = \frac{G_{ij} * P_t(j)}{P_n(i)} \geq SNR\_Thresh_i \tag{5}$$

where $P_n(i)$ is the total noise node i observes on the channel. Again, this implies that node j can successfully transmit to i when j with its maximum transmission power satisfies Eq. (5).

To summarize, if the above two equations (Eqs. (4) and (5)) are satisfied when one node receives a RREQ from another node, these two nodes are considered to be able to communicate directly with each other and hence have a bi-directional link. Otherwise, it can be concluded as having a unidirectional link between them. With this modification, we have our scheme work better in more realistic scenarios, with its improved estimation accuracy. Nevertheless, there is a clear tradeoff between accuracy and complexity in estimation. Furthermore, the RREQ packet size needs to be increased to include additional information. The transmitter node i of a RREQ packet is now required to include more information (i.e., its transmitted power $P_t(i)$, observed total noise $P_n(i)$, minimum received power threshold RX_Thresh_i, and minimum signal-to-interference ratio SNR_Thresh_i) on the RREQ packet.

5 Performance Evaluation

Although the EUDA framework can be applied to any ad hoc routing protocol, for performance evaluation purposes, we add the EUDA framework to AODV to simulate AODV-EUDA and compare it with the Basic-AODV and AODV with Black Listing (AODV-BL) protocols. As explained in the previous section, Basic-AODV does not include any technique for handling unidirectional links, whereas AODV-BL reactively avoids unidirectional links by using a blacklist set. In AODV-BL, the next hop of a failed

RREP packet is inserted by a node detecting the RREP delivery failure. We performed a simulation study using an extended version of the network simulator *ns-2*. *ns-2* is a discrete event-driven network simulator with extensive support for simulation of TCP, routing, and multicast protocols. The extensions implemented by CMU Monarch project were used for our simulations. Their extensions enable simulation of multi-hop wireless ad hoc networks. Extensions include simulation modules for the IEEE 802.11 MAC protocol and the two-ray radio propagation model. For AODV-EUDA, we simulated both approaches (i.e., the network layer approach and the cross-layer approach) and they gave the same performance as the ns-2 simulator only provides an ideal environment where the node location and path loss information are always correct.

5.1 Simulation Environment

In our simulation model, initial node locations (X and Y coordinates) are obtained using a uniform distribution. All 100 nodes in the network move around in a rectangular region of size 1500 $m \times 300$ m according to the following mobility model. Each node chooses a direction, moving speed, and distance of move based on a predefined distribution and computes its next position *P* and the time instant *T* of reaching that position. Once the node reaches this destined position, it stays there for *pause time* and repeats the process. We always use *zero* pause time (i.e., continuous mobility), and two maximum speeds: 1 m/s and 20 m/s. Thus, each node is assumed to move in a continuous fashion, at a random speed chosen from the interval [0 m/s, (1 or 20) m/s]. Our total simulation time is 900 seconds and we repeated each scenario ten times with different random seed numbers. In our experiments, the transmission range of a node is defined as either one of the two different values (250 meters corresponding to long ranges and 125 meters corresponding to short ranges). We modified the *ns-2* to implement these variable transmission ranges and model unidirectional links between nodes. The wireless link bandwidth is 2 Mb/s. Traffic pattern we used consists of 10 CBR connections running on UDP. Each CBR source generates four 512-byte data packets every second.

5.2 Simulation Results

Figure 2 shows the *number of unidirectional links* in the entire network as a function of varying fraction of low power nodes. The most number of unidirectional links exist when the fraction is 0.5 (thus, 50 nodes with transmission range of 250 m and the other 50 nodes with range 125 m). A similar observation was also made in [18]. We performed experiments with different maximum speed of mobile nodes at 1 m/s and 20 m/s, but the results are nearly identical. This result shows that the number of unidirectional links are influenced more by the low power node fraction rather than mobility, and the fraction remains the same between scenarios with different maximum speeds at a pause time of zero.

Figure 3 presents the *packet delivery ratio* of the three schemes we simulate, as a function of fraction of low power nodes. Packet delivery ratio is defined as the ratio of the number of data packets *received* by the CBR sinks and the number of data packets *originated* by the application layer CBR sources. We report the average over 10 CBR connections between source-destination pairs. Note that the y-axis scale in these figures

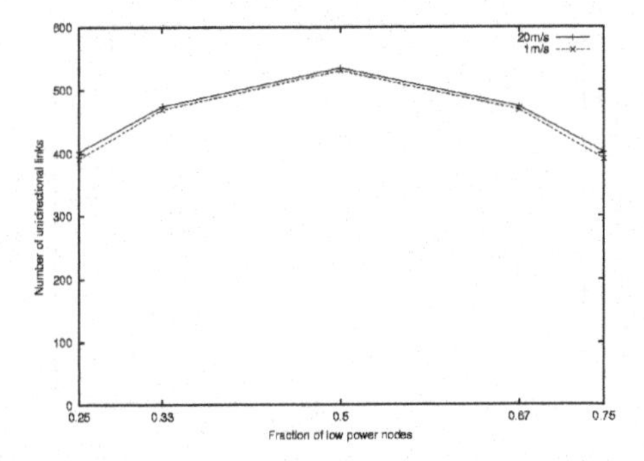

Fig. 2. Number of unidirectional links with varying fraction of low power nodes.

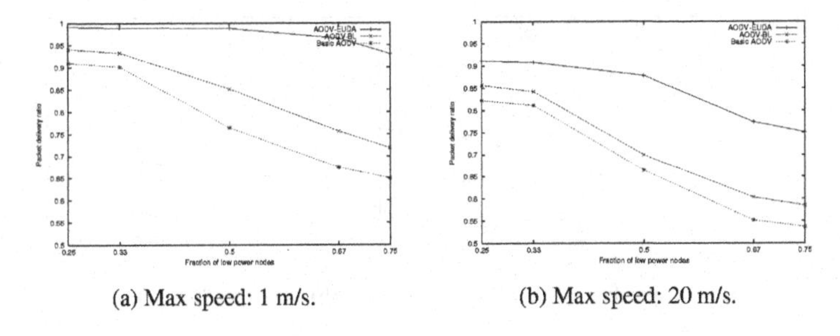

(a) Max speed: 1 m/s. (b) Max speed: 20 m/s.

Fig. 3. Packet delivery ratio with varying fraction of low power nodes.

range from 0.5 to 1. In both subfigures with different mobility speeds, as the fraction of low power nodes increases, the packet delivery ratio decreases for all protocols. Even the delivery ratio for the Basic-AODV, which does not handle unidirectional links, drops monotonically, even though there are the most unidirectional links when the fraction is 0.5. This is somewhat unexpected, because with the increase in unidirectional links, Basic-AODV scheme is expected to experience more route discovery failures and show less packet delivery success rate. By analyzing the traces, we found that the network connectivity becomes poor with the increase in fraction of low power nodes as they have short transmission range. With a shorter transmission range, the likelihood of being connected with other nodes will decrease. Consequently, with the decrease in connectivity, the packet delivery ratio decreases as well because there are less number of routes available. Figure 4 shows the total number of neighbors per node as a function of low power node fraction. The total number of neighbors reflects the level of connectivity in the network—the larger the number of neighbors per node, the higher the number of paths between each nodes.

Back to Figure 3 with this observation in mind, we see that the drop in packet delivery ratio is much less drastic and the success delivery rate is consistently higher for AODV-EUDA compared with the other two AODV schemes. This improvement of AODV-EUDA is due to efficient and fast detection of unidirectional links. With AODV-EUDA, a route search failure will not occur even when there is unidirectional path from a source

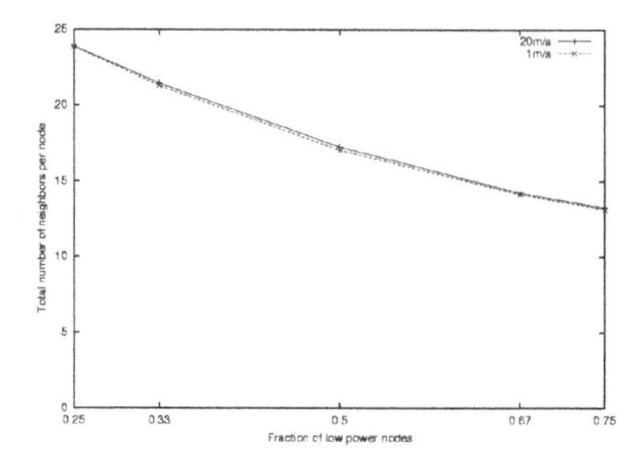

Fig. 4. Total Number of Neighbors per Node with varying fraction of low power nodes.

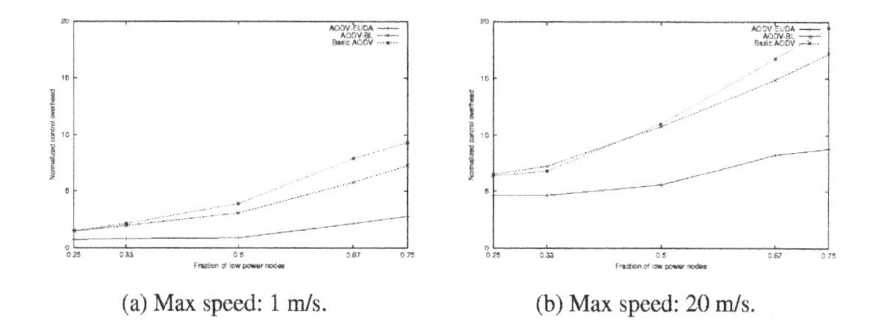

Fig. 5. Normalized control overhead with varying fraction of low power nodes.

to a destination. The Basic-AODV performs poorly in most cases as it does not take notice of unidirectional links and repeatedly performs route re-discoveries. As the Basic-AODV always chooses the shortest path between a source/destination pair, it cannot transmit any data packets when the shortest path includes one or more unidirectional links. The AODV-BL delivered less data compared with AODV-EUDA, but it still performed better than the Basic-AODV. Although AODV-BL detects unidirectional links, it only does so after a delivery failure and hence requires another route discovery process. AODV-EUDA on the other hand, finds unidirectional links during the RREQ propagation phase and avoids including them in the route in the first route discovery attempt.

Mobility also affects the protocol performance. As expected, packet delivery fraction degrades for all protocols with increase in mobility as there are more route breaks (see Figure 3 (a) and (b), at the maximum speeds of 1 m/s and 20 m/s, respectively). AODV-EUDA continues to perform significantly better than the other two schemes.

Figure 5 shows the *normalized routing overhead* with varying fraction of low power nodes. We define the normalized routing overhead as the ratio between the total number of routing control packets *transmitted* by all nodes and the total number of data packets *received* by the destinations. Overall, AODV-EUDA has the lowest overhead compared with AODV-BL and the Basic-AODV. AODV-BL and the basic AODV perform excessive flooding as they can neither detect unidirectional links or detect them in a timely fashion. Such an excessive flooding clearly contributes to a larger routing overhead.

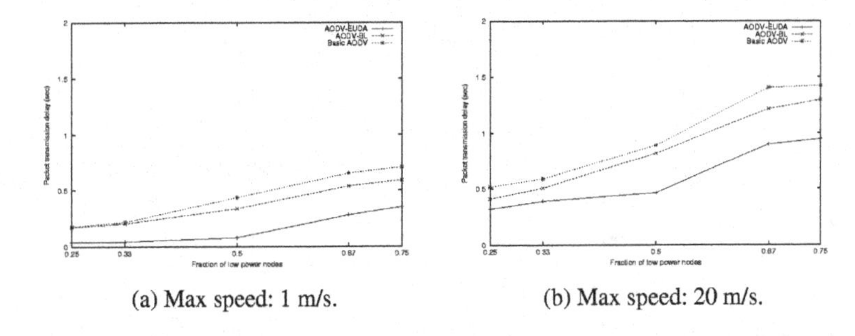

(a) Max speed: 1 m/s.

(b) Max speed: 20 m/s.

Fig. 6. End-to-end delay with varying fraction of low power nodes.

AODV-EUDA shows better efficiency because of its unique ability of early detection and avoidance of unidirectional links. As can be seen in Figure 5 (a) and (b), the routing overhead is also affected by mobility. The number of routing packet transmissions increases for all routing protocols with the increase in the speed of mobile nodes. With more dynamic mobility, route recoveries are more frequent as wireless links will break more often. We can still observe that AODV-EUDA provides a lower rate of increase than the other two schemes.

In Figure 6, we report the *average end-to-end delay* of successfully delivered data packets. The end-to-end delay is measured for the time from when a source generates a data packet to when a destination receives it. Therefore, this value includes all possible delay such as a buffering delay during route discovery, queuing and MAC delay during packet transmission, and propagation delay. The result again shows that AODV-EUDA yields a significantly better performance (i.e., smaller end-to-end latency) than other protocols for both cases of two different maximum node speeds. This shows that AODV-EUDA effectively overcomes unidirectional links. By exploring the early unidirectionality detection and avoidance feature, AODV-EUDA is able to shorten the route discovery latency and hence the overall end-to-end delay. Note that route (re)discovery latency may dominate the total end-to-end delay. For this reason, the Basic-AODV and AODV-BL consistently showed poor delay performance. When any shortest path includes a unidirectional link, the sources running the Basic-AODV experience significant amount of route discovery delay as they cannot receive corresponding RREP packets until such unidirectional links disappear by some network topology change. AODV-BL also shows a poor delay performance compared with AODV-EUDA because it produces more number of RREQ packets to find bidirectional routes.

The effect of varying the moving speed of nodes is depicted more in detail in the following figures. The fraction of low power nodes is fixed to 0.5 as this value produces the largest number of unidirectional links. From Figures 7 to 9, we again see that AODV-EUDA gives the best performance for all moving speeds in terms of packet delivery, routing overhead, and latency. With higher mobility, the frequency of route breaks increases and consequently packet delivery ratio becomes lower while routing overhead increases. AODV-EUDA has a lower increasing rate compared with the other schemes as its number of route requests is significantly reduced by preventing route discovery failures. With the Basic-AODV and AODV-BL, the existence of unidirectional links

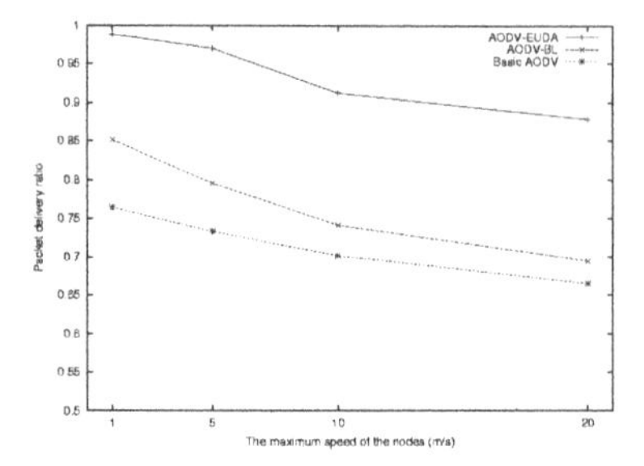

Fig. 7. Packet delivery ratio as a function of the maximum speed of the nodes.

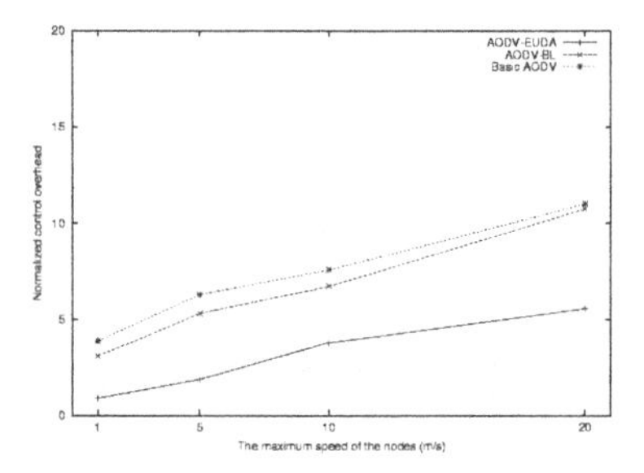

Fig. 8. Normalized control overhead as a function of the maximum speed of the nodes.

cause the failure of RREP delivery and hence the number of route requests is increased. It is interesting to see that AODV-BL has similar normalized routing overhead to that of the Basic-AODV (see Figure 8). This is due to the fact that AODV-BL transmits additional RREP-ACK packets to detect unidirectional links. Moreover, if the AODV-BL scheme fails to deliver RREP-ACK, it works exactly as the Basic-AODV. The delay of the AODV-EUDA is 100% to 300% smaller than that obtained by the Basic-AODV and AODV-BL schemes.

6 Conclusion and Future Work

We discussed node heterogeneity of real ad hoc network settings and studied a specific problem of unidirectional links that result from variance in radio transmission power. We addressed the limitations of existing routing protocols in the presence of unidirectional links. We proposed EUDA that detects unidirectional links in a timely fashion and excludes such links from being part of end-to-end communication paths. Two approaches have been considered. The first scheme uses node location information to estimate uni- or bi-directionality of a link while the second scheme utilizes a path loss model. Our simulation results have shown that when EUDA is applied to AODV, compared with AODV and AODV blacklisting, it gives superior throughput, less overhead with high efficiency, and shorter latency. Early detection of unidirectional links enables EUDA to use less messaging overhead and deliver more data packets with less delay.

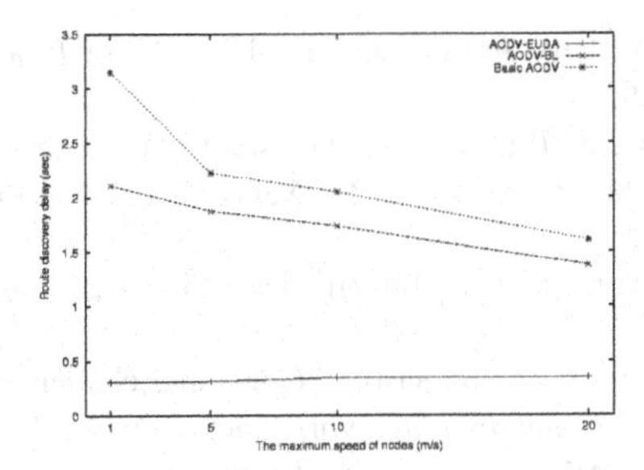

Fig. 9. End-to-end delay as a function of the maximum speed of the nodes.

Although EUDA is shown to work well, we could not truely evaluate our scheme as the ns-2 simulator provides rather optimistic channel propagation model. Our future work includes performing simulations with a more realistic path loss model and investigating further on the cross-layer approach where new MAC scheme is applied. Consideration of node heterogeneity other than radio transmission power, such as power source, antenna type, computing power are also ongoing. We are also building a network testbed with heterogeneous mobile devices for further performance study.

References

1. A. Avudainayagam, W. Lou, and Y. Fang, "DEAR: A device and energy aware routing protocol for heterogeneous ad hoc networks," *Journal of Parallel and Distributed Computing*, vol. 63, no. 2, pp. 228–236, Feb. 2003.
2. L. Bao and J. J. Garcia-Luna-Aceves, "Link state routing in networks with unidirectional links," in *Proc. of IEEE ICCCN*, Boston, MA, Oct. 1999.
3. J. Broch, D. A. Maltz, and D. B. Johnson, "Supporting hierarchy and heterogeneous interfaces in multi-hop wireless ad hoc networks," in *Proc. of the IEEE Workshop on Mobile Computing, (I-SPAN)*, Perth, Western Australia, June 1999.
4. J. Broch, D. A. Maltz, D. B. Johnson, Y.-C. Hu, and J. Jetcheva, "A performance comparison of multi-hop wireless ad hoc network routing protocols," in *Proc. of ACM MobiCom*, Dallas, TX, Oct. 1998, pp. 85–97.
5. S. Capkun, M. Hamdi, and J.-P. Hubaux, "GPS-free positioning in mobile ad-hoc networks," in *Proc. of HICSS*, Hawaii, Jan. 2001, pp. 3481–3490.
6. IETF mobile ad hoc networks (MANET) working group charter. [Online]. Available: http://www.ietf.org/html.charters/manet-charter.html
7. P. Johansson, T. Larsson, N. Hedman, B. Mielczarek, and M. Degermark, "Scenario-based performance analysis of routing protocols for mobile ad-hoc networks," in *Proc. of ACM MobiCom*, Seattle, WA, Aug. 1999, pp. 195–206.
8. D. Johnson and D. Maltz, "Dynamic source routing in ad hoc wireless networks," in *Mobile Computing*, T. Imielinski and H. Korth, Eds. Kluwer Publishing Company, 1996, ch. 5, pp. 153–181.
9. C. E. Jones, K. M. Sivalingam, P. Agrawal, and J. C. Chen, "A survey of energy efficient network protocols for wireless networks," *ACM/Kluwer Wireless Networks*, vol. 7, no. 4, pp. 343–358, 2001.
10. R. E. Kahn, S. A. Gronemeyer, J. Burchfiel, and R. C. Kunzelman, "Advances in packet radio technology," *Proc. IEEE*, vol. 66, no. 11, pp. 1468–1496, Nov. 1978.

11. G. S. Lauer, "Hierarchical routing design for SURAN," in *Proc. of IEEE ICC*, Toronto, Canada, June 1986, pp. 93–102.
12. S.-J. Lee, M. Gerla, and C.-K. Toh, "A simulation study of table-driven and on-demand routing protocols for mobile ad hoc networks," *IEEE Network*, vol. 13, no. 4, pp. 48–54, July/Aug. 1999.
13. B. M. Leiner, D. L. Nielson, and F. A. Tobagi, "Issues in packet radio network design," *Proc. IEEE*, vol. 75, no. 1, pp. 6–20, Jan. 1987.
14. B. M. Leiner, R. J. Ruth, and A. R. Sastry, "Goals and challenges of the DARPA GloMo program," *IEEE Personal Commun. Mag.*, vol. 3, no. 6, pp. 34–43, Dec. 1996.
15. C. R. Lin and M. Gerla, "Adaptive clustering for mobile wireless networks," vol. 15, no. 7, pp. 1265–1275, Sept. 1997.
16. A. Lindgren and O. Schelen, "Infrastructured ad hoc networks," in *Proc. of IWAAN*, Vancouver, Canada, Aug. 2002, pp. 64–70.
17. H. Lundgren, E. Nordstrom, and C. Tschudin, "Coping with communication gray zones in ieee 802.11b based ad hoc networks," in *Proc. of ACM WoWMoM*, Atlanta, GA, Sept. 2002, pp. 49–55.
18. M. K. Marina and S. R. Das, "Routing performance in the presence of unidirectional links in multihop wireless networks)," in *Proc. of ACM MobiHoc*, Lausanne, Switzerland, June 2002, pp. 12–23.
19. M. Mauve, J. Widmer, and H. Hartenstein, "A survey on position-based routing in mobile ad hoc networks," *IEEE Network*, vol. 15, no. 6, pp. 30–39, Nov. 2001.
20. J. Monks, V. Bharghavan, and W.-M. Hwu, "A power controlled multiple access protocol for wireless packet networks," in *Proc. of IEEE INFOCOM*, Anchorage, AK, Mar. 2001.
21. C. Perkins, E. Belding-Royer, and S. Das, "Ad hoc on-demand distance vector (AODV) routing," IETF, RFC 3561, July 2003.
22. C. Perkins and E. Royer, "Ad-hoc on-demand distance vector routing," in *Proc. of IEEE WMCSA*, New Orleans, LA, Feb. 1999, pp. 90–100.
23. N. Poojary, S. V. Krishnamurthy, and S. Dao, "Medium access control in a network of ad hoc mobile nodes with heterogeneous power capabilities," in *Proc. of IEEE ICC*, Helsinki, Finland, June 2001.
24. R. Prakash, "A routing algorithm for wireless ad hoc networks with unidirectional links," *ACM/Kluwer Wireless Networks*, vol. 7, no. 6, pp. 617–625, Nov. 2001.
25. R. Ramanathan and M. Steenstrup, "Hierarchically-organized, multihop mobile wireless networks for quality-of-service support," *ACM/Baltzer Mobile Networks and Applications*, vol. 3, no. 1, pp. 101–119, June 1998.
26. R. Ramanathan and R. Rosales-Hain, "Topology control of multihop wireless networks using transmit power adjustment," in *Proc. of IEEE INFOCOM*, Tel Aviv, Israel, Apr. 2000, pp. 404–413.
27. V. Ramasubramanian, R. Chandra, and D. Mosse, "Providing a bidirectional abstraction for unidirectional ad hoc networks," in *Proc. of IEEE INFOCOM*, New York, NY, June 2002.
28. F. A. Tobagi and L. Kleinrock, "Packet switching in radio channels: Part II-the hidden terminal problem in carrier sense multiple-access and the busy-tone solution," *IEEE Trans. Commun.*, vol. COM-23, no. 12, pp. 1417–1433, Dec. 1975.
29. Y.-C. Tseng, S.-L. Wu, W.-H. Liao, and C.-M. Chao, "Location awareness in ad hoc wireless mobile networks," *IEEE Computer*, vol. 34, no. 6, pp. 46–52, June 2001.
30. K. Xu, X. Hong, and M. Gerla, "An ad hoc network with mobile backbones," in *Proc. of IEEE ICC*, New York, NY, Apr. 2002.
31. Z. Ye, S. V. Krishnamurthy, and S. K. Tripathi, "A framework for reliable routing in mobile ad hoc networks," in *Proc. of IEEE INFOCOM*, San Francisco, CA, Apr. 2003.

802.11 Link Quality and Its Prediction – An Experimental Study

Gregor Gaertner, Eamonn ONuallain, Andrew Butterly, Kulpreet Singh, and Vinny Cahill

Distributed Systems Group
Department of Computer Science
Trinity College, Dublin, Ireland
gregor.gaertner@cs.tcd.ie

Abstract. Reliable link quality prediction is an imperative for the efficient operation of mobile ad-hoc wireless networks (MANETs). In this paper it is shown that popular link quality prediction algorithms for 802.11 MANETs perform much more poorly when applied in real urban environments than they do in corresponding simulations. Our measurements show that the best performing prediction algorithm failed to predict between 18 and 54 percent of the total observed packet loss in the real urban environments examined. Moreover, with this algorithm between 12 and 43 percent of transmitted packets were lost due to the erroneous prediction of link failure. This contrasts sharply with near-perfect accuracy in corresponding simulations. To account for this discrepancy we perform an in-depth examination of the factors that influence link quality. We conclude that shadowing is an especially significant and hitherto underestimated factor in link quality prediction in MANETs.

1 Introduction

With the deployment of MANETs having just begun, practical experience of wireless protocol performance is limited. A-priori performance analysis of MANET protocols is difficult due to multi-hop communication with intractable and environment-specific signal propagation effects. Hence provably reliable communication is still in its infancy [1] and network performance evaluations are largely based on simulations. Recent studies however have cast serious doubts on the reliability of results obtained from such simulations. Discrepancies between simulated and real-world network performance has been primarily attributed due to the inappropriate level of detail at which these simulations are performed including the use of overly-simplified propagation models [2,3,4].

The propagation models commonly used in such simulations are based on very 'benign' obstacle-free environments. Obstructing static and dynamic objects (e.g. buildings, people, cars, etc.) as well as important radio propagation effects (diffraction, scattering and transmission [5]) are not considered. These are major shortcomings since the primary deployment areas for MANETs are urban environments where such effects are pronounced. Since the network topology

I. Niemegeers and S. Heemstra de Groot (Eds.): PWC 2004, LNCS 3260, pp. 147–163, 2004.

and environment change rapidly in MANETs, the communication quality varies considerably with time. Link quality prediction is a key technique to alleviate link quality degradations by pro-actively adapting the network to the operating environment.

We now evaluate the performance of some of the more popular link quality prediction algorithms for MANETs in two real urban environments. Our results indicate that these algorithms perform poorly in such a setting despite having exhibited near-perfect performance in simulations. While some performance degradation is to be expected when such algorithms are applied in real environments, the observed discrepancy is much larger than expected. This motivates us to systematically identify the factors that influence link quality (henceforth called 'influencing factors') and assess their importance for link quality prediction so that these discrepancies can be accounted for.

The remainder of this paper is organized as follows: Section 2 discusses current work on link quality prediction and its applications. Section 3 evaluates the accuracy of current link quality prediction algorithms in a real-world case study and compares their performance with simulations. Improvements for future algorithms are recommended. Section 4 identifies the factors influencing link quality and assesses their importance in prediction accuracy. We conclude our work in Sect. 5.

2 Related Work

It has been shown in numerous simulations that link quality prediction is important for routing in ad-hoc networks. In wired networks the shortest distance criterion is often used to select optimal paths for routing. However, as a result of the significantly higher number of link failures in wireless ad-hoc networks, a more sophisticated means is required for route optimisation in these networks. Using link quality prediction methods, the protocols of [6,7,8,9,10,11,12] can select routes with higher lifetimes. To reduce traffic on shared low bandwidth channels on-demand routing protocols, which only initiate route discovery for active paths, are used. If a path breaks, high latency is the costly outcome due to multiple timeouts. Proactive route discovery enabled by link quality prediction is utilized in [8,9,12,13,14] to alleviate this problem by initiating route discovery before the path breaks. This approach reduces packet loss and jitter. Link quality prediction may also be used to dynamically cluster the network into groups of stable nodes. This reduces the update propagation time of changes in network topology [15]. Another application of pro-active failure detection is enabling consistency maintenance in group communication [16].

Link quality prediction algorithms can be categorized as follows: deterministic approaches [6,7,8,9,10,12,13,14] which give a precise value for link quality and stochastic approaches [11,17,18] which give a probabilistic measure. All approaches (with the exception of [17]) employ a deterministic signal propagation model either explicitly or implicitly. The signal propagation models in common usage are the simple Radial Model of [16,18]; the assumption that the transmitted power and the distance separating the nodes alone determines the signal strength [6,7,10]; the Free Space Propagation Model [19, pg. 107-109] used in [9,12] and the Two-Ray Ground Model [19, pg. 120-125] used in [8,13,14]. At runtime [6,7,10,13,14] use signal strength criteria to provide an estimate for link quality whereas [8,9,12,20] use node location for this purpose. All approaches are evaluated by simulation with either NS-2 [21] or GloMoSim [22] except for [8] which is evaluated experimentally.

It should be noted that in the above link quality prediction algorithms the link quality metric is determined primarily from the distance between nodes. Though this may hold for very benign open-space environments, it breaks down in real urban environments where there are substantial signal fading effects due to manifold signal propagation phenomena. In the above works only simple techniques, if any, have been suggested for dealing with fast signal fading (e.g. exponential average [6], linear regression [14] or the ping-pong mechanism of [13]). Consequently these approaches exhibit a significantly poorer performance when applied in real-world urban scenarios over simulations (see Sect. 3).

In [6] the signal strength is measured and the link is classified simply as being either strong or weak. An 'affinity metric' proposed in [7] and later used in [10] gives in contrast a continuous measure for the link quality which is determined primarily from the trend of the most recent samples of the signal strength.

In [13] a pre-emptive threshold is compared with the current signal strength. If the signal strength is lower than this threshold, the possibility of link failure is considered. This leads to an exchange of a pre-set number of messages called ping-pong rounds. If the signal strength of greater than a certain number of these packets is under the threshold, then a link failure is predicted. This mechanism aims to reduce the number of link failures which are predicted erroneously due to fast signal fading. The pre-emptive threshold is calculated using the node transmission range and the Two-Ray Ground Model so that there is enough time to establish a potential alternative route before the link fails. Regrettably an appropriate threshold is difficult to calculate in a real-world environment since the range of the nodes is environment-specific and unknown.

In [14] the Law of Cosines is used to derive the remaining time to link failure. This prediction algorithm is based on the last three received signal strength samples, the Two-Ray Ground Model and assumptions of the Random Waypoint Mobility Model [23]. Linear regression is suggested to pre-process the signal strength values in order to counter the effects of fast signal fading. The algorithm assumes that nodes have a constant velocity and that the signal strength is affected only by the distance between nodes.

Node location and velocity are used with the Free-Space Propagation Model in [9,12] to predict the time to link failure. This propagation model is inappropriate even for open-space environments (see [19, pg. 120]) and the need for location and velocity information makes the algorithm costly to implement. The prediction method suggested in [8] consists of modules for mobility prediction, signal strength prediction and an environment map. The future signal strength is determined from the predicted distance between the nodes, an experimentally-determined site-specific coefficient for shadowing and the Two-Ray Ground Model. The concept is evaluated experimentally in a parking lot. Unfortunately the graphs given in the paper use a time scale that precludes a direct evaluation of the prediction accuracy in the order of seconds. The algorithm is critically dependent on the distance between nodes determined via the absolute node positions without tolerating missing position data. This is a shortcoming since the suggest positioning system GPS is often unreliable in dense urban environments (this is confirmed by our measurements in Sect. 3.1). Furthermore it is not clear whether memory consuming maps of the environment with their associated computationally intensive algorithms justify their cost since unaccounted for moving objects have a strong bearing on link quality (see Sect. 4.2).

Three stochastic approaches have been proposed. [11,18] predict link failures based on the Radial Propagation Model and the mobility pattern of the Random Ad-hoc Mobility Model [18]. It is unclear how well these approaches work in practice, especially since very simple propagation models are employed. In [17] links with an expected higher remaining lifetime are chosen based on previous link lifetimes. It is unclear weather this concept can be used for link failure prediction.

The handoff problem (see the survey article [24] for example) in cellular communication networks is analogous to the link quality prediction problem in MANETs. Based on the current mobility pattern it is attempted to execute optimal handoff from one cell to another such that potential consequent disruptions are minimized. While this problem has been well studied, the suggested solutions are only partially transferable to link quality prediction in MANETs. For a detailed discussion see Sect. 4.2.

Since the literature on link quality prediction focuses on simulations as the means for the evaluation of algorithms, it is clear that the accuracy of the used simulators is crucial. [2] compares popular simulators by their physical layer models, their implementations and a case study. This study revealed that the evaluation of protocols in different simulators may give different absolute and even different relative performance measures. In [3] similar results were reported for a simple broadcast protocol implemented by flooding. Since broadcast protocols are used as basic building blocks for many wireless protocols, [3] concludes that "finally, beside simulations and according to the feeling of the MANET community, there is an important lack of real experiments that prove the feasibility of wireless protocols".

3 Accuracy of Link Quality Prediction in the Real-World

In this section we evaluate the accuracy of some popular link quality prediction algorithms experimentally in two typical urban environments. We then compare the results obtained with those derived by simulations.

3.1 Case Study and Test Bed Description

The urban environments we chose include a variety of static (e.g. buildings, trees, etc.) and dynamic objects (people, cars). The selected locations in Dublin's city centre are fairly typical of European cities. Grafton Street is a long narrow pedestrian precinct overshadowed by moderately high (three-storey) buildings. There is no vegetation and it is commonly quite full of people. The second location, O'Connell Street, is a wide avenue-like street. There are two lanes of traffic in both directions. These are separated by a verge with some trees. Traffic is heavy throughout the day. On both sides of the street are sidewalks which are normally full of pedestrians. We conducted our experiments at busy times. We will refer to Grafton Street as 'Street 1' and O'Connell Street as 'Street 2' in this paper.

Shadowing and propagation effects are the prominent factors bearing on link quality in urban environments over the occurrence of collisions in the presence of a moderate number of nodes. Hence, working with two nodes, we consider the former effects only. In our study the bearers of both nodes moved between randomly chosen shops, their movement interspersed with pauses of about two seconds as might be expected of some pedestrians on these streets. This mobility pattern corresponds to the widely used Random Waypoint Mobility Model with a maximum speed of 1.5 m/s (walking speed) using pauses of two seconds. The movement area of the nodes was confined to the main street and its immediate precinct to ensure a certain level of connectivity.

Both nodes were broadcasting messages of 100 bytes at regular intervals (heartbeats) directly over the MAC layer without using any higher-level protocol at a transmission rate of 2 Mbit. The coherence time of 17.63 ms for the speed of our nodes suggests a sampling rate of greater than 28.36 Hz if multi-path effects are to be fully captured by the measurements. However, the experience of the cellular network community shows that the incorporation of multi-path fading is impractical for prediction algorithms (see [25] for example) and the prediction algorithms evaluated in this work do not incorporate such information. Consequently, in order to reduce the amount of data, we use a sample rate of 10 Hz which allows for prediction in convenient 100 ms intervals. We conducted three trials of 30 minutes duration for each street. All sent and received packets were recorded together with their associated average signal strength.

Two Dell Latitude C400 notebooks with Lucent Orinoco Gold wireless 802.11b cards acted as nodes. The transmission power was set to the maximum of 15 dBm with disabled power management. We used Redhat Linux 7.3 together with the GPL orinoco_cs driver by David Gibson in version 0.12 [26], which we extended to send and receive packets directly over the MAC layer. During the trials it was attempted to record the current node positions with either a Magellan GPS 315 or a Garmin ETrax receiver. However, the number of satellites in view was mostly insufficient to obtain a position.

For the simulations NS-2 was configured according to the real world study. The movements of the two nodes followed the Random Waypoint Mobility Model with a maximum speed 1.5 m/s, 2 seconds pause and an area of 500 m squared.

The Two-Ray Ground Model and the parameters of a Lucent Wavelan card with a maximum range of 250 m were employed. Our choice of mobility and propagation models corresponds to most MANET simulations (see [2,10,13,14] for examples).

3.2 Algorithm Implementations

We restrict our evaluation to algorithms with the following properties:

1. The algorithm predicts at time t if the link is either available or unavailable at time $t+T_{\mathrm{p}}$ (T_{p} is called the prediction horizon). Most pro-active applications in routing or group communication (see [7,13,14,16] for examples) require this very weak property ('weak' since only a binary value for the link quality at only one point in time is used).
2. The algorithms must tolerate missing node locations for long time periods since GPS signal reception can be very unreliable in dense urban environments.
3. The signal propagation model used must be at least as accurate as the Two-Ray Ground Model. Even this model does not account for the manifold propagation phenomena in urban environments.

These criteria that aim to identify the most likely successful algorithms in urban environments reduce our evaluation to [7,13,14].

In order to unify the notation for the algorithms considered, we use the following symbols: a lower case 't' denotes a point in time, 'T' denotes a time interval. 'P_{t}' denotes the signal strength of a packet received at time t. 'P_{maxrange}' denotes the minimum signal strength at which a packet at the maximum distance *maxrange* can be received. Threshold values used in the algorithms are denoted using the symbol 'δ'. The symbol 'Δ' is used to denote the difference between successive values of a variable.

The Affinity Algorithm. The Affinity Algorithm originating in [7] and applied in [10], associates a link with an affinity metric a. The algorithm assumes that the mobility pattern remains constant and that the signal strength increases if and only if the nodes move closer and decreases if the nodes move further apart. Thus, if the average signal trend $\Delta P_{[t-n+1,t]}$ over n samples remains positive, the nodes are assumed to move closer and the affinity is classified as being 'high'. The link is predicted to remain available in this case. However, if the signal trend is negative, the nodes are assumed to be moving apart and the affinity at time t for a horizon T_{p} is calculated via $a_t = \frac{P_{\mathrm{maxrange}} - P_{\mathrm{t}}}{\Delta P_{[t-n+1,t]}}$. Under the condition that all assumptions are true the affinity value is a direct measure of the time to link failure.

Our implantation predicts a link failure at t for $t + T_{\mathrm{p}}$ if the affinity value is below a threshold a_δ. For best performance in the real world we determine the parameter values a_δ and n using an optimization routine (see Sect. 3.3 for details). An analytical determination of these parameters based on the original work gives good performance results in simulations only.

The Law of Cosines (LoC) Algorithm. This algorithms [14] calculates the time to link failure based on the signal strength values of three consecutive received packets. The algorithm assumes that the signal strength follows strictly the Two-Ray Ground Model and that nodes maintain a constant velocity until the predicted link failure occurs. The algorithm is based on a simple geometric evaluation using Law of Cosines. A link is predicted to be available if the signal strength over three consecutive received packets is either constant or increasing. Otherwise a time to link failure is predicted.

Fast signal fading effects are attempted to be masked using linear regression. However, the signal strength may still differ greatly with that given by the Two Ray Ground model even when using an optimum window-size and so the quadratic equation used to give link failure time occasionally gives an imaginary solution. This problem was not observed during simulation.

The Pre-emptive Threshold Algorithm. The above algorithms use the recent signal trend as their link quality metric. In contrast, [13] uses a combination of the signal strength itself and the recent packet loss. If the current signal strength is under P_δ, a link failure is suspected. As result packets are exchanged in n ping-pong rounds between the two nodes. If $k \leq n$ packets are below the signal strength threshold P_δ, a link failure is predicted. This method attempts to overcome erroneous prediction of link failures due to fast signal fading by tolerating adverse link quality fluctuations over $k-1$ packets. The threshold of the signal strength P_δ is determined from the so called pre-emptive threshold as $\delta = (\frac{maxnoderange}{maxnoderange - maxrelspeed\, T_{\mathrm{p}}})^4$. The threshold attempts to enable a communication task to be completed even if two nodes move away with the maximum speed *maxrelspeed*. However, this is only achieved if the signal strength behaves strictly according to the Two-Ray Ground Model and the maximum transmission range *maxrange* is known.

If in our implementation the signal strength value associated with one packet is below P_δ, a node counts how many out of the next n received heartbeat packets have also a signal strength value below P_δ. If this number is equal or greater than k, a link failure is predicted. We calculated the threshold values for the simulation as outlined in the original work and set $n = k = 3$. However, for the real-world study these parameters are unknown, especially the environment specific average maximum transmission range. Consequently we determined all parameter values by optimization as described in the next section.

3.3 Evaluation Metrics and Results

In current simulators link failures are modelled as being 'sharp'. This contrasts with the real world behaviour, where a link typically oscillates between being available and unavailable over a transitional period before eventually remaining unavailable. Hence a definition of link stability is necessary. We define a link to be stable in a system with a constant heartbeat rate, if at least p percent of packets are received during a time span T_{s}. T_{s} should be small enough to reflect fast changes of the link status but large enough to tolerate fast-fading effects (i.e., to prevent unnecessary route discoveries in routing protocols). p must be small enough to meet the minimum link quality requirements of an application and large enough to tolerate fast signal fading. Our results suggest $p = 70\%$ and $T_s = 2\,\mathrm{s}$ are reasonable values. Since all algorithms considered here are based on the implicit assumption that a link fails sharply, we only measure the prediction accuracy outside the transitional period.

We define our accuracy metrics for link quality prediction in like manner with those used in handoff algorithms in cellular networks (i.e., service failure and unnecessary handoffs [25]): the percentage of packets that are lost while the link was predicted to be available (henceforth called 'missed packet loss') and the percentage of packets that are received while the link was predicted to be unavailable (henceforth called 'unnecessary packet loss'). These metrics capture the fundamental trade-off in link failure prediction: if the prediction is too pessimistic, the unnecessary packet loss is high but the missed packet loss is low. If the prediction is too optimistic the reverse is the case. Since both metrics have an equal bearing on performance, an overall accuracy metric must include both. We define the degree of prediction inaccuracy $i = m^2 + u^2 + (m - u)^2$ as the sum of the squares of missed packet loss m and unnecessary packet loss u with an additional term that accounts for an imbalance of these two metrics.

As stated earlier, suitable parameter values for the prediction algorithms must be determined for the environment in question. Clearly the way to do this is to minimise the inaccuracy i. This is, however, a difficult task since the inaccuracy function associated with a prediction algorithm and dataset is not analytically differentiable and also highly non-linear. The parameter space is cleft so that even classical direct search algorithms perform inadequately since they get stuck at local minima as a result of the greedy criterion. We resolved these problems using a robust evolutionary optimization algorithm – The Differential Evolution Algorithm [27].

Table 1 displays the performance of all the algorithms considered under different criteria. 'MPL' stands for the missed packet loss, 'UPL' – the unnecessary packet loss and 'OI' – the overall inaccuracy. The table is divided into separate sections corresponding to real-world experiments and simulation. The experimental section is further partitioned into data sets on which the optimization of the algorithm parameters was conducted. This distinction is required since the optimum parameters are environment-specific. Optimization (Opt.) 1 and 2 refers to minimizing the inaccuracy based on either the data set for Street 1 or 2 respectively. Opt. 3 refers to minimization of the inaccuracy on a compound data set of Street 1 and 2 that aims to find 'compromise' parameter values for both environments. In the simulation study optimization is not required since the parameters are determined analytically.

The table shows that in our real-world measurements the Pre-Emptive Threshold Algorithm achieves the best accuracy by far. However 22.85/28.39% of missed packet loss and 28.07/30.50% of unnecessary packet loss was measured in Street 1/2 despite optimal parameters having been used for the environment in

Table 1. Prediction accuracy in real-world urban environments and in simulations

	Affinity Algorithm			LoC Algorithm			Pre-Emptive Threshold Algorithm		
Experimental Trial – Street 1 (Grafton Street)									
	Opt. 1	Opt. 2	Opt. 3	Opt. 1	Opt. 2	Opt. 3	Opt. 1	Opt. 2	Opt. 3
MPL	48.72%	49.63%	49.45%	58.17%	64.48%	58.17%	22.85%	53.71%	31.73%
UPL	34.73%	34.77%	34.77%	27.35%	22.71%	27.35%	28.07%	11.72%	22.62%
OI	3775.53	3892.91	3869.76	5081.64	6418.15	5081.64	1337.30	4785.28	1601.45
Experimental Trial – Street 2 (O'Connell Street)									
	Opt. 1	Opt. 2	Opt. 3	Opt. 1	Opt. 2	Opt. 3	Opt. 1	Opt. 2	Opt. 3
MPL	55.59%	54.19%	54.19%	61.75%	60.59%	61.75%	17.81%	28.39%	19.43%
UPL	36.89%	35.90%	35.91%	27.40%	24.62%	27.40%	43.40%	30.50%	39.06%
OI	4800.81	4559.89	4560.24	5743.75	5571.13	5743.75	2855.60	1740.69	2288.55
Simulation									
	Calculated			Calculated			Calculated		
MPL	0.00%			6.67%			0.00%		
UPL	0.00%			0.00%			0.00%		
OI	0.00			44.44			0.00		

question. This performance decreases with the use of non-optimum parameters that have been obtained in similar environments. For example, the accuracy of the Pre-Emptive Threshold Algorithm deteriorated in Street 1 from 1337.30 to 4785.28 when the algorithm parameters used were the optimum parameters for Street 2. These results indicate that an online learning is necessary that adapts the parameters to the current operating environment.

The Affinity and LoC Algorithms performed much more poorly on all counts than the Pre-emptive Threshold algorithm. Both algorithms failed to predict approximately half the packet loss while a quarter to one third of packets was unnecessarily predicted to be lost. We attribute this poor accuracy to these algorithms' reliance on signal trend data which is highly sensitive to fast-fading effects. The mechanism of linear regression employed in the LoC Algorithm and the window-size used in the Affinity Algorithm seem both to be ineffective in dealing with fast signal fading. The Pre-Emptive Threshold algorithm shows a greater robustness in this regard since it uses the current signal strength values and not signal trend. This observation demonstrates that signal strength is a more powerful predictor of packet loss if the path loss in the prediction interval is insignificant like it is the case for low node speeds.

In the simulation all algorithms show near-perfect accuracy, both in terms of missed and unnecessary packet loss. Only the LoC Algorithm fails to predict some lost packets. This occurs in situations where the node speed changed during the sampling interval of the three consecutive packets used for prediction (as described in [14]). We would have liked to have compared the results of our simulations directly with those of the original authors. However [7,13] assessed the prediction accuracy indirectly using metrics that show the enhancement of the routing protocol performance due to link failure prediction. [14], nevertheless, gives explicit prediction accuracies where it is reported that more than 90 percent of lost packets were predicted successfully for various mobility patterns. We observed a value of 93 percent in our simulation.

The near-perfect accuracy in simulations can be attributed to the fact that the prediction algorithms and simulation programs use the same deterministic radio propagation model. It is well known that the propagation models used by such simulators are simple and yield only to real-world approximations (see [5] for example). Thus simulation accuracy is perceived as being optimistic (see [14] for example). However, the large discrepancy between link quality prediction accuracy given by simulators and those obtained in real-world trials has apparently not been appreciated in the literature to date. To the best of our knowledge, we are unaware of any work in which link quality prediction algorithms for MANETs are specifically designed for and evaluated in real-world urban environments. Our results emphasize the need for further research in this area, especially with regard to adaptive prediction models and techniques with which to deal with fast signal fading.

4 Assessing Influencing Factors On Link Quality

The poor performance of prediction algorithms in urban environments motivated us to examine systematically the factors that influence link quality and its prediction in 802.11 MANETs. The results are summarised in Table 2.

Table 2. Influence of various factors on link quality

Factor	Model of wireless card	Type of ground	Height of nodes	Orientation without shadowing body	Orientation with shadowing body	Shadowing by person
Influence	High	Low	High	Low	High	High

Factor	Shadowing by car	Small scale movements	Large scale movements at different speeds	Communication load without collisions	Message length	Payload pattern
Influence	High	None	None	None	None	None

4.1 Experimental Setup

The manifold signal propagation phenomena observed in urban environments make the separation of potential and actual 'influencing factors' cumbersome in these environments. Hence we chose a beach on a deserted island as our 'benign' obstacle-free environment to conduct our experiments. We used the same hardware as for the urban environment study. Accurate markings and

GPS receivers determined the distances between the two nodes that were always facing each other.

Three types of experiments were designed to assess potential and actual influencing factors on link quality:

- Type I One node was motionless while the other node moved away at constant walking speed until returning after 300 m. This type of experiment exhibits path loss and is for example suitable for comparing the theoretical Two-Ray Ground Model with actual measurements.
- Type II Both nodes were placed at a fixed distance. One node was motionless while the other node oscillated between being in motion (e.g., node was rotated, shaken, etc.) and motionless. The comparison of the signal strength between the two different states reveals the influence of a factor.
- Type III Both nodes were placed at a fixed distance and were motionless. An obstacle was moved in the line of sight between the nodes, paused and moved out. The difference in signal strength of the two states measures the influence.

All experiments were conducted three times and the mean was used for evaluation where possible. As described in Sect. 3.1, both nodes transmitted messages periodically. All diagrams display the signal strength that was measured at one node along the ordinate. If a packet was lost no value for the signal strength is given.

4.2 Results

Most simulations for MANETs (see [13,14] for examples) are based on the Two-Ray Ground Model with a maximum node transmission radius of 250 m for a transmission rate of 2 Mbit at maximum transmission power. Nevertheless, a recent experimental outdoor study from [28] claims that 250 m transmission range is far too optimistic and that the actual range would be 90 to 100 meters in an open environment. However, our measurements (Fig. 1, Type I experiment) confirm the widely accepted view of the literature that the transmission range is indeed around 250 m. Our measurements confirm further that the Two-Ray Ground Model is a good approximation for open environments with a measured average model error of 1.49 dBm (standard deviation 1.53 dBm).

These contradictory results may be attributed to the model of wireless card. In Fig. 2 (Type I experiment) we compare signal strength measurements of a Lucent Orinoco card with a card from a different brand. Both comply with the same 802.11 specification and have the same transmission power but show different behaviour. The observed performance discrepancy of the cards may be caused by different chip sets, the design of radio frequency components, different antennas and the applied signal processing of the cards' onboard processors. Unfortunately, no manufacturer provided us with the necessary information to enable an in-depth analysis. However, it should be noted that our comparison of

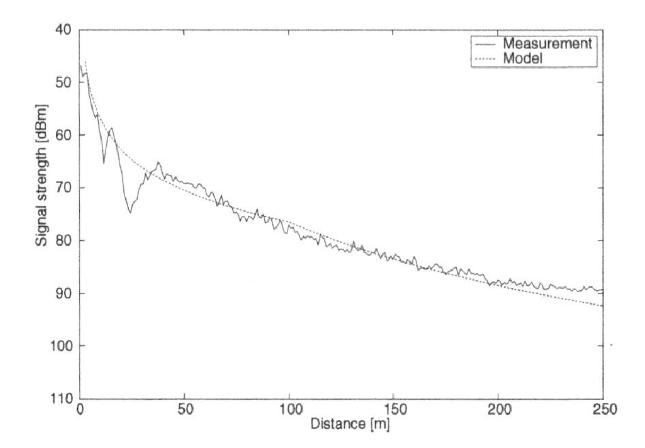

Fig. 1. Theoretical Two-Ray Ground Model and real-world measurements

simulation and real-world performance for different prediction algorithms in Sect. 3.3 is fair, even without detailed knowledge about the applied post-processing algorithms and hardware designs. As we have shown above, the measurements obtained using Lucent Orinoco cards correspond in an open environment very well with those given by the theoretical Two-Ray Ground Model on which the simulations are based on.

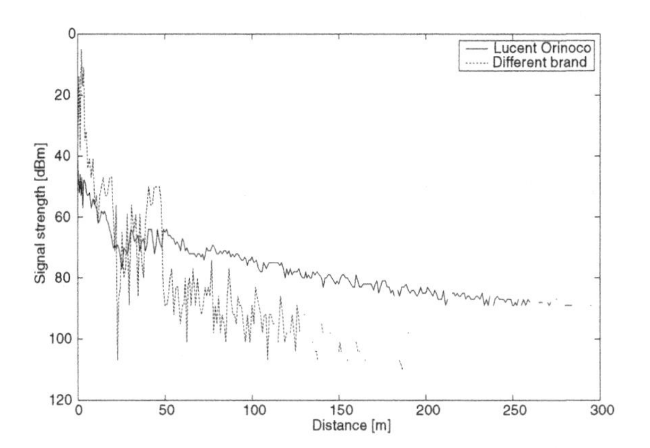

Fig. 2. Different card brands may show different behaviour

In all our trials we noted the persistent occurrence of a single but substantial 'dip' in the signal strength over approximately 10 m at distances ranging from 30 m to 50 m from the source (see for example Fig. 1). This is most likely due to coupling between the ground and the source (antenna) [29].

The influence of node height was assessed by placing two nodes 50 m apart while one node was stationary at navel height and the other changed its height as shown in Fig. 3 (Type II experiment). The signal strength dropped significantly when one node was placed close to the ground and no relevant difference was observable between navel and head height. Since we assume that node heights change infrequently in ubiquitous computing scenarios, we rate the importance for link quality predictions in these applications to be low. However, link quality fluctuations caused by varying node heights may pose a problem for military or disaster recovery applications due to their mobility patterns.

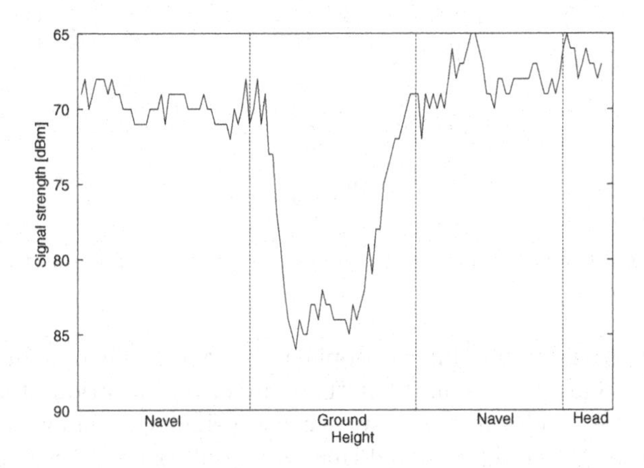

Fig. 3. Influence of node height on link quality

In personal wireless networks, the orientations of nodes change with the movements of the person (e.g. by turning) that holds the node. If standard hardware has a truly omni-directional antenna, this does not seem to pose a problem. However, we identified that the person's body can shadow the link severely based on its orientation and the node positions. This effect is illustrated at 150 m distance in Fig. 4 (Type II experiment). At around 180 degrees the body's shadowing caused a significant drop in the signal strength (a node rotation without a shadowing body caused no relevant changes in the signal strength). This leads to a link failure although the node could move 100 m further away at a different orientation without loosing the connection. Since the orientation changes frequently and suddenly in personal wireless networks, inevitable prediction errors in even the most benign open-space environment are the consequence.

People and cars are common in urban environments. We evaluate their influence on link quality by moving a person or car into the line of sight between two 100 m apart nodes (Type III experiment). Figure 5 and Fig. 6 present the influence of shadowing at different distances from the sender for a person and a car. At 20 m distance a person exhibits virtually no influence while a significant drop of link quality can be noticed 1 m away. A person directly in front of a node, which is common in crowds, leads even to a link failure at 100 m node distance. Similar but stronger link quality degradations can be observed for cars, where the connection was already lost at a shadowing distance of 1 m.

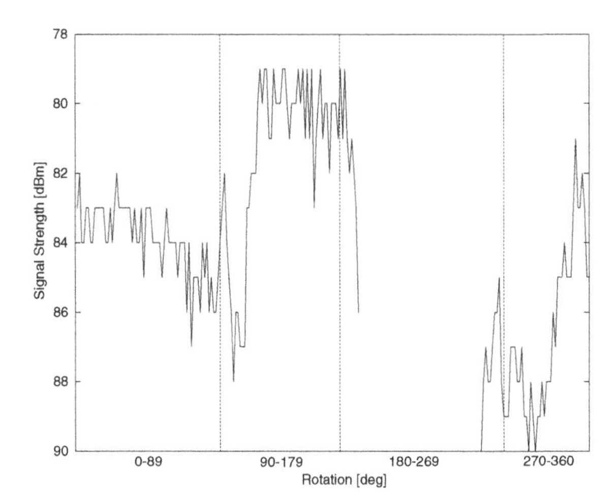

Fig. 4. Person shadowing a node with the own body while changing the orientation

We noticed that shadowing due to people and cars may degrade link quality by up to 30 dBm. Since we assume that the number and location of these objects are unpredictable, shadowing will always cause a significant amount of inevitable prediction errors. It should be noted that this result is not trivial since shadowing is a very important factor in 802.11 MANETs with more adverse effects on link quality than in traditional wireless networks (e.g. cellular phone networks). These traditional networks rely on an infrastructure with single-hop communication to base stations with antennas high over the ground (e.g. on building tops). Therefore the number of people in the line of sight is generally less than in 802.11 MANETs, where the antennas are usually located at either belt or car height. The infrastructure of cellular networks also provides a well defined coverage and well known reference points while in 802.11 MANETs no such reference points exists due to the mobility of all nodes. Moreover, cell phones can dynamically increase the transmission power to a much higher level for compensation of shadowing than 802.11 devices. 802.11 devices are restricted in their maximum transmission power to 15 dBm due to the license free band. Furthermore the diffracting ability of 802.11 signals is lower due to the much higher operating frequency.

Other factors that influence signal strength were also examined but found to be insignificant. These are also listed in Table 2.

5 Conclusions

We have observed that popular link quality prediction algorithms for 802.11 MANETs achieve only a much lower accuracy in real-world urban environments than in simulations. The best performing algorithm failed to predict 18 to 54 percent of total packet loss while 12 to 43 percent of packets were erroneously

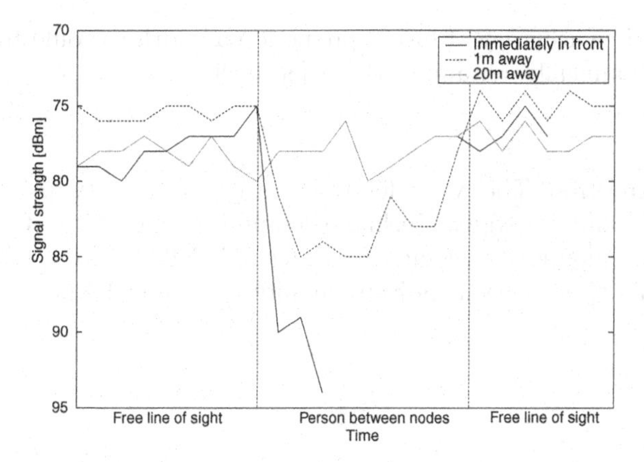

Fig. 5. Person shadowing a link at different distances

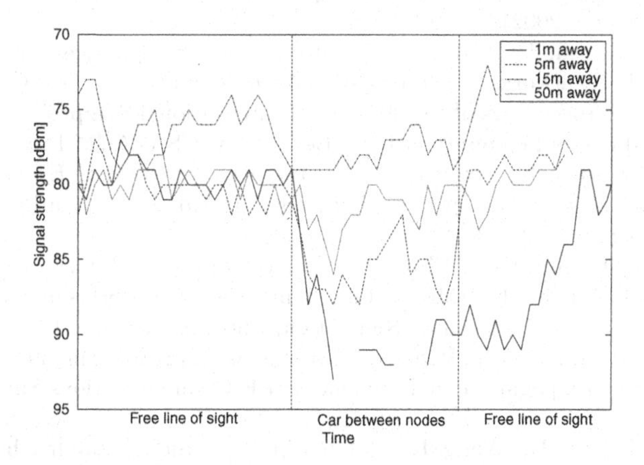

Fig. 6. Car shadowing a link at different distances

predicted to be lost. This contrasts sharply with near-perfect accuracy in simulations. The magnitude of this discrepancy has apparently not been appreciated in the literature to date.

We have observed that link quality prediction is especially difficult due to shadowing effects caused by user orientation, people and cars. The influence of such shadowing on link quality is much greater than in traditional cellular phone networks since in 802.11 MANETs there is no supporting infrastructure, transmission power is more limited and there is greater signal attenuation and poorer diffracting ability due to the higher operating frequencies used.

We identified, for low speed mobility scenarios, that algorithms based on the current received signal strength perform better than their trend-based counterparts. Current techniques to deal with fast signal fading were shown to be insufficient. Further research in this area is necessary. Moreover, we have determined that algorithm parameters are very specific to the actual operating environment. The accuracy of future prediction algorithms could be greatly improved by dynamically obtaining these parameters for the current operating environment.

Acknowledgments. The work described in this paper was partly supported by the Future and Emerging Technologies programme of the Commission of the European Union under research contract IST-2000-26031 (CORTEX – CO-operating Real-Time sentient objects: architectures and EXperimental evaluation).

References

1. Scheideler, C.: Models and techniques for communication in dynamic networks. In: 19th Symposium on Theoretical Aspects of Computer Science, Antibes Juan-les-Pins, France (2002) 27–49
2. Takai, M., Martin, J., Bagrodia, R.: Effects of wireless physical layer modeling in mobile ad hoc networks. In: International Conference on Mobile Computing and Networking, Proceedings of the 2001 ACM International Symposium on Mobile ad hoc networking and computing, Long Beach, CA, USA, ACM Press (2001) 87–94
3. Cavin, D., Sasson, Y., Schiper, A.: On the accuracy of MANET simulators. In: Proceedings of the Workshop on Principles of Mobile Computing (POMC'02), ACM Press (2002) 38–43
4. Heidemann, J., Bulusu, N., Elson, J., Intanagonwiwat, C., Lan, K., Xu, Y., Ye, W., Estrin, D., Govindan, R.: Effects of detail in wireless network simulations. In: SCS Multiconference on Distributed Simulation, Phoenix, Arizona, USA (2001) 3–11
5. Neskovic, A., Neskovic, N., Paunovic, G.: Modern approaches in modeling of mobile radio systems propagation environment. IEEE Communications Surveys **3** (2000) 2–12
6. Dube, R., Rais, C.D., Wang, K.Y., Tripathi, S.K.: Signal stability-based adaptive routing (SSA) for ad hoc mobile networks. IEEE Personal Communications **4** (1997) 36 –45
7. Paul, K., Bandyopadhyay, S., Mukherjee, A., Saha, D.: Communication-aware mobile hosts in ad-hoc wireless network. In: International Conferenceon Personal Wireless Communications (ICPWC'99), Jaipur, India (1999) 83–87
8. Punnoose, R.J., Nikitin, P.V., Broch, J., Stancil, D.D.: Optimizing wireless network protocols using real-time predictive propagation modeling. In: IEEE Radio and Wireless Conference (RAWCON), Denver, CO, USA (1999)
9. Lee, S.J., Su, W., Gerla, M.: Ad hoc wireless multicast with mobility prediction. In: IEEE ICCCN'99, Boston, MA, USA (1999) 4–9
10. Agarwal, S., Ahuja, A., Singh, J.P., Shorey, R.: Route-lifetime assessment based routing (RABR) protocol for mobile ad-hoc networks. In: IEEE International Conference on Communications. Volume 3., New Orleans (2000) 1697–1701
11. Jiang, S., Hez, D., Raoz, J.: A prediction-based link availability estimation for mobile ad hoc networks. In: IEEE INFOCOM, The Conference on Computer Communications, Twentieth Annual Joint Conference of the IEEE Computer and Communications Societies. Volume 3., Anchorage, Alaska, USA, IEEE (2001) 1745–1752
12. Su, W., Lee, S.J., Gerla, M.: Mobility prediction and routing in ad hoc wireless networks. International Journal of Network Management **11** (2001) 3–30

13. Goff, T., Abu-Ghazaleh, N.B., Phatak, D.S., Kahvecioglu, R.: Preemptive routing in ad hoc networks. In: MOBICOM 2001, Proceedings of the seventh annual international conference on Mobile computing and networking, Rome, Italy, ACM Press (2001) 43–52
14. Qin, L., Kunz, T.: Pro-active route maintenance in DSR. ACM SIGMOBILE Mobile Computing and Communications Review **6** (2002) 79–89
15. McDonald, A.B., Znati, T.: A mobility-based framework for adaptive clustering in wireless ad-hoc networks. IEEE Journal on Selected Areas in Communication **17** (1999)
16. Roman, G.C., Huang, Q., Hazemi, A.: Consistent group membership in ad hoc networks. In: IEEE 23rd International Conference in Software Engineering (ISCE), Toronto, Canada (2001) 381–388
17. Gerharz, M., Waal, C.d., Frank, M., Martini, P.: Link stability in mobile wireless ad hoc networks. In: 27th Annual IEEE Conference on Local Computer Networks (LCN'02), Tampa, Florida (2002) 30–42
18. McDonald, A.B., Znati, T.: A path availability model for wireless ad-hoc networks. In: IEEE Wireless Communications and Networking Conference 1999 (WCNC '99), NewOrleans, LA, USA (1999)
19. Rappaport, T.S.: Wireless communications – principles and practice. 2nd edn. Prentice Hall Communications Engineering and Emerging Technologies Series. Prentice Hall (2002)
20. He, D., Shengming, J., Jianqiang, R.: A link availablity prediction model for wireless ad hoc networks. In: Proceedings of the International Workshop on Wireless Networks and Mobile Computing, Taipei, Taiwan (2000)
21. McCanne, S., Floyd., S.: UCB/LBNL/VINT network simulator – ns (version 2). http://www.isi.edu/nsnam/ns (1999)
22. Bajaj, L., Takai, M., Ahuja, R., Tang, K., Bagrodia, R., Gerla., M.: Glomosim: A scalable network simulation environment. Technical report, UCLA Computer Science Department (1999)
23. Johnson, D., Maltz, D.: Dynamic source routing in ad hoc wireless networks. In Imelinsky, Korth, H., eds.: Mobile computing. Kluwer Academic Publishers (1996) 153–181
24. Tripathi, N.D., Reed, J.H., Van Landingham, H.F.: Handoff in cellular systems. IEEE Personal Communications **5** (1998) 26–37
25. Veeravalli, V.V., Kelly, O.E.: A locally optimal handoff algorithm for cellular communications. IEEE Transactions on Vehicular Technology **46** (1997) 603–609
26. Gibson, D.: Orinoco_cs. http://ozlabs.org/people/dgibson/dldwd/ (2003)
27. Storn, R., Price, K.: Differential evolution a simple and efficient heuristic for global optimisation over continuous spaces. Journal of Global Optimization **11** (1997) 341–359
28. Anastasi, G., Borgia, E., Conti, M., Gregori, E.: IEEE 802.11 ad hoc networks: performance measurements. In: Workshop on Mobile and Wireless Networks (MWN 2003) in conjunction with ICDCS 2003. (2003)
29. Janaswamy, R.: A fredholm integral equation approach to propagation over irregular terrain. In: Antennas and Propagation Society International Symposium. Volume 2. (1992) 765–768

BER Estimation for HiperLAN/2

Lodewijk T. Smit, Gerard J.M. Smit, Johann L. Hurink, and Gerard K. Rauwerda

Department of Electrical Engineering, Mathematics & Computer Science
University of Twente, Enschede, the Netherlands
L.T.Smit@utwente.nl

Abstract. This paper presents a method to estimate the bit error rate (BER) of the wireless channel based on statistical analysis of the soft output of the receiver only. In HiperLAN/2 several modulation schemes can be used. The system should select the most suitable modulation scheme dependent on the quality of the wireless link and the Quality of Service requirements of the user. Our BER estimation method can be used to estimate the current quality of the wireless link and the quality when another modulation mode is considered. With this information, it is possible to select the most suitable modulation scheme for the current situation.

1 Introduction

This paper presents a method to estimate the bit error rate (BER) of the wireless channel based on statistical analysis of the soft output of the receiver. The method does not use pilot symbols and does not require knowledge of the properties of the channel.

The physical layer of HiperLAN/2 (and also IEEE 802.11a) can use four modulation schemes: BPSK, QPSK, 16QAM and 64QAM. A modulation scheme with more bits per symbol allows a higher throughput, but requires a better channel to receive the bits with the same quality.

HiperLAN/2 uses a forward error correction (FEC) Viterbi decoder after the receiver to correct the incorrectly received bits in a frame. Figure 1 depicts this configuration. In most cases, the Viterbi decoder can still correct a frame with a high BER (e.g. up to 10%). It is attractive to switch to a modulation scheme with as many symbols per bit as possible because of the higher throughput, as long as the used Viterbi decoder is able to correct most of the received frames. To operate such a mechanism at run-time, an accurate estimation of the current wireless channel (BER) is required to select the optimal modulation mode. Furthermore, it would be nice to be able to predict what will occur with the current quality when we consider to change the modulation scheme. Our BER estimation method can be used to estimate the quality for the current modulation scheme and the quality when we consider to change to other modulation schemes. With this information, the control system in Figure 1 can select the best modulation scheme for the current situation and given the requested QoS.

I. Niemegeers and S. Heemstra de Groot (Eds.): PWC 2004, LNCS 3260, pp. 164–179, 2004.

The paper is organized as follows. Section two describes related work. Section three describes a method to obtain detailed information about the quality of the wireless link. Instead of using pilot bits to obtain this quality information, we introduce another method based on statistical analysis of the received data. Section four evaluates the performance of this method. Section five discusses some implementation issues of our proposed method.

2 Related Work

Khun-jush [5] states that the Packet Error Rate (PER), determined using the checksum of a packet, is a suitable measure of the link performance. Although the PER gives an indication of the quality, this quality metric is rather coarse. Besides that, a slightly higher PER might be desirable compared to a lower PER with a considerable lower thoughput (e.g. due to another modulation scheme). A disadvantage is the introduced latency, because the PER should be calculated using enough packets to get an accurate estimation of the PER. Furthermore, it gives no indication what we could do to improve the current situation.

A commonly used method to estimate the quality of a wireless link is to compute the BER using pilot symbols. Pilot symbols represent a predefined sequence of symbols, which are known at the transmitter and the receiver side. Therefore, the BER can be computed from these pilot symbols. For example, HiperLAN/2 as well as third generation telephony uses pilot symbols [1]. This approach has two disadvantages. First, the transmission of the pilot symbols introduces overhead. Second, the BER is only computed over a small amount of the total bits that are transmitted.

Another approach is to model the channel with all the known effects, e.g. [7]. A state of the art article on this area is [3]. Using this method it is possible to achieve accurate BER estimations for the modeled channel. However, the actual properties of the channel and the modeled effects can differ significantly from the constructed model. Also, effects that are not modeled can happen in real situations. In practice, it is not possible to model all the different effects that

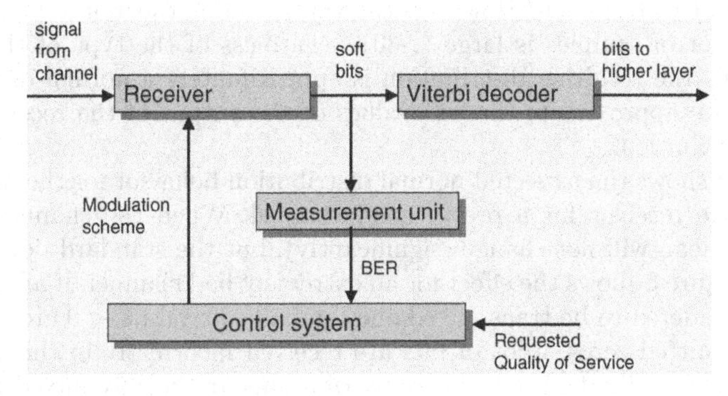

Fig. 1. The Control System of the HiperLAN/2 Terminal

cause the disturbance of the wireless channel. Estimation of the exact quality of the signal of the wireless channel is therefore impossible.

Our approach differs significantly from the mentioned approaches. We only use the soft output from the receiver, and require no additional information about the channel. Furthermore, no pilot symbols are used. In our opinion, it does not matter which physical effect is responsible for the degradation of the signal to determine the BER. Therefore, information of the channel is not required to make an estimation of the BER. The advantage is that an accurate estimation can be made independent of the unpredictable dynamic changing external environment. Furthermore, our method provides the possibility to estimate the quality resulting from a planned adaptation of a parameter.

3 BER Estimation

This section explains how to estimate the BER of the output of the receiver so that we can predict the probability that the Viterbi decoder can correct a frame using the results of the previous section. Our method uses only the soft output of the receiver and thus no pilot symbols. Although pilot symbols are used for different purposes in HiperLAN/2 and therefore still need to be transmitted, this reason may be important for other applications of our BER estimation algorithm.

We start with an explanation of the method in detail for BPSK modulation followed by a shorter explanation of the method for the QPSK, 16-QAM and 64-QAM modulation schemes.

3.1 BPSK

In an ideal situation, without disturbance of the channel, the output of the soft value (also called symbol) of the receiver is equal to the transmitted symbol value. In case of BPSK modulation this means 1 or -1. In case of disturbance of the channel, the sampled values are no longer exactly equal to 1 or -1, but can be higher of lower. Figure 2 depicts this situation. A lot of external causes may be responsible for this disturbance. Most effects that change the signal can be modeled by a normal distribution. Other effects, e.g. fading, do not behave like a normal distribution. However, the central limit theorem [6] states that, if the number of samples is large (>30), regardless of the type of the original distributions the resulting distribution is approximately a normal distribution. Therefore, we approximate the soft values of the output of the receiver with a normal distribution.

Figure 2 shows the expected normal distribution behavior for the soft output values of the receiver for a pretty good channel. When the channel becomes worse, the mean will not change (significantly), but the standard deviation will increase. Figure 3 shows the effect for an extremely bad channel. If all soft values > 0 are considered to be transmitted ones and all soft values < 0 are considered to be transmitted zeros, a lot of bits are received incorrectly in this figure. As can be seen from the figure, the two distributions are heavily mixed up. Every bit with value 1 that is received with a negative soft output is received badly

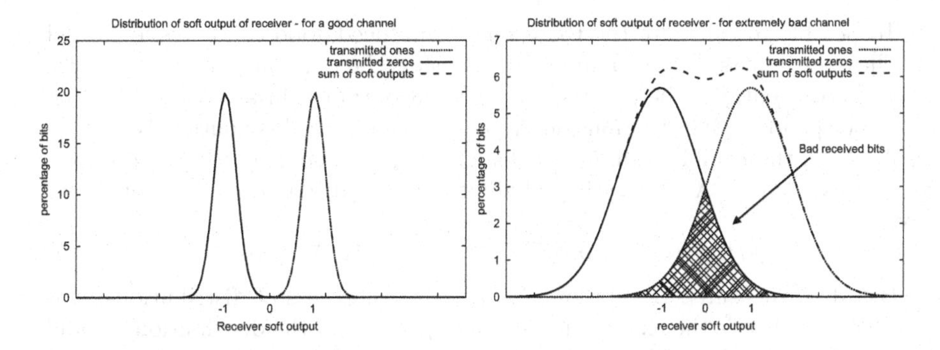

Fig. 2. Good Channel

Fig. 3. Very Bad Channel

and also the positive soft output for a transmitted bit with value -1 is received badly. Thus, the marked area is the probability that a bit is received incorrectly.

Unfortunately, the receiver can not determine whether a soft value belongs to the 1-distribution or to the -1-distribution. The soft output of the receiver is the addition of the 1-distribution and the -1-distribution, which is also plotted in Figures 2 and 3 as a dotted line.

Our goal is to predict the bit error rate (BER), i.e. the size of the marked area in Figure 3. Let X(Y) denote the distribution of the soft output values of the transmitted -1 (1). Using these distributions, the BER can be expressed by:

$$BER = pP(X \geq 0) + (1 - p)P(Y \leq 0). \tag{1}$$

where
X: denotes the soft value of a transmitted -1.
Y: denotes the soft value of a transmitted 1.
p: denotes the probability that a -1 is transmitted.
Since both distributions are mirrored at the zero axis and due to the mentioned assumption that we can model these distributions with a normal distribution, X and Y can be expressed in terms of a standard normal distribution:

$$X = \sigma Z - \mu. \tag{2}$$

$$Y = \sigma Z + \mu. \tag{3}$$

where Z denotes the standard normal distribution, μ the mean and σ the standard deviation.
Using this, the BER reduces to:

$$BER = P(X \geq 0) = P(Z < \frac{\mu}{\sigma}) = \Phi(-\frac{\mu}{\sigma}). \tag{4}$$

where $\Phi(z)$ is the function that gives the area of the standard normal distribution to the right of z, i.e. the probability that a value is smaller than z. The function $\Phi(z)$ is widely available in tabular form. Note that $\Phi(-\frac{\mu}{\sigma})$ is equal to $Q(\frac{\mu}{\sigma})$, with Q being the complementary error function that is commonly used in communication theory.

To be able to calculate the BER via (4), we need good estimates for μ and σ. These estimates $\widehat{\mu}$ and $\widehat{\sigma}$ are derived using the soft output values of the receiver. As mentioned before, the received soft output values of the receiver do not correspond to the distribution X and Y, but to a distribution W, which results from the combination of the distributions X and Y (with probability p we get distribution X and with probability $(1-p)$ distribution Y). For W we have:

$$P(W \leq w) = pP(X \leq w) + (1-p)P(Y \leq w). \tag{5}$$

Based on measured results for W and using moments of distributions, it is possible to estimate the characteristic values μ and σ of the distributions X and Y, which together form distribution W (see [10]). If r is a positive integer, and if X is a random variable, the rth moment of X is defined to be $m_r(X) = E(X^r)$, provided the expectation exists, see [4]. For a standard normal distribution, the moments of Z are shown in Table 1. The first and third moment of Z are zero and can not be used to compute the two unknown variables $\widehat{\mu}$ and $\widehat{\sigma}$. Therefore the second and fourth moment of W are used.

Table 1. Moments of Z

$m_1(Z)$	0
$m_2(Z)$	1
$m_3(Z)$	0
$m_4(Z)$	3

The second moment of W is:

$$m_2(W) = p(E(X^2)) + (1-p)(E(Y^2)). \tag{6}$$

The scrambling used in HiperLAN/2 ensures that approximately an equal number of ones and zeros are transmitted. This means that $p \approx \frac{1}{2}$. Setting $p = \frac{1}{2}$, and using equations (2), (3) and Table 1, equation (6) becomes:

$$m_2(W) = \mu^2 + \sigma^2. \tag{7}$$

therefore,

$$\widehat{\sigma}^2 = m_2(W) - \widehat{\mu}^2. \tag{8}$$

The fourth moment of W is:

$$m_4(W) = p(E(X^4)) + (1-p)(E(Y^4)). \tag{9}$$

With $p = \frac{1}{2}$, this equation becomes:

$$m_4(W) = \mu^4 + \binom{4}{2}\mu^2\sigma^2 E(Z^2) + \sigma^4 E(Z^4). \tag{10}$$

Substituting the moments of Z gives:

$$m_4(W) = \mu^4 + 6\mu^2\sigma^2 + 3\sigma^4. \tag{11}$$

Replacing σ^2 with (7) and simplifying yields:

$$\mu^4 = \frac{3}{2}(m_2(W))^2 - \frac{1}{2}m_4(W). \tag{12}$$

So,

$$\mu = \sqrt[4]{\frac{3}{2}(m_2(W))^2 - \frac{1}{2}m_4(W)}. \tag{13}$$

With the moment estimators of equations (8) and (14), the mean $\widehat{\mu}$ and standard deviation $\widehat{\sigma}$ can be computed from the individual samples $W_1..W_n$ by:

$$\widehat{\mu} = \sqrt[4]{\frac{3\left(\frac{\sum_{i=1}^{n} W_i^2}{n}\right)^2 - \frac{\sum_{i=1}^{n} W_i^4}{n}}{c_1}} \tag{14}$$

$$\widehat{\sigma} = \sqrt{\frac{\sum_{i=1}^{n} W_i^2}{n} - c_2\,\widehat{\mu}^2} \tag{15}$$

where c_1=2, c_2=1 and c_3=12. We have introduced these constants to be able to use the same formulas for the other modulation schemes.

Sometimes, there exists no (real) solution for μ and/or σ. In this case μ^4 and/or σ^2 in Equations (12) and (8) respectively are negative. So, this should always be checked before μ and σ are computed.
Finally, the BER estimation can be computed with:

$$\widehat{BER} = \frac{c_3}{12}\Phi\left(-\frac{\widehat{\mu}}{\widehat{\sigma}}\right) \tag{16}$$

3.2 Extention to QPSK, 16-QAM, and 64-QAM Modulation Schemes

HiperLAN/2 allows four modulation schemes: BPSK, QPSK, 16QAM and 64QAM. Therefore, the method for BER estimation for BPSK and QPSK schemes, that was presented in Chapter 3.1, should be extended for 16QAM and 64QAM modulation schemes. The 16QAM modulation scheme uses complex symbols. The real as well as the imaginary part of the complex symbol can have four different values: -3,-1,1 and 3. Therefore, the complex symbol can have 16 values, representing 4 bits. Figure 4 shows the possible complex values of a transmitted symbol for 16QAM.

The 64QAM modulation scheme uses also complex symbols. The real as well as the imaginary part of the complex symbol can have eight different values: -7,-5,-3,-1,1,3,5 and 7. Therefore, the complex symbol can have 64 values, representing 6 bits.

Section 3.3 describes the derivation of the estimators for μ and σ for 16QAM modulation. These estimators will be used in Section 3.4 to derive an estimator for the BER for 16QAM modulation. Similarly, section 3.5 describes the derivation of the estimators for μ and σ for 64QAM modulation and Section 3.6 derives an estimator for the BER for 64QAM modulation.

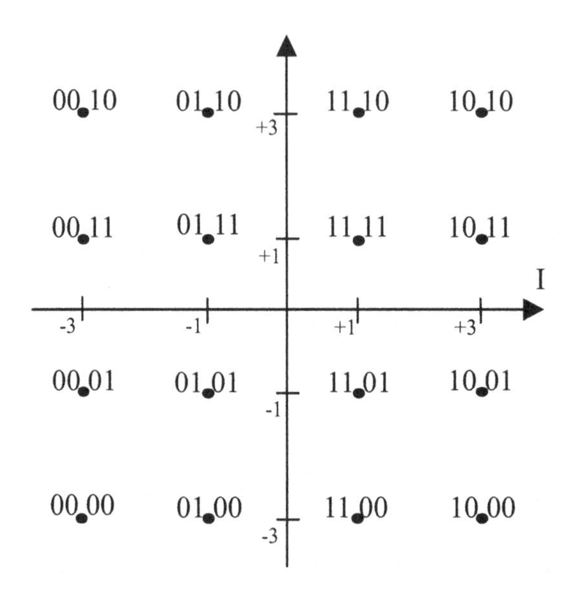

Fig. 4. 16-QAM Constellation Bit Encoding Scheme

3.3 Estimators for 16QAM

Figure 4 shows the possible complex values of a transmitted symbol for 16QAM. We consider the real and the complex part separate, because they are independent. So, the real part of a symbol that is transmitted can have four different values (-3,-1,1 and 3), compared to two possible values (-1,1) in the UMTS case with BPSK/QPSK modulation. Therefore, new estimators should be derived for $\widehat{\mu}$ and $\widehat{\sigma}$.
First, we define four stochastic variables for the four possible values, with:

X: denotes the distribution of the soft values of the transmitted minus ones.
Y: denotes the distribution of the soft values of the transmitted ones.
Q: denotes the distribution of the soft values of the transmitted minus threes.
R: denotes the distribution of the soft values of the transmitted threes.

We assume that the distribution have the same variance σ and that the means are $-\mu$, μ, -3μ and 3μ respectively. Using this, the stochastic variables X,Y,Q and R are expressed in terms of a standard normal distribution Z:

$$X = \sigma Z - \mu. \tag{17}$$

$$Y = \sigma Z + \mu. \tag{18}$$
$$Q = \sigma Z - 3\mu. \tag{19}$$
$$R = \sigma Z + 3\mu. \tag{20}$$

The received output values of the receiver do not correspond to these four individual distributions, but to a distribution V, which results from the combination of the four distributions. We assume that these four separate stochastic variables occur with the same frequency. In other words, the probability for each value is equal to $\frac{1}{4}$. For HiperLAN/2 the scrambling ensures that this property is fulfilled. For V we have:

$$P(V < v) = \frac{1}{4}\big(P(X < v) + P(Y < v) + P(Q < v) + P(R < v)\big). \tag{21}$$

Based on measured results of V and using the moments of the distributions, it is possible to estimate the characteristic values μ and σ of the four distributions, which form together distribution V.

The second moment m_2 of V is:

$$m_2(V) = \frac{1}{4}(E(X^2) + E(Y^2) + E(Q^2) + E(R^2)). \tag{22}$$

Using the following expressions for the second moments of the stochastic variable X,Y,Q and R

$$m_2(X) = E(X^2) = \sigma^2 + \mu^2. \tag{23}$$
$$m_2(Y) = E(Y^2) = \sigma^2 + \mu^2. \tag{24}$$
$$m_2(Q) = E(Q^2) = \sigma^2 + 9\mu^2. \tag{25}$$
$$m_2(R) = E(R^2) = \sigma^2 + 9\mu^2, \tag{26}$$

the second moment of V becomes:

$$m_2(V) = \sigma^2 + 5\mu^2. \tag{27}$$

This results in the following estimator for the variance:

$$\widehat{\sigma}^2 = m_2(V) - 5\mu^2. \tag{28}$$

To get an estimator for the mean μ, we again use the fourth moment of V:

$$m_4(V) = \frac{1}{4}(E(X^4) + E(Y^4) + E(Q^4) + E(R^4)). \tag{29}$$

Using the fourth moments of the different stochastic variables (the computation of the fourth moments is done in a similar way as in Section 3.1

$$m_4(X) = E(X^4) = E\left((\sigma Z - \mu)^4\right) = \mu^4 + 6\sigma^2\mu^2 + 3\sigma^4 \tag{30}$$
$$m_4(Y) = E(Y^4) = E\left((\sigma Z + \mu)^4\right) = \mu^4 + 6\sigma^2\mu^2 + 3\sigma^4 \tag{31}$$
$$m_4(Q) = E(Q^4) = E\left((\sigma Z - 3\mu)^4\right) = 81\mu^4 + 54\sigma^2\mu^2 + 3\sigma^4 \tag{32}$$
$$m_4(R) = E(R^4) = E\left((\sigma Z + 3\mu)^4\right) = 81\mu^4 + 54\sigma^2\mu^2 + 3\sigma^4, \tag{33}$$

the fourth moment m_4 of V becomes:

$$m_4(V) = 41\mu^4 + 30\sigma^2\mu^2 + 3\sigma^4. \tag{34}$$

With substitution of $\widehat{\sigma}$ in $m_4(V)$, we derive an estimator for μ:

$$\widehat{\mu} = \sqrt[4]{\frac{3m_2(V)^2 - m_4(V)}{34}} \tag{35}$$

With substitution of $\widehat{\mu}$ in $m_2(V)$, the estimator for σ becomes:

$$\widehat{\sigma} = \sqrt{m_2(V) - \widehat{\mu}^2}. \tag{36}$$

With this derivations, estimators $\widehat{\sigma}$ and $\widehat{\mu}$ can be computed from the individual samples V_1 .. V_n by:

$$\widehat{\mu} = \sqrt[4]{\frac{3\left(\frac{\sum_{i=1}^{n} V_i^2}{n}\right)^2 - \frac{\sum_{i=1}^{n} V_i^4}{n}}{34}} \tag{37}$$

$$\widehat{\sigma} = \sqrt{\frac{\sum_{i=1}^{n} V_i^2}{n} - 5\widehat{\mu}^2} \tag{38}$$

Simulations within Matlab show that both estimators work well if the four individual distributions are normal distributions.

3.4 BER Estimation for 16QAM

A received symbol is mapped to the nearest symbol in the bit constellation encoding diagram for 16QAM (see Figure 4). This symbol may be another symbol than the transmitted symbol resulting in one or more bit errors.

Based on the estimators $\widehat{\mu}$ and $\widehat{\sigma}$ we derive an estimation of the bit error rate $\widehat{BER}$. First, we investigate the error probabilities in one dimension. Figure 5 shows the four distributions of the four stochastic variables. The probability that the real part of a complex transmitted symbol with value (i,q) is received wrong is denoted by: $P_{e\_re(i)}$.
These probabilities are given by:

$$P_{e\_re(-3)} = P(Q \geq -2) \tag{39}$$

$$P_{e\_re(-1)} = P(X \leq -2) + P(X \geq 0) \tag{40}$$

$$P_{e\_re(1)} = P(Y \leq 0) + P(Y \geq 2) \tag{41}$$

$$P_{e\_re(3)} = P(R \leq 2), \tag{42}$$

and can be calculated by using the mean μ and the standard deviation σ:

$$P_{e\_re(-3)} = \Phi(\frac{-\mu}{\sigma}) \tag{43}$$

$$P_{e\_re(-1)} = 2 * \Phi(\frac{-\mu}{\sigma}) \tag{44}$$

$$P_{e\_re(1)} = 2 * \Phi(\frac{-\mu}{\sigma}) \tag{45}$$

$$P_{e\_re(3)} = \Phi(\frac{-\mu}{\sigma}) \tag{46}$$

The probability P_{e_re} that the real part of a symbol is received incorrect is now given by:

$$P_{e\_re} = \frac{1}{4}P_{e\_re(-3)} + \frac{1}{4}P_{e\_re(-1)} + \frac{1}{4}P_{e\_re(1)} + \frac{1}{4}P_{e\_re(3)} = \frac{6}{4}\Phi\left(-\frac{\mu}{\sigma}\right). \tag{47}$$

The same holds for the probability P_{e_im} for the imaginary part of the symbol.

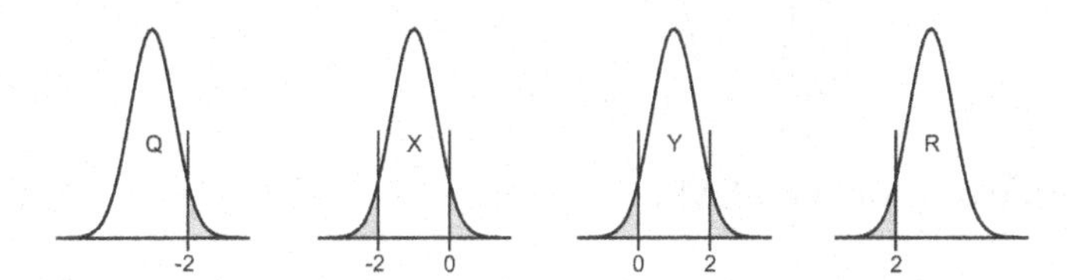

Fig. 5. BER Areas for 16-QAM Modulation Scheme in One Dimension

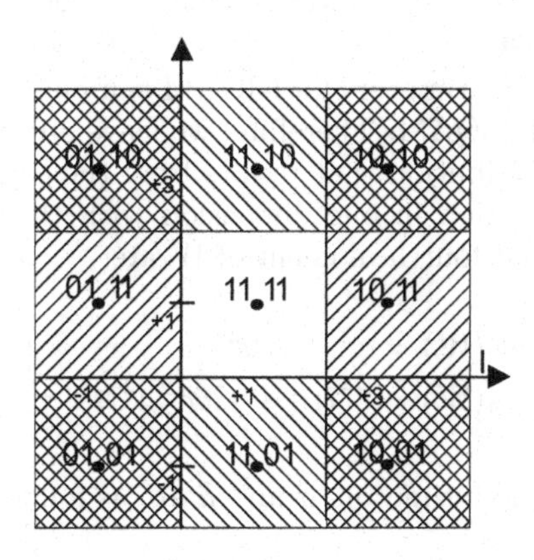

Fig. 6. Bit Error Probability for '1111' for 16-QAM

To calculate the BER, the mapping between symbol and bit sequence has to be considered. Figure 4 shows this mapping. Due to the used Gray coding, only one out of four bits is in error, if one of the direct neighbor symbols (of the transmitted symbol) is received in one dimension. If a symbol is received incorrectly, we assume that instead of the correct symbol, one of the direct neighbors is received. So, we assume that an incorrectly received real or complex part of a symbol leads to precisely one bit error. This assumption however, introduces a small estimation error. For now this error is ignored. In [8] this small error is investigated. affects the estimation. The probability of a complex bit error (in two dimensions) is the addition of the real part and the complex part. Note that the overlap of the two probability areas are counted twice. This is correct, due to the fact that indeed two bits are incorrect, instead of one. Figure 6 shows this effect for the reception of symbol '1111'. The probability of a bit error becomes:

$$\begin{aligned}\widehat{BER} &= (P_{e\_re} + P_{e\_im})/(\text{number of bits per symbol}) \\ &= \left(\frac{6}{4}\Phi\left(-\frac{\widehat{\mu}}{\widehat{\sigma}}\right) + \frac{6}{4}\Phi\left(-\frac{\widehat{\mu}}{\widehat{\sigma}}\right)\right)/4 \\ &= \frac{3}{4}\Phi\left(-\frac{\widehat{\mu}}{\widehat{\sigma}}\right) \end{aligned} \tag{48}$$

3.5 Estimators for 64QAM

The 64QAM modulation scheme uses also complex symbols. The real as well as the imaginary part of the complex symbol can have eight different values: -7,-5,-3,-1,1,3,5 and 7. We consider the real and the complex part separate, because they are independent. So, the real part of a symbol that is transmitted can have eight different values (-7μ,-5μ,-3μ,-1μ,1μ,3μ,5μ and 7μ), compared to four values for the 16QAM case.

The derivation of the estimators for 64QAM can be done in the same way as for BPSK (QPSK) and 16QAM. Therefore, we leave out all intermediate steps. We define W as the distribution that results from the combination of the eight separate normal distribution in one dimension with the eight means:-7,-5,-3,-1,1,3,5,7. The second and fourth moment of W are:

$$m_2(W) = \sigma^2 + 21\mu^2 \tag{49}$$

$$m_4(W) = 777\mu^4 + 126\sigma^2\mu^2 + 3\sigma^4 \tag{50}$$

With these moments, the estimators $\widehat{\sigma}$ and $\widehat{\mu}$ are as follows:

$$\widehat{\mu} = \sqrt[4]{\frac{3\left(\frac{\sum_{i=1}^{n} W_i^2}{n}\right)^2 - \frac{\sum_{i=1}^{n} W_i^4}{n}}{546}} \tag{51}$$

$$\widehat{\sigma} = \sqrt{\frac{\sum_{i=1}^{n} W_i^2}{n} - 21\widehat{\mu}^2} \tag{52}$$

where $W_1...W_n$ denote the n individual samples.

3.6 BER Estimation for 64QAM

In a similar was as done in Section 3.4, we can derive the probability of a bit error for 64-QAM modulation:

$$\begin{aligned}\widehat{BER} &= (P_{e\_re} + P_{e\_im}) / (\text{number of bits per symbol}) \\ &= \left(\frac{14}{8}\Phi\left(-\frac{\widehat{\mu}}{\widehat{\sigma}}\right) + \frac{14}{8}\Phi\left(-\frac{\widehat{\mu}}{\widehat{\sigma}}\right)\right)/6 \\ &= \frac{7}{12}\Phi\left(-\frac{\widehat{\mu}}{\widehat{\sigma}}\right)\end{aligned} \tag{53}$$

3.7 Summary for BPSK, QPSK, 16-QAM, and 64-QAM Modulation Scheme

The BER estimation for the different types of modulation differs only by the constants c_1, c_2 and c_3 in the Formulas (14) to (16). Table 2 summarizes the constants for the different types of modulation that should be used for the Formulas (14) to (16). It may be a bit surprisingly that also in the BER estimation with Formula (16) the function Φ has to be weighted. This results from the fact that symbols that have neighbour symbols on both sides have a higher probability on a bit error.

Table 2. Constants for Different Modulation Schemes

	c_1	c_2	c_3
BPSK	2	1	12
QPSK	2	1	12
16-QAM	34	5	9
64-QAM	546	21	7

4 Results BER Estimation

We have chosen to show the test results of the BER estimation method for the 64-QAM modulation scheme. More bits per symbol makes the BER estimation more difficult. For these tests we used a HiperLAN/2 simulator [11]. Figure 8 shows the results for the BER estimation for 64-QAM modulation, which is the most difficult case. The figure shows the BER on the x-axis and the error in the estimation on the y-axis. The estimation error is equal to the (estimated BER - real BER) * 100%. The line with label "1 term" shows the results obtained by using the Formulas (14) to (16) and Table 2. As can be seen from the figure, there is an under estimation when the BER increases. The reason is that Formula (16) accounts for 1 bit error only when the received symbol is not the transmitted symbol. However, 16-QAM and 64-QAM modulation schemes are not modeled with two normal distributions, but with four or eight distributions respectively. This means that a symbol error also can be caused by two interfering distributions that are not directly next to each other, resulting in more bit errors. For example, if the received symbol is the neighbor of the neighbor of the transmitted symbol (denoted with distance 2), then it will account for 2 bit errors, as can be seen from Table 3. We can correct for this case with the addition of an additional term to Formula (16). This term is the probability that a symbol has neighbors with distance 2 multiplied with the error introduced if indeed a neighbor with distance 2 is received instead of the correct symbol. So, the correction term for a received neighbor with distance 2 is: $\frac{6}{12}\left(\Phi(\frac{5\mu}{\sigma}) - \Phi(\frac{3\mu}{\sigma})\right)$, see Figure 7. A similarly correction term for a received neighbor with distance 3 is: $\frac{5}{12}\left(\Phi(\frac{7\mu}{\sigma}) - \Phi(\frac{5\mu}{\sigma})\right)$.

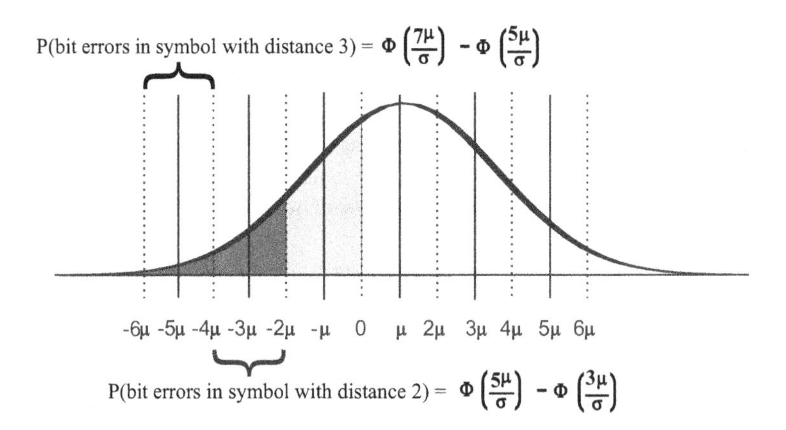

Fig. 7. More Bit Errors

Dependent on the desired accuracy and computational complexity of the BER estimation we can add one or more correction terms to Formula (16). Figure 8 shows the results of these correction factors. The line with the label "2 terms" is the result of the simulation for the original Formula (16) plus the correction term for received neighbors with distance 2. The line with the label "3 terms" shows the performance of the BER estimation with Formula (16) plus the correction factors for received neighbors with distance 2 and 3. Since the Viterbi decoder can not correct frames with a high BER, it is not usefull to add the correction term(s) in our case.

Figure 8 shows that the error of the BER estimation algorithm is smaller than 2%. Additional simulations for different channels gave a similar result.

Table 3. 64QAM Constellation Bit Encoding

000100	001100	011100	010100	**7**	110100	111100	101100	100100
000101	001101	011101	010101	**5**	110101	111101	101101	100101
000111	001111	011111	010111	**3**	110111	111111	101111	100111
000110	001110	011110	010110	**1**	110110	111110	101110	100110
-7	**-5**	**-3**	**-1**		**1**	**3**	**5**	**7**
000010	001010	011010	010010	**-1**	110010	111010	101010	100010
000011	001011	011011	010011	**-3**	110011	111011	101011	100011
000001	001001	011001	010001	**-5**	110001	111001	101001	100001
000000	001000	011000	010000	**-7**	110000	111000	101000	100000

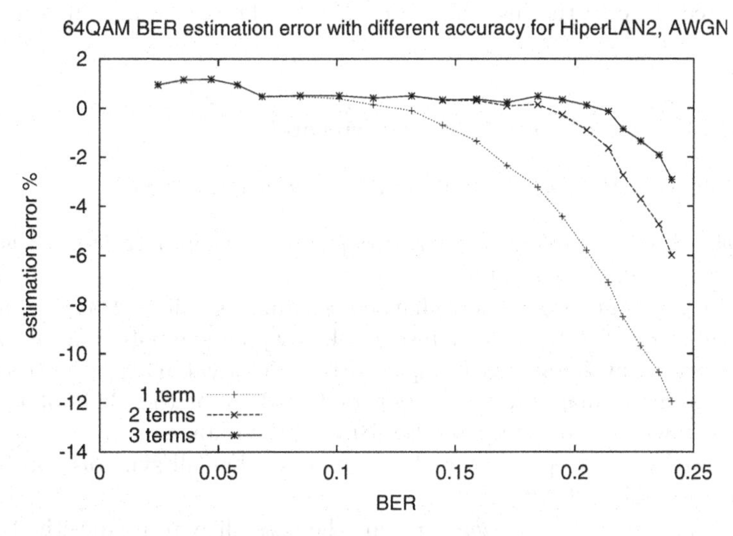

Fig. 8. BER Estimation Error with Different Number of Terms for 64QAM in HiperLAN/2 Simulation with a AWGN Channel

5 Discussion

This section dicusses the implementation of the presented in hardware and the advantages of using this method.

5.1 Implementation

The basic idea of the proposed method is to express the quality of the channel (BER) in terms of a normal probability function. This function can be read from a z-table, or certain functions exist that approximate the c.d.f. of a standard normal distribution. For example, $P(x) = 1 - \frac{1}{2}(1+c_1x+c_2x^2+c_3x^3+c_4x^4)^{-4} + \epsilon(x)$, with $|\epsilon| < 2.5 * 10^{-4}$ [2].

The estimators for the μ and the σ of the BER (Formulas (14 and 15) require the summation of W_i^2 and W_i^4 of the incoming bits. This computation has to be done at the incoming bit rate. These computations can be done with very simple hardware support. Because the incoming values of the receiver are always quantized with a limited number of bits, the values of the power of two and four can be stored in a look-up table (LUT). This LUT in combination with an adder and a register are sufficient to compute the sum. The rest of the computation has to be performed only once per frame and can be done e.g. with a general purpose processor.

The estimators do not contain a correction factor for the bias. When the number of samples is large enough (>30), the difference is so small that this can be neglected. Due to the high bit rate (Mbit/s) the bias is not an issue in our case.

5.2 Advantages of the Presented Method

The presented method has several attractive properties, such as:

- *simplicity* – The presented method requires little computation and is therefore easy to implement in a receiver.
- *accuracy* – Simulations show that the accuracy is within 2%. The method uses all received data symbols instead of only pilot symbols. This means that more data is available which improves the statistical analysis. Furthermore, most methods make an estimation of the BER of the BER of the pilot symbols, which can differ from the BER of the data symbols.
- *low overhead* – No pilot symbols are used so that all symbols can be used for transmission of data.
- *parameter prediction possible* – Beside the possiblitiy to predict the BER for the current situation, it is also possible to predict the BER after changes of parameters. For example, it is possible to predict the BER when the modulation is changed.
- *no assumptions about environment* – Some analytical methods use assumptions about the environment to make a model for prediction of the BER. The presented method does not make assumptions about the environment.
- *generality* – The presented method is not only useful for HiperLAN/2, but also for other wireless communication methods. A detailed example for wideband code division multiple access (WCDMA) is given in [9]. WCDMA differs significantly from the OFDM transmission technology used in HiperLAN/2.

6 Conclusion

The presented method to estimate the BER in HiperLAN/2 is simple and effective. With a few formulas, we can describe the complete behaviour of the system in terms of quality. The BER estimation requires no overhead (such as pilot symbols) and has an accuracy of about 2% with respect to the real BER. A key advantage of our method is that we can also predict what will happen with the BER when we consider to change the modulation type.

Acknowledgements. This research is conducted within the Chameleon project (TES.5004) supported by the PROGram for Research on Embedded Systems & Software (PROGRESS) of the Dutch organization for Scientific Research NWO, the Dutch Ministry of Economic Affairs and the technology foundation STW. We would like to thank dr. W.C.M. Kallenberg for his support.

References

1. `http://www.3gpp.org`.
2. M. Abramowitz and I. A. Stegun. *Handbook of Mathematical Functions.* General Publishing Company, Ltd., seventh edition, 1970. ISBN: 0-486-61272-4.
3. J. Cheng and N. C. Beaulieu. Accurate DS-CDMA bit-error probability calculation in Rayleigh fading. *IEEE transactions on wireless communications*, 1(1):3–15, 2002.
4. E. J. Dudewicz and S. N. Mishra. *Modern Mathematical Statistics.* John Wilsey & Sons, Inc., 1988. ISSN: 0271-6232.
5. J. Khun-Jush, P. Schramm, G. Malmgren, and J. Torsner. HiperLAN2: Broadband wireless communications at 5 GHz. *IEEE Communications Magazine*, pages 130–136, June 2002.
6. P. S. Mann. *Introductory Statistics.* John Wiley & Sons, second edition, 1995. ISBN: 0-471-31009-3.
7. R. K. Morrow. Accurate CDMA BER calculations with low computational complexity. *IEEE Transactions on Communications*, pages 1413–1417, Nov. 1998.
8. L. T. Smit. *Energy-Efficient Wireless Communication.* PhD thesis, University of Twente, Dec. 2003. ISBN: 90-365-1986-1.
9. L. T. Smit, G. J. M. Smit, J. L. Hurink, and A. B. J. Kokkeler. Soft output bit error rate estimation for WCDMA. In *Proceedings of Personal Wireless Conference 2003*, pages 115–124, Sept. 2003. ISBN: 3-540-20123-8; ISSN: 0302-9743.
10. W. Y. Tan and W. C. Chang. Some comparisions of the method of moments and the maximum likelihood in estimating parameters of a mixture of two normal densities. *Journal of the American Statistical Association*, 67(33):702–708, Sept. 1972.
11. L. F. W. van Hoesel. Design and implementation of a software defined HiperLAN/2 physical layer model for simulation purposes. Master's thesis, University of Twente, Aug. 2002.

An Efficient Backoff Scheme for IEEE 802.11 DCF

Ho-Jin Shin, Dong-Ryeol Shin[1], and Hee-Yong Youn

School of Information and Communication Engineering,
Sungkyunkwan University,
300 cheoncheon-dong, jangan-gu, suwon, Korea
{hjshin, drshin, youn}@ece.skku.ac.kr

Abstract. This paper proposes a new backoff algorithm for a CSMA/CA protocol and an analytical model computing the saturation throughput and delay of the scheme. The scheme differs from the standard backoff schemes in that the backoff time is not uniformly chosen in the contention window interval which results in reduced collision probability under high loads. Numerical analysis shows that saturation throughput and delay of the proposed scheme outperform the earlier approach.

1 Introduction

In IEEE 802.11 carrier sensing is performed both at the physical layer and MAC layer, which is also referred to as physical carrier sensing and virtual carrier sensing, respectively. The PCF is a polling-based protocol, which was designed to support collision free and real time services. This paper focuses on the performance and delay analysis and modeling of a new scheme for DCF in IEEE 802.11 wireless LAN.

There are two techniques used for packet transmission in DCF. The default one is a two-way handshaking mechanism called basic access method. The optional one is a four-way handshaking mechanism, which uses RTS/CTS frame to reserve the channel before data transmission. This mechanism has been introduced to reduce performance degradation due to hidden terminals. However, it has some drawback of increased overhead for short data frames.

According to the CSMA/CA protocol, if the medium is busy, the station has to wait until the end of the current transmission. It then waits for an additional DIFS time and then generates a random backoff interval before transmitting a frame. The backoff counter is decremented as long as the channel is sensed idle. The station transmits a frame when the backoff time counter reaches zero. The initial value of the backoff counter is randomly chosen in the contention window (0, w-1). At the first attempt, $w = W_0$, the minimum size of the contention window. After each unsuccessful transmission, w is doubled, up to a maximum value within which the backoff counter value is again randomly selected with uniform distribution.

As more stations are added, the new contention window size due to collisions is doubled and chosen randomly in (0, w-1). In such a case, there is a good chance for the collided station to select smaller value and thus increasing the possibilities of

[1] This paper was supported by Faculty Research Fund from Sungkyunkwan University.

I. Niemegeers and S. Heemstra de Groot (Eds.): PWC 2004, LNCS 3260, pp. 180–193, 2004.

collisions with existing nodes with small values of backoff counters, which results in degrading performance.

With this observation in mind, we propose a new scheme deciding the contention window size and procedure decrementing the backoff counter, which gives preferences to the active nodes over colliding nodes to reduce the collision probability and thereby improving the throughput. This is the main contribution of the paper, which is to be detailed later.

In recent year, modeling of IEEE 802.11 has been a research focus since the standard has been proposed. In the literature, performance evaluation of IEEE 802.11 has been carried out either by means of simulation [3][12][13] or analytical models with simplified backoff rule assumptions. In particular, constant or geometrically distributed backoff window was used in [4][5], while [6] considered an exponential backoff limited to two stages by employing a two dimensional Markov chain analysis. This paper investigates the performance of the new proposed scheme in terms of throughput and delay. Then comparing with other models, we demonstrate the effect of the proposed mechanism on network performance.

The rest of the paper is organized as follows. Section 2 briefly describes the DCF of IEEE 802.11 MAC protocols and an analytical model computing the saturated throughput and delay of 802.11 is proposed. Section 3 evaluates the proposed scheme and compares it with earlier models. Finally, Section 4 concludes the paper.

2 Analysis of DCF

In this paper we concentrate on the analytical evaluation of the saturation throughput and delay, which is described in [7][8][10][11][12][14]. These fundamental performance figures are defined as the throughput limit reached by the system as the offered load increases, and the corresponding delay with the load that the system can carry in stable condition.

One of key contributions of this paper is to derive an analytical model of the saturation throughput and delay, in the assumption of ideal channel conditions (i.e., no hidden terminals and capture [9]). In addition, the Markov model in [8] does not consider the frame retries limits, and thus it may overestimate the throughput of IEEE 802.11. With this observation, our analysis is based on a more exact Markov chain model [11] taking into the consideration of the maximum frame retransmission limit. We assume a fixed number of wireless stations, each always having a packet available for transmission. In other words, the transmission queue of each station is assumed to be always nonempty. To ease comparison with the model in [11], we use the same symbols and variables as in [11]. The analysis is divided into two parts. First, we examine the behavior of a single station with a Markov model, and obtain the stationary probability τ that the station transmits a packet in a generic slot time. This probability does not depend on the access mechanism employed as shown in [8]. Then, we express the throughput and delay of both Basic and RTS/CTS access method as a function of the computed value τ.

2.1 Markov Chain Model

We employ the same assumptions in [11] for our analysis. Consider a fixed number n of contending stations. In saturation condition, each station has a packet immediately available for transmission after the completion of each successful transmission. Moreover, being all packets consecutive, each packet needs to wait for a random backoff time before transmission.

Let $b(t)$ be the stochastic process representing the backoff time counter for a given station at slot time t. A discrete and integer time scale is adopted: t and t+1 correspond to the beginning of two consecutive slot times and the backoff time counter of each station decrements at the beginning of each slot time [8]. Let $s(t)$ be the stochastic process representing the backoff stage (0, … , m) of the station at slot time t. As in [8][11], the key conjecture in this model is that the probability p that a transmitted packet collides is independent on the stage $s(t)$ of the station. Thus, the bi-dimensional process $\{s(t), b(t)\}$ is a discrete-time Markov chain, which is shown in Fig. 1.

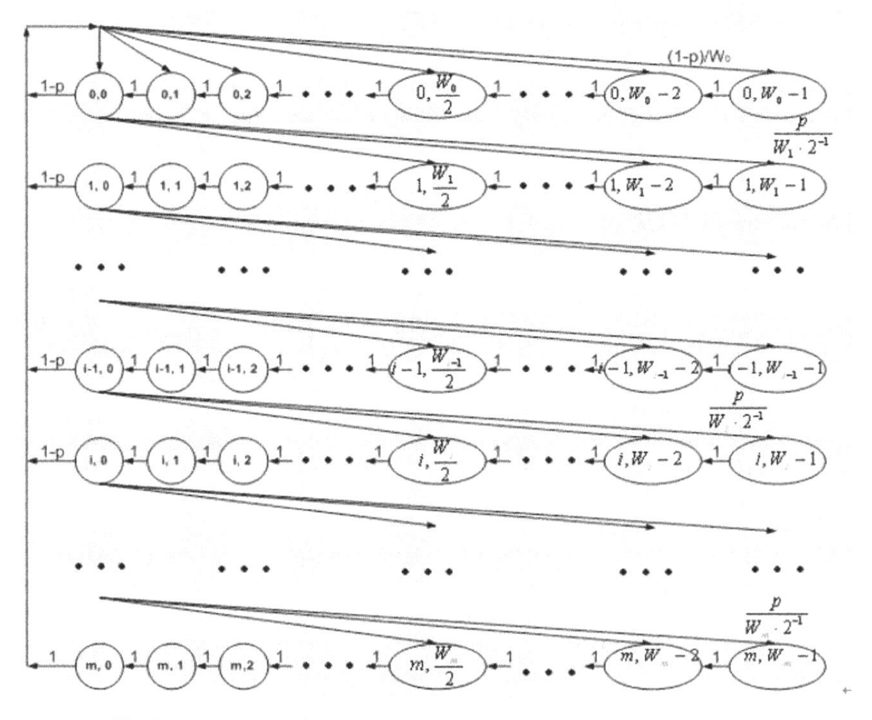

Fig. 1. The Markov Chain model of the new backoff window scheme.

This paper uses all the parameters assigned for Direct Sequence Spread Spectrum (DSSS) PHY in 802.11, for the comparison with the model in [11]. In DSSS, CW_{min} and CW_{max} are set to 31 and 1023, respectively. Therefore, we have

$$\begin{cases} W_i = 2^i W & i \le m' \\ W_i = 2^{m'} W & i > m' \end{cases} \tag{1}$$

Where $W = (CW_{min} + 1)$, and $2^{m'} W = (CW_{max} + 1)$. Thus for DSSS, we have $m' = 5$.

In the IEEE 802.11 backoff procedure, backoff time is randomly chosen between [0, W_i-1], where W_i is a contention window at backoff stage-i. Fig. 2 and Fig. 3 show the configuration of backoff counter for backoff stage i=0 and i=1, respectively. This scheme is to reduce the possibility of collision by letting each station get a different range of contention window. The motivation behind choosing a non-uniformly distributed contention window counter value is as follows. For example, a station which transmits a packet successfully generates new backoff time counter value from CW_0 (0~31). If a station cannot transmits a packet due to collision, it moves to the next backoff stage and generates a backoff time counter value from CW_1(0~63) as shown in [1][8][11]. As a result, there is still a good chance for two stations having the same backoff time counters in which collision may occur when counter reaches zero again since the window size values are integers and multiple of some basic time slot value.

We thus propose new backoff time counter selection scheme. Unlike the schemes in [1][8][11], we give preferences to the stations actively competing with the colliding stations within current contention window for it, the contention window is constructed non-uniformly as follows:

$$CW_i = [R_i, W_i - 1] \qquad 0 \le i \le m$$

Here, if a station transmits a packet successfully in backoff stage-0, R_0 = 0. Otherwise, there is a busy channel to incur collision and thus sets a contention window to [W_i/2, W_i-1] rather than [0, W_i-1]. In other words,

$$R_i = W_i/2 \qquad 1 \le i \le m$$

Comparing with the example above, the proposed scheme adapts a different mechanism for selecting backoff time counter value. The station then moves to the next backoff stage, generating backoff time counter from CW_1 (32~63). This approach will reduce the probability of each station choosing the same backoff time counter value. Fig. 4 shows exponential increment of CW for the proposed backoff time counter scheme.

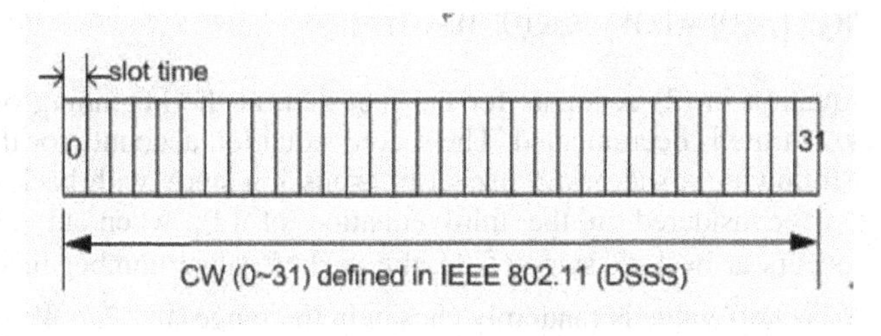

Fig. 2. CW_i when backoff stage i=0.

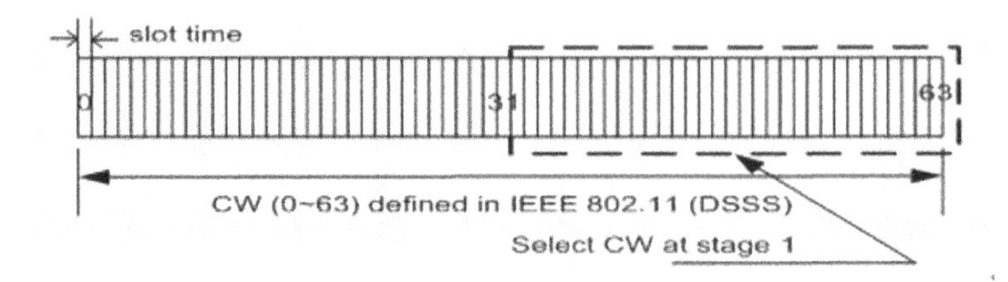

Fig. 3. CW_i when backoff stage i =1

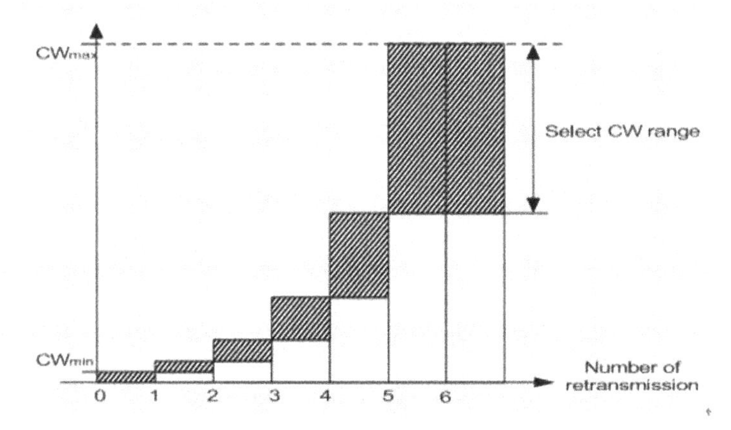

Fig. 4. The backoff counter range with the proposed scheme.

In order to validate and analyze the proposed scheme, we employ the same Markov chain models and assumptions in [11].

Here, the only non-null one-step transition probabilities are

$$\begin{cases} P\{i,k \mid i,k+1\} = 1 \quad k \in (0,\ W_i - 2) \quad i \in (0,\ m) \\ P\{0,k \mid i,0\} = (1-p)/W_0 \quad k \in (0,\ W_0 - 1) \quad i \in (0,\ m) \\ P\{i,k \mid i-1,0\} = p/(W_i/2) \quad k \in (W_i/2,\ W_i - 1) \quad i \in (1,\ m) \\ P\{0,k \mid m,0\} = 1/W_0 \quad k \in (0,\ W_0 - 1) \end{cases} \tag{2}$$

The first equation in (2) accounts for the fact that, at the beginning of each slot time, the backoff time is decremented. The second equation accounts for the fact that a new packet following a successful packet transmission starts with backoff stage-0. In particular, as considered in the third equation of (2), when an unsuccessful transmission occurs at backoff stage-(i-1), the backoff stage number increases, and the new initial backoff value is randomly chosen in the range $(W_i/2,\ \ W_i - 1)$. This is different part from others in [8][10][11]. Finally, the fourth case models the fact that when the backoff stage reaches the maximum backoff stage, the contention window is reset if the transmission is unsuccessful or restart the backoff stage for new packet if the transmission is successful.

The stationary distribution of the Markov chain is defined as $b_{i,k} = \lim_{t\to\infty} P\{s(t)=i, b(t)=k\}$. Using chain regularities in steady state, we can derive the following relations.

$$b_{i-1,0} * p = b_{i,0} \quad 0 < i \le m \tag{3}$$

$$b_{i,0} = p^i b_{i,0} \quad 0 \le i \le m \tag{4}$$

If a backoff time value k of a station at stage-i is selected among $CW_i = [W_i/2,\ W_i - 1]$, the following relation can be derived.

$$\begin{cases} b_{i,k} = \dfrac{W_i - k}{W_i}(1-p)\sum_{j=0}^{m-1} b_{j,0} + b_{m,0} \quad i = 0,\ k \in (0,\ W_i - 1) \end{cases} \tag{5}$$

for each $k \in (0,\ \frac{W_i}{2} - 1)$

$$b_{i,k} = p b_{i-1,0} \quad 0 < i \le m$$

each $k \in (\frac{W_i}{2},\ W_i - 1)$

$$b_{i,k} = \frac{W_i - k}{W_i/2} p b_{i-1,0} \quad 0 < i \le m$$

With (4) and transitions in the chain, equation (5) can be simplified as

$$b_{i,k} = \gamma * b_{i,0} \quad 0 \le i \le m \tag{6}$$

By using the normalization condition for stationary distribution, we have

$$\begin{aligned} 1 &= \sum_{i=0}^{m}\sum_{k=0}^{W_i-1} b_{i,k} = \sum_{i=0}^{m}\sum_{k=0}^{W_i-1} \gamma b_{i,0} = \sum_{i=0}^{m} b_{i,0} \sum_{k=0}^{W_i-1}\gamma \\ &= b_{0,0}\sum_{k=0}^{W_i-1}\gamma + \sum_{i=1}^{m} b_{i,0}\sum_{k=0}^{W_i-1}\gamma = b_{0,0}\cdot\frac{W_0+1}{2} + \sum_{i=1}^{m} b_{i,0}\sum_{k=0}^{W_i-1}\gamma \end{aligned} \tag{7}$$

Let τ be the probability that the station transmits a packet during a generic slot time. A transmission occurs when the backoff window is equal to zero. It is expressed as:

$$\tau = \sum_{i=0}^{m} b_{i,0} = \frac{1-p^{m+1}}{1-p} b_{0,0} \tag{8}$$

Then, using equation (1), (6) and (7), $b_{0,0}$ is computed as

$$b_{0,0} = \begin{cases} \dfrac{4(1-2p)(1-p)}{\alpha} & m \le m' \\ \dfrac{4(1-2p)(1-p)}{\beta} & m > m' \end{cases} \tag{9}$$

$$\alpha = 2(W+1)(1-2p)(1-p) + 3W\,2p(1-(2p)^{m}(1-p) + p(1-p^{m})(1-2p)$$

$$\beta = 2(W+1)(1-2p)(1-p) + 3W\,2p(1-(2p)^{m'}(1-p) + p(1-p^{m})(1-2p)$$
$$+3W\,2m'p^{m'+1}(1-p^{m-m'})(1-2p)$$

In the stationary state, a station transmits a packet with probability τ, so we have

$$p = 1-(1-\tau)^{n-1} \tag{10}$$

Therefore, equation (8), (9) and (10) represent a nonlinear system with two unknowns τ and p, which can be solved using numerical method. We must have $\tau \in (0,1)$ and $p \in (0,1)$.

Since the Markov chain of Fig. 1 is a little different from that of [8] and [11], the result obtained for $b_{0,0}$ is different from that in [8] and [11]. Similarly, different τ and p.

2.2 Throughput Analysis

Once τ is known, the probability P_{tr} that there is at least one transmission in a slot time given n wireless stations, and the probability P_s that a transmission is successful are readily obtained as

$$P_{tr} = 1-(1-\tau)^{n} \tag{11}$$

$$P_s = \frac{n\tau(1-\tau)^{n-1}}{P_{tr}} = \frac{n\tau(1-\tau)^{n-1}}{1-(1-\tau)^{n}} \tag{12}$$

We are finally in the position to determine the normalized system throughput, S, defined as the fraction of time the channel is used to successfully transmit payload bits, and express it as a fraction of a slot time

$$S = \frac{E[\text{payload infomation transmitted in a slot time}]}{E[\text{length of a slot time}]} = \frac{P_s P_{tr} E[P]}{(1-P_{tr})\sigma + P_s P_{tr} T_s + (1-P_s) P_{tr} T_c} \tag{13}$$

Where $E[P]$ is the average packet length, T_s is the average time the channel is sensed busy because of a successful transmission, T_c is the average time the channel is sensed busy by the stations during a collision, and σ is the duration of an empty slot time. The times $E[P]$, T_s and T_c must be measured in slot times.

Let packet header be $H = PHY_{hdr} + MAC_{hdr}$ and propagation delay be δ. Then we must have the following expression for ACK timeout effect, which are same as in [11].

$$\begin{cases} T_s^{bas} = DIFS + H + E[P] + \delta + SIFS + ACK + \delta \\ T_c^{bas} = DIFS + H + E[P^*] + SIFS + ACK \end{cases} \tag{14}$$

Where *bas* means basic access method and $E[P^*]$ is the average length of the longest packet payload involved in a collision. In our cases, all the packets have the same fixed size, $E[P] = E[P^*] = P$.

For the RTS/CTS access method,

$$\begin{cases} T_s^{rts} = DIFS + RTS + SIFS + \delta + CTS + SIFS + \delta \\ \qquad\quad + H + E[P] + \delta + SIFS + ACK + \delta \\ T_c^{rts} = DIFS + RTS + SIFS + CTS \end{cases} \tag{15}$$

We suppose collision occurs only between RTS frames and consider only the CTS timeout effect.

2.3 Delay Analysis

The delay discussed here refers to the medium access delay; it includes the total time from a station beginning to contend the channel for a transmission to the data frame being transmitted successfully. The delay represents the interval during which two contiguous data frames in a station are transmitted successfully.

Let D be the delay defined above. Then D is computed as

$$D = T_s + D_s + D_c + T_{slot} \tag{16}$$

In equation (16), T_s is the time for a successful transmission, D_s is the average time the channel is sensed busy because of successful transmission by other stations, D_c is the average time the channel is sensed busy because of collisions, T_{slot} is the total time of idle slots, which includes the total backoff time of successful transmissions and collisions by each station.

During the interval of two contiguous successful transmissions in a station, the time for a successful transmission in each other station is $T_s N_s$, where N_s is the number of successful transmissions by other stations. For a long enough period time, each station transmits data frame successfully with the same probability. Therefore, during two contiguous successful transmissions in one station, each station must have a successful transmission. We have $N_s = n\text{-}1$. Then we obtain

$$D_s = T_s(n-1) \tag{17}$$

Let P_c be the probability that a collision occurs on the channel, which is

$$P_c = 1 - P_s \tag{18}$$

Let N_c be the number of contiguous collisions, and we have

$$P\{N_c = i\} = P_c^i P_s = (1 - P_s)^i P_s \qquad i = 0, 1, 2, \cdots. \tag{19}$$

The mean of N_c is

$$E[N_c] = \sum_i iP\{N_c = i\} = \frac{1 - P_s}{P_s} = \frac{1 - (1-\tau)^n - n\tau(1-\tau)^{n-1}}{n\tau(1-\tau)^{n-1}} \tag{20}$$

Now consider the whole network. There are $E[N_c]$ contiguous collisions between two random contiguous successful transmissions. According to the analysis above, there are n successful transmissions in the period of a time D. Therefore, we have,

$$D_c = nE[N_c]T_c = \frac{1 - (1-\tau)^n - n\tau(1-\tau)^{n-1}}{\tau(1-\tau)^{n-1}} T_c \tag{21}$$

Let N_{slot} be the number of contiguous idle slots in a backoff interval, then

$$P\{N_{slot} = i\} = (1 - P_{tr})^i P_{tr} \qquad i = 0, 1, 2, \cdots. \tag{22}$$

The mean of N_{slot} is

$$E[N_{slot}] = \frac{1 - P_{tr}}{P_{tr}} = \frac{(1-\tau)^n}{1 - (1-\tau)^n} \tag{23}$$

Since there is a backoff interval before each successful transmission or collision and n successful transmissions and N_c collisions in the period of time of D according to the analysis above, the total time of idle slot is

$$T_{slot} = (E[N_c] + n)E[N_{slot}]\sigma = \frac{1-\tau}{\tau}\sigma \tag{24}$$

Using equation (14)-(17), (21) and (24), we can get the medium access delay as

$$D = nT_s + \frac{1 - (1-\tau)^n - n\tau(1-\tau)^{n-1}}{\tau(1-\tau)^{n-1}} T_c + \frac{1-\tau}{\tau}\sigma \tag{25}$$

3 Numerical Results

We assume each station has enough data to send to obtain the saturation throughput with the new backoff scheme. We change the number of stations to check the effect of throughput degradation due to increased collision probability.

All the parameters used in the analysis follow the same parameters in [11] for DSSS summarized in Table I. We assume the application data payload is 1000 bytes, IP header and UDP header are 20 and 8 bytes, respectively. Thus the packet payload at MAC layer becomes 1028 bytes.

Table 1. System parameters for MAC and DSSS PHY Layer

Packet payload	8224 bits
MAC header	224 bits
PHY header	192 bits
ACK	112 bits + PHY header
RTS	160 bits + PHY header
CTS	112 bits + PHY header
Channel bit rate	1Mbps
Propagation delay	1us
Slot time	20 us
SIFS	10 us
DIFS	50 us

Fig. 5 shows the results of basic access method. From the figure, observe that the proposed scheme shows better saturation throughput than that of [11]. The results with RTS/CTS access method are similar to those of basic access method as shown in Fig 6. Note that the scale of vertical axis of Fig. 6 is slightly different from that of Fig. 5. From the comparisons, the proposed scheme is verified to reduce the chance of collision.

Fig. 7 shows that the delay of the basic access method highly depends on the initial contention window size, W, while Fig. 8 shows that RTS/CTS access method does not. In addition, the proposed scheme allows lower delay than [11]. The improvement in the delay is due to reduced collision probability between active stations and collided stations by using the proposed backoff time counter scheme.

Fig. 9 and 10 show the delay versus the number of wireless stations for the basic access method and RTS/CTS access method, respectively. The initial contention window size is set to 32. The delay for the basic access method is larger than that for the RTS/CTS access method under the same condition. This may be caused by the lager collision delay in basic access method. Note that delay for both the access methods strongly depends on the number of stations in the network.

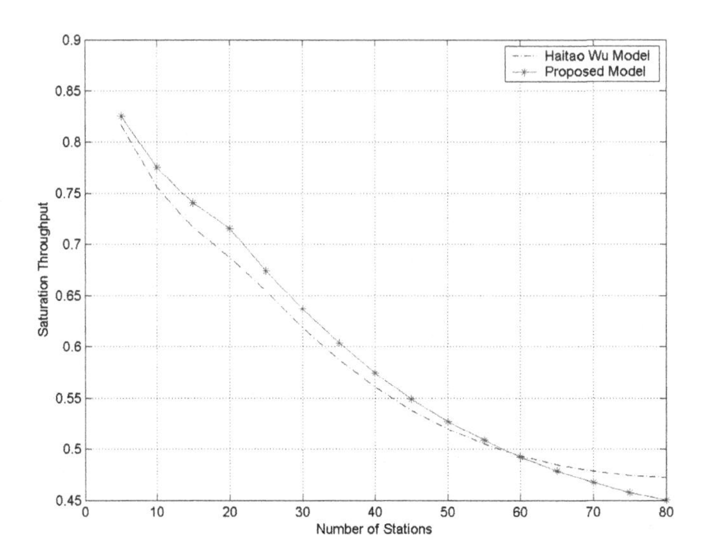

Fig. 5. Throughput: basic access method.

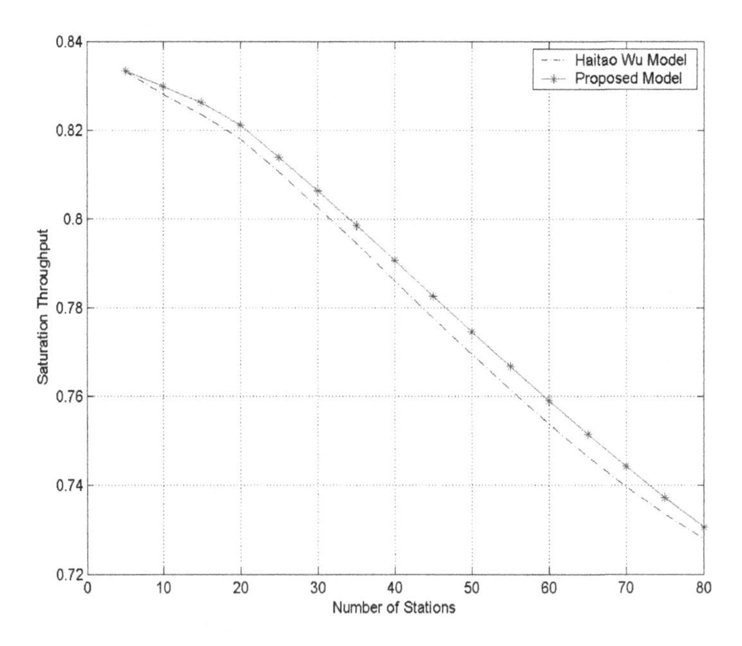

Fig. 6. Throughput: RTS/CTS access method.

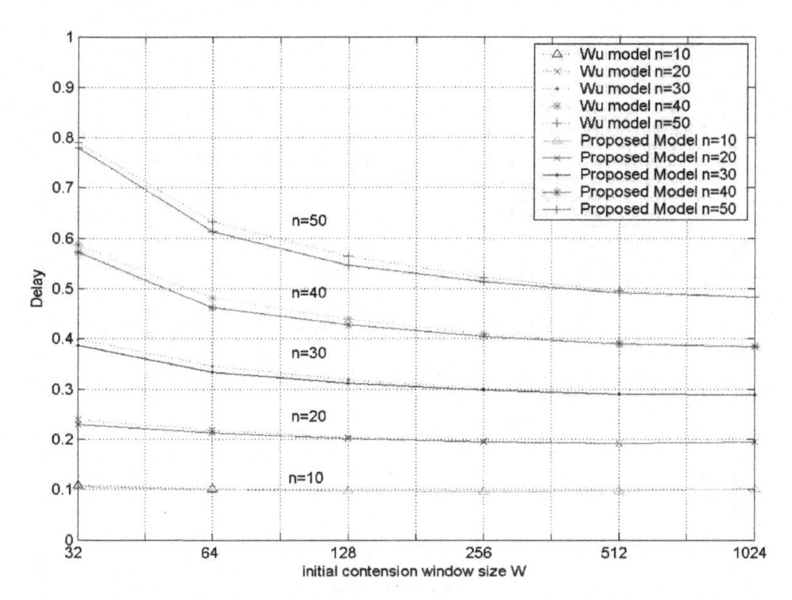

Fig. 7. Delay verse initial contention window size W for basic access method.

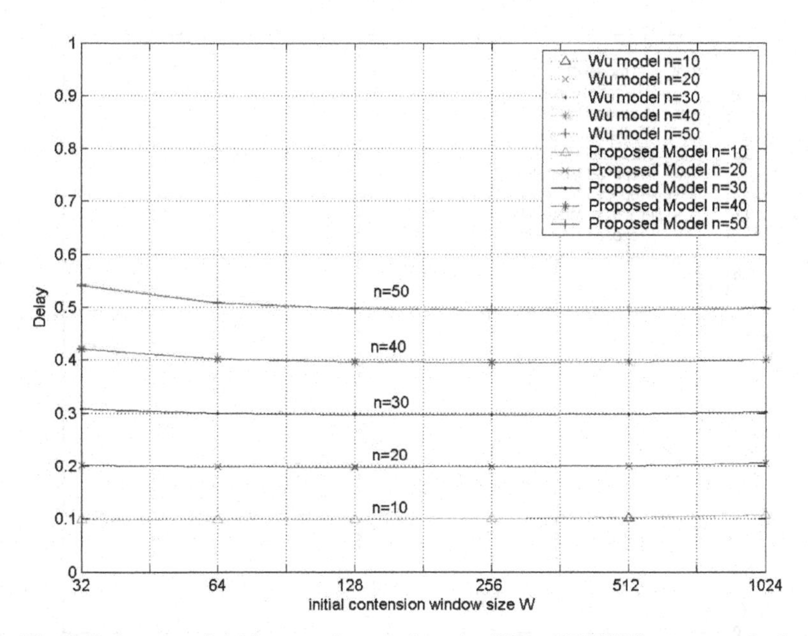

Fig. 8. Delay verse initial contention window size W for RTS/CTS access method.

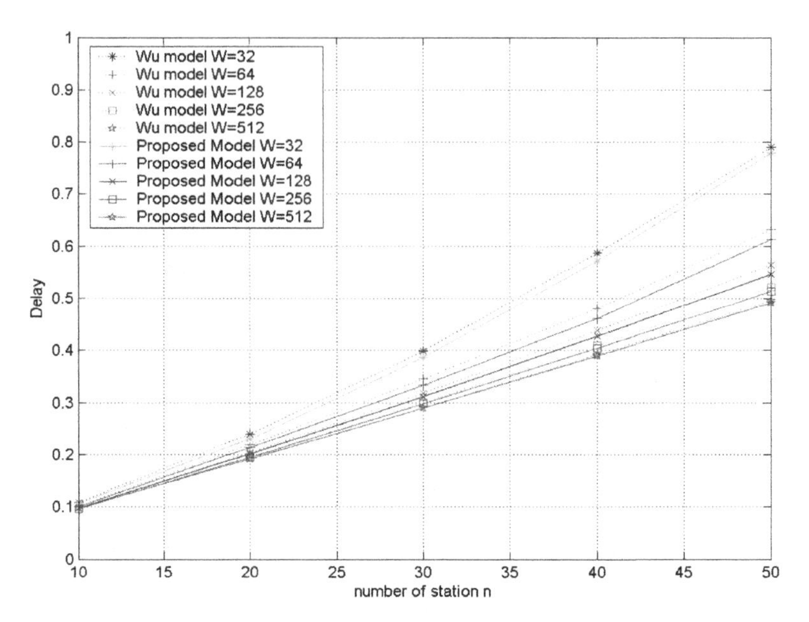

Fig. 9. Delay verse the number of stations for basic access method

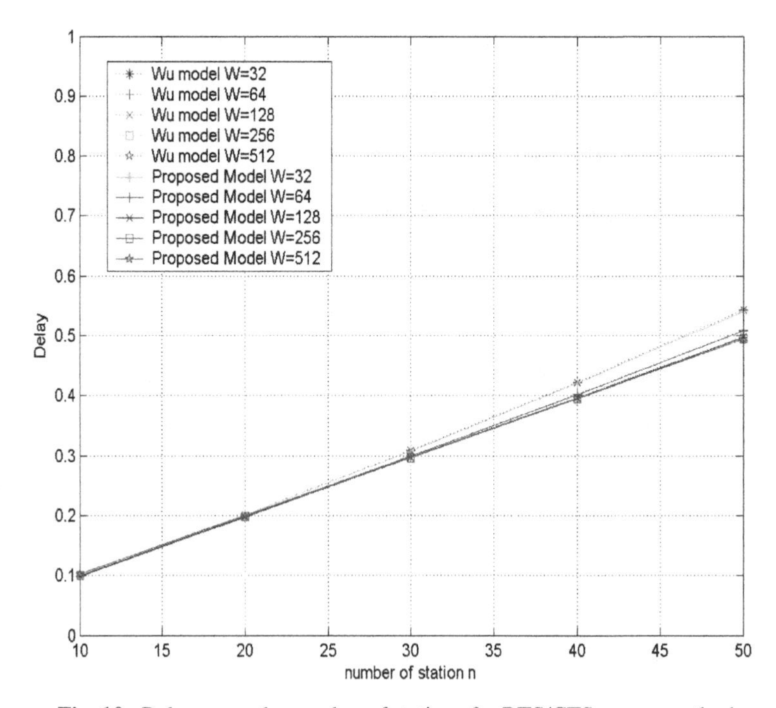

Fig. 10. Delay verse the number of stations for RTS/CTS access method

4 Conclusions

This paper has proposed a new backoff scheme for CSMA/CA protocols, analyzed the throughput and delay, and based on a bi-dimensional discrete Markov chain. The analytical results showed that the proposed scheme allows higher throughput and lower delay compared to the legacy mechanism [11] at high traffic conditions. Comparing with the model in [11], even though the proposed scheme shows better performance, fairness remains to be solved. The reason for unfairness in the new scheme is due to the fact that the selection is not made within the full range of the congestion window. Future works will be focused on the enhancement of the proposed scheme and the corresponding analytical model by guaranteeing the fairness.

References

[1] IEEE standard for Wireless LAN Medium Access Control (MAC) and Physical Layer (PHY) specifications, ISO/IEC 8802-11: 1999 Edition, Aug. 1999
[2] B. P. Crow, J. G. Kim, et. "IEEE 802.11 Wireless Local Area Networks", IEEE Communication magazine. Sept. 1997
[3] J. Weinmiller, M. Schlager, A. Festag and A. Wolisz, "Performance study of access control in wireless LANs IEEE 802.11 DFWMAC and ETSI RES 10 HIPERLAN", Mobile Networks and Applications, pp. 55-67, Vol. 2, 1997
[4] F. Cali, M. Conti and E. Gregori, "IEEE 802.11 wireless LAN: Capacity analysis and protocol enhancement", INFOCOM'98, 1998
[5] H. S. Chhaya and S. Gupta, "Performance modeling of asynchronous data transfer methods of IEEE 802.11 MAC protocol", Wireless Networks, pp. 217-234, Vol. 3, 1997
[6] G. Bianchi, "IEEE 802.11 – Saturation Throughput Analysis", IEEE Communications Letters, pp. 318-320, Vol. 2, No. 12, Dec. 1998
[7] G. Bianchi, L. Fratta and M. Oliveri, "Performance analysis of IEEE 802.11 CSMA/CA medium access control protocol", Proc. IEEE PIMRC, pp. 407-411. Oct. 1996
[8] G. Bianchi, "Performance Analysis of the IEEE 802.11 Distributed Coordination Function", IEEE Journal on Selected Area in Communication. pp. 514-519, Vol. 18, No. 3, Mar. 2000
[9] K. C. Huang K. C. Chen, "Interference analysis of nonpersistent CSMA with hidden terminals in multicell wireless data networks", Proc. IEEE PIMRC, pp. 907-911. Sep. 1995
[10] H. Wu, Y. Peng, K. Long and S. Cheng, "A Simple model of IEEE 802.11 Wireless LAN", Proceedings. ICII 2001 - Beijing. 2001
[11] H. Wu, Y. Peng, K. Long and J. Ma, "Performance of reliable transport protocol over IEEE 802.11 wireless LAN: analysis and enhancement", Proc. IEEE Inforcom'02. 2002
[12] G. Wang, Y. Shu, and L. Zhang, "Delay Analysis of the IEEE 802.11 DCF", Proc. IEEE PIMRC, pp. 1737-1741, 2003
[13] K. H. Choi, K.S. Jang and D.R. Shin, "Delay and Collision Reduction Mechanism for Distributed Fair Scheduling in Wireless LANs", ICCSA 2004 Italy Perugia
[14] H. J. Shin and D. R. Shin, "Throughput Analysis Based on a New Backoff Scheme", CIC2004, pp. 79-82, LasVegas USA, 2004

Selective Channel Scanning for Fast Handoff in Wireless LAN Using Neighbor Graph

Sang-Hee Park, Hye-Soo Kim, Chun-Su Park, Jae-Won Kim, and Sung-Jea Ko

Department of Electronics Engineering, Korea University,
Anam-Dong Sungbuk-Ku, Seoul, Korea
Tel: +82-2-3290-3672
{jerry, hyesoo, cspark, jw9557, sjko}@dali.korea.ac.kr

Abstract. Handoff at the link layer 2 (L2) consists of three phases: scanning, authentication, and reassociation. Among the three phases, scanning is dominant in terms of time delay. Thus, in this paper, we propose an improved scanning mechanism to minimize the disconnected time while the wireless station (STA) changes the associated access points (APs). According to IEEE 802.11 standard, the STA has to scan all channels in the scanning phase. In this paper, based on the neighbor graph (NG), we introduce a selective channel scanning method for fast handoff in which the STA scans only channels selected by the NG. Experimental results show that the proposed method reduces the scanning delay drastically.

1 Introduction

In recent years, wireless local area network (WLAN) with wide bandwidth and low cost has emerged as a competitive technology to adapt the user with strong desire for mobile computing. The main issue of mobile computing is handoff management. Especially, for real-time multimedia service such as VoIP, the long handoff delay has to be reduced. Many techniques [1]-[4] have been proposed by developing new network protocols or designing new algorithms. Their approaches are categorized as network layer (L3), L2, and physical layer (PHY).

One of the previous works based on L3 uses the reactive context transfer mechanism [1],[2]. This mechanism is designed solely for access routers (ARs) and is reactive rather than pro-active. On the other hand, Nakhjiri [5] proposed a general purpose context transfer mechanism, called SEAMOBY, without detailing transfer triggers. In SEAMOBY, a generic framework for either reactive or pro-active context transfer is provided, though the framework does not define a method to implement either reactive or pro-active context transfer.

To reduce the L2 handoff delay in WLAN using inter access point protocol (IAPP) [7], an algorithm on the context transfer mechanism utilizing the NG [6] was suggested in [7]. But originally, IAPP was only reactive in nature and creates an additional delay in handoff. Thus, this algorithm can not shorten the original L2 handoff delay.

I. Niemegeers and S. Heemstra de Groot (Eds.): PWC 2004, LNCS 3260, pp. 194–203, 2004.

One approach on PHY is the method using two transceivers. In this method, an STA has two wireless network interface cards (WNICs), one for keeping connection to current AP and the other for scanning channels to search alternative APs [8].

In this paper, we propose a selective channel scanning mechanism using the NG to solve the L2 handoff delay. In the proposed mechanism, the STA scans only channels selected by the NG without scanning all the channels. And on receiving a *ProbeResponse* message, the STA scans the next channels without waiting for the pre-defined time. Therefore, the delay incurred during the scanning phase can be reduced.

This paper is organized as follows. Section 2 describes the operations in IEEE 802.11 [9]. In Section 3, the proposed algorithm is presented. Finally, Section 4 shows the results experimented on our test platform and presents brief conclusion comments.

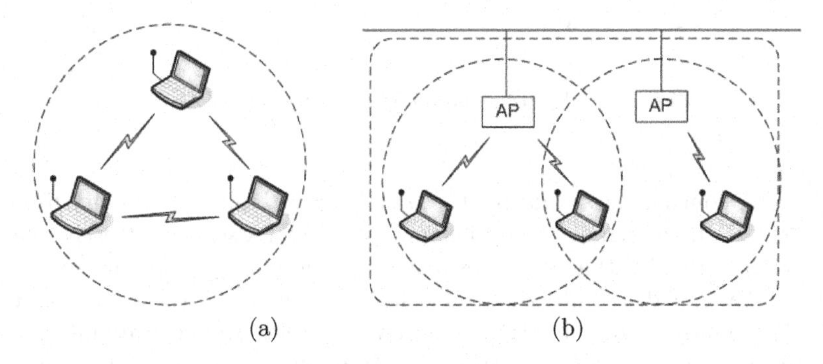

Fig. 1. Types of WLAN (a) ad hoc mode(b) infrastructure mode.

2 IEEE 802.11

As shown in Fig. 1, IEEE 802.11 MAC specification allows for two modes of operation: ad hoc and infrastructure modes. In the ad hoc mode, two or more STAs recognize each other through beacons and establish a peer-to-peer relationship. In this configuration, STAs communicate with each other without the use of an AP or other infrastructure. The ad hoc mode connects STAs when there is no AP near the STAs, when the AP rejects an association due to failed authentication, or when the STAs are explicitly configured to use the ad hoc mode. In the infrastructure mode, STAs not only communicate with each other through the AP but also use the AP to access the resource of the wired network which can be an intranet or internet depending on the placement of the AP. The basic building block of IEEE 802.11 network is a basic service set (BSS) consisting of a group of STAs that communicate with each other. A set of two or more APs connected to the same wired network is known as an extended service set (ESS) which is identified by its service set identifier (SSID).

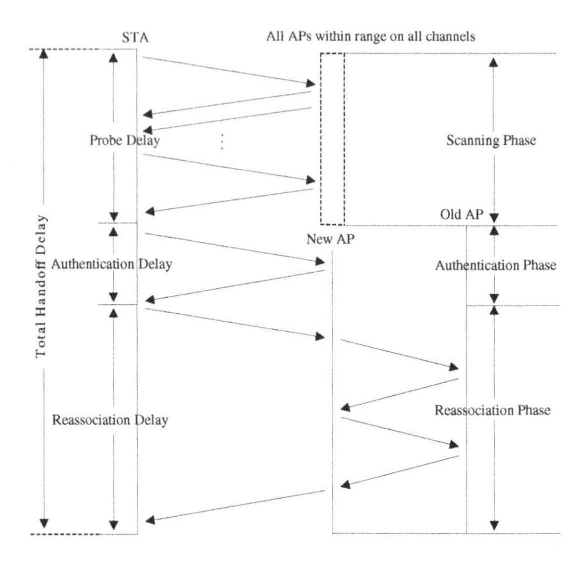

Fig. 2. IEEE 802.11 handoff procedure with IAPP.

The STA continuously monitors the signal strength and link quality from the associated AP. If the signal strength is too low, the STA scans all the channels to find a neighboring AP that produces a stronger signal. By switching to another AP, the STA can distribute the traffic load and increase the performance of other STAs. During handoff, PHY connectivity is released and state information is transferred from one AP to another AP. Handoff can be caused by one of the two types of transition including BSS and ESS transitions. In the BSS transition, an STA moves within an extended service area (ESA) that constructs an ESS. In this case, an infrastructure does not need to be aware of the location of the STA. On the other hand, in the ESS transition, the STA moves from one ESS to another ESS. Except for allowing the STA to associate with an AP in the second ESS, IEEE 802.11 does not support this type of transition. In this paper, we focus on BSS transition.

2.1 Handoff Procedure

The complete handoff procedure can be divided into three distinct logical phases: scanning, authentication, and reassociation. During the first phase, an STA scans for APs by either sending *ProbeRequest* messages (Active Scanning) or listening for *Beacon* messages (Passive Scanning). After scanning all the channels, the STA selects an AP using the received signal strength indication (RSSI), link quality, and etc. The STA exchanges IEEE 802.11 authentication messages with the selected AP. Finally, if the AP authenticates the STA, an association moves from an old AP to a new AP as following steps:

(1) An STA issues a *ReassociationRequest* message to a new AP. The new AP must communicate with the old AP to confirm that a previous association existed;
(2) The new AP processes the *ReassociationRequest*;
(3) The new AP contacts the old AP to finish the reassociation procedure with IAPP;
(4) The old AP sends any buffered frames for the STA to the new AP;
(5) The new AP begins processing frames for the STA.

The delay incurred during these three phases is referred to as the L2 handoff delay, that consists of probe delay, authentication delay, and reassociation delay. Figure 2 shows the three phases, delays, and messages exchanged in each phase.

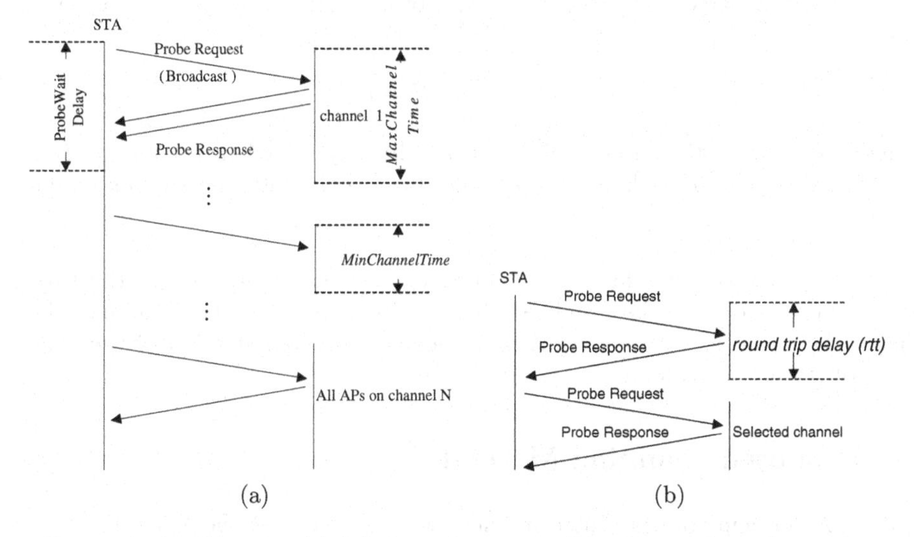

Fig. 3. Active Scanning (a) full channel scanning (b) selective channel scanning.

2.2 Passive and Active Scanning Modes

The STA operates in either a passive scanning mode or an active scanning mode depending on the current value of the *ScanMode* parameter of the MLME-SCAN.request primitive. To become a member of a particular ESS using the passive scanning mode, the STA scans for *Beacon* messages containing SSID indicating that the *Beacon* message comes from an infrastructure BSS or independent basic service set (IBSS). On the other hand, the active scanning mode attempts to find the network rather than listening for the network to announce itself. STAs use active scanning mode with the following procedure:

(1) Move to the channel and wait for either an indication of an incoming frame or for the *ProbeTimer* to expire. If an incoming frame is detected, the channel is in use and can be probed;

(2) Gain access to medium using the basic distributed coordination function (DCF) access procedure and send a *ProbeRequest* message;
(3) Wait for the *MinChannelTime* to elapse.
  a. If the medium has never been busy, there is no network. Move to the next channel.
  b. If the medium was busy during *MinChannelTime* interval, wait until *MaxChannelTime* and process any *ProbeResponse* messages.

When all the channels in the *ChannelList* are scanned, the MLME issues an MLME-SCAN.confirm primitive with the *BSSDescriptionSet* containing all information gathered during the active scanning. Figure 3 (a) shows messages, *MaxChannelTime*, and *MinChannelTime* for the active scanning.

During the active scanning, the bound of scanning delay can be calculated as

$$N \times T_b \leq t \leq N \times T_t, \tag{1}$$

where N is the total number of channels which can used in a country, T_b is *MinChannelTime*, T_t is *MaxChannelTime*, and t is the total measured scanning delay.

Mishra [10] showed that the scanning delay is dominant among the three delays. Thus, to solve the problem of the L2 handoff delay, the scanning delay has to be reduced. Therefore, our paper focuses on the reduction of the delay time of the active scanning.

3 Proposed Scanning Method

Before introducing our proposed method, we briefly review the NG. The NG is an undirected graph where each edge represents a mobility path between APs. Therefore, for a given edge, the neighbors of the edge become the set of potential next APs. Figure 4 shows the concept of the NG with five APs.

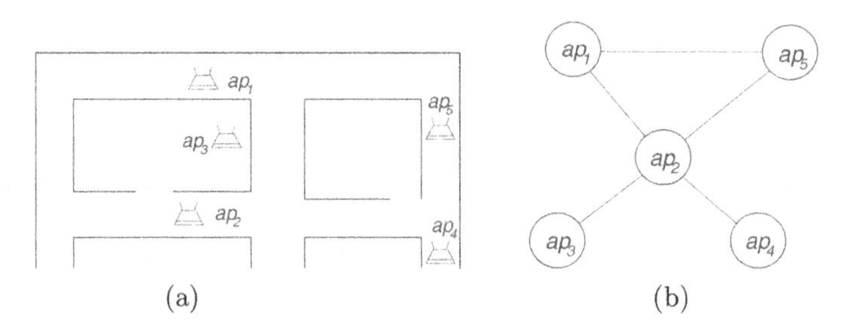

Fig. 4. Concept of neighbor graph (a) placement of APs (b) corresponding neighbor graph.

The undirected graph representing the NG is defined as

$$\begin{aligned} &G = (V, E), \\ &V = \{ap_1, ap_2, \ldots, ap_i\}, \\ &e = (ap_i, ap_j), \\ &N(ap_i) = \{ap_{ik} : ap_{ik} \in V, (ap_i, ap_{ik}) \in E\}, \end{aligned} \tag{2}$$

where G is the data structure of NG, V is a set containing all APs, E is a set consisting of edges, e's, and N is the neighbor APs near an AP.

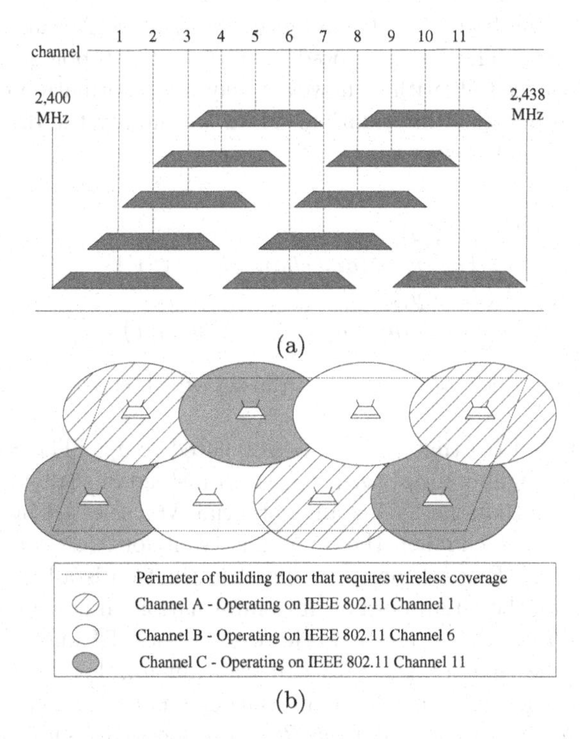

Fig. 5. Selecting channel frequencies for APs (a) channel overlap for 802.11b APs (b) example of channel allocation.

The NG is configured for each AP manually or is automatically generated by an individual AP over time. The NG can be automatically generated by the following algorithm with the management messages of IEEE 802.11.

(1) If an STA sends *Reassociate Request* to AP_i with *old-ap* $= AP_j$, then create new neighbors (i, j) (i.e. an entry in AP_i, for j and vice versa);
(2) Learn costs only one '*high latency handoff*' per edge in the graph;

(3) Enable mobility of APs which can be extended to wireless networks with an ad hoc backbone infrastructure.

In general, the STA scans all channels from the first channel to the last channel, because it is not aware of the deployment of APs near the STA. However, as shown in Fig. 5 (a), the STA can use only three channels at the same site (in case of United States, channel 1, 6, 11) because of the interference between adjacent APs. Thus, all channels are not occupied by neighbor APs to permit efficient operation although multiple APs are operating at the same site. If channels of APs existing near the STA are known, the STA does not need to scan all channels as shown in Fig. 5 (b). In this section, based on the NG, we introduce a selective channel scanning method for fast handoff in which an STA scans only channels selected by the NG. The NG proposed in [6] uses the topological information on APs. Our proposed algorithm, however, requires information on channels of APs as well as topological information. Thus, we modify the data structure of NG defined in (2) as follows:

$$\begin{aligned}
&G' = (V', E), \\
&V' = \{v_i : v_i = (ap_i, channel), v_i \in V\}, \\
&e = (ap_i, ap_j), \\
&N(ap_i) = \{ap_{ik} : ap_{ik} \in V', (ap_i, ap_{ik}) \in E\},
\end{aligned} \tag{3}$$

where G' is the modified NG, and V' is the set which consists of APs and their channels.

Assume in Fig. 4 (b) that an STA is associated to ap_2. The STA scans only 4 channels of its neighbors (ap_1, ap_3, ap_4, ap_5) instead of scanning all channels. Figure 3 (b) shows that the STA scans potential APs selected by the NG.

In order to scan a channel, the STA must wait for *MaxChannelTime* after transmitting a *ProbeRequest* message whose destination is all APs as shown in Fig. 3 (a). Because the STA does not have information on how many APs would respond to the *ProbeRequest* message. However, if the STA knows the number of APs occupying the channel, the STA can transmit another *ProbeRequest* message to scan the next channel without waiting for *MaxChannelTime* when it receives the number of predicted *ProbeResponse* messages. Our mechanism can be summarized as follows:

(1) Retrieve the NG information (neighbor APs and their channels) according to the current AP;
(2) Select one channel;
(3) Transmit a *ProbeRequeset* message to the selected channel and start *ProbeTimer*;
(4) Wait for *ProbeResponse* messages not until *ProbeTimer* reaches *MaxChannelTime*, but until the number of the received *ProbeResponse* messages becomes the same as the number of potential APs in the channel.
  a. If the channels to be scanned remain, process (1).
  b. Otherwise, start the authentication phase.

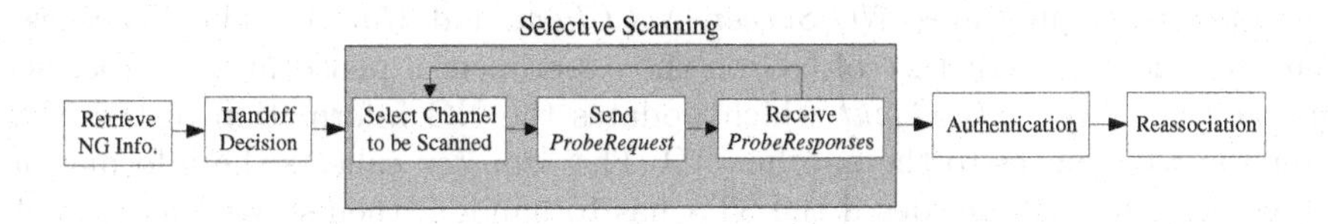

Fig. 6. Block diagram of the proposed method.

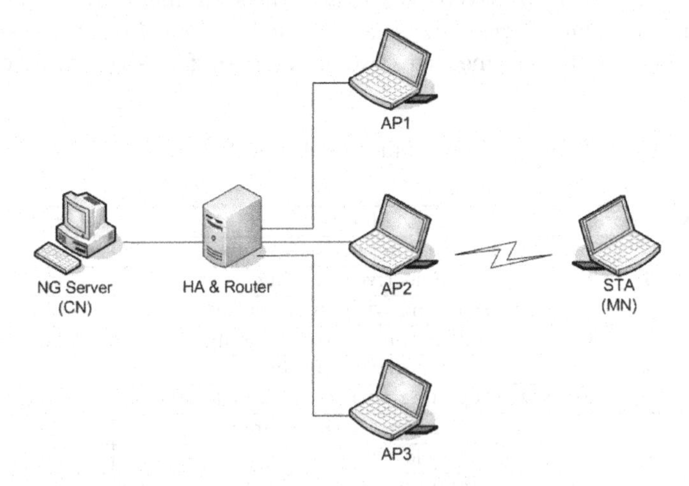

Fig. 7. Experimental platform.

Figure 6 shows the block diagram of the proposed method described above. With the proposed scanning algorithm, the scanning delay can be expressed as

$$\begin{aligned} t &= N' \times (t_{proc} + rtt) + (N - N')/2 \times (t_{proc} + rtt) + \alpha \\ &= (N + N')/2 \times (t_{proc} + rtt) + \alpha, \end{aligned} \tag{4}$$

where N is the number of the potential APs, N' is the number of channels selected by the NG, t_{proc} is the processing time of MAC and PHY headers, rtt is the round trip time, and α is the message processing time. While *MinChannelTime* and *MaxChannelTime* in (1) is tens of milliseconds, t_{proc} is a few milliseconds.

4 Experimental Results

Figure 7 shows our experimental platform consisting of an STA, APs, router, and correspondent node (CN). To exchange the NG information, socket interface is used, and Mobile IPv6 is applied to maintain L3 connectivity while experimenting. The device driver of a common WNIC was modified such that the STA operates as an AP. And emulated APs perform the foreign agents on our experimental platform. To simulate the operation of the proposed mechanism, we developed three programs: *NG Server*, *NG Client*, and *Monitor*. The *NG Server* manages the data structure of NG on the experimental platform and processes the request of the *NG Client* which updates the NG information of the STA after the STA moves to the another AP. The *Monitor* collects the information on wireless channel, decides if the STA has to handoff, then starts the handoff. In Fig. 7, *NG Client* and *Monitor* are deployed at the STA and *NG Server* is deployed at the CN, respectively. We defined the scanning delay as the duration between the first *ProbeRequest* message and the last *ProbeResponse* message. To capture these two kinds of messages, we used the sniffer program, AiroPeek.

Table 1. Average probe delay of each method.

Neighbors	Method	delay [ms]
1	Full channel active scanning	322
	Selective scanning	55
	Selective scanning without *MaxChannelTime*	12
2	Full channel active scanning	322
	Selective scanning	109
	Selective scanning without *MaxChannelTime*	21
3	Full channel active scanning	322
	Selective scanning	145
	Selective scanning without *MaxChannelTime*	30
4	Full channel active scanning	322
	Selective scanning	190
	Selective scanning without *MaxChannelTime*	39

To evaluate the proposed algorithm, we experimented with three mechanisms: full channel active scanning mechanism; selective scanning; and selective scanning without *MaxChannelTime* for the number of neighbors. As shown in Table 1, the scanning delay by the selective scanning is shorter than the full channel active scanning. Note that the selective scanning without *MaxChannelTime* produces a smallest delay time among the three mechanisms.

5 Conclusions

we have introduced the selective channel scanning method for fast handoff in which an STA scans only channels selected by the NG. The experimental results showed that the proposed method can reduce the overall handoff delay drastically. Therefore, we expect that the proposed algorithm can be an useful alternative to the existing full channel active scanning due to its reduced delay time.

References

1. Koodli, R., Perkins, C.: Fast Handovers and Context Relocation in Mobile Networks. ACM SIGCOMM Computer Communication Review. Vol. 31 (2001)
2. Koodli, R.: Fast Handovers for Mobile IPv6. Internet Draft : draft-ietf-mipshop-fast-mipv6-01.txt (2004)
3. Cornall, T., Pentland, B., Pang, K.: Improved Handover Performance in wireless Mobile IPv6. ICCS. Vol. 2 (2003) 857–861
4. Park, S. H., Choi, Y. H.: Fast Inter-AP Handoff Using Predictive Authentication Scheme in a Public Wireless LAN. IEEE Networks ICN. (2002)
5. Nakhjiri, M., Perkins, C., Koodli, R.: Context Transfer Protocol. Internet Draft : draft-ietf-seamoby0ctp-09.txt. (2003)
6. Mishra, A., Shin, M. H., Albaugh, W.: Context Caching using Neighbor Graphs for Fast Handoff in a Wireless. Computer Science Technical Report CS-TR-4477. (2003)
7. IEEE: Recommended Practice for Multi-Vendor Access Point Interoperability via an Inter-Access Point Protocol Across Distribution Systems Supporting IEEE 802.1 Operation. IEEE Standard 802.11. (2003)
8. Ohta, M.: Smooth Handover over IEEE 802.11 Wireless LAN. Internet Draft : draft-ohta-smooth-handover-wlan-00.txt. (2002)
9. IEEE: Part 11: Wireless LAN Medium Access Control (MAC) and Physical Layer (PHY) Specifications. IEEE Standard 802.11. (1999)
10. Mishra, A., Shin, M. H., Albaugh, W.: An Empirical Analysis of the IEEE 802.11 MAC Layer Handoff Process. ACM SIGCOMM Computer Communication Review. Vol. 3 (2003) 93–102

Throughput Analysis of ETSI BRAN HIPERLAN/2 MAC Protocol Taking Guard Timing Spaces into Consideration

You-Chang Ko[1], Eui-Seok Hwang[2], Jeong-Shik Dong[2], Hyong-Woo Lee[3], and Choong-Ho Cho[2]

[1] LG Electronics Inc. UMTS Handsets Lab,
Korea
yckoe@lge.com

[2] Dept. of Computer Science & Technology,
KOREA UNIV., Korea
{ushwang, bwind, chcho}@korea.ac.kr

[3] Dept. of Electronics & Information Engineering,
KOREA UNIV., Korea
hwlee@korea.ac.kr

Abstract. In this paper we examine the effects of the required portions of guard timing spaces in a MAC frame of ETSI BRAN HIPERLAN/2 system such as inter-mobile guard timing space in UL(Up Link) duration, inter-RCH(Random CHannel) guard timing space, sector switch guard timing space. In particular, we calculate the number of OFDM(Orthogonal Frequency Division Multiplexing) symbols required for these guard timing spaces in a MAC frame. We then evaluate the throughput of HIPERLAN/2 system as we vary parameter such as the guard time values defined in [2], the number of DLCCs(Data Link Control Connections), and the number of RCHs. Finally we show by numerical results that the portions for the total summation of required guard timing spaces in a MAC frame are not negligible, and that they should be properly considered when trying to evaluate the performance of MAC protocol of HIPERLAN/2 system and also when determining the number of RCHs as well as the number of DLCCs in UL PDU trains at an AP/CC(Access Point/Central Controller).

1 Introduction

HIPERLAN/2, providing high-speed communications between mobile terminal and various broadband infrastructure networks and representing a centralized access of MAC protocol, is an ETSI BRAN standard with a 2ms duration of MAC frame. There have been studies which analyze throughput of HIPERLAN/2 MAC layer[4][5]. Their first step to analyze the throughput of MAC layer is to calculate the length of the LCHs(Long transmit CHannels) in a number of OFDM symbols in a MAC frame in such a manner that the overhead durations(preambles and signals) for each channels and PDU trains are subtracted from the whole duration of a MAC frame. In these studies, however, they omit to subtract from the total length of a MAC frame some of

I. Niemegeers and S. Heemstra de Groot (Eds.): PWC 2004, LNCS 3260, pp. 204–214, 2004.

non-negligible components of overheads in a MAC. Firstly, they do not take into account some guard timing spaces in order to cope with propagation delay such as inter-mobile guard timing space in UL duration, inter-RCH guard timing space, and sector switch guard timing space which is needed in a multi-sector environment. Secondly, they omit to subtract the overhead portions of DL(Down Link) PDU train as well as SCHs(Shot transmit CHannels) of UL PDU train from a MAC frame duration. Because of these reasons, the previous works tend to overestimate the possible length of LCHs as the pure user data path, and consequently may lead to inflating result of its throughput. Having considered the missing points of the previous works and further taken into account multi-sector environment and DiL(Direct Link) PDU train as optional, what we try to show in this paper is the effects on the throughput of the required guard timing spaces imposed on a MAC frame based on [1][2]. Specifically we observe the results as varying the number of active user's DLCC as well as guard time values defined in [2]. Finally we show by numerical results that the whole required portions of the guard timing spaces are not negligible within a MAC frame duration of 500 OFDM symbols, as the number DLCC and RCH is increased. These overheads should be properly considered when evaluating the performance of HIPERLAN/2 system, and also when determining the number of RCHs as well as the number of DLCCs in UL PDU trains at an AP/CC.

In section II we introduce system parameters to evaluate throughput of HIPERLAN/2 system in this paper. In section III we analyze the length of each PDU parts regarding guard timing spaces for throughput analysis in a MAC frame. Section IV deals with various numerical results examining the effects of the number of DLCCs as well as guard time values. Finally we end with conclusions in section V.

2 System Parameters

The PHY(physical) layer of HIPERLAN/2 is based on the modulation scheme OFDM with 52 sub-carriers whose possible modulations are BPSK(Binary Phase Shift Keying), QPSK(Quaternary Phase Shift Keying), and 16QAM(Quadrature Amplitude Modulation) for mandatory, and 64QAM for optional. In order to improve radio link capacity due to different interference situations and distances of MTs to the AP or CC, a multi-rate PHY layer is applied, where the appropriate mode can be selected by the link adaptation scheme. The mode dependent parameters are listed in TABLE I. We assume that the preambles for UL PDU trains and RCH PDUs are long preamble i.e. four OFDM symbols and each active user has two DLCCs; one is for UL and the other is for DL. We further assume that every DLCC has one SCH, which is only used to convey control user data, and also that the guard time value for RCH specified in BCCH(Broadcast Control CHannel) is concurrently applied to that for UL inter-PDU trains. We define and assume system parameters for numerical calculation as TABLE II.

Table 1. Phy layer mode dependent parameters

Modulation	Coding rate	Nominal bit rate[Mbps]	Data bits/octets per OFDM symbol
BPSK	1/2	6	23/3
BPSK	3/4	9	36/4.5
QPSK	1/2	12	48/6
QPSK	3/4	18	72/9
16QAM	9/16	27	108/13.5
16QAM	3/4	36	144/18
64QAM	3/4	54	216/27

Table 2. Parameters for numerical calculations

Parameters	Meaning	Value
L_X	The length of X PDU in a MAC frame.	Variable
BpS_X	The number of bytes coded per OFDM symbols for X PDU train.	Depend on PHY layer modes
N_{sec}	The number of sectors per AP.	1 ~ 8
N_{IE}	The number of IE blocks in whole sectors.	Variable
NX_{SCH}	The total number of SCHs in X PDU train.	1
NX_{MT}	The number of active MTs in X PDU train.	Variable
$NDiL_{MT\_Diff}$	The number of different index of transmitter between two consecutive MTs in DiL PDU train.	Variable (Optional)
S_g	Sector switch guard time	800ns
P_g	Propagation delay guard time	2.0μs ~ 12μs
UD_{OFDM}	Unit Duration of an OFDM symbol	4μs
Δ(t)	Delta step function	0, t≤1 and t, elsewhere

3 Throughput Analysis

3.1 BCH PDU Trains Length

The BCH(Broadcast CHannel) has the size of 15 octets long and shall be transmitted using BPSK with coding rate 1/2. The size of preamble shall consist of four OFDM symbols(16μs). According to Fig.1 we can calculate the length of BCH PDU train as (1) where sector switch guard timing space should be imposed on every interval between two BCHs if multi-sector is used. No sector switch guard timing space is inserted if omni-sector is used.

In the case of multi-sector, the total length of sector switch guard timing spaces in a number of OFDM symbols is calculated by dividing the sum of the number of guard times by UD_{OFDM}.

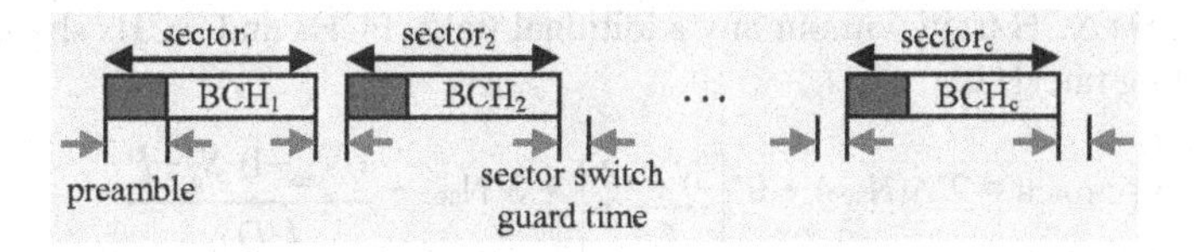

Fig. 1. The structure of BCH PDU trains

$$L_{BCH} = (PRE_{BCH}+15bytes/3)*N_{sec}+\frac{\Delta(N_{sec})\cdot S_g}{UD_{OFDM}} = 9*N_{sec}+\frac{\Delta(N_{sec})\cdot S_g}{UD_{OFDM}} \quad (1)$$

3.2 FCH+ACH PDU Trains Length

A FCH(Frame CHannel) shall be built of fixed size IE(Information Element) blocks. Every IE block shall contain three IEs, each with a length of eight octets and a CRC(Cycle Redundancy Check) of length 24 bits which shall be calculated over the three consecutive IEs.

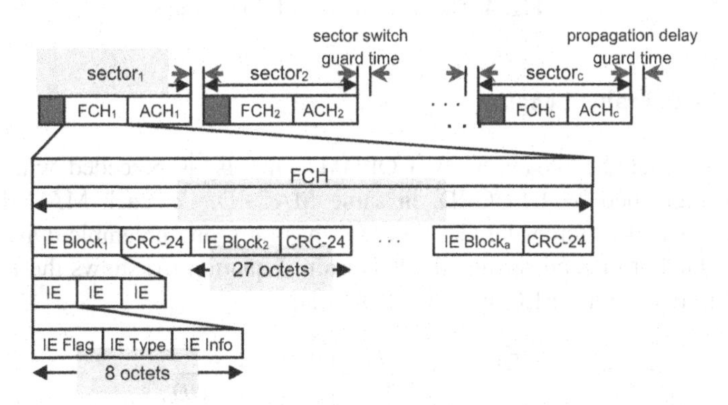

Fig. 2. The structure of FCH+ACH PDU trains

Thus a FCH shall consist of multiple of 27 octets as Fig.2 depicts. The length of ACH(Access feedback CHannel) consists of nine octets. One preamble shall be added in the beginning of each FCH+ACH PDU train if multiple sectors are used in AP and its size shall be two OFDM symbols. We assume one IE carries the signal information for one DLCC. Equation (2) shows the length of the FCH+ACH PDU train in a MAC frame. In the same fashion as in BCH PDU train, the sector switch guard timing space is set on every interval between FCH+ACH PDUs, and the propagation delay guard timing space is also added at the end of the FCH+ACH PDU train. If omni-sector is used only one preamble, size of four OFDM symbols, shall be imposed on the

BCH+FCH+ACH train without any additional ticks. FCHs and ACHs shall use BPSK with coding rate 1/2.

$$L_{FCH+ACH} = 2*\Delta(N_{sec}) + 9*\left\lceil \frac{N_{IE}\cdot 8}{24} \right\rceil + 3*N_{sec} + \frac{(N_{sec}-1)\cdot S_g + P_g}{UD_{OFDM}} \quad (2)$$

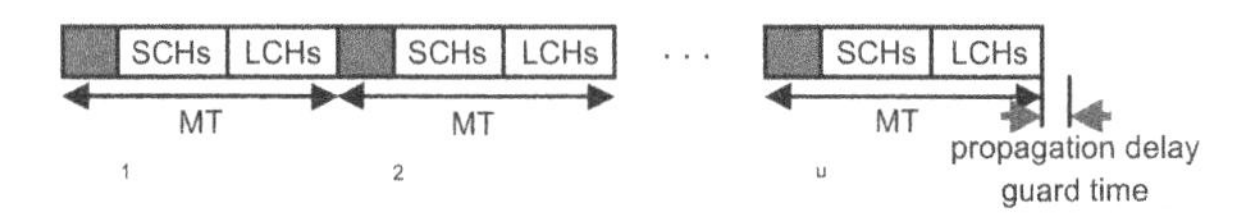

Fig. 3. The structure of DL PDU trains

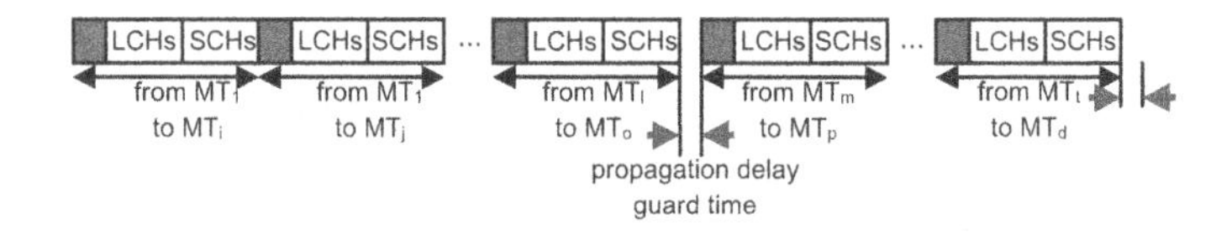

Fig. 4. The structure of DiL PDU trains

3.3 DL PDU Trains Length

Fig.3 shows that the preamble, two OFDM symbols, is preceded with every first PDU, distinguished by DLCC-ID, in same MAC-ID, i.e. each MAC-ID GROUP, which consists of different DLCC-ID PDUs has only one preamble. One guard time shall be added at the end of the DL PDU train. Equation (3) shows the length of the DL PDU train except for LCHs in this PDU train.

$$L_{DL\text{-}LCH} = 2*NDL_{MT} + \left\lceil \frac{9}{BpS_{SCH}} \right\rceil *NDL_{SCH} + \frac{P_g}{UD_{OFDM}} \quad (3)$$

3.4 DiL PDU Trains Length as Optional

Fig.4 shows that a guard timing space is needed where different TX mobile-id is positioned between two consecutive PDU trains.

Equation (4) shows the length of DiL PDU train except for LCHs in this PDU train.

$$L_{DiL\text{-}LCH} = 4*NDiL_{MT} + \left\lceil \frac{9}{BpS_{SCH}} \right\rceil *NDiL_{SCH} + \frac{(NDiL_{MT-Diff}+1)\cdot P_g}{UD_{OFDM}} \quad (4)$$

3.5 UL PDU Trains Length

Being different from DL PDU trains, UL PDU train always needs a guard timing space between two PDU trains whose MAC-IDs are different each other as in Fig.5. Equation (5) depicts the length of the UL PDU train except for LCHs duration. The guard timing spaces are directly proportional to the number of DLCCs used by active users in UL PDU trains as (5) shows.

$$L_{UL\text{-}LCH} = 4*NUL_{MT} + \left\lceil \frac{9}{BpS_{SCH}} \right\rceil *NUL_{SCH} + \frac{\Delta(NUL_{MT})\cdot P_g}{UD_{OFDM}} \quad (5)$$

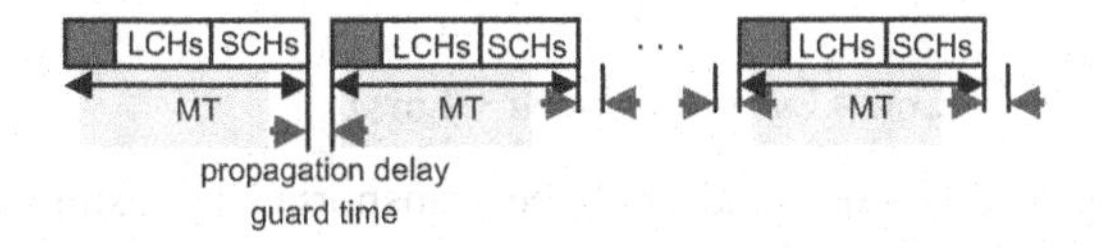

Fig. 5. The structure of UL PDU trains

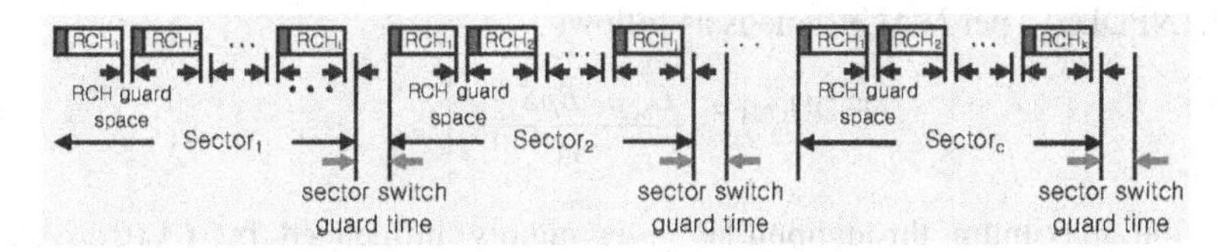

Fig. 6. The structure of RCH PDUs

3.6 RCH PDUs Length

At least one RCH shall be allocated per sector in each MAC frame and the RCHs for all sectors shall be clustered together. Between RCHs shall be space for preamble and guard timing space as in Fig.6. The RCH consists of nine octets and its format is identical to that of the SCH. The RCHs shall use BPSK with coding rate 1/2. The size of RCH's preamble is same as that of UL, i.e. four OFDM symbols for long preamble or three OFDM symbols for short one. Equation (6) gives the total length of RCHs.

$$L_{RCH} = 7*N_{RCH} + \frac{N_{RCH}\cdot P_g + \Delta(N_{sec})\cdot S_g}{UD_{OFDM}} \quad (6)$$

3.7 Total Required Guard Timing Spaces and Maximum Throughput in a MAC Frame

Using equations from (1) to (6), we can specify their durations into two parts where the first is the total length for the preambles and the signaling overheads($L_{PRE+SOH}$), and the second is the total length for the guard timing spaces(L_{GTS}). Subtracting $L_{PRE+SOH}$

and L_{GTS} from the total length of a MAC frame(L_{MF}) in a number of OFDM symbols, we can now calculate the total length of LCHs for DL, DiL, and UL PDU trains in a MAC frame as (9).

$$L_{PRE+SOH} = 12*N_{sec} + 2*\Delta(N_{sec}) + 9*\left\lceil \frac{N_{IE} \cdot 8}{24} \right\rceil + 2*(NDL_{MT} + 2*NDiL_{MT} + 2*NUL_{MT}) + \left\lceil \frac{9}{BpS_{SCH}} \right\rceil *(NDL_{SCH} + NDiL_{SCH} + NUL_{SCH}) + 7*N_{RCH} \quad (7)$$

$$L_{GTS} = \left\lceil \frac{\{2\cdot\Delta(N_{sec})+N_{sec}-1\}\cdot S_g + \{3+NDiL_{MT\_Diff}+\Delta(NUL_{MT})+N_{RCH}\}\cdot P_g}{UD_{OFDM}} \right\rceil \quad (8)$$

$$L_{LCH} = L_{MF} - (L_{PRE+SOH} + L_{GTS}) \quad (9)$$

The LCH consists of 54 octets and shall be transported by using any modulation scheme in Table 1. Here one of our main concerns is L_{GTS} imposed on MAC frame duration and its effects will be discussed in section IV. In order to get maximum throughput we use (10) as given in [4]. The total number of PDUs for LCHs($NPDU_{LCH}$) per MAC frame is as follows :

$$NPDU_{LCH} = \left\lfloor \frac{L_{LCH} \cdot BpS_{LCH}}{54} \right\rfloor \quad (10)$$

Then the maximum throughput(TP_{max}) is mainly influenced by CL(Convergence Layer) and is finally given by

$$TP_{max} = NPDU_{LCH} \cdot \frac{x}{\left\lceil \frac{x+4}{48} \right\rceil} \cdot \frac{8}{t_{frame}} \quad (11)$$

where x is the length of the user data packets in bytes.

4 Numerical Results

In this section we examine the effects on the total required guard timing spaces and the maximum throughput of the number of RCHs, the number of DLCCs, and guard time values under various modulation schemes of HIPERLAN/2 system. In Fig.7, we observe the effect of the sector switch guard timing space and inter-RCH guard timing space by increasing the number of sectors whose maximum value is eight as specified in BCCH. We assume that one RCH is included in each sector. Except for the case that maximum guard time value(12μs) is used, at most 13 OFDM symbols are required to carry the whole guard timing spaces even if the maximum number of sectors is hired. Referring to (8) we can see more clearly the effect of guard timing spaces required to support RCHs and NUL_{MT} as shown in Fig.8. We assumed that $NDiL_{MT\_Diff} = 0$ for the experiments in this paper, as it is optional.

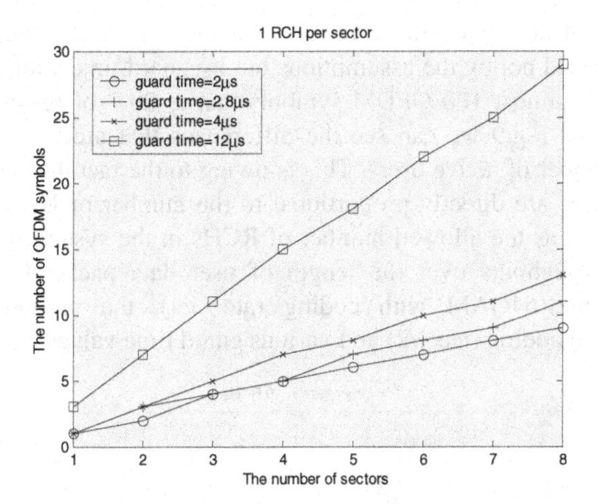

Fig. 7. The number of OFDM symbols by L_{GTS} for N_{sec}s

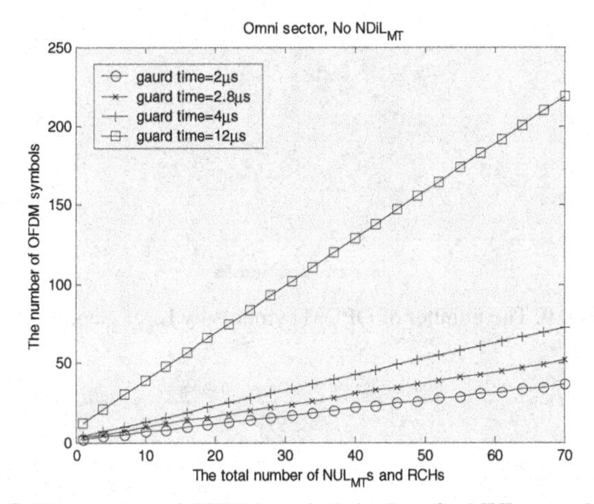

Fig. 8. The number of OFDM symbols by L_{GTS} for NUL_{MT}s and RCHs

When the sum of number becomes 70, then the required number of OFDM symbols to carry their total guard timing spaces is 37 for minimum guard time value(2μs), and 219 for maximum guard time value(12μs). Increasing the number of active users in the system we have plotted the total required number of OFDM symbols according to (7), (8) in Fig.9. We can realize the differences of how many additional number of OFDM symbols there are needed when various guard time values defined in [2] are adapted. Under the assumptions and parameters we have made in Section II, when the number of active users reaches 20 the required number of OFDM symbols is over the total size of a MAC frame for the case that maximum guard time value is used. By changing the assumptions, especially adapting the use of SCH more sporadically, maximum active users can be increased in a MAC frame or the required number of OFDM symbols used by $L_{PRE+SCH}$ and L_{GTS} could be reduced. However the differences between the case of no guard time value and the cases of various guard time values are mainly influenced not by the assumptions but by guard time values, and its maximum difference is almost 100 OFDM symbols that is 20% of resources in a MAC frame. In Fig.8 and Fig.9 we can see the differences that are becoming bigger by increasing the number of active users. This is owing to the fact that the total required guard timing spaces are directly proportional to the number of UL DLCCs used by active users as well as the allowed number of RCHs in the system. Fig.10 compares the maximum throughputs over the length of user data packet by the maximum modulation schemes(64QAM with coding rate 3/4), the minimum modulation scheme(BPSK with coding rate 1/2) and various guard time values.

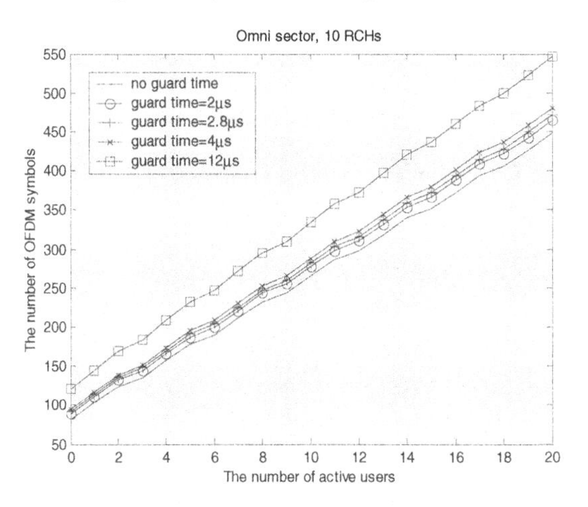

Fig. 9. The number of OFDM symbols by $L_{PRE+SCH}$ and L_{GTS}

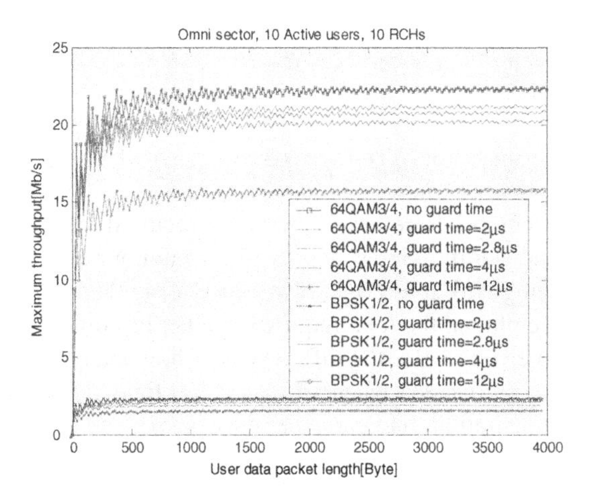

Fig. 10. Throughput by modulations and guard time values

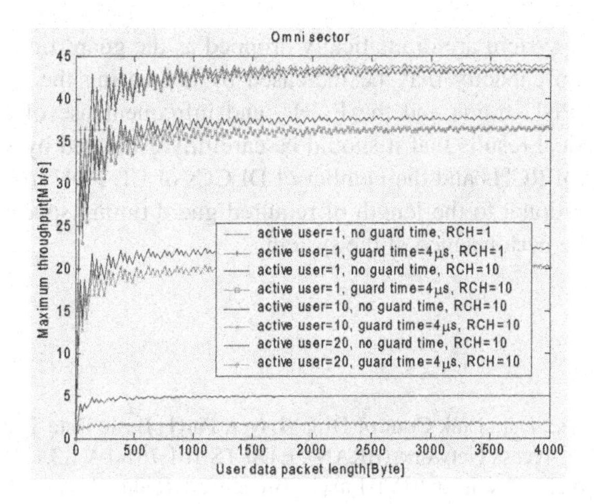

Fig. 11. Throughputs by 64QAM3/4 modulation for LCH

We can see that the system throughputs are seriously degraded in comparison to the case with no guard time value in both maximum and minimum modulations as the guard time value has increased. The graphs of BPSK with coding rate 1/2 for 2μs guard time value and BPSK with coding rate 1/2 for 2.8μs guard time value are completely overlapped with the third curve from the last one due to the property of floor function in (10). Fig.11 shows the throughput performances under the various number of active users and the number of RCHs for 64QAM modulation with coding rate 3/4 of LCH. The first graph of each figure shows the maximum throughput that it can achieve according to each modulation scheme. In that case only one active user and one RCH are used for simulation.

Note that the gap of throughput loss between two cases with and without guard time value becomes larger as we add the number of active users. This phenomenon is similar to that in Fig.8 and Fig.9.

5 Conclusions

In this paper we have discussed the effects of guard timing spaces in a MAC frame of ETSI BRAN HIPERLAN/2 system on the throughput performance. These guard timing spaces are needed between RCHs and different MAC-ID PDU trains in UL, and different sectors. We estimated throughputs under various guard time values defined in [2], the number of RCHs and number of DLCCs used by active users. We analyzed the required guard timing spaces in every PDU trains according to [1][2] and calculated their lengths in a number of OFDM symbol imposed on MAC frame depending on the number of RCHs and the number of DLCCs. The numerical results showed that the required length for the sectorized system is at most 13 OFDM symbols except for maximum guard time value, and that is relatively insignificant comparing to those for multi-DLCC of UL PDU trains and multi-RCH. It is obvious that the throughputs of system are dramatically dropped as the guard time value becomes bigger. The system capacity may be increased by decreasing the guard time value between the UL PDU trains and the RCHs, and infrequent use of SCHs. Also we showed by numerical results that it should be carefully estimated by AP/CC to determine the number of RCHs and the number of DLCCs of UL PDU trains because they are directly proportional to the length of required guard timing spaces and may, otherwise, degrade the performance of the system.

References

1. HIPERLAN Type2; Data Link Control(DLC) Layer; Part1: Basic Data Transport Functions, Broadband Radio Access Networks(BRAN), ETSI TS 101 761-1 V1.3.1, Dec(2001).
2. HIPERLAN Type2; Physical(PHY) Layer: Broadband Radio Access Networks(BRAN), ETSI TS 101 475 V1.3.1, Dec(2001).
3. HIPERLAN Type2; Packet based Convergence Layer; Part1: Common Part, Broadband Radio Access Networks(BRAN), ETSI TS 101 493-1 V1.1.1, Apr.(2000).
4. Bernhard H. WALKE, et al. "IP over Wireless Mobile ATM-Guaranteed Wireless QoS by HiperLAN/2", Proc. IEEE, vol.89, no.1, Jan. (2001).
5. Andreas Hettich, Arndt Kadelka, "Performance Evaluation of the MAC Protocol of the ETSI BRAN HIPERLAN/2 Standard", Proc. European. Wireless'99, Munich Germany, Oct. (1999).
6. Andreas Hettich, Matthias Schrother, "IEEE 802.11 or ETSI BRAN HIPERLAN/2: Who will win the race for a high speed wireless wireless LAN standard", Proc. European. Wireless'99, Munich Germany, Oct. (1999).

Reconfigurable Wireless Interface for Networking Sensors (ReWINS)

Harish Ramamurthy, B.S. Prabhu, and Rajit Gadh

Wireless Internet for the Mobile Enterprise Consortium (WINMEC)
The Henry Samueli School of Engineering and Applied Science, UCLA
38-138/A, 420 Westwood Plaza,
Los Angeles, California 90095
{harish, bsp, gadh}@wireless.ucla.edu

Abstract. Remote Monitor & Control systems are increasingly being used in security, transportation, manufacturing, supply chain, healthcare, biomedical, chemical engineering, etc. In this research, attempt is to develop a solution of wireless monitoring and control for such industrial scenarios. The solution is built on two components – a generic wireless interface for remote data collection/actuation units and control architecture at the central control unit (CCU) for smart data processing. The data collection/actuation unit (sensors/actuators) is intelligent by virtue of smart-reconfigurable-microcontroller based wireless interface, which is reconfigurable using Over-the-Air (OTA) paradigm. The RF link is also reconfigurable to accommodate a variety of RF modules (Bluetooth, 802.11 or RFID) providing plug-n-play capability. These capabilities make the interface flexible and generic. The control architecture supports services such as naming, localization etc., and is based on JavaBeans, which allows a component level description of the system to be maintained, providing flexibility for implementing complex systems.

1 Introduction

The current generation automation systems, control & monitoring systems, security systems, etc., all have the capability to share information over the network and are being increasingly employed to aid real-time decision support. The inter-device communication in such systems can be leveraged to maximize the efficiency and convenience in a variety of situations. Intelligent wireless sensors based controls have gained significant attention due to their flexibility, compactness and ease of use in remote unattended locations and conditions. These wireless sensor modules can be designed to combine sensing, provide in-situ computation, and contact-less communication into a single, compact device, providing ease in deployment, operation and maintenance. Already large-scale wireless sensor networks having different capabilities are being used to monitor real-time application needs.

Different types of sensors (thermal, photo, magneto, pressure, accelerometers, gyros, etc.) having different capabilities, interfaces and supporting different protocols

I. Niemegeers and S. Heemstra de Groot (Eds.): PWC 2004, LNCS 3260, pp. 215–229, 2004.

are used for different applications. For remote data collection, RF communication links with different characteristics, such as frequencies, data carrying capacity, bandwidth, susceptibility to interference, power needs, etc., have been employed to transmit the data. The appropriate design of the wireless sensing device depends on the end application needs such as data bandwidth, range, interference, etc.

In the proposed work the attempt is to provide a single comprehensive architecture to support the diverse control automation needs of a variety of industrial applications. The data collection/actuation units will be made intelligent by equipping them with smart microcontroller based wireless interface. Utilizing the plug-n-play modularity of the system architecture, appropriate sensor interfaces and RF communication interfaces can be chosen according to the application requirements. Further, once the interfaces are chosen, the (developed) application interface could be used to implement the specifics like description, placement, interaction of sensors and actuators, etc. For example; the interaction could be a "closed loop" control of a motor where the sensor is an encoder and actuator a motor.

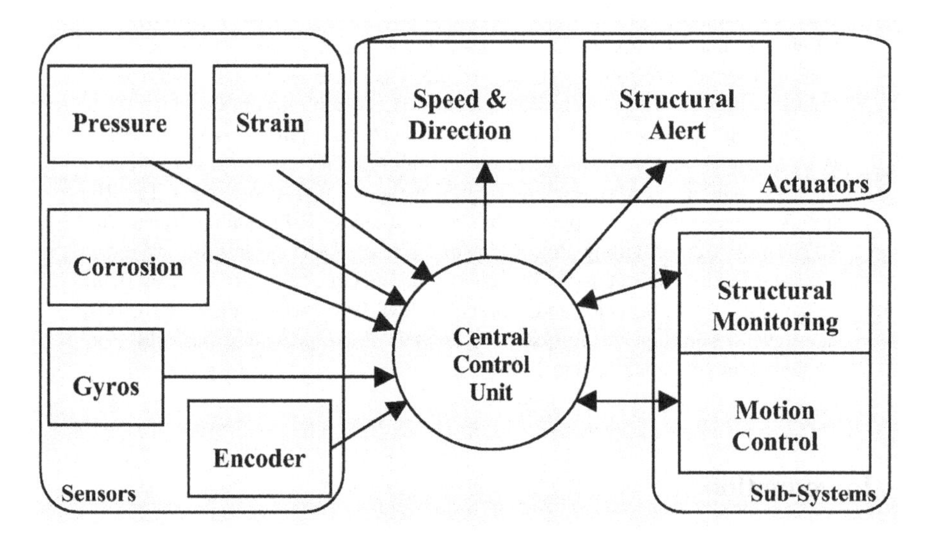

Fig. 1. Application Scenario – Aircraft health monitoring; the arrow direction denotes the direction of signal flow

1.1 Typical Application Scenario

A typical application scenario for our system can be aircraft sub-systems' monitoring.

As depicted in Fig. 1 monitoring the operational status and condition of a number of aircraft sub-systems such as structural monitoring, fuel level, motion control, etc., involves data retrieval and processing from a high-density of sensors, and provide real-time decision support. Each such subsystem may have multiple sensors installed, which continually update the status of the aircraft. The CCU is responsible for continuously aggregating real-time information from sensors and disseminates it to each of the respective subsystem for appropriate response (to actuators). Benefits of wireless communication such as significant space reduction, lower maintenance costs (as armored cables are done away with) and flexibility of deployment can be suitably leveraged. Further, utilizing the proposed reconfigurable and plug-n-play functionality in this research, the sensing system can be upgraded or updated with lesser effort.

1.2 Organization of the Paper

The various sections of the paper are as under. Review of the current published work done in this research field constitutes section 2. A brief overview of the system in explained in Section 3. Section 4 and 5 describe the intelligent wireless sensor architecture and the structure of a micro-controller that provides the intelligent interface between the sensor and RF module. In Section 6, the developed control architecture or application interface is explained. Implementation details, snapshots and some simulation results of the current implementation will be presented in Sections 7 and 8. Finally Section 9 reports the conclusions on the work done so far.

2 Related Work

Early work in the field of wireless sensor networks finds its roots in DARPA's military surveillance and distributed sensor network project, low-power wireless integrated micro-sensor (LWIM) and the SenseIT project. Through these initiatives DARPA has been quite instrumental in pushing the field from concept to a deliverable implementation form. Some of the relevant related research is described here in brief.

Wireless Integrated Network Sensors or the WINS project and NIMS project [1, 2] at University of California, Los Angeles is about ad-hoc wireless sensor network research – dealing mainly with building micro-electronic mechanical sensors (MEMS), efficient circuit design, and design of self-organizing wireless network architecture. Though these projects have been successful in demonstrating a network of self-organized sensor wireless nodes, they seem to have a bias towards environmental and military applications. Also they use proprietary RF communication technology and hence the solutions are restrictive for wide scale deployments in industries.

Motes and Smart Dust project [3] – at University of California, Berkeley involved creating extremely low-cost micro-sensors, which can be suspended in air, buoyed by currents. Crossbow Inc. has commercialized the outcome of this project. Here again the solution is restrictive, as proprietary communication technologies have been used to achieve inter-device communication. Further, the focus has been on development of sensors and their interaction rather than how the sensors will be integrated to form systems (simple or complex). This is generally termed as the "bottom-up" [4] approach, which may not be suitable for building complex systems.

Pico-Radio [4] – A group headed by Jan Rabaey at University of California, Berkeley is trying to build a unified wireless application interface called Sensor Network

Service Platform. An attempt is to develop an interface that will abstract the sensor network and make it transparent to the application layer. A preliminary draft describing the application interface has been recently released [4]. They believe in a "top-down approach" (from control to sensor nodes) for building sensor networks, which is probably more suitable for building complex systems.

Recently, there have been several initiatives like TinyDB [5], Cornell's Cougar [6] etc. to develop a declarative SQL-like language to query sensors and define certain standard query services. Here the implementation is sensor-interface specific and not a generic or abstracted sensor-networking platform. These query services can be implemented with ease on top of our (developed) wireless interface and sensor-networking platform and can be made generic by extending them for other sensors.

Other research initiatives in this field include MIT's μAMPS [7], Columbia University's INSIGNIA [8], Rice University's Monarch [9]. For a more detailed literature survey readers are encouraged to refer [10].

Though there have been a lot of research efforts in developing ad-hoc wireless networks, the focus has been on developing smart wireless sensor interfaces and not much attention has been paid to the actual application integration. Typical approach has been to develop powerful smart wireless interfaces, which supports the important features/requirements for a particular class of applications (like military, environment sensing or more focused applications like fuel-level control in automobiles). The result is a plethora of wireless interfaces appropriate for a certain class of application; but almost no interoperability between them. We believe that the deployment of wireless infrastructure in industries will occur in incremental stages and thus interoperability (between different sensor-networks) and extendibility (according to application needs) will form the basic requirements of any prospective solution. A prospective good solution would be an end-to-end solution, which is modular and extendable and hence capable of addressing the needs of majority of industrial applications, if not all. The ReWINS research initiative is an attempt to develop such an end-to-end solution with support for incremental deployment through a transparent lower layer implementation and control architecture and a user-friendly application interface.

3 System Overview

In this section, we present a brief overview of the system architecture. The system consists of a network of sensors, actuators and aggregators communicating with the central control unit using standard RF-links. The basic scenario is shown in Fig. 2. Here D represents the devices, which can be sensors or actuators and A is the aggregator. The scenario we consider is an infrastructure based deployment. In a typical industrial scenario like aircraft systems monitoring or automotive control, the sensors are typically stationary and have fixed wireless communication links, unlike mobile ad-hoc scenario. Fixed wireless communication links are more reliable than mobile links, which have inherent problems such as loss of connection, varying data rates, etc. Fixed links help in providing certain performance guarantees in terms of data-rate, mean delay, etc., which are essential for deployment in industrial scenarios. We thus believe an infrastructure-based deployment (i.e. fixed wireless deployment) is more suitable in such scenarios.

We term sensor as any kind of transducer which is capable of exchanging information in the form of electrical signals and similarly actuator is any kind of device which will accept data in the form of electrical signals and perform a measured action. Sensors and actuators will be referred generically as devices henceforth. The main function of the aggregator is to collect the data and signals from/to the devices and using a backhaul link (Wi-Fi) to transmit it back to the CCU.

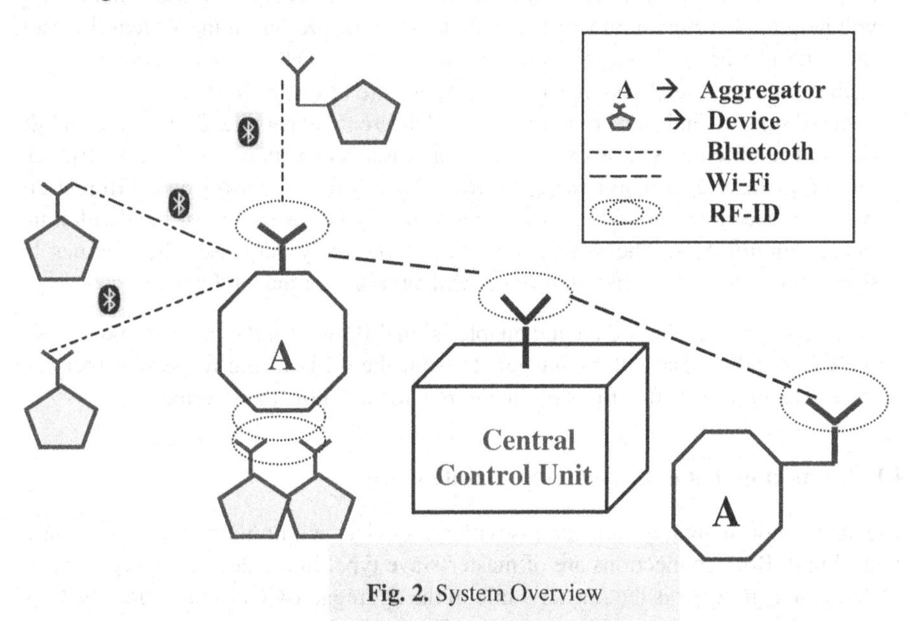

Fig. 2. System Overview

The features of the proposed system include:

1. Reconfigurability: Central Control Unit can set the run-time parameters of the device and/or update/upgrade the firmware of the system over the air (OTA).
2. Plug-n-play: Depending on the application needs/requirements, different infrastructure and their configurations can be deployed quickly.
3. Self-calibration: the intelligent adaptive sensors can accurately measure data and self-calibrate without significant user intervention.
4. Wireless connectivity: provide bi-directional communication over a wireless connection.
5. Lower installation and maintenance cost as no wiring/cables, etc., is required and therefore enabling greater acceptance and speedy deployment.

4 Aggregator – Functions and Description

The Aggregator has been introduced in the architecture to deal with the case of a highly dense network of distributed sensors and where it may not be possible for all sensors to directly communicate with the central control unit. The problems encountered in a direct sensor-control unit network can be summarized as:

1. Paucity of communication channels – The number of RF communication channels available in a given space are limited; thus limiting the number of sensors that can be present in the system. The aggregator with spatial channel re-use functionality will help improve the number of sensors that can be present in the system. Further, this will also bring down the power requirements (smaller transmission range) at each device and thereby increasing the battery life of the device.
2. Diverse sensors and their requirements – Different sensors have diverse capabilities and requirements. For example, a corrosion sensor may not need a stringent monitor/control as a motor rotary system. Also different sensors and different application needs warrant wireless communications with a variety of bandwidth and range requirements. The aggregator model efficiently addresses these issues by supporting multiple wireless interfaces and aggregating the payload.

An aggregator collects data and enables signal flow from the devices and sends it to the CCU and forwards the data/signals from the CCU to the respective recipient devices. The important functions of the aggregator will now be described.

4.1 Connection Establishment and Maintenance

The aggregator maintains connections with the devices within its purview and central control unit. Both connections are of master-slave type. In the device-aggregator connection, an aggregator is the master while in the aggregator-CCU connection the CCU is the master. In this type of connections, the master initiates connections and is responsible for maintaining them. For each type of device, the aggregator maintains the list of devices (i.e. device IDs – Section 5) connected to it. For the backhaul connection i.e. aggregator-CCU, Wi-Fi is used; the devices can use different RF technologies to communicate with aggregator, which supports multiple wireless interfaces.

4.2 Data Forwarding

The aggregator receives data from both the devices & CCU and forwards it to its intended recipients. Data from devices are marked with their specific device IDs at the aggregator and sent to CCU. Data from CCU is marked with device IDs and the aggregator extracts this information and forwards the data to the respective-device.

A typical payload is shown in Fig. 3. The data payload, i.e. the particular order in which bits will be transmitted, is device-specific and can be tailored to application needs. For example; the data payload could be a trigger or acknowledgement as speci-

Aggregator – Central Control Unit Communication

Start	Device ID	Data Payload	Stop

Fig. 3. Payload over the wireless link

fied in IEEE 1451.2 standard or it could be a simple data measurement byte. The format of data payload is fixed during initialization of the device (Section 6.1) and can be modified while the device is in Command mode.

4.3 Exception Handling

In the event of wireless connection drop between the aggregator and any of the devices, the aggregator reports to the CCU about the loss of connection. This helps differentiate between device inactivity and connection loss. The CCU then takes appropriate steps to compensate for the loss of device. This for example includes: entering a "safe-state", instructing other aggregators to start looking for the device, raising an exception, etc. In the event of connection severance with CCU, the aggregator instructs all the devices to enter the "safe-state" and wait for the CCU to respond.

5 Device Architecture

Architecture of the wireless intelligent data collection device is shown in Fig. 4. The device is composed of three components: a Sensor/Actuator, Microcontroller interface and RF interface. Microcontroller acts like an intelligent interface between the sensor or the data collection unit and the RF module. It handles tasks related to data collection, data processing and wireless transmissions. The RF trans-receiver communicates with the aggregator or CCU over the RF-link.

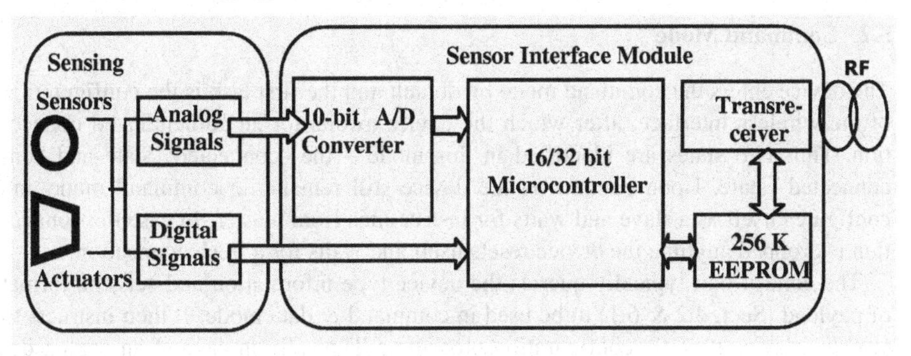

Fig. 4. Hardware design of Intelligent Sensor

5.1 Software Architecture

The device is implemented as an asynchronous finite state machine as shown in the Fig. 5. The device while in operation will be in any one of the two modes: Command Mode and Data Mode. This design facilitates flexible configuration i.e. forming networks, setting up device specific parameters, etc., while the device is in command mode. Further, once the device is configured, the device data (i.e. sensed data or actuator instructions) can be communicated without overhead in the data mode utilizing the wireless link efficiently. We now describe each of the modes in detail.

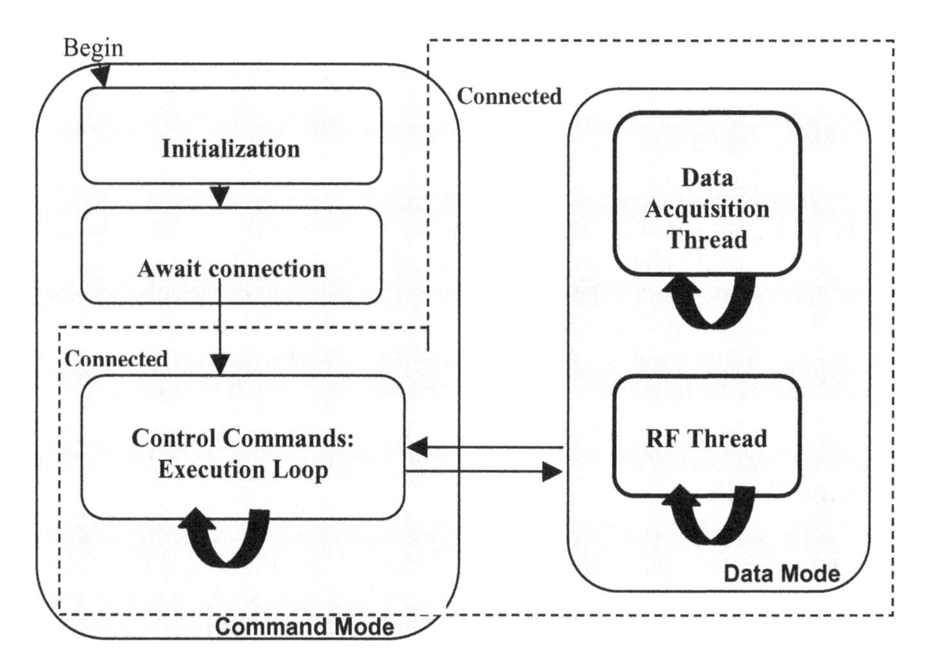

Fig. 5. Software Architecture of the Device (State Machine)

5.2 Command Mode

The device enters the command mode by default and the first task is the configuration of the wireless interface, after which the device awaits for an authenticated connection. Thus two states are identified in this mode – the "connected" state and "unconnected" state. Upon connection the device still remains in command mode and configures itself as a slave and waits for instructions from master. In case the connection is dropped anytime the device resets itself and waits for a fresh connection.

The control unit typically queries the device type information and sets the format of payload (Sect. 4.2 & 6.1) to be used in command & data mode. It then instructs to start the device specific configuration procedure. Typically in case of an analog device, the A/D or D/A converters are initialized. Further, the control unit may instruct the device to switch to data mode. The device can be reconfigured in this mode.

5.3 Data Mode

Once in Data mode the device will transmit/receive the device data using the RF-link. The device can be switched back to command mode using an escape sequence through the wireless interface. Otherwise the device will remain in data mode and transmit/receive data in the prescribed format. In the data mode, the device has to perform the following two tasks – 1) data collection in case of sensor or issuing instructions in case of actuator, 2) Interfacing and communicating with RF-link. As real-time processing of these tasks is required, separate threads are run for each task on the device.

5.4 Connection Drop and Exception Handling

The device periodically checks the status of the wireless link. In case the connection is dropped and cannot be reinstated; the device executes an "emergency routine" where the device is set to a "safe-state", which is particularly necessary in case of actuators. For example, a safe-state for a motor would be a complete stop. After executing the emergency routine the device just waits for the connection from the aggregator/CCU. The device thus can be only in "connected" state in this mode.

5.5 Memory Organizations and Over-the-Air Reconfiguration

The data stored in the microcontroller/EEPROM is divided into five types and each stored in different portion of memory. The memory allocation is shown in Fig. 6. This kind of memory organization helps in supporting reconfigurability, upgrade & update of firmware over-the-air (OTA) and fast real-time data processing in the device.

Self-identification information contains the unique device ID and type information.

The main program is the basic "monitor" program which initializes the system i.e. device hardware (sensor and wireless) interfaces. The main program cannot be altered and is permanent. Services like naming, service discovery, functionality of reconfigurability and other such core services are supported in the main program.

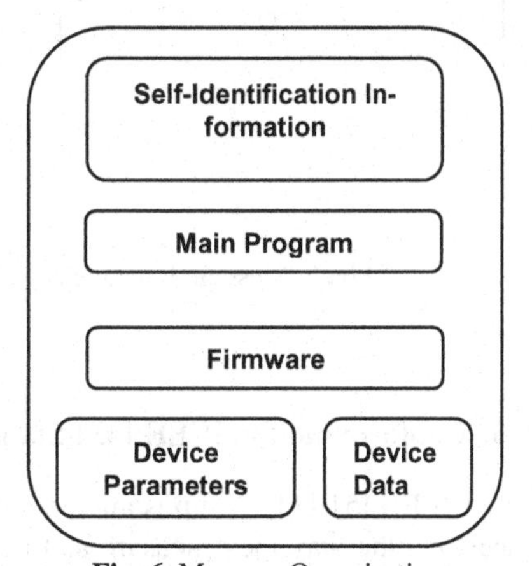

Fig. 6. Memory Organization

The firmware program takes care of data handling, parsing, taking actions and setting run-time parameters. The main program can modify the firmware.

Device Parameters contain the device-specific information like bit accuracy, sampling rate, aggregator identification etc and functions (names) supported by the device like sampling rate modification, toggle switch support etc. The attribute-value pair characterizes these parameters. For function names, the value corresponds to the function version. The device parameter characterization is essential for supporting querying services over the device. Reconfiguration Over-The-Air (OTA) is initiated by the CCU and is carried out through a set of messages exchange between the CCU and device as shown in Fig. 7.

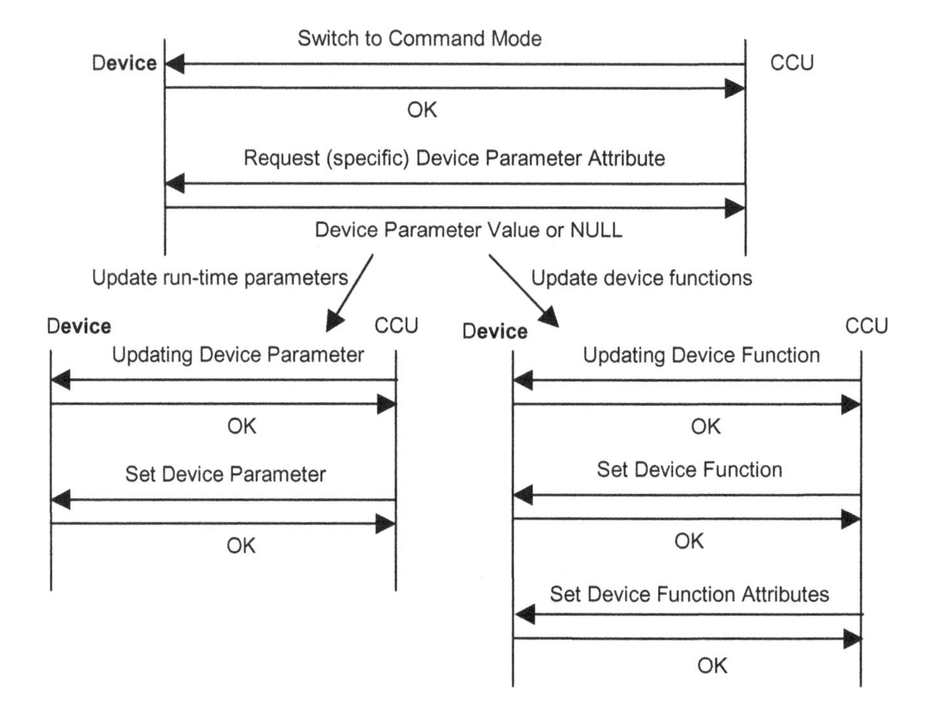

Fig. 7. Message Exchange Flowchart for Device Reconfiguration (OTA)

5.6 Conformation to IEEE 1451 Standard

The IEEE 1451 [11] group is attempting to standardize the sensor (or actuator) interfaces, i.e. the way they measure and report (or get instructions) parameters and values. More specifically, the standard defines the physical interface – Smart Transducer Interface Module (STIM), the Transducer Electronic Data Sheet (TEDS) and Network Capable Application Processors (NCAPS) – which derives information and controls the transducers. Utilizing the modularity and reconfigurability functionality of our device interface, adhering to IEEE 1451 standard is envisioned in future. The Control unit coupled with aggregator will be the network capable application process (NCAPs) as specified in the preliminary draft of IEEE 1451. The developed sensor interface can be configured to adhere to the STIM specifications and TEDS will be available from each device through a query service (defined by the control architecture).

6 Central Control Unit (CCU) – Software Architecture

This section provides a brief overview of the software architecture of the CCU. The developed system is targeted to be used with complex systems such as aircraft subsystem monitoring where a number of devices have to be identified, probed and instructed after performing some complicated logical computation such as closed loop control, wing balancing, etc. In order to implement complex systems, we had identified that the following features need to be supported by the CCU:

1. Flexibility of implementation – No restrictions on how device should be placed (geographically and logically) and connected to the system
2. Flexibility of control – Allow inclusion of complex control algorithms over devices & the sub-system formed by these devices
3. Friendly user and control interface – A framework by which the system can be described with ease; i.e. using a declarative language like XML.

The software architecture is what drives the central control unit. Thus in order for the system to support the aforementioned features, the software architecture of the control unit has to address the following issues:

1. Device detection, representation and characterization in the system
2. Inter-device interaction & information sharing; formation of sub-systems by these devices and aggregation of subsystems to form the complete system

6.1 Device Detection and Representation

The devices configure themselves as slaves and wait for connection from the central control unit or the aggregator. After initialization the aggregator will start probing for other devices and the control unit. Since the aim is to maximize device detection, coverage and communication, if possible, all device connections will be routed through an aggregator. Devices are represented as entities with data interfaces in the control system software architecture. The data interface provisions the exchange of information among devices.

6.2 Device Query Services

We are currently developing a standard query service model which will support IEEE 1451 and provide a query platform to implement query services like Cornell's Cougar or Berkeley's PicoRadio and Tiny DB, etc. Standard query services like service discovery, identification, device parameters modification etc. are currently supported.

6.3 Formation of Subsystems

A subsystem can be defined as a collection of logically connected devices; for example a simple motor control subsystem will have a motor (actuator) and an encoder (sensor). The logical connection specifies how data present in the device can be exchanged as information, for e.g., the control logic like closed-loop control of a motor control system. Further only those devices, which have some information to share, can be connected. Here we will like to differentiate between "data" and "information". Information is what devices can extract from data to self-adjust and modify their behavior. For example for a motor, rotary position from an encoder is "information" but linear position from a sensor may be just "data". The complete system has a hierarchical architecture and hence facilitates a very user-friendly control interface. The control architecture has been implemented using the Java Beans API [12], where each entity of the system is represented as a component and a reusable software unit. Further using the application builder tools of Java Beans these software units can be visually composed into composite systems. This makes the user interface extremely friendly but still gives the power of building complex systems.

7 Experimental Setup and the Current Status of the Project

Currently the proof-of-concept implementation of the networking platform has been demonstrated. In the current version of the system different type of sensors, viz., rotary and linear have been interfaced. Class -1 Bluetooth has been used for the RF communication link. Work on supporting other RF technologies like UWB and RFID is currently being done. In the current implementation, as the aggregator is just a logical component, the aggregator resides on the CCU itself. Here the aim was to support both "open-loop" and "closed-loop" control on a motor as actuator. An encoder was used to sense the position of motor. Fig. 8 shows the block-diagram and snapshots of the current implementation.

8 Results

Preliminary tests of the system were carried out in certain scenarios and the results are presented below:

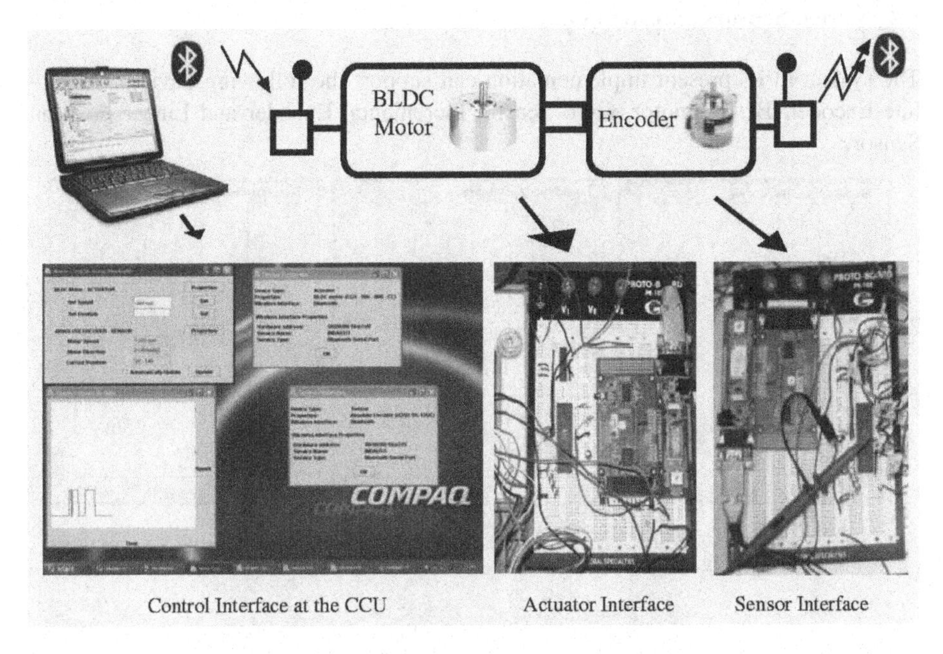

Fig. 8. Snapshots of the Current Implementation

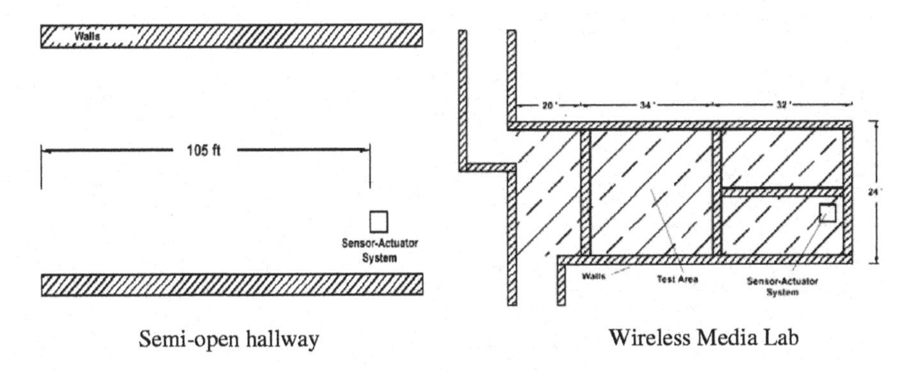

Fig. 9. System Test Scenarios

8.1 Geographical Range Tests

The system was tested in two scenarios as shown in Fig. 9. Here the walls are shown as hatched rectangles and the rectangular box is the sensor-actuator system. The laptop (control-unit) was able to control the sensor-actuator system from 105ft without degradation in the performance. In the second scenario, the hatched area shows where system performance didn't degrade. The system performed well even in presence of obstacles like walls (2).

8.2 Varied Sensors and Actuators

The system in its present implementation can support the following devices - Absolute Encoder, BLDC motor, Gyro Sensor, Incremental Encoder and Linear Position Sensor.

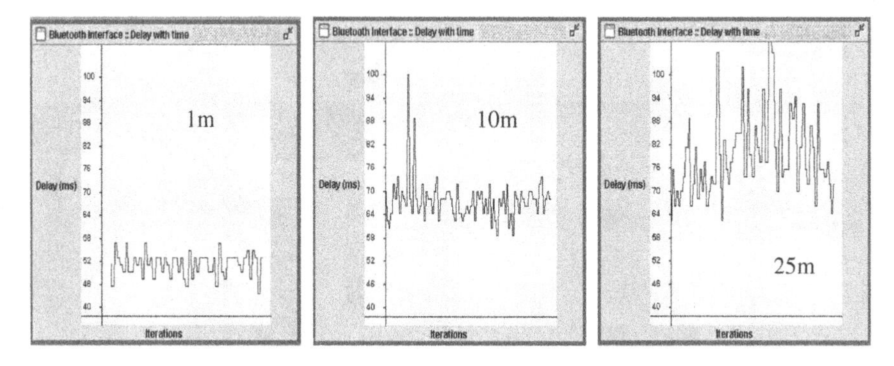

Fig. 10. Delay Performance with varying testing distance

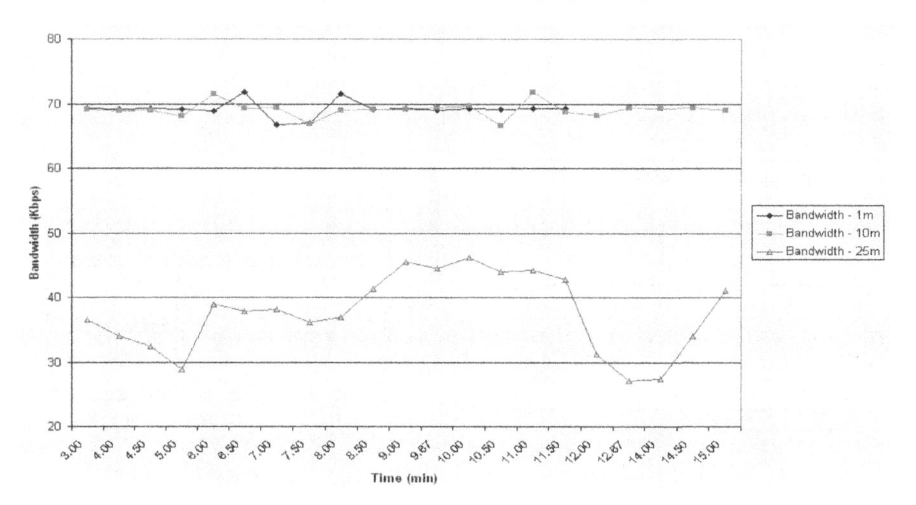

Fig. 11. Bandwidth Performance with distance

8.3 System Performance Tests

Since these systems will be used for real-time monitoring and control, performance studies were carried out. Delay and bandwidth have a significant effect on the fidelity and responsiveness of the system. To characterize the parameters, simulations were performed in an "echo-scenario", i.e. whenever the CCU sends a packet to the device; the device simply echoes back the packet. Measuring the delay from start of transmission from CCU to end of reception at CCU gives the round-trip-delay of the link

Bandwidth is measured by the data rate. Simulations were carried out for different inter-device distances and the results are shown in Fig. 10 and 11.

As can be seen from simulation results the round-trip-delay increases with distance. Further, the delay becomes jittery as the distance is increased. This is attributed to the increased interference and fading at longer distances. Similarly bandwidth decreases with distance and becomes jittery due to the same reason

9 Conclusion

In this research, an end-to-end solution for wireless monitoring & control in industrial scenarios has been proposed. It was shown how traditional sensors could be transformed into intelligent wireless sensors, capable of making real-time decisions using the developed generic wireless interface. A proof-of-concept working model of the solution is also demonstrated with successful integration of a variety of sen sors/actuators by using the developed application interface. To illustrate, a suitable application scenario was also discussed for such a remote data collection system.

Acknowledgements. We would like to thank Xiaoyong Su and other members of Wireless Media Lab and WINMEC [http://winmec.ucla.edu] for their valuable comments and suggestions.

References

1. Wireless Integrated Network Sensors project at University of California at Los Angeles, URL: http://www.janet.ucla.edu/WINS/
2. NIMS – Networked Info-Mechanical Systems project at University of California at Los Angeles, URL: http://www.cens.ucla.edu
3. Smart Dust and motes project at University of California, Berkeley, URL: http://robotics.ee-cs.berkeley.edu/~pister/SmartDust/
4. Pico-Radio project at University of California, Berkeley, URL: http://bwrc.eecs.berkeley.ed-u/Research/Pico_Radio/
5. TinyDB project at University of California, Berkeley, URL: http://telegraph.cs.berkeley.edu /tinydb/
6. Cougar project at Cornell University, URL: http://www.cs.cornell.edu/database/couga-r/index.htm
7. Micro-Adaptive Multi-domain Power-aware Sensors (μAMPS) project at University of California, Berkeley, URL: http://www-mtl.mit.edu/research/icsystems/uamps/
8. Insignia project at Columbia University, URL: http://comet.ctr.columbia.edu/insignia/
9. Monarch project at Rice University, URL: http://www.monarch.cs.rice.edu/
10. Wireless Ad-hoc Network Links at NIST webpage, URL: http://www.antd.nist.gov-/wctg/manet/adhoclinks.html
11. IEEE 1451.2 Standard, URL: http://grouper.ieee.org/groups/1451/2/
12. Java Beans specification, URL: http://java.sun.com/products/javabeans/

Mission-Guided Key Management for Ad Hoc Sensor Network

Shaobin Cai, Xiaozong Yang, and Jing Zhao

Harbin Institute of Technology, Department of Computer, Mail-box 320 Harbin Institute of Technology, Harbin, China, 150001
Phone: 86-451-6414093
csb@ftcl.hit.edu.cn

Abstract. Ad hoc Sensor Networks (ASNs) are ad-hoc mobile networks that consist of sensor nodes with limited computation and communication capabilities. Because ASNs may be deployed in hostile areas, where communication is monitored and nodes are subject to be captured by an adversary, ASNs need a cryptographic protection of communications and sensor-capture detection. According to that the ASN is deployed to carry out some certain tasks, we present a mission-guided key-management scheme. In our scheme, a key ring, which consisting of randomly chosen k keys from a sub-pool of a large offline-generated pool of P keys, is pre-distributed to each sensor node of a group. Compared with Laurent's scheme, our scheme improves the probability that a shared key exists between two sensor nodes of the same group, and doesn't affect its security.

Keywords. Ad hoc Sensor Networks, Security, Key Management, Mission-guarded

1 Introduction

The last decade of last century has seen the advances in micro-electro-mechanical systems (MEMS) technology, wireless communications, and digital electronics have enabled ad hoc sensor networks (ASN) to monitor the physical world. When they are deployed in the hostile environments, their open architectures make potential intruder easy to intercept, eavesdrop and fake messages. Therefore, the ASN need strong security services.

Although some significant progress has been made in many aspects of ASN, which include topology management, routing algorithms, data link protocol and sensor data management [1], very little work is done on the security of ASN. Since proposals addressing security in general ad hoc networks [2][3][4] aren't suitable for ASN, the research into authentication and confidentiality mechanisms designed specifically for ASN is needed.

Most of the security mechanisms require the use of some kind of cryptographic keys that need to be shared between the communicating parties. The purpose of key

I. Niemegeers and S. Heemstra de Groot (Eds.): PWC 2004, LNCS 3260, pp. 230–237, 2004.

management is to [5]: Initialize system users within a domain; Generate, distribute and install keying material; Control the use of keying material; Update, revoke and destroy keying material; Store, backup, recover and archive keying material. Key management is an unsolved problem in ASN.

The hardware resources of the sensor are so scarce that it is impractical for it to use typical asymmetric (public-key) cryptosystems to secure communications. For example, the Smart Dust sensors [6, 7] only have 8Kb of program and 512 bytes for data memory, and processors with 32 8-bit general registers that run at 4 MHz and 3.0V. Carman, Kruus, and Matt [8] report that on a mid-range processor, such as the Motorola MC68328, the energy consumption for a 1024-bit RSA encryption (signature) operation is much higher than that for a 1024-bit AES encryption operation. Hence, symmetric-key ciphers, low-energy, authenticated encryption modes [9, 10, 11], and hash functions become the tools for protecting ASN communications.

In order to reduce the usage of hardware resource, Laurent's scheme [12] distributes a key ring, which consisting of randomly chosen k keys from a large offline-generated pool of P keys, to each sensor node. Although, Lauren's scheme saves some hardware resources, its possibility that there is a secure link between any pair sensor nodes is low. According to that the sensors are deployed to perform certain tasks, we propose a mission-guided key management scheme. In our scheme, the sensors, which are deployed to perform a certain tasks, form a group. The scheme randomly chooses a key sub-pool from the large pool of P keys for the group according to the size of the group. And then, it distributes a ring of keys, which consists of randomly chosen k keys from the sub-pool, to each sensor node off-line. In the sensor network, most communications among sensors are among the sensors, which cooperate to accomplish assigned tasks. Therefore, our mission-guided key management scheme improves probability that a shared key exists between two sensor nodes. By the random graph analysis and simulation, we analyze the performance of both key management schemes.

The rest of the paper is organized as follow. First, we give an overview of our scheme in section 2. Secondly, we setup a mathematic model and analyze its performance in section 3. Thirdly, we analyze its performance by simulations in section 4. Finally, we draw a conclusion in section 5.

2 Overview of Our Scheme

Our scheme modifies Laurent's scheme according to that the actions of the sensor nodes are mission-guided. The difference between our scheme and Laurent's scheme is their key pre-distribution. Their shared-key discovery, path-key establishment, key revocation, re-keying and resiliency to sensor node capture are identical. In this section, we present how the keys are pre-distribute to the sensors according to the mission-guided scheme.

In the Laurent's scheme, a key ring, which consisting of randomly chosen k keys from a large offline-generated pool of P keys, is pre-distributed to each sensor node.

According to the usage of ASN, most sensors of the ASN are deployed at the same time and to the same place for a special mission. Therefore, the sensors can be divided into groups for the sub-missions. Since most tasks of the mission are completed by the cooperation of the group members, most communications among sensors are happened among the number of a group. Therefore, our scheme can improve secure connectivity among the sensors, and reduce the path length between any pair of sensor nodes. The key pre-distribution phase of our scheme consists of the following six steps:

1. It first generates a large pool of P keys, which normally has 2^{17} - 2^{20} keys in Laurent's scheme, and their key identifiers offline;
2. It selects P_i ($P_i = \frac{l}{n} \times P$, l is the number of the sensors of the group and n is the number of the sensors of the ASN) keys from the pool to form a sub-pool, $Subpool_i$, for group G_i;
3. It randomly selects k keys from $Subpool_i$ to establish the key ring of a sensor of group G_i;
4. It loads the key ring into the memory of each sensor;
5. Its saves the key identifiers of a key ring and associated sensor identifier on a trusted controller node of the group;
6. It loads the ith controller node with the key shared with the controller.

In the step 6, the key shared by a node with the ith controller node, K^{ci}, can be computed as $K^{ci} = E_{K_x}(ci)$, where $K_x = K_1 \oplus, \ldots, \oplus K_k$, K_i are the keys of the node's key ring, ci is the controller's identity, and E_{K_x} denotes encryption section, with node key K_x. Hence, the keys shared by a node with controllers, which are only infrequently used, need not take any space on the key ring. However, in this case, the K^{ci} changes upon any key change on a ring.

Compared with Laurent's scheme, the advantage of our scheme is that the sensor nodes, which cooperate to accomplish some tasks, get their key ring from a sub-pool. Since the sub-pool is smaller than the pool, the possibility that there is a secure link between any pair sensor nodes of the same group increased. On the other hand, the key selection from the sub-pool doesn't affect randomicity of the key selection. Therefore, this scheme doesn't increase the possibility that an adversary decrypts a key.

3 Analysis

In this section, we compare the probability that a shared key exists between two sensor nodes in Laurent's scheme and our scheme.

In the sensor network, not only the security considerations but also the limits of the wireless communication ranges of sensor nodes preclude that the ASNs are fully connected by shared-key links between all sensor nodes. Therefore, it is impossible for the shared-key discovery phase to guarantee full connectivity for a sensor node with all its neighbors. Let p be the probability that a shared key exists between two sensor nodes, n be the number of network nodes, and $d = p \times (n-1)$ be the expected degree of a node of a fully connected network, in which d is the average number of edges connecting that node with its graph neighbors.

Since the wireless connectivity constraints limit sensor node neighborhoods, the a node has n' ($n' \leq$ n-1) neighbor nodes, which implies that the probability of sharing a key between any two neighbors becomes $p' = \frac{d}{n'} \geq p$. Hence, we set the probability that two nodes share at least one key in their key rings of size k chosen from a given pool of P keys to p', and then derive p' as a function of k. In the derivation, the size of the key pool, P, isn't a sensor-design constraint; the size of the key ring, k, is sensor design constraint.

The probability that two key rings share at least a key is 1 – Pr [two nodes do not share any key]. To compute the probability that two key rings do not share any key, we first compute the number of the possible key rings. Since each key of a key ring is drawn out of a pool of P keys without replacement, the number of possible key rings is:

$$\frac{P!}{k!(P-k)!}$$

After picking out the first key ring, the total number of possible key rings that do not share a key with this key ring is the number of key rings that can be drawn out of the remaining P - k unused key in the pool, namely:

$$\frac{(P-k)!}{k!(P-2k)}!$$

Therefore, the probability that no key is shared between the two rings is the ratio of the number of rings without a match by the total number of rings. Thus, the probability that there is at least a shared key between two key rings is:

$$p' = 1 - \frac{k!(P-k)!}{P!} \times \frac{(P-k)!}{k!(P-2k)}!$$

and thus

$$p' = 1 - \frac{((P-k)!)^2}{(P-2k)!P!}$$

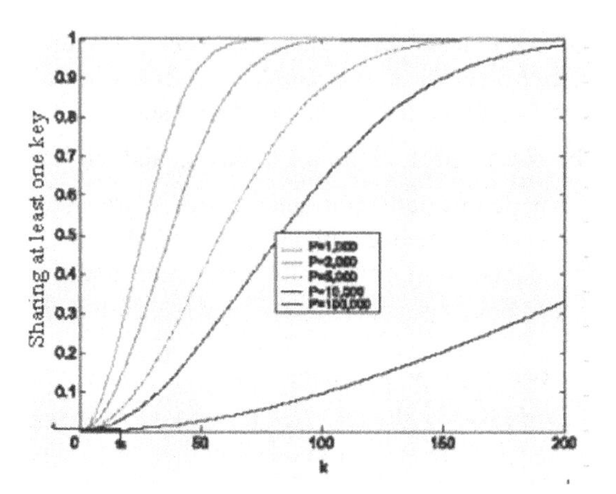

Fig. 1. Probability of sharing at least one key when two nodes choose k keys from a pool of size P (cited from Laurent's scheme)

In order to simplify the analysis of our scheme, we assume that there are only sensor nodes of the same group in the ASN. Therefore, the p' of our scheme between two sensors of the same group, whose sub-pool has P_i keys, is

$$p' = 1 - \frac{((P_i - k)!)^2}{(P_i - 2k)!P_i!}$$

Since P is very large, we use Stirling's approximation for

$$n! \approx \sqrt{2\pi}\, n^{n+\frac{1}{2}} e^{-n}$$

to simplify the expression of p', and obtain:

$$p' = 1 - \frac{(1 - \frac{k}{P})^{2(P-k+\frac{1}{2})}}{(1 - \frac{2k}{P})^{(P-2k+\frac{1}{2})}}$$

Since the size of sub-pool is much smaller than that of pool, the probability that a shared key exists between two sensor nodes of the same group is improved by our scheme. But, our scheme doesn't affect the probability that a shared key exists between two sensor nodes of the different group.

Figure 1 illustrates a plot of this function for various values of P. When a pool size P is 10,000 keys, and 75 keys are distributed to any sensor nodes, Laurent's scheme only makes any two nodes have the probability $p' = 0.5$ that they share a key in their key ring, our scheme can make any two nodes of the same group have the probability $p' \approx 1$ that they share a key in their key ring (we assume that there are 10 groups in the ASN, and each sub-pool has 1,000 keys).

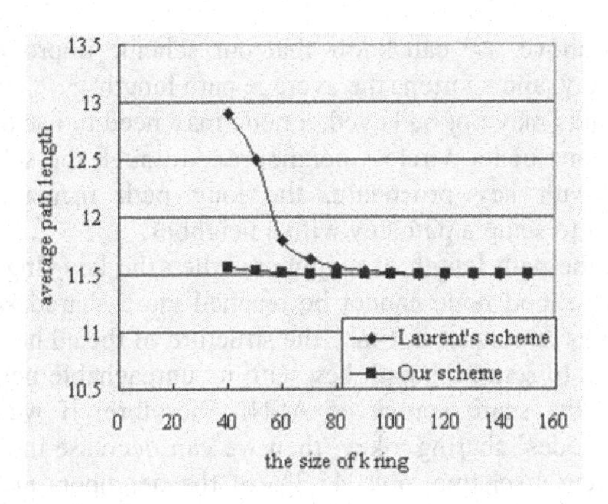

Fig. 2. Average path length at network layer

4 Simulations

We investigate the effect of the various parameters on ASN by the simulations. In our simulations, the ASN is made up of 1,000 nodes, each node averagely has 40 neighbor sensor nodes, and the pool of key has 10,000 keys. In our scheme, the sensor nodes are divided into 10 groups, and the movement of the nodes is guided by mission-guided mobility model. In the mission-guided mobility model, the movements of the nodes have three characters:

1. When a task is assigned to a group, all nodes of the group move to the assigned area at similar speed;
2. When a node of the group arrived at the assigned area, the node move according to the "random waypoint" in the assigned area;
3. When a group finishes its assigned task, it waits a new task in the previous assigned area.

4.1 Effect on the Network Topology

Whether two neighbor nodes share a key during the shared-key discovery phase means that whether a link exists or not between these two nodes from a network routers' point of view. Therefore, the probability that two nodes share a key in their key ring has an effect on the average path length between two nodes after shared-key discovery.

Figure 2 shows the relationship between the path length and the sizes of the key ring. From the figure we can see that the average path length of the network depends on the size of the key ring. The smaller k is, the higher the probability that a link does not have a key and, therefore, the longer paths need to be found between nodes. When k is too small, the network is disconnected. Since, the sensor group is deployed to perform certain tasks, most of them are in an assigned zone, and most links between two sensor nodes are between two sensor nodes of the same group.

From the analysis above, we can know that our scheme improves the probability, which a link has a key, and shortens the average path length.

Because some links may not be keyed, a node may need to use a multi-link path to communicate with one of its wireless neighbors. Although the schemes can encrypt this link by the path key procedure, the long path increases the delay and communication cost to setup a path key with a neighbor.

Figure 3 shows the path length of neighbors when the key ring of sensors has 75 keys. When neighborhood node cannot be reached via a shared key, the node must take at least two links to contact it. Since the structure of the ad hoc sensor network is unstable, a node has to setup the path key with its unreachable neighbors constantly. The effects waste the scare source of ASN. Therefore, if we can improve the probability of two nodes' sharing a key, then we can decrease the usage of path key procedure. In Laurent's scheme, only 45.3% of the neighbors are reachable over a single link, and other 17.8% of the neighbors are reachable over two-link paths. In our scheme, almost 100% of the neighbors of the same group are reachable over a single link. Although, some neighbors of different group aren't reachable over a single link, the 98.7% of the neighbors are reachable over a single link, and other 1.3% of the neighbors are reachable over two-link paths.

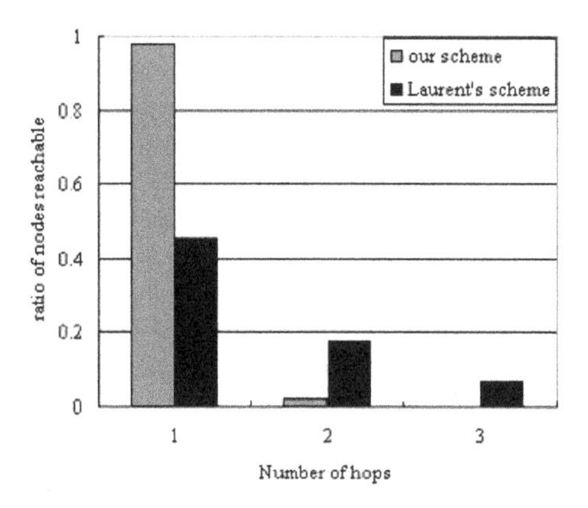

Fig. 3. Path length to neighbors

4.2 Resiliency to Sensor Node Capture

If an adversary captures a node, then they can acquire k keys, and the adversary can attack $\dfrac{k \times number\ of\ links}{P}$ links. Therefore, how many links are secured with the same key is an important factor that affects the resiliency to sensor node capture. From the simulation results, we can know that the usage of the keys of our scheme is similar with that of Laurent's scheme. Out of the pool of 10,000 keys, only about 50% of the keys are used to secure links, only about 30% are used to secure one link, about

10% are used to secure two links, and only about 5% are used to secure 3 links. Therefore, the ability of both schemes to stand against the node capture is similar.

5 Conclusions

In this paper, we presented a new mission-guided key management scheme for large scale ASNs. In our scheme, the sensors, which are deployed to perform a certain tasks, form a group. The scheme randomly chose a key sub-pool from the large pool of P keys for the group according to the size of the group, and distributes a ring of keys, which consists of randomly chosen k keys from the sub-pool, to each sensor node off-line. By the analysis and simulations, we compare the difference between our scheme and Laurent's scheme. The results show that our scheme outperforms the Laurent's scheme.

References

1. I. Akyildiz, W. Su, Y. Sankarasubramaniam, E. Cayirci, "A survey on sensor networks", IEEE Commun. Mag. (August) (2002).
2. N. Asokan, P. Ginzboorg, "Key agreement in ad hoc networks", Comp. Commun., 23, pp. 1627 - 1637.
3. L. Zhou, Z.J. Haas, "Securing ad hoc networks", IEEE Networks 13 (6) (1999).
4. J. Kong, P. Zerfos, H. Luo, S. Lu, L. Zhang, "Providing robust and ubiquitous security support for mobile ad-hoc networks", IEEE ICNP (2001).
5. A. Menezes, P. van Oorschot, S. Vanstone, "Handbook of Applied Cryptography", CRC Press, 1997, ISBN 0849385237.
6. J. Hill, R. Szewczyk, A. Woo, S. Hollar, D. Culler, K. Pister, "System architecture directions for network sensors," Proc. of ASPLOS-IX, Cambridge, Mass. 2000.
7. J. M. Kahn, R. H. Katz and K. S. J. Pister, "Mobile Networking for Smart Dust", ACM/IEEE Intl. Conf. on Mobile Computing and Networking (MobiCom 99), Seattle, WA, August 17-19, 1999, pp. 271 - 278.
8. D. W. Carman, P. S. Kruus and B. J. Matt, "Constraints and Approaches for Distributed Sensor Network Security", dated September 1, 2000. NAI Labs Technical Report #00-010, available at http://download.nai.com /products/media/nai/zip/nailabs-report-00-010-final.zip
9. V.D. Gligor and P. Donescu, "Fast Encryption and Authentication: XCBC Encryption and XECB Authentication Modes", Fast Software Encryption 2001, M.Matsui (ed), LNCS 2355, Springer Verlag, April 2001.
10. C.S. Jutla, "Encryption Modes with Almost Free Message Integrity", Advances in Cryptology EUROCRYPT 2001, B. Pfitzmann (ed.), LNCS 2045, Springer Verlag, May 2001.
11. P. Rogaway, M. Bellare, J. Black, and T. Krovetz, "OCB: A Block-Cipher Mode of Operations for Efficient Authenticated Encryption", Proc. of the 8th ACM Conf. on Computer and Communication Security, Philadelphia, Penn., November 2001.
12. Laurent Eschenauer, Virgil D. Gligor "A Key Management Scheme for Distributed Sensor Networks" CCS'02, November 18–22, 2002, Washington, DC, USA.

Aggregation-Aware Routing on Wireless Sensor Networks*

Antonio J.G. Pinto, Jorgito M. Stochero, and José F. de Rezende

Grupo de Teleinformática e Automação (GTA)
COPPE/PEE – Universidade Federal do Rio de Janeiro
Caixa Postal 68504,
21945-970 Rio de Janeiro, RJ, Brazil
{antonio,stochero,rezende}@gta.ufrj.br

Abstract. In wireless sensor networks, data aggregation is critical to network lifetime. It implies that data will be processed in an efficient flow from multiple sources to a specific node named sink. However, there are trade-offs for fusing multisensor data and creating a path between source and sink that increase the likelihood of aggregation. In this work, we propose a decentralized mechanism using parametric-based techniques, such as Bayesian Inference and Dempster-Shafer Method, for data aggregation in wireless sensor networks. Moreover, we propose an extension to an existing data-centric routing protocol in order to favor aggregation. Our approach is evaluated by means of simulation.

Keywords: Sensor networks, routing, in-networking aggregation

1 Introduction

Recently, the research interest in Wireless Sensor Networks (WSNs) is considerable because of its wide range of potential applications, from weather data-collection to vehicle tracking and physical environment monitoring [1]. Communicating wirelessly consumes much more power at the nodes than processing or sensing [2]. Thus, it is preferred that data be processed *in-network* as opposed to a centralized processing. Sensors data are transmitted from multiple acquisition sources toward one or more processing points, which may be connected to external networks. Since sensors monitor a common phenomenon, it is likely to appear significant data redundancy, which may be exploited to save transmission energy, throughout in-network filtering and data aggregation procedures. Techniques employed range from suppressing duplicated messages and performing distributed basic functions, such as *max*, *min*, *average*, and *count*, to data fusion.

In this paper we deal with two different issues: aggregation tree construction in data-centric routing protocols and aggregation mechanism using parametric techniques, in order to save energy with a low cost to the network. The main

* This work was supported by CNPq, CAPES, SUFRAMA, and FAPERJ/Brazil.

I. Niemegeers and S. Heemstra de Groot (Eds.): PWC 2004, LNCS 3260, pp. 238–247, 2004.

motivation behind this work is to exploit the colaborative characteristics of sensor nodes to jointly estimate about a certain common sensed phenomena. The remainder of the paper is organized as follows. In section 2 we present a review of related work in routing for aggregation and aggregation techniques themselves. Section 3 describes the routing protocol and the multisensor aggregation techniques proposed. Simulation results are presented in Section 4. Finally, the paper is concluded in Section 5.

2 Data-Centric Routing and Aggregation

Energy is a factor of utmost importance in WSNs. To increase network lifetime, energy must be saved in every hardware and software solution composing the network architecture. According to the radio model proposed in [2], data communication is responsible for the greatest weight in the energy budget when compared with data sensing and processing. Therefore, to save energy it is better to increase data processing in order to avoid data transmission. Energy-aware routing protocols can exploit two particularities of WSNs: data redundancy and a many-to-one (sources to sink) association.

In order to *in-network aggregation* to be effective, routing protocols should use the most suitable addressing and forwarding schemes. In data-centric routing protocols [3,4], intermediate nodes build forwarding tables based on interests sent by the sink and on the information provided by the sources. Source nodes are recognized based on the sort of data content they can provide (temperature, smoke presence, etc.) and in their current geographical location (if available). Thus, instead of assigning a Global Unique ID to sensor nodes, network protocols may use *attributed-based addressing.*

Other routing protocols aim at building aggregation trees from multiple sources to the sink. The NP-hard Minimum Steiner Tree gives the optimal aggregation path in terms of number of transmissions [5]. Authors in [6] analyze three suboptimal aggregation schemes as function of two positioning models and an increasing number of source nodes. Hierarchical cluster-based routing favors aggregation in the cluster-heads and is analyzed in [2]. Building aggregation paths creates a trade-off between delay and energy saving.

Concerning the data aggregation mechanisms, the breadth of applications and diversity of applicable techniques [1] make their study very complex. These techniques correspond to a mixture of mathematical and heuristic methods drawn from statistics, artificial intelligence, and decision theory [7]. The most trivial data aggregation function is to suppress duplicated messages. The aggregation techniques involve *how* to aggregate data. The basic functions of aggregation (such as *max* and *min*) are appropriate to aggregate data from multiple inputs of a same type of sensor, while fusing data from multiple kinds of sensors requires more complex methods. Some of these methods have been exploited in diverse mathematical developments, such as those called parametric techniques, like Bayesian and Dempster-Shafer Inferences. These techniques compose the data fusion model created by *Joint Directors of Laboratories Date Fusion Sub-*

panel (JDL DFS) of the American Department of Defense [7]. JDL DFS model has been the target of several research and developments involving traditional sensors (for instance, radar and ELINT sensors). However, WSN literature has been limited only to cite the potential of these techniques. In this work, we apply parametric techniques on a data-centric routing protocol, establishing an aggregation mechanism that minimizes the negative aspects of traditional approaches.

3 Aggregation-Centric Routing and Multisensor Data Aggregation

This work addresses different aspects of aggregation in WSNs. These aspects involve the use of routing protocols and filtering mechanisms implemented in the sensor nodes to save energy. In this paper, we propose an extension to an existing data-centric routing protocol in order to increase the likelihood of aggregation without incurring in the overhead related to aggregation tree building and maintenance. In addition, an aggregation mechanism, using parametric techniques, with the minimum of losses in terms of delay, scalability and robustness.

3.1 Privileged Aggregation Routing (PAR)

The data-centric routing protocol used is the Directed Diffusion proposed in [3], which allows application specific processing within the network. Data is named using attribute-value pairs. The sink performs *interests* dissemination to assign sensing tasks to the sensor nodes. This dissemination sets up *gradients* in order to draw data that match the interest. Events (data) start flowing towards the originators of interest along multiple paths. The sink *reinforces* one particular neighbor from which it receives the first copy of the data message. Intermediate nodes using the same criterion reinforce their respective neighbor. This approach privileges lower delay paths, while it does not necessarily lead to the establishment of better aggregation paths.

We propose a new scheme to reinforce neighbors. The idea is to use the same localized interactions of Directed Diffusion and exploit its filtering architecture, described in [8], to build an empirical aggregation tree in the network. The diffusion filter architecture is a software structure for a distributed event system that allows an external software module, called *filter*, to interact with the Directed Diffusion *core* and modify its routing capabilities to influence how data is moved through the network. In this work, Directed Diffusion is modified to reinforce intermediate nodes with greater potential to combine data from different sources. After these nodes are found and included in the source-to-sink path, distributed estimation algorithms are used to efficiently aggregate network traffic into a reduced number of higher delay data messages.

A sensor node has potential for aggregation if it receives, in a timely manner and from different sources, data messages with the same attributes (*e.g.* temperature), but not necessarily the same values. Data fusion will depend on the aggregation function used. We consider a path efficient if the number of aggregation nodes is higher and these nodes are closer to the sources. By selecting these paths, we trade an increase in delay and in processing time for a reduced number of transmissions, while limiting measurement accuracy loss. To identify and select these better paths, we added a software module to the Directed Diffusion implementation available on ns-2 [9]. We named this module *PAR filter* (Privileged Aggregation Routing filter). Our filter acts on the data messages used for route discovery and path setup, i.e. *exploratory* data messages.

Instead of immediately forwarding the first received exploratory data message toward the interest message originator, the PAR filter sets up a timer associated to this message and compares it with other messages received before the timer expires. To perform the comparison, three attributes were added to the message header, namely *distance_to_source*, *aggreg_nodes* and *aggreg_ID*. The first contains the number of hops between the source and the first aggregation node, indicating how far is the aggregation point from that specific source of information. The second stores the number of aggregation nodes in the path followed by that exploratory data message. Finally, *aggreg_ID* keeps an indication of which of the sources in the network are included in the aggregation path.

Messages received from a particular neighbor with the highest *aggreg_nodes* and the lowest *distance_to_source* values are chosen to be forwarded, even if they are received later than other messages, which are discarded. The last field (*aggreg_ID*) is used to avoid computation of the same source twice in the aggregation path, which would give erroneous numbers for *aggreg_nodes* and could jeopardize our mechanism for route selection. Because a node does not have global knowledge of the topology, an exploratory data message from a specific source could have been aggregated two hops away, and the current node has no means of knowing this. The only way to provide the information of which sources are included in the *aggreg_nodes* computation is to pass this information along with the message. If the *aggreg_ID* of the received message matches the messages already in the node, then the current message is not aggregated.

The route selection mechanism works in two phases. The first one occurs when an exploratory data message arrives. It is compared with similar messages already stored in the filter to check if it is a candidate for aggregation or to be discarded. Messages are candidate for aggregation if they respond to the same interest come from different sources and have not been previously aggregated, as shown by their *aggreg_ID*. When two candidates are found, their aggregation parameters are updated to reflect this. To discard a message, the PAR filter checks aggregation parameters of messages that come from the same source, preserving messages with better parameters and discarding the others. The comparison is made using the *aggreg_nodes*(higher is better) and *distance_to_source* (lower is better) values, in this order. If a message is to be kept, a timer is set before transmission. Otherwise, it is deleted. When the timer for a specific message expires, the second phase begins. Now, a double check is performed to identify any modification in the message queue, and then the better message is sent back to the diffusion core.

The Directed Diffusion routing module will act as it normally does, except that now it receives different exploratory data messages than before, with better aggregation parameters, thus different neighbors are reinforced. These neighbors are the ones we chose to privilege aggregation. For instance, suppose a node has two neighbors, a and b, from which it receives similar messages, using the criteria discussed above, and the message from a arrives first. Normally, the a message (low latency) would be forwarded and the b message would be discarded. With the PAR processing, it could be identified that the *aggreg_nodes* value of a is 1, indicating it conveys aggregated information from two distinct sources, where the *aggreg_nodes* value of b is 2, indicating an aggregation of three sources. The PAR filter would then discard a and forward b to the next node.

This process continues in every intermediate node in the path until the sink is reached and the desired (aggregated) routes are reinforced. Empirically, intermediate nodes in the reinforced path receive similar data messages, candidates for aggregation, from different sources. These intermediate nodes are the closest to these sources, which presents the greatest gains in terms of a reduced number of transmissions. The aggregation timer can be used to adjust the desired latency to the number of transmissions ratio. A small timer implies low latency, while a larger timer reduces the number of transmissions with the cost of higher latency. After the desired routes were established, another software module must be attached to the nodes of the sensor field to act on the *data* messages and to aggregate and suppress redundant information. In the remainder of this work, we present two aggregation filters, one using Bayesian Inference (Bayes Filter) and another the Dempster-Shafer Inference (DS Filter).

3.2 Distributed Multisensor Aggregation

In order to reduce the impact of multisensor data aggregation in the WSN, we propose a distributed aggregation mechanism. In such mechanism, aggregation occurs locally at each node on the basis of available information (local observations and information communicated from neighboring nodes). Data does not need to be held at any node in order to be combined, and in this sense it differs from traditional store-and-forward techniques. The result is a distributed processing in the network, which favors scalability and a reduction of transmissions, without occurring in increase of the latency.

The aggregation mechanism starts after the definition of source-to-sink routes, as illustrated in the figure 1. The aggregation procedure consists of keeping a copy of the last evidence received in each node (in the source-to-sink path). This evidence can be used for aggregation for a period T_v, as illustrated in the figure 1(a), which defines the valid time of the evidence. To prevent data aggregation of the same sensor, the parameter T_v must be inferior to the lesser interval of data transmission (I_t) from the sources. Another parameter to be adjusted is the discard time of the messages T_d, illustrated in figure 1(b). This parameter defines the period in which aggregate messages can be considered redundant, and therefore discarded. Finally, in the case of Dempster-Shafer aggregation, the

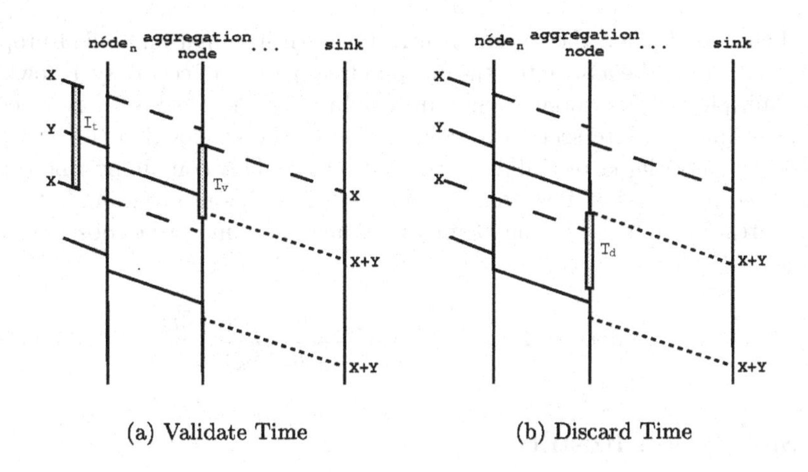

(a) Validate Time (b) Discard Time

Fig. 1. Parameters of the Aggregation Mechanism.

parameter C defines the admitted maximum variation for the degree of certainty of the messages (in an interval T_d).

When a data type message arrives in the node it is forwarded to the aggregation filter. The first step consists of verifying valid evidences in the cache. This is done comparing the difference between current time and the evidences arrival time with T_v. After that, the message's content is read to identify which evidences are being sent. After these procedures, the evidences (local and received) are combined, generating a new message. Then, the cache is updated with the most recent evidences. The new message could be forwarded to Gradient filter or simply discarded, if identical (or without significant changes) when compared with the previous message, within interval T_v.

The nodes carry out aggregation by combining data fusion based on parametric techniques with removal of redundant messages. Such techniques allow a direct mapping between the evidences detected by sensors with events of interest, which is simplified by the data centric addressing schema of Directed Diffusion. The aggregation methods have been implemented in the Bayes (Bayesian Inference) and DS (Dempster-Shafer Inference) Filters, which were software modules added to the protocol.

The Bayesiana Inference [7] was implemented in the **Bayes Filter**. Bayesian Inference is a statistical-based data fusion algorithm based on Bayes' rule (equation 1) that provides a method for calculating the conditional or a posteriori probability of a hypothesis being true given supporting evidence detected by sensors.

$$p(H_i|E) = \frac{p(E|H_i)p(H_i)}{\sum_i [p(E|H_i)p(H_i)]} \tag{1}$$

DS Filter implements data aggregation based on Dempster-Shafer Inference [10]. This technique can be considered as a generalization of the Bayesian Inference, being used when the sensors contribute with information (called propositions) that cannot be associated the one hundred percent of certainty. Knowledge from multiple sensors about events are combined using Dempster' rule (equation 2) to find the intersection or conjunction of the proposition (for instance, e_1) and the associated probability, denoted by m_i. This combining rule can be generalized by iteration if we treat m_j not as sensor S_j's proposition, but rather as the already combined (using Dempster-Shafer combining rule) observation of sensors.

$$(m_i \oplus m_j)(e_1) = \frac{\sum_{E_k \cap E_l = e_1} m_i(E_k) m_j(E_l)}{1 - \sum_{E_k \cap E_l = \emptyset} m_i(E_k) m_j(E_l)} \tag{2}$$

4 Simulation Results

In this section, we describe the results obtained with our simulation model implemented in the ns-2 simulator [9]. In this evaluation, we used two different scenarios. In the first one, a certain number of sensor nodes are randomly distributed over a sensor field of size 1000m x 1000m. The second scenario uses a grid topology where each node is spaced 150m from its neighbors. In both scenarios, all nodes have a transmission range of 250m and the interval between data messages sent by the source nodes is of 1 second. Due to space constraints, some of more detailed simulations and analysis have been omitted.

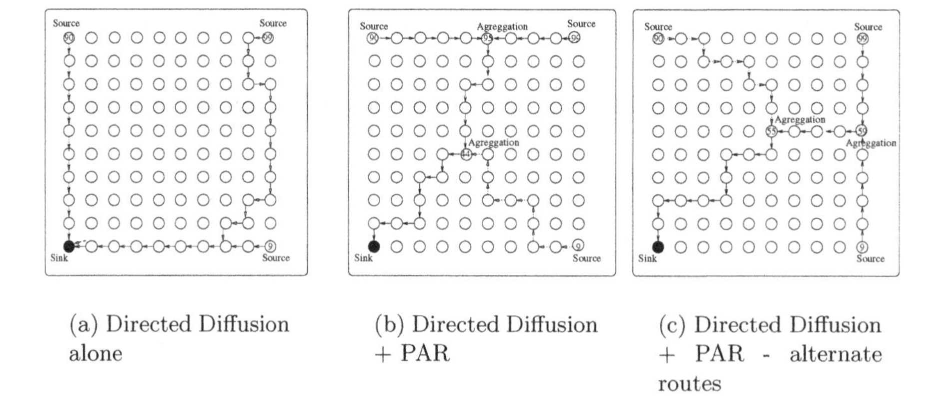

(a) Directed Diffusion alone

(b) Directed Diffusion + PAR

(c) Directed Diffusion + PAR - alternate routes

Fig. 2. Preferred and Alternate Routes.

The first step in the simulation involves route discovery. We tested a 100-nodes grid topology with and without the PAR filter. We found, as expected, that routes discovered by directed diffusion, without PAR, followed independent paths, where preferred and alternate routes can be observed. Figure 2(a) shows the preferred routes for each source. The alternate paths can be explained as a way to keep the energy balance in the network, avoiding early depletion of the nodes involved in the preferred path. When the PAR filter was added, we found new preferred routes. These routes alternate always looking for aggregation nodes, as we can observe in Figures 2(b) (node 95) and 2(c) (node 55).

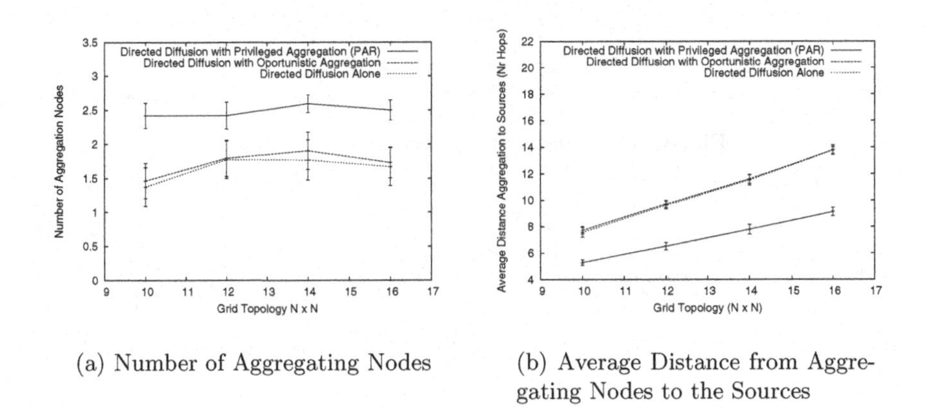

(a) Number of Aggregating Nodes

(b) Average Distance from Aggregating Nodes to the Sources

Fig. 3. Aggregation in the N x N grid.

Figure 3 shows the impact of these new routes in the aggregation parameters. Figure 3(a) shows the number of aggregating nodes for different grid sizes, and Figure 3(b) their average distance to the sources. We may observe that the PAR filter increased the number of aggregating nodes for all grid sizes, while reducing their distance to the sources. As discussed in section 3, we expect that the larger number of aggregation nodes will reduce the number of messages received at the sink, while the small distance from the aggregation point to the sources will reduce the number of hops in the network.

To verify the impact of the aggregation timer in the route discovery, we increased its value in successive rounds of simulation. The timer is important for the PAR filter to collect neighborhood information to decide on best aggregation path. However, it increases the route discovery latency, calculated by the difference between the time of arrival of the first positive reinforcement in the source node and the first corresponding exploratory data sent by the same source. Figure 4 shows this metric as function of the aggregation timer. In larger grids, the timer influence is even larger since the route discovery messages pass through a greater number of nodes. A timer comprised around 0,5s proved to be a good balance between efficient aggregation trees and route discovery latency.

Energy savings are compared through the use of Bayes and DS filters in two simulation scenarios. In the first one, we use the Bayes filter in the grid topology and we vary the grid size. The second scenario uses the DS filter and a random topology for a varying number of nodes. In both scenarios, the number of sources

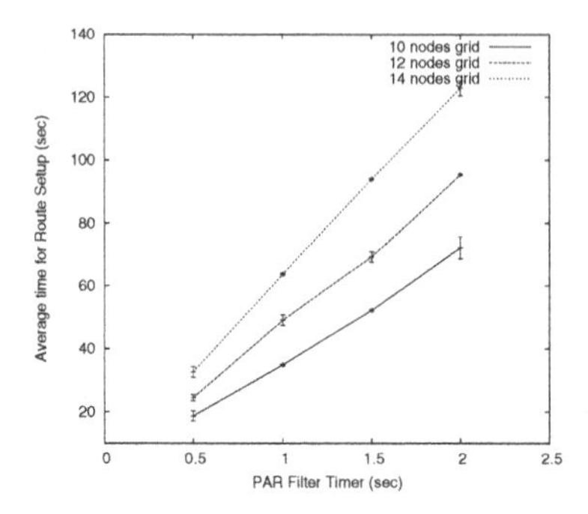

Fig. 4. Route Discovery Latency Versus Timer.

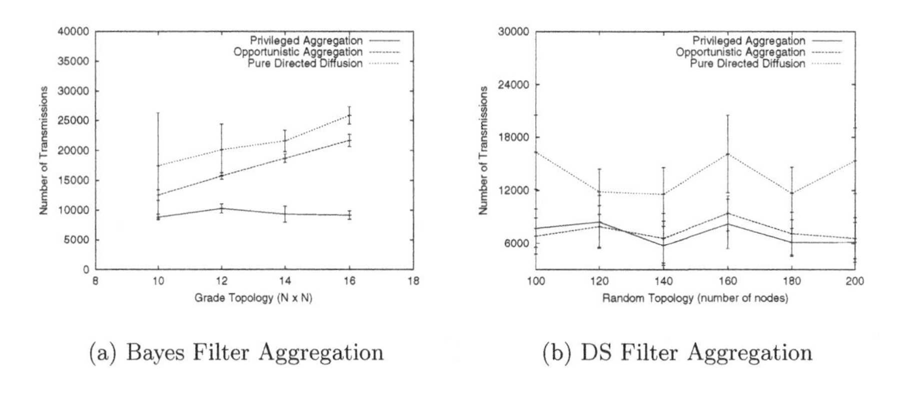

(a) Bayes Filter Aggregation

(b) DS Filter Aggregation

Fig. 5. Total Number of Hop-by-Hop Transmissions.

is constant and equal to five. The total number of transmissions used to send the same amount of information from the source to the sink as function of grid size and the number of nodes is shown in Figure 5. The longer is the route the larger is the number of hops even with the same number of messages arriving at the sink. Normally, the routes obtained with the PAR filter are longer than the ones with the pure Directed Diffusion since the former moves away from shorter routes in order to find aggregation nodes. So, it is important to find efficient aggregation trees to reduce the number of hops. The results of Figure 5 show the impact of bringing the aggregation points near to the sources, which produced a considerable gain in this performance metric. Once the sources were far away from the sink, for bigger grids the aggregation benefits increased, because much less hops were necessary to convey information to the sink.

5 Conclusions

In this paper, we propose a multisensor-aggregation-centric routing protocol that increases sensors network lifetime. In such scheme, two filters were added to the directed diffusion protocol. The first filter is called Privileged Aggregation Routing (PAR) and it acts during route setup to identify intermediate nodes with greater potential to combine data from different sources. These nodes are included in the source-to-sink path in order to favor data aggregation, but with the overhead of increasing route discovery time. The second filter is responsible for the aggregation mechanism. In this paper, we explore two techniques, Bayesian Inference (Bayes Filter) and Dempster-Shafer Inference (DS Filter). Both were implemented in order to minimize the aggregation cost in WSNs, since it occurs locally at each node on the basis of the available information (local observations and information received from neighboring nodes). We show by means of simulation that our aggregation mechanism in a privileged fusion path between sources and sink can reduce drastically the total number of transmissions needed to accomplish the task.

References

1. I. F. Akyildiz, W. Su, Y. Sankarasubramaniam, and E. Cayirici, "A survey on sensor networks," *Communications of the ACM*, pp. 102–114, Aug. 2002.
2. W. R. Heinzelman, A. Chandrakasan, and H. Balakrishnan, "Energy-efficient communication protocol for wireless microsensor networks," *Proceedings of the 33rd IEEE Hawaii International Conference on System Sciences (HICSS)*, pp. 1–10, Jan. 2000.
3. C. Intanagonwiwat, R. Govindan, and D. Estrin, "Directed diffusion: A scalable and robust communication paradigm for sensor networks," *Proceedings of ACM MobiCom'00, Boston, MA*, pp. 56–67, Aug. 2000.
4. J. Kulik, W. R. Heinzelman, and H. Balakrishman, "Negotiation-based protocols for disseminating information in wireless sensor networks," *Proceedings of the 5th Annual ACM/IEEE International Conference on Mobile Computing and Networking (MobiCom'99)*, 1999.
5. M. R. Garey and D. S. Johnson, *Computers and Intractability: A Guide to the Theory of NP-Completeness.* Freeman, San Francisco, 1979.
6. B. Krishnamachari, D. Estrin, and S. Wicker, "Modeling data-centric routing in wireless sensor networks," *IEEE INFOCOM*, June 2002.
7. D. L. Hall, *Mathematical Techniques in Multi-Sensor Data Fusion.* Artech House, 1992.
8. J. Heidemann, F. Silva, Y. Yu, D. Estrin, and P. Haldar, *Diffusion Filters as a Flexible Architecture for Event Notification in Wireless Sensor Networks.* USC/ISI Technical Report 2002-556, 2002.
9. http://www.isi.edu/nsnam/ns, *The Network Simulator NS-2.*
10. H. Wu, R. S. Mel Siegel, and J. Yang, "Sensor fusion using dempster-shafer theory," *In Proceedings of the IEE Instrumentation and Measurement Technology Conference, Anchorage, AK, USA*, May 2002.

A Self-Organizing Link Layer Protocol for UWB Ad Hoc Networks*

N. Shi and I.G.M.M. Niemegeers

Center for Wireless and Personal Communication
Delft University of Technology, Delft, the Netherlands
{n.shi, i.niemegeers}@ewi.tudelft.nl

Abstract. Ultra Wide Band (UWB) impulse radio is a promising technology for future short-range, low-power, low cost and high data rate ad hoc networks. The technology is being explored in a number of research projects. While most UWB research for this class of networks is concentrating on the physical layer, little research has been published on link layer protocols which exploit the specifics of UWB impulse radio. In this paper, we focus on the self-organization concept and the peculiarities of UWB technology from a physical and a link layer point of view. A novel self-organizing link layer protocol based on time hopping spread spectrum is proposed in this paper. This protocol promises to be an efficient and collision-free mechanism that enables the devices to discover neighbor nodes and arrange the access to communication resources shared among the nodes. The adjustable parameters of the protocol enable the network to adapt to a dynamic environment.

1 Introduction

A range of services supporting future mobile applications are expected to require high data rates, high communication quality and efficient network access. A case in point is mobile interactive gaming, where fast transmission of image and voice in dynamic environments is a prerequisite.

Wireless networks that meet these expectations will have a hybrid character, consisting mainly of ad hoc networks with occasional access to infrastructures, in order to reach remote nodes or infrastructure-based servers. They will have to operate completely automatically without the intervention of system administrators, and therefore will have to be self-organizing. Self-organization in this context implies the automatic finding of neighbor nodes, the creation of connections, the scheduling of transmissions and the determining of routes. This should be performed in a distributed manner so that all nodes in the network are able to exchange information and reconfigure the network when nodes join or leave, or when radio links are broken or established.

A promising but, because of implementation difficulties, not well explored radio technology is UWB impulse radio. This technology has a lot of potential for high-

* This research is part of the AIRLINK project funded by the Dutch Ministry of Economic Affairs under the Freeband Impulse Program.

I. Niemegeers and S. Heemstra de Groot (Eds.): PWC 2004, LNCS 3260, pp. 248–261, 2004.

bandwidth, low-power, low-cost and short-range communication in self-organizing ad hoc networks. Self-organization is helped by flexible radio resource management. UWB impulse radio has this intrinsic flexibility [1] because of the parameters of the physical layer that can be explicitly controlled by the upper layers. Furthermore it offers the potential for achieving high data rates with low spectral density and immunity to interference [2]. Its radar like pulses can in principle also be used for high precision positioning.

The research on UWB-based ad hoc networks is still in its infancy. Currently, IEEE 802.15.3a [3] is considering UWB technology for the physical layer. A debate is going on whether this should be based on OFDM or on impulse-based radio. OFDM technology is well understood, for impulse-based radio on the other hand, research is needed to understand the practical limitations and to come up with technical solutions. This paper intends to contribute to this goal.

An ongoing project, which explores UWB impulse radio technology for use in short-range ad hoc networks, is the AIR-LINK project [4]. The research reported on here was carried out in this project. We approach the UWB ad hoc network for the applications that need to efficiently establish connections and exchange data at high speed within a short range. We consider the scenarios where the transmission distance is about 10 to 100 meters, the data rate is at least 100 Mbps and the time needed for establishing a network is of the order of 0.1s.

In this paper the key issues involved in UWB ad hoc and self-organization are analyzed and a new link layer device discovery protocol is proposed to discover disconnected nodes within radio range and establish links among them. We assume that the network layer and higher layers will be based on IETF protocols. We also assume that the communicating devices are energy constrained, since many of them will be portable and battery powered.

The paper is organized as follows. In Section 2, we introduce the concept of self-organization. In Section 3, we describe an UWB ad hoc network and address the physical and link layer issues. In Section 4, a new self-organization device-discovery protocol (SDD) is proposed and specified. A conclusion is made in Section 5.

2 Self-Organization

Self-organization refers to the ability of a system to achieve the necessary organizational structures without human intervention [5]. In our case it is the process network nodes go through to autonomously organize and maintain a network topology either at network initialization or during operation. During the latter the topology may change due to nodes joining or leaving the network, and links appearing and disappearing. Note that, given a set of connected nodes, the topology may change because extra (radio) links become operational, making the topology more interconnected. One should also envisage the partitioning of a given topology into two, no longer connected networks and the merger of two networks when a radio link is established between them.

Self-organization plays a role both at link and network layer. The issues that need to be addressed are neighbor discovery and connection setup, link scheduling and channel assignment, network topology formation and re-configuration, control and routing information adaptation and mobility management [6].

Self-organization has been investigated in ad hoc networks, Bluetooth scatternets, and sensor networks. Recently a number of protocols have been proposed in these areas. An example of a device discovery protocol for short-range wireless ad hoc networks is the DD protocol proposed in [7]. It has been designed to be time-efficient when the number of devices in the network is large. The protocol is designed to perform well when devices that are in-range of each other become active in a short time interval. Some Bluetooth scatternet construction protocols [8] contain self-organization procedures that cater for the case of nodes frequently joining and leaving the scatternet, leading to dynamic topologies. However these protocols rely on the Bluetooth random-access inquiry mechanisms, which are reputed to be slow; delays of the order of 10s [9] are not unusual. Therefore, these protocols are unsuitable for very dynamic environments. Sensor networks [10] are formed by collections of distributed sensor nodes in order to sense the environment and inform users. The self-organization process in this type of network is designed with an emphasis on prolonging the lifetime of the network without the need for human assistance. This is achieved by initiating a communication-link schedule among the nodes and maintaining this schedule periodically in the long-term. Protocols belonging to this class have been proposed in [11], e.g., SMACS and EAR. Because of the static nature of the sensor nodes, the self-organization process is mostly a one-time initialization effort, except in case nodes fail.

3 UWB Ad Hoc Networks

The UWB ad hoc network we envisage is composed of a set of nodes, each of which is assumed to be equipped with an UWB transceiver. The nodes are personal digital devices such as notebooks, PDAs, mobile phones or, in principle, any device with computing and UWB communication capabilities. An example is shown in Fig. 1.

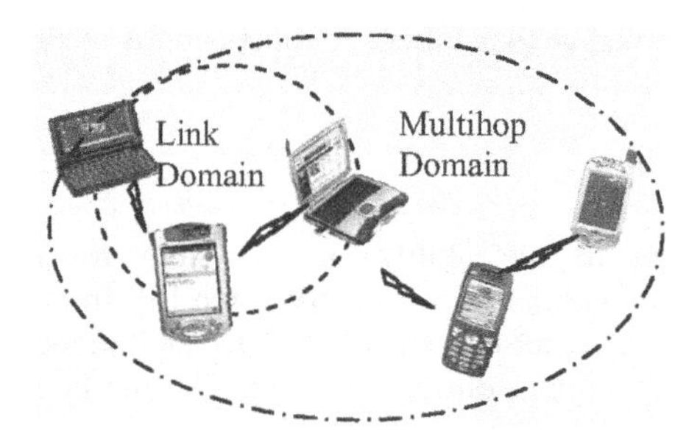

Fig. 1. UWB ad hoc network

3.1 Communication Architecture

We assume that the network has a multi-hop architecture. This implies that we can distinguish two domains: the link domain corresponding to sets of nodes that are within radio range of each other, and the network domain corresponding to multi-hop interconnection of nodes in different link domains (Fig. 1). Protocols for device discovery, the establishment of UWB physical connections between a node and its neighbors and MAC protocols belong to the link domain. The network domain is where end-to-end multi-hop connectivity is realized. In this domain, each node is assumed to have equal functionality. In particular, each node is able to initiate end-to-end communication and to forward packets on behalf of other nodes.

3.2 Impulse Radio

The UWB impulse radio signal consists of a spreading pulse train in a framed period. The duration of each frame is equal and is divided into multiple time bins. Each pulse belonging to one signal is repeated per frame in a randomly positioned time bin (Fig. 2). The position sequence, which is called *time hopping sequence* or *TH code* [12], is based on a pseudorandom (PN) process. The pulses are further dithered within a bin based on the information bit of the signal.

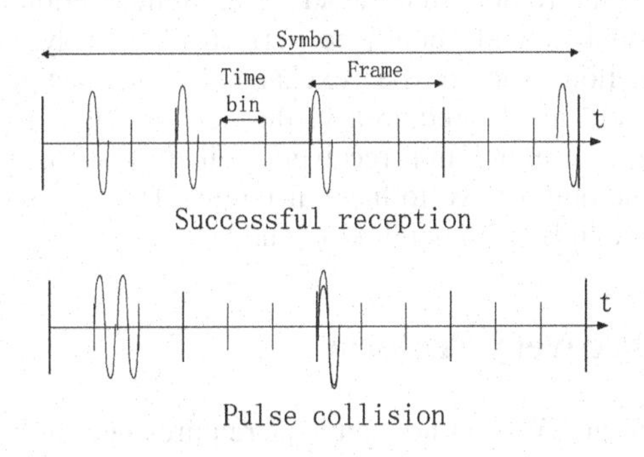

Fig. 2. Impulse radio signal

The reception of an impulse radio signal is performed by a correlator [13]. A template signal is applied to the received signal to retrieve the original pulse train. The expected TH code is used to generate the template signal. When pulses of two signals (Fig. 2) arrive in one time bin, a pulse collision happens. When two signals encoded by the same TH code are received simultaneously, a collision happens if they overlap exactly.

Impulse radio offers a flexibility that can be exploited in the design of a self-organizing ad hoc network. First the parameters, e.g., the TH code, can be used to identify the network nodes and to indicate the required radio resource. Second, because a TH code is applied to the original pulse train, only a small fraction of the frame is occupied (typically one percent or less) [14]. Thus, a collision hardly happens in multiple packets transmission. If a device uses two orthogonal TH codes for transmitting and receiving, the signals can be received without interfering with the signals that are being transmitted; if a device simultaneously transmits or receives multiple packets which are transmitted on orthogonal TH codes, collisions are avoided.

However, the potential drawback of impulse radio is its long signal acquisition time. This is the time required for a receiver to collect sufficient energy for bit synchronization. Synchronization between two devices requires one or more signal acquisition times. Because the transmitted pulse train has an extremely low power spectral density, impulse radio receivers inherently have a longer signal acquisition time than conventional receivers [15]; this can be of the order of milliseconds. Our physical layer design aims at a value of 0.1ms.

3.3 Link Layer Issues

UWB-based medium access control (MAC) can in principle be achieved by contention-based and by explicit methods. Both methods are impacted by the long acquisition time. The impact is more obvious in the contention-based method. For instance, the CSMA/CA protocol has a high packet delay because the signal acquisition occurs twice per packet transmission [16].

Since we consider a dynamic environment, access schemes with fixed periodical schedules, e.g., TDMA as used for sensor networks [11] are clearly not applicable. Consistent with the requirement that the ad hoc communication should be completely distributed, link establishment should be performed in a symmetrically distributed way; link information should be exchanged with neighbor nodes. Hence, synchronization should not be required in the network domain but only in the link domain between transmitters and receivers. Finally, since time hopping spread spectrum is used, the nodes have to have different TH codes so that they are able to initialize conflict-free links with neighboring nodes.

4 The Device Discovery Protocol

Self-organization of an UWB ad hoc network requires new link layer protocols to be able to automatically discover the in-range nodes, form a distributed link-layer topology and dynamically arrange the access to communication resources shared among the nodes.

4.1 Protocol Assumptions

We consider the situation that a node has no prior knowledge about its surroundings. When a node is powered on, the node is able to establish direct links with in-range nodes after it has exchanged its time hopping pattern with these nodes. When a node loses most of its direct links, this is an indication that the topology of the network has significantly changed; the node then needs to discover its neighbors again. We also need a link level mechanism for data transmission over established links. An RTS-CTS handshake is used for this purpose.

In the sequel we describe a self-organizing distributed protocol on top of the TH-based impulse radio physical layer. We also describe the essentials of the protocol for data transmission.

We consider a scenario where a variable number of nodes within radio range of each other are expected to form and maintain a single-hop ad hoc network. It is assumed that the nodes can move at most at walking speed. Let $K(t)$ denote the number of the nodes in-range at time t. $K(t)$ increases when nodes that are powered on or move into the radio range join; $K(t)$ decreases when nodes that are powered off or move out of the radio range leave. Each node keeps a list of the TH codes of its neighbors. This list is denoted by $L_i(t)$ (i can be any node, i <= $K(t)$). We assume that the TH codes are sufficient to identify a node. $L_i(t)$ is updated whenever a node discovers a new TH code or detects the disappearance of a neighbor; it records the TH codes of its neighbors.

A node is not aware of the value of $K(t)$; it only knows $L_i(t)$. When a node i initially joins, it has no information about the other $K(t)$-1 nodes, i.e., $L_i(t)$ is empty, unless it contains outdated information about its surroundings. If a node j other than i leaves an established network, the information in each $L_i(t)$ is updated only after node i detects the absence of node j. In addition, in order to avoid an explicit removal procedure, a TH code in $L_i(t)$ expires after it has not been used for a period longer than a timeout T_{lmax}.

We assume omni-directional antennas, a negligible propagation delay and errorless channels. The nodes know their own TH codes. We assume that the UWB device consumes very little power when it is continually monitoring a TH code.

We define a packet collision on a TH code if two or more packets transmitted on the same TH code arrive at one node at the same moment; the physical layer detects this. When multiple packets transmitted on different TH codes arrive at a node simultaneously, the node is able to decode the information on its own TH code while the information on other codes is perceived as noise.

4.2 Self-Organizing Device Discovery Protocol (SDD)

The device discovery protocol, SDD, is based on a RTS-CTS type of signaling dialog on the link layer.

4.2.1 Overall Protocol Procedure

The SDD protocol is a link layer protocol. It directly interfaces with the physical layer and the network layer. Via the physical layer interface the following information is communicated:

Physical layer to link layer:

- A collision indication.
- The TH code of the received packet, i.e., the identification of the sender.
- The information decoded from the packet received on a particular TH code.

Link layer to physical layer:

- The information to be sent out via the physical layer

The performance parameters of the SDD protocol, in particular, the discovery time, the node join-time, the node leave-time and the data transmission time, are impacted by the parameters of the physical layer, in particular by the signal acquisition time.

Via the network layer interface the following information is communicated.

Link layer to network layer:

- Current link information used for routing
- Service data unit that needs to be forwarded

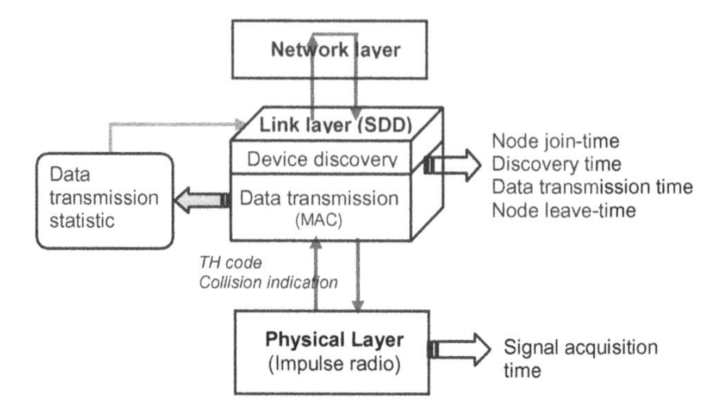

Fig. 3. SDD overall protocol procedure

Network layer to link layer:

- Service data unit when data transmission is requested from the application layer of the device
- Service data unit when data is requested to be forwarded

The SDD protocol is initialized when the node is powered on and terminates when the node is powered off. The protocol consists of two types of sub-processes: *device discovery* and *data transmission* (Fig. 3).

A device discovery process can be driven by a discovery process timer with an adaptive time interval, denoted by $T_{interval}$. The short time interval is used when a device is powered on to form a network and join a network or, in case it has reason to believe it has lost its connectivity with its surroundings, e.g., due to movement. In the former case, the neighbor list $L_i(t)$ is empty. A node could assume the latter has happened, e.g., when the data transmissions to most of its neighbors on $L_i(t)$ keep failing persistently. In addition when a destination node one-hop away from node i is not included in $L_i(t)$, a device discovery process has to be executed before the data transmission process can be initialized. The discovery process will execute in a long interval, $T_{interval}$, set to a large value, when the number of the discovered nodes $L_i(t)$ is larger than a maximum value, L_{MAX}, or no more neighbors are discovered in three continuous discovery processes.

The data transmission process in a node associated to one of its neighbors executes whenever there is a request from the higher layers of the node to send data to its neighbor.

The device discovery and data transmission processes contend in using the physical channel resources. For example a node could be attempting to discover other nodes while a data-sending request arrives from upper layer or one neighbor node. We will show later that both actions require the use of the same TH code. However one process cannot preempt another one and has to wait until the other process is completed and resources are released again.

4.2.2 Definitions

We assume each node is assigned three basic types of TH codes [17]. The codes are used in the device discovery process and the data transmission process.

- *Common code C*: a fixed short TH access code known by all nodes. It is used to initialize device discovery and data transmission through a broadcast operation.
- *Receiver-based code* $C_{i,R}$: the TH code of node i for receiving packets from neighboring nodes. It is used in device discovery for an inquiring node to receive responses from the nodes inquired by it.
- *Transmitter-based code* $C_{i,T}$: It is the TH code of node i for transmitting packets to neighboring nodes. This code is used in both device discovery and data transmission. In *half-duplex* communication (e.g., i to j), the code in use is always associated with $C_{i,T,}$ which is the transmitter-based TH code of the node that needs to send data. *Full-duplex* communication between nodes i and j is initialized by exchanging the transmitter-based codes of both nodes, i.e., $C_{i,T}$ and $C_{j,T}$.

The receiver-based and transmitter-based TH codes are generated by a pseudo random process. The unique 48-bit IEEE MAC address of a device can be used as the key of the process.

The protocol defines seven PDU types defined as follows:

- *IS packet* (Inquiry): The purpose of this packet is to inquire the as yet undiscovered neighboring devices. It contains packet type, the TH codes of the inquiring node, i.e., $C_{i,R}$ and $C_{i,T}$, and the *response scan window* size $N_i(t)$. It is transmitted on the C code.
- *IR packet* (Inquiry Reply): The purpose of this packet is to reply to the IS packet. Apart from the packet type, it contains the transmitter-based code of the inquired node $C_{j,T}$. It is sent using the receiver-based TH code $C_{i,R}$ of the inquiring node.
- *LRS packet* (Low Rate Synchronization): This packet is used to keep synchronization between the inquiring node and the undiscovered neighbors it is inquiring. It is a short packet with packet type information. It is sent by the inquiring node using code $C_{i,T.}$
- *RTS packet* (Request to Send): The purpose of this packet is to synchronize and inform the potential receiving node to be ready for data transmission. It contains packet type, the transmitter-based TH codes of the receiving node(s) and transmitting node, e.g., $C_{j,T}$ and $C_{i,T}$, assuming that the data should be transmitted from i to j. It is sent using the C code.
- *CTS packet* (Clear to Send): The purpose of this packet is to confirm the request for data transmission. It contains the code of the transmitting node, i.e., $C_{j,T}$, to match to the RTS. It is sent using the C code.
- *Data packet*: This packet carries the data that needs to be sent to the receiver. The packet is transmitted using code $C_{i,T}$ assuming node i is the sender.
- *ACK packet*: The purpose of this packet is to confirm the successful reception of a Data packet. The packet is sent using the $C_{j,T}$ code.

An operational node has eight states. These states correspond to different phases of the protocol, which are explained as follows:

- *Inquiry*: The node is initiating an inquiry request in order to discover the as yet unknown neighboring nodes.
- *Inquiry response scan*: An inquiring node stays in this state while waiting for the inquiry responses from its not yet discovered neighboring nodes.

- *Inquiry response*: An inquired node stays in this state until it replies to the inquiring node.
- *Data send request*: The node is broadcasting a request to send a packet.
- *Data request response*: A node is replying to the sender to be clear to send.
- *Data send*: The data is being sent from this node.
- *Data receive*: The node is receiving data.
- *Idle:* The node state is idle, when there is no ongoing operation on the node. At this time, common code is always free and available for initiating new operations.

4.2.3 Device Discovery Process

The function of the device discovery process is to automatically discover new in-range nodes and quickly form a distributed link-layer topology. The discovery process adapts itself to the number of nodes that are in-range.

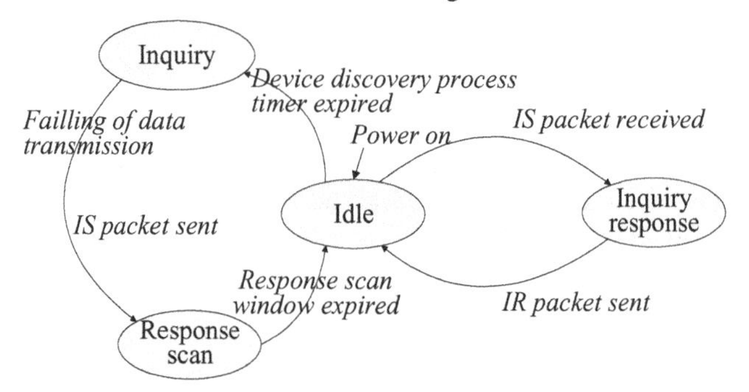

Fig. 4. State transition diagram of the device discovery process

The device discovery process includes the operations of *inquiry* and *response scan* running on the inquiring node, and *inquiry response* running on the inquired nodes. Fig. 4 shows the state transition diagram of the device discovery process.

After a node is powered on, it enters the *Idle* state and waits for a random period T_{start}. Then it moves to the *Inquiry* state. A transition from the *Idle* state to the *Inquiry* state can also occur when either the device discovery process timer expires or the node detects during data transmission that the number of consecutive failing transmissions to is above a threshold or the destination node is not a neighbor of the node. In *Inquiry* state an IS packet is generated by the node and is broadcast. After sending the IS packet, the inquiring node moves to the *Response scan* state to receive the responses from its neighbor nodes. All neighbor nodes in *Idle* state will receive the IS packet and synchronize with the inquiring node. Then, each inquired node gets into the *Inquiry response* state and schedules an IR packet at a randomly chosen time to reduce the probability of packet collisions. After these operations, the inquiring and inquired nodes return to the *Idle* state.

We use an example with two nodes i and j to illustrate how the process works. As shown in Fig. 5, when node i is switched on, node i first stays in *Idle* state for a random period, T_{start}. The discovery process starts when node i broadcasts an IS packet on code C. The acquisition header of the IS packet allows node j, which is in range and listening, to synchronize with node i. After sending the IS packet, node i starts a response scan operation for receiving responses from the inquired nodes. The number of the slots during the response scan operation is called the *response scan window size* $N_i(t)$. During this operation, node i will periodically send LRS control packets on code $C_{i,T}$ to keep all inquired nodes synchronized.

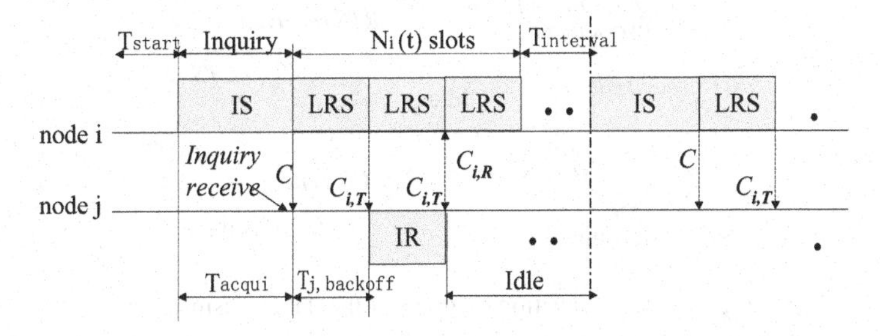

Fig. 5. Device discovery process

If the link i-j already exists, j doesn't have to reply and returns to *Idle* state monitoring code C. Otherwise, j starts listening on code $C_{i,T}$ to receive periodically LRS packets, which keep it synchronized with node i. After a random number of time slots, denoted by $T_{j,backoff}$, node j responds with an IR packet on code $C_{i,R}$. $T_{j,backoff}$ is determined by node i's scanning window size $N_i(t)$. It is randomly chosen in the range $[0, N_i(t)]$. Afterwards, node j stops listening to code $C_{i,T}$ and returns to Idle state. If node j is successful in sending the IR packet containing node j's TH code $C_{j,T}$, node i and j have discovered each other and are ready for data transmission.

Adaptation Algorithm of Response Scan Window

A node i can adjust the response scan window size $N_i(t)$ that will be used for the next device discovery process based on the reception statistics during the response scan operation. The result of every time slot as observed by node i can be *null*, *success* or *collision*; null implies that no IR packet was received nor a collision occurred, success means that an IR packet was successfully received and collision means that a collision was detected. Node i counts the number of the three results during $N_i(t)$ time slots. At the end of the response scan operation, if the number of *null* results is much larger than the other two, it means that the response scan window size $N_i(t)$ is too large for the number of in-range nodes and can be reduced without significantly increasing the probability of collisions. If the number of *success* results is dominating, it implies that $N_i(t)$ is a suitable number. Finally, if the number of *collision* results is much larger than the others, it indicates that $N_i(t)$ is too small and should be enlarged for the next device discovery process.

4.2.4 Data Transmission Process

The data transmission process is initiated by an RTS-CTS handshake using code C. Unicast and multicast are both supported. In multicast operation, multiple TH codes of the destination nodes are included in one RTS packet. Our design allows simultaneously transmission and reception on several orthogonal TH codes. Multicast operation is done by encoding the same data information into the multiple TH codes and sending them at the same time. A successful multicast operation is confirmed if the source node receives the acknowledgments from all the destination nodes.

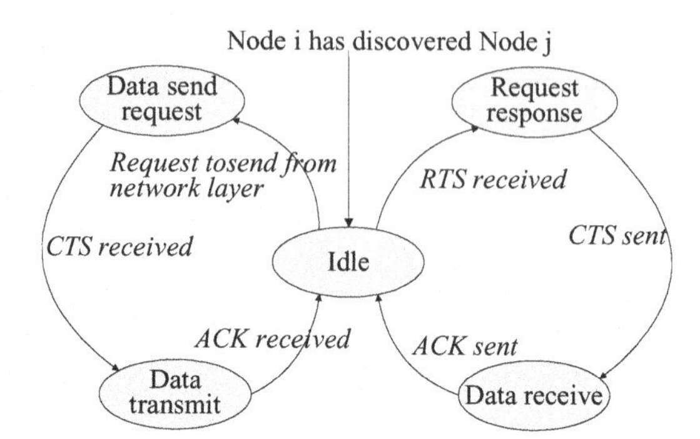

Fig. 6. State transition diagram of the data transmission

Data transmission can only happen after the destination node has been discovered by the source node. Data transmission is invoked by a request from the network layer or a reception of a RTS packet. In Fig. 6, when a node is in *Idle* state and needs to send data to a neighbor node, it makes a transition to the *Data send request* state and sends an RTS packet. If the destination node is in *Idle* state, it responds with a CTS packet and moves to the *Data receive* state. The source node then begins sending data and moves to the *Data send* state. After the source node receives confirmation of successful data transmission by receiving an ACK from the destination node, both nodes make a transition to the Idle state.

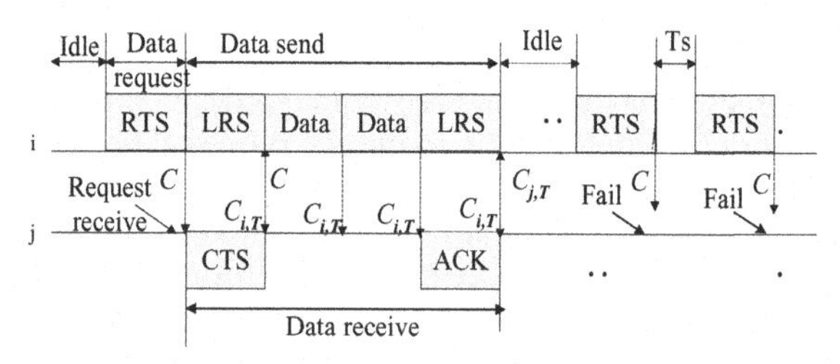

Fig. 7. Data transmission

Fig. 7 illustrates this process. Combining two half-duplex links, i.e., by having a handshake in both directions, forms a full-duplex link. Let us assume that node i needs to send data to node j. As shown in Fig. 6, if node i is *Idle*, it first broadcasts an RTS packet on code C. Without waiting for the response, node i continues to send an LRS packet on code $C_{i,T}$ to keep node j synchronized. Right after node j receives the RTS packet, it replies to node i by sending a CTS on code C. In the meantime, node j begins to monitor code $C_{i,T}$. Subsequently, the data packets are transmitted on code $C_{i,T}$ from node i to node j. After all the data packets are transmitted, node j replies to node i with an ACK packet on code $C_{j,T}$. Finally both nodes i and j return to *Idle* state.

When node i does not receive a CTS packet from node j, node i will periodically attempt to resend an RTS packet with period T_s on code C for at most m times. This is illustrated in Fig. 7. If none of the RTS packets are replied to, it means that either node j has left the radio range of node i, has been powered off or radio communication between i and j is disturbed for a longer time.

4.2.5 Collision Analysis in Multiple Nodes Example

In the SDD protocol, we assume that a packet collision only happens when two or more packets arrive at the receiver at the exact same moment on the same TH code. The collision problems of SDD protocol on the common code and on other codes can be illustrated by using multiple nodes examples of device discovery process.

Collision Avoidance on the Common Code

If a node successfully accesses code C, the other nodes in its radio range are aware of this. Every node uses code C only in the beginning of the device discovery process and the data transmission process. A collision happens on code C only when two or more IS or RTS packets arrive at a node at the same time. When the packets partly overlap, collisions don't happen. The random backoff procedure (with backoff time $T_{start)}$ has been introduced in the SDD protocol (see Section 4.2) to avoid potential collisions when all the in-range nodes are powered on at the same time. Therefore, the probability of a collision on the common code C is minimized.

Collision on Code $C_{i,R}$

In the Inquiry Response state of the device discovery process, the number of successful transmissions depends on the mismatch between the response scan window size $N_i(t)$ and the number of neighbor nodes that receive the inquiry, denoted by $N_{i,inquired}(t)$. When $N_{i,inquired}(t)$ is much larger than $N_i(t)$, i.e., the maximum number of responses node i can receive, there is a higher probability of collisions. In Fig. 8, node i sends an IS packet on code C and starts to scan code $C_{i,R}$. The in-range nodes j and k successfully receive node i's inquiry. Node j and k only scan code $C_{i,T}$ to stay synchronized but do not scan code C any more. Thus, the IS or RTS packets from other nodes, e.g., node l, cannot be received by node j or k. Once their random back-off periods, i.e., $T_{j,backoff}$ and $T_{k,backoff}$, are expired, they send IR packets to node i on code $C_{i,R}$. When $T_{j,backoff}$ is equal to $T_{k,backoff}$, a collision happens between their IR packets (see Fig. 8). Node i detects this collision on $C_{i,R}$ and sets the reception result of the time slot as *collision*. Afterwards, node j and k return to monitor code C.

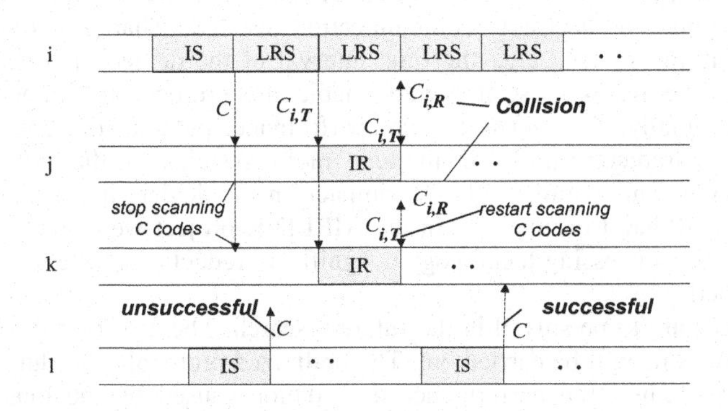

Fig. 8. Collision on $C_{i,R}$

4.3 System Parameters and Variables

The following system parameters and variables are used and need to de determined in the SDD protocol:

- $K(t)$: The number of nodes in radio range at a time.
- $L_i(t)$: The list of the TH codes that node i has found at a time.
- L_{MAX}: The maximum number of neighbors or TH codes that a node can store.
- $T_{interval}$: The time interval between two adjacent device discovery processes.
- T_{lmax}: The maximum time that a TH code can be kept unused in $L_i(t)$ before it expires.
- T_{start}: The random period that a node has to wait before moving to the Inquiry state after it is powered on. It is used to avoid collisions caused by simultaneous power on.
- $N_i(t)$: Response scan window size at time t.
- T_{slot}: The duration of a time slot, a typical value could be 0.05ms
- $N_{i,inquired}(t)$: The number of nodes that receive an inquiry from node i as a result of a single inquiry.
- $T_{i,backoff}$: A random number of time slots that a node waits before sending a response to an inquiry.
- T_s: The repetition period of an RTS message, in case a neighbor does not answer with a CTS packet.
- m: The maximum number of times that a device repeats an RTS packet, in case a neighbor does not answer.

We are presently investigating the performance of the protocol by means of a Glomosim [18] simulation model. The most important performance parameters are the node join-time, the discovery time of the device discovery process, the data transmission time of a packet, the node leave-time and the throughput.

5 Conclusion

In this paper we have presented a novel self-organizing link layer protocol for short-range ad hoc networks based on impulse radio UWB. The protocol exploits the specific features of time hopping spread spectrum. This protocol promises to be an efficient and collision-free mechanism that enables the devices to discover neighbor nodes and arrange the access to communication resources shared among the nodes. The adjustable parameters, e.g., the time interval of the device discovery processes and the adaptive response scan period enable the protocol to adapt to dynamic environment. Finally, the main system and performance parameters were identified.

A critical parameter in UWB impulse radio systems is the potentially long acquisition time. This parameter has an impact on the device discovery process and the efficiency of data transmission. In the AIRLINK project we expect that by using optimized signal processing technology a significant reduction of the acquisition time can be obtained.

Some issues are to be solved in the future research. The effort on the reduction of the acquisition time will be carried out. The optimized protocol procedure will also be worked out to reduce the performance degradation caused by the long acquisition time.

The protocol is being implemented in a Glomosim simulator, which will be used to analyze the performance and to optimize the system parameters.

References

1. M.G. Di Benedetto, P. Baldi, "A Model for Self-organizing Large-scale Wireless Networks", Invited paper, Proceedings of the International Workhsop on 3G Infrastructure and Services, Athens, July 2001, pp. 210-213
2. J. Foerster, E. Green, S. Somayazulu, D. Leeper, "Ultra-Wideband Technology for Short- or Medium-range Wireless Communications", Intel Technology Journal Q2, 2001
3. http://grouper.ieee.org/groups/802/15/pub/TG3a.html
4. http://www.freeband.nl/projecten/airlink/ENindex.html
5. H. Balakrishnan, S. Seshan, P. Bhagwat, M.F. Kaashoek, "Self-organizing Collaborative Environments", (Position paper) NSF/DARPA/NIST Workshop on Smart Environments, Atlanta, GA, July 1999.
6. A.N. Zadeh, B. Jabbari, "Self-organizing Packet Radio Ad Hoc Networks with Overlay (SOPRANO)", Communications Magazine, IEEE, vol. 40, issue 6, June 2002, pp. 149 - 157
7. P. Popovski, T. Kozlova, L. Gavrilovska, R. Prasad, "Device Discovery in Short-Range Wireless Ad Hoc Networks", Proc. 5th Conf. Wireless Personal Multimedia Communications (WPMC), October 2002
8. T. Salonidis, P. Bhagwat, L. Tassiulas, R. LaMaire, "Distributed Topology Construction of Bluetooth Personal Area Networks", INFOCOM 2001, Proceedings. IEEE, vol. 3, April 2001, pp. 1577 - 1586
9. D. Groton, J.R. Schmidt, "Bluetooth-based Mobile Ad Hoc Networks: Opportunities and Challenges for a Telecommunication Operator," Vehicular Technology Conference, Spring 2001, vol. 2, pp. 1134-1138.
10. I.F. Akyildiz, Weilian Su; Y, Sankarasubramaniam, E. Cayirci, "A Survey on Sensor Networks", Communications Magazine, IEEE, vol. 40, Aug 2002, pp. 102 – 114
11. K. Sohrabi, J. Gao, V. Ailawadhi, G.J. Pottie, "Protocols for Self-organization of a Wireless Sensor Network", IEEE Personal Communications, October 2000
12. M.Z. Win, R.A. Scholtz, "Ultra-Wide Bandwidth Time-Hopping Spread-Spectrum Impulse Radio for Wireless Multiple-Access Communications", IEEE Transactions on Communications, vol. 48, no. 4, April 2000
13. M.Z Win, R.A. Scholtz, "Impulse Radio: How It Works", IEEE Comm. Letters, vol. 2, no. 2, February 1998, pp. 36–38
14. F. Cuomo, C. Martello, A. Baiocchi, "Radio Resource Sharing for Ad Hoc Networking With UWB", IEEE Journal on Selected Areas In Communications, vol. 20, no. 9, December 2002
15. S.S.Kolenchery, J.K.Townsend, J.A. Freebersyser, "A Novel Impulse Radio Network for Tactical Military Wireless Communications", Proceedings of IEEE Milcom '98, Boston, October 1998.
16. J. Ding, L. Zhao, S.R. Medidi, K.M. Sivalingam, "MAC Protocols for Ultra-Wide-Band (UWB) Wireless Networks: Impact of Channel Acquisition Time", ITCOM, 2002
17. M. Joa-Ng, I.T. Lu, "Spread Spectrum Medium Access Protocol with Collision Avoidance in Mobile Ad-hoc Wireless Network", INFOCOM '99. Proceedings. IEEE, vol. 2, March 1999, pp. 776 - 783
18. http://pcl.cs.ucla.edu/projects/glomosim/

Quality of Service Support in Wireless Ad Hoc Networks Connected to Fixed DiffServ Domains

M.C. Domingo and D. Remondo

Telematics Eng. Dep., Catalonia Univ. of Technology (UPC)
Av. del Canal Olímpic s/n.
08860 Castelldefels (Barcelona), Spain
{cdomingo, remondo}@mat.upc.es

Abstract. This paper analyzes the provision of end-to-end Quality of Service between nodes in a mobile ad hoc network and a fixed IP network that supports Differentiated Services. The ad hoc network incorporates the Stateless Wireless Ad Hoc Networks (SWAN) model to perform admission control for real-time traffic flows. We propose a new protocol, named DS-SWAN (Differentiated Services-SWAN), where end-to-end delays and loss rates of real-time traffic are monitored continuously at the destination nodes in the fixed network and at the edge routers respectively. In this way, nodes in the ad hoc network are warned when congestion is excessive for the correct functioning of a real-time application (specifically, Variable Bit Rate Voice-over-IP), so that the nodes restrain best-effort traffic in order to favour real-time flows. The results indicate that DS-SWAN significantly improves end-to-end delays without starvation of background traffic, adapting itself to changing traffic and network conditions in a relatively small ad hoc network. Besides, we compare different notification procedures in DS-SWAN aimed to improve scalability.

1 Introduction

Ad hoc networks [1] are formed by mobile devices that are able to communicate without having to resort to a pre-existing network infrastructure. In an ad hoc network, terminals can communicate with each other even if they are out of range because they can reach each other via intermediate nodes acting as routers.

At first glance, it may seem incoherent to deal with Quality of Service (QoS) support in such dynamic systems with unreliable wireless links. However, some authors have presented proposals to support QoS in wireless ad hoc networks including QoS oriented MAC protocols [2], QoS aware routing protocols [3] and resource reservation protocols [4]. Moreover, a flexible QoS model for mobile ad hoc networks has been proposed in [5]. This paper explores the dynamics of a system where a resource reservation mechanism within the ad hoc network co-operates with the Differentiated Services (DiffServ) domain of the fixed network to which the ad hoc network is attached. The aim of this work is to investigate whether aiding resource reservation mechanisms at the ad hoc network by DiffServ based QoS support could yield satisfactory end-to-end QoS properties.

I. Niemegeers and S. Heemstra de Groot (Eds.): PWC 2004, LNCS 3260, pp. 262–271, 2004.

Specifically, we consider a scenario where an ad hoc network is connected via a single gateway to a fixed IP network that supports DiffServ. The ad hoc network incorporates the SWAN [10] scheme to provide QoS. The authors in [10] study the behavior of CBR voice traffic in this context but voice transmission of Variable Bit Rate (VBR) real-time traffic has not yet been analyzed. There are also some works related to voice transmission in IEEE 802.11, but only very few in the ad hoc mode [6]. To our knowledge, there has been little or no prior work on analyzing voice transmission between an ad hoc network and a fixed IP network providing end-to-end QoS for real-time traffic that shares resources with background traffic.

The paper is structured as follows: Section 2 describes related work about how to support QoS in mobile ad hoc networks. Section 3 presents the protocol that supports end-to-end QoS in the mentioned context, which we have named DS-SWAN (DiffServ-SWAN). Section 4 presents and shows our simulation results. Finally, Section 5 concludes this paper.

2 QoS in Mobile Ad Hoc Networks

In a mobile environment it is difficult to provide a certain QoS because the network topology changes dynamically and in wireless networks the packet loss rates are much higher and more variable than in wired networks. Some authors have adapted the DiffServ [7] model for mobile ad hoc networks [8]. However, when DiffServ is compared with the SWAN model in an isolated ad hoc network, SWAN clearly outperforms DiffServ in terms of throughput and delay requirements [9]. For this reason, we will concentrate on the SWAN scheme.

2.1 SWAN

SWAN is a stateless network scheme that has been specifically designed to provide end-to-end service differentiation in wireless ad hoc networks employing a best-effort distributed wireless MAC [10]. It distinguishes between two traffic classes: real-time UDP traffic and best-effort UDP and TCP traffic.

A classifier (see Fig. 1) differentiates between real-time and best-effort traffic. Then, a leaky-bucket traffic shaper handles best-effort packets at a previously calculated rate, applying an AIMD (Additive Increase Multiplicative Decrease) rate control algorithm. Every node measures the per-hop MAC delays locally and this information is used as feedback for the rate controller. Every T seconds, each device increases its transmission rate gradually (additive increase with increment rate of c bit/s) until the packet delays at the MAC layer become excessive. As soon as the rate controller detects excessive delays, it reduces the rate of the shaper with a decrement rate (multiplicative decrease of r %).

Rate control restricts the bandwidth for best-effort traffic so that real-time applications can use the required bandwidth. On the other hand, the bandwidth not used by real-time applications can be efficiently used by best-effort traffic. The total best-effort and real-time traffic transported over a local shared channel is limited below a certain 'threshold rate' to avoid excessive delays.

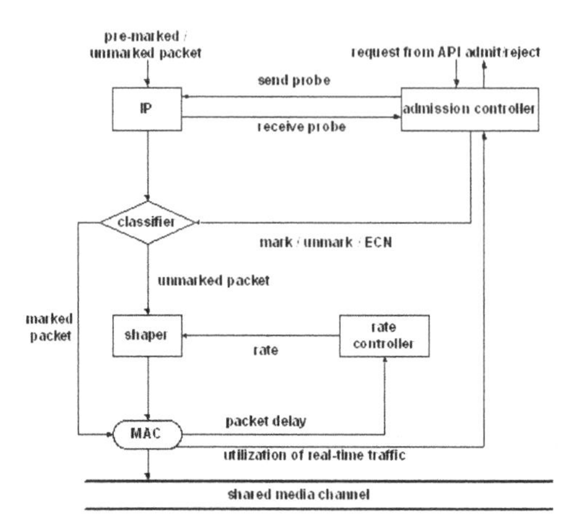

Fig. 1. SWAN model

SWAN also uses sender-based admission control for real-time UDP traffic. The rate measurements from aggregated real-time traffic at each node are employed as feedback. This mechanism sends an end-to-end request/response probe to estimate the local bandwidth availability and then determine whether a new real-time session should be admitted or not. The source node is responsible for sending a probing request packet toward the destination node. This request is a UDP packet containing a "bottleneck bandwidth" field. All intermediate nodes between the source and destination must process this packet, check their bandwidth availability and update the bottleneck bandwidth field in the case that their own bandwidth is less than the current value in the field. The available bandwidth can be calculated as the difference between an admission threshold and the current rate of real-time traffic. The admission threshold is set below the maximum available resources to enable that real-time and best-effort traffic are able to share the channel efficiently. Finally, the destination node receives the packet and returns a probing response packet with a copy of the bottleneck bandwidth found along the path back to the source. When the source receives the probing response it compares the end-to-end bandwidth availability and the bandwidth requirement and decides whether to start a real-time flow accordingly. If the flow is admitted, the real-time packets are marked as RT (Real-Time packets) and they bypass the shaper mechanism at the intermediate nodes and are thus not regulated.

Since the traffic load conditions and network topology change dynamically, real-time sessions might not be able to maintain the bandwidth and delay bound requirements and they will have to be rejected or readmitted. For this reason, it is said that SWAN offers soft QoS. SWAN incorporates the Explicit Congestion Notification mechanism (ECN), which regulates real-time sessions as follows. When a mobile node detects congestion or overload conditions, it starts marking the ECN bits in the IP header of the real-time packets. The destination monitors the packets with the marked ECN bits and informs the source sending a 'regulate' message. Then the source node tries to re-establish the real-time session with its bandwidth needs accordingly.

In SWAN, intermediate nodes do not keep any per-flow information and thus avoid complex signaling and state control mechanisms. This makes the system relatively simple and scalable.

3 DS-SWAN (Differentiated Services-SWAN)

To support end-to-end QoS it is not only necessary to provide service differentiation inside the ad hoc network: the fixed IP network must use a QoS architecture, such as DiffServ, to provide scalable service differentiation in the Internet. In the DiffServ architecture [7], each priority class is associated to a different PHB (Per-Hop Behavior). The PHB defines how packets are forwarded by the routers. Each packet carries a particular marking ('codepoint') that is unique for each PHB. Edge routers perform the marking of the incoming packets and the core routers only need to examine the packets' codepoints and forward them acccording to the associated PHBs. One DiffServ service class corresponds to the EF (Expedited Forwarding) PHB, that provides low loss, low latency, low jitter and end-to-end assured bandwidth service. It provides a Premium Service. In our study the EF aggregates correspond to real-time traffic and are policed with a token bucket meter. Some bursts are tolerated but the traffic that exceeds the profile is marked with a different codepoint and then it is dropped. The number of dropped packets at the edge router and the end-to-end delay of the real-time connections are associated with the QoS parameters of the SWAN model in the ad hoc network. We observe that if the rate of the best-effort leaky bucket traffic shaper is lower then best-effort traffic is more efficiently rate controlled and real-time traffic is not so much influenced by best-effort traffic and it is able to maintain the required QoS parameters. For this reason, it is necessary that the SWAN model co-operates with the DiffServ model in the ad hoc network.

We propose a new protocol that enables the co-operation between the described DiffServ architecture at the fixed network and the explained SWAN scheme in the ad hoc network to improve end-to-end QoS support. We consider a scenario where best-effort CBR background traffic and real-time VBR traffic are transmitted as the mobile nodes in the ad hoc network communicate with one the fixed hosts located in the Internet through the gateway. In the proposed model, DS-SWAN, the edge router that is close to the gateway periodically monitors the number of packets of EF (real-time) traffic that are dropped because they are out of the established profile for this kind of traffic. Besides, the destination nodes in the wired IP network periodically monitor the average end-to-end delays of the real-time flows that have been established. It is thus required that the real-time application provides time-stamps in the data packets. Specifically, we use an interesting real-time VBR application: VBR Voice-over-IP (VoIP). In this context, if the end-to-end delay of one or more VBR VoIP flows is larger than 140 ms, then the destination nodes send a QoS_LOST packet to the edge router near the gateway to warn it. We have chosen this value because the ITU-T (International Telecommunication Union) recommends in its standard G.114 that the end-to-end delay should be kept below 150 ms to maintain an acceptable conversation quality in VoIP [11].

For PCM encoding with the G. 711 codec, the VoIP packet loss should never drop over a percentage of 5% of all generated frames to prevent significant losses in quality [6]. We have observed from initial simulation runs that the number of dropped

VoIP packets in the ad hoc network is always kept under 1%. Therefore, we establish that if the number of dropped VoIP packets at the edge router is less than 4 % and the edge router has received a notification that the end-to-end delays for VoIP flows are excessive, then the edge router must send a QoS_LOST message to the nodes in the ad hoc network to inform them that the system is too congested to maintain the desired QoS. In this way, we can change the parameter values of SWAN dynamically according to the traffic conditions not only in the ad hoc network but also in the fixed IP network.

The nodes in the ad hoc network use a queue to store packets at the MAC layer waiting for medium access. This queue uses priority scheduling to prioritize routing packets. QoS_LOST packets are treated as routing packets because they are warnings and must arrive to their destinations as soon as possible.

When the mobile nodes in the ad hoc network are warned, they will react by modifying the parameter values in the SWAN's AIMD rate control algorithm mentioned above. In DS-SWAN, every time that a QoS_LOST message is received, the node decreases the value of c by Δc- with a certain minimum value. When no QoS_LOST message is received during T seconds, the node increases the value of c by Δc+ bits/s unless the initial value has been reached. This is done to prevent starvation of best-effort traffic.

When a node receives a QoS_LOST message, it increases the value of r by Δr+ up to a maximum value. When no QoS _LOST message has been received in the period T, the value of r is decreased by Δr- up to the initial value.

SWAN has a minimum rate m for the best-effort leaky bucket traffic shaper. In DS-SWAN nodes are also allowed to reduce m. When a node receives a QoS_LOST message, it reduces the minimum rate by Δm- Kbit/s. However, this parameter value is kept above a minimum value of m_0 Kbit/s and is increased Δm+ bits/s every second up to the initial value when the mobile nodes do not receive a warning message in T seconds. Table 1 shows the specific parameter values that we have selected for the simulations. However, operators and users are free to set these values according to their own needs, based on the characteristics of the targeted network.

Table 1. Parameter values in our simulations

Paremeters	Initial value of c	Δc-	Δc+	Minimum value of c	Initial value of r	Δr+	Δr-	Maximum value of r	Initial minimum rate	Δm-	Δm+	m_0
Values in our simulations	41 Kbit/s	10 Kbit/s	50 bits/s	11 Kbit/s	50 %	10%	1%	90%	31 Kbit/s	10 Kbit/s	50 bits/s	11 Kbit/s

4 Simulations

The aim of DS-SWAN is that real-time traffic can satisfy its bandwidth and delay requirements and best-effort traffic can use the remaining bandwidth effectively. The end-to-end delays of individual real-time flows will be reduced if there is congestion due to excess of best-effort traffic. However, it is important to remark that, on the contrary, if end-to-end delays become excessive because of reasons such as failures of physical links, then the question remains whether our algorithm will be able to maintain end-to-end QoS requirements for these real-time flows. Therefore, we have run simulations with the NS-2 [12] tool in order to investigate the performance of DS-SWAN with a relatively realistic system model that incorporates effects of all relevant communication layers.

The system framework is shown in Fig. 2. We consider a single DiffServ domain (DS-domain) covering the whole network between the wired corresponding hosts and the gateway. The chosen scenario consists of 20 mobile nodes, 1 gateway, 3 fixed routers and 3 fixed hosts. The mobile nodes are distributed in a square region of 500 m by 500 m

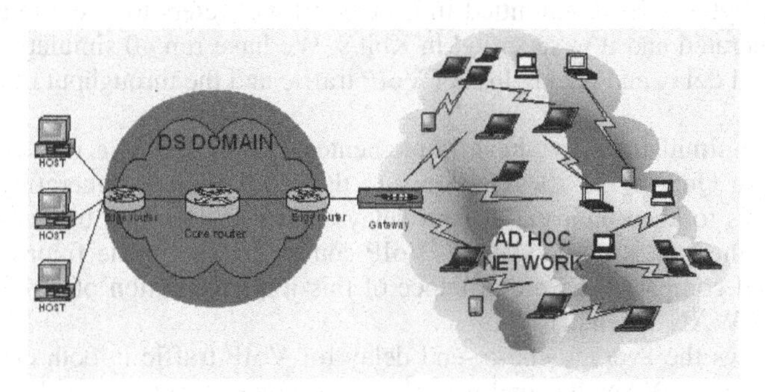

Fig. 2. Simulation framework

We assume that two traffic classes are transmitted: best-effort CBR background traffic and real-time VBR VoIP traffic. The mobile nodes communicate with one of the three fixed hosts located in the Internet through the gateway. Thus, the destination of all the CBR and VBR VoIP traffic is one of the three hosts in the wired network and some nodes in the ad hoc network will act as intermediate nodes or routers forwarding the packets from other nodes. In order to represent best-effort background traffic, 13 of the 20 mobile nodes are selected to act as CBR sources and fifteen nodes are selected to send VBR VoIP traffic.

The CBR best-effort packets that need to be sent are first processed by a leaky bucket traffic shaper so that they are delayed accordingly to a rate determined by the shaper. Afterwards, they are put in a queue at the MAC layer and should wait for medium access. The VBR VoIP packets are put in the same queue as well. This queue uses priority scheduling to prioritize routing packets and QoS_LOST packets. The rest of the traffic (VBR VoIP and CBR packets) are served without priorities so that always the oldest request is handled first.

The dynamic routing algorithm is AODV [13] and the mobile hosts use IEEE 802.11b. Each node selects a random destination within the area and moves toward it at a velocity uniformly distributed between 0 and 3 m/s. Upon reaching the destination the node pauses a fixed time period of 20 seconds, selects another destination and repeats the process.

To avoid synchronization problems due to deterministic start time, background traffic is generated with CBR traffic sources whose starting times are drawn from a uniform random distribution in the range [15 s, 20 s] for the first source, [20 s, 25 s] for the second one and so on. They have a rate of 48 Kbit/s with a packet size of 120 bytes. The VBR mode is used for VoIP traffic. We employ a silence suppression technique in voice codecs so that no packets are generated in silence period. For the voice calls, we use the ITU G.711 a-Law codec [14]. The VoIP traffic is modelled as a source with exponentially distributed on and off periods with 1.004 s and 1.587 s average each. Packets are generated at a constant inter-arrival time during the on period. Fifteen VoIP connections are activated at a starting time chosen from a uniform distribution in the range [10 s, 15 s]. Packets have a constant size of 128 bytes.

Shaping of EF (VoIP) and BE (Best-Effort) (CBR) traffic is done in two different drop tail queues of size 30 and 100 packets respectively. The EF and BE aggregates are policed with a token bucket meter with CBS = 1000 bytes and CIR = 200 Kbit/s. CBS (Committed Burst Size) refers to the maximum size of the token bucket and it is measured in bytes. CIR (Committed Information Rate) refers to the rate at which tokens are generated and it is specified in Kbit/s. We have run 40 simulations to assess the end-to-end delay and packet loss of VoIP traffic and the throughput of background traffic.

In the first simulations, we have implemented DS-SWAN in a way that the edge router sends a QoS_LOST message only to the VoIP sources generating flows that have problems to keep their end-to-end delays under 150 ms and to the intermediate nodes along the routes ("DS-SWAN- VoIP sources" label in the figures). We have evaluated and compared the performance of this implementation of DS-SWAN with the existing SWAN scheme.

Fig. 3 shows the average end-to-end delay for VoIP traffic in both cases. We observe that using SWAN the end-to-end delays increase progressively because the system is congested due to the large number of VoIP flows and background VoIP traffic. From the second 115 until the end of the simulation the end-to-end delays are too high for an acceptable conversation quality [11]. In DS-SWAN there exist flows that suffer end-to-end delays larger than 140 ms; hence, the destination nodes warn the edge router, which checks the percentage of lost packets and after verifying that it is less than 4%, it warns the nodes in the wireless ad hoc network to react accordingly. Then the nodes in the ad hoc network increase or decrease the pertinent parameters following the already explained DS-SWAN implementation and thus the system prevents the end-to-end delay to become larger than 150 ms.

Fig. 4 shows the average throughput for background traffic. In DS-SWAN, the average throughput for this kind of traffic is lower than in SWAN because the nodes in the ad hoc network react by decreasing the rate of the best-effort traffic shaper when they receive a warning. We must recall that a node may be part of a real-time and a background route at the same time. In any case, DS-SWAN functions correctly because there is no starvation of background traffic.

All simulations indicate that the packet loss rate for VoIP is well below the required 5%.

Now we have evaluated and compared the performance of the DS-SWAN protocol in the already explained scenario using two different implementations:

- The already explained implementation, where warnings are sent only to the VoIP sources having problems to keep their end-to-end delays under 150 ms and to the intermediate nodes along their respective routes ("Case 1, DS-SWAN - VoIP sources").

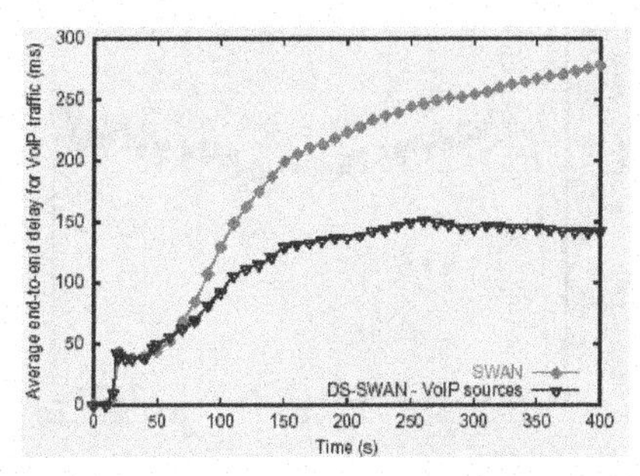

Fig. 3. Average end-to-end delay for VoIP traffic: DS-SWAN vs. SWAN

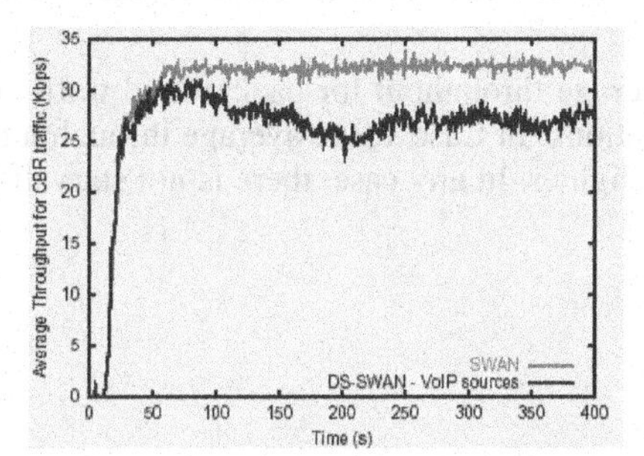

Fig. 4. Average throughput for CBR traffic: DS-SWAN vs. SWAN

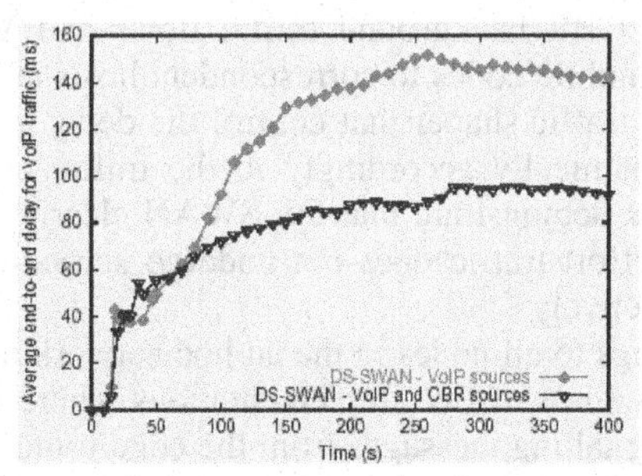

Fig. 5. Average end-to-end delay for VoIP traffic in the two DS-SWAN implementations (Case 1 and Case 2)

- Where warnings are sent to the VoIP sources having problems to keep their end-to-end delays under 150 ms, to all the CBR sources and to the intermediate nodes along the routes ("Case 2, DS-SWAN - CBR and VoIP sources").

Fig. 5 shows the average end-to-end delay for VoIP traffic in the two cases. Average end-to-end delays are kept well below 150 ms in both cases, but in Case 2 it is significantly smaller. This is because two neighbouring nodes that belong to two different routes, each carrying a different type of traffic, may still compete to access the medium. In Case 2, nodes carrying best-effort traffic that are in the proximity of a VoIP route and may contend with it for the medium access, are forced to reduce their data rates.

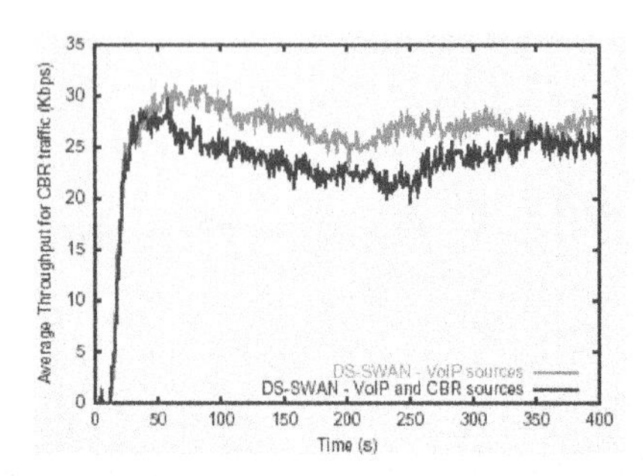

Fig. 6. Average throughput for CBR traffic (Case1 versus Case 2)

Fig. 6 shows the average throughput for background traffic obtained with the two DS-SWAN implementations. In Case 1, the average throughput is larger than in Case 2 because signalling is lighter. In any case, there is not starvation of background traffic in Case 2.

5 Conclusions

This work presents simulations of DS-SWAN in a relatively small mobile ad hoc network connected to a DiffServ domain. We have analyzed the functioning of the systems when multiple CBR background traffic flows and VBR VoIP flows have been established from mobile nodes to correspondent hosts at the fixed network. The parameter values of the traffic shaper that control the delay undergone by best-effort traffic are changed dynamically accordingly to the traffic conditions in the whole route. Simulation results demonstrate that DS-SWAN clearly outperforms SWAN in this scenario and best-effort traffic does not undergo starvation and can use the remaining bandwidth effectively.

Since sending warnings to all nodes in the ad hoc network may not be scalable, we have studied the performance of two implementations where only a selection of the mobile nodes receive signalling messages from the edge router. The two implementations show similar performance, and the choice of one over the other depends on the trade-off between end-to-end delay of real-time flows and throughput of background traffic. It still remains to be seen what the performance will be for larger networks.

Acknowledgements. This work was partially supported by the "Ministerio de Ciencia y Tecnología" of Spain under the project TIC2003-08129-C02, which is partially funded by FEDER, and under the programme Ramón y Cajal.

References

[1] D. Remondo and I. G. Niemegeers, "Ad hoc networking in future wireless communications", *Computer Communications*, vol 26, no. 1, Jan. 2003, pp. 36-40.

[2] M. Benveniste, G. Chesson, M. Hoeben, A. Singla, H. Teunissen, and M. Wentink, "EDCF proposed draft text", *IEEE working document 802.11-01/131r1,* March 2001.

[3] A. Iwata, C-C. Chiang, G. Yu, M. Gerla, and T-W. Chen, "Scalable routing strategies for ad hoc wireless networks", *IEEE Journal on Selected Areas in Communications,* 17(8):1369-1379, August 1999.

[4] S. B. Lee and A. Campbell, "INSIGNIA", *Internet Draft,* May 1999.

[5] H. Xiao, K.G. Seah, A. Lo and K.C. Chua, "A flexible quality of service model for mobile ad hoc networks", *IEEE Vehicular Technology Conference (VTC Spring 2000)*, Tokyo, Japan, May 2000, pp. 445-449.

[6] P.B. Velloso, M. G. Rubinstein and M. B. Duarte, "Analyzing Voice Transmission Capacity on Ad Hoc Networks", *International Conference on Communications Technology - ICCT 2003*, Beijing, China, April 2003.

[7] S. Blake, D. Black, M. Carlson, E. Davies, Z. Wang, and W. Weiss, "An architecture for differentiated service", *Request for Comments (Informational) 2475, Internet Engineering Task Force,* December 1998.

[8] H. Arora and H. Sethu, "A Simulation Study of the Impact of Mobility on Performance in Mobile Ad Hoc Networks," *Applied Telecommunications Symposium*, San Diego, Apr. 2002.

[9] H. Arora, Ll. Greenwald, U. Rao and J. Novatnack, "Performance comparison and analysis of two QoS schemes: SWAN and DiffServ", *Drexel Research Day Honorable Mention,* April 2003.

[10] G.-S. Ahn, A. T. Campbell, A. Veres and L.-H. Sun, "SWAN", *draft-ahn-swan-manet-00.txt*, February 2003.

[11] ITU-T Recommendation G.114, "One way transmission time", May 2000.

[12] NS-2: Network Simulator, http://www.isi.edu/nsnam/ns.

[13] C.E. Perkins, E.M. Royer, "Ad-hoc On-demand Distance Vector routing," in *Proc. of the 2nd IEEE Workshop on Mobile Computing Systems and Applications,* New Orleans, U.S.A., Feb. 1999.

[14] D. Chen, S. Garg, M. Kappes and K.S. Trivedi, "Supporting VBR Traffic in IEEE 802.11 WLAN in PCF Mode," in *Proc. OPNETWORK'02*, Washington D.C., Aug. 2002.

Scheduling of QoS Guaranteed Services in DS-CDMA Uplink Networks: Modified Weighted G-Fair Algorithm

Young Min Ki, Eun Sun Kim, and Dong Ku Kim

Yonsei University, Dept. of Electrical and Electronic Engineering
134 Shinchon Dong, Seodaemun Gu, Seoul, Korea
{mellow, esunkim, dkkim}@yonsei.ac.kr
http://mcl.yonsei.ac.kr

Abstract. In this paper, the modified weighted g-fair (MWGF) scheduling scheme with rise over thermal (RoT) filling is proposed for QoS guaranteed services in DS-CDMA uplink networks, in which RoT is directly related to the loading factor of a wireless network. The proposed scheme computes priorities of QoS guaranteed traffic for each slot based upon the weighted values of the QoS factor, channel factor, and fairness factor. Once priorities are assigned, power and rate used for each user are computed within a limited RoT by using RoT filling, which adjusts the transmit rate of the traffic assigned least priority. The proposed algorithm increases more throughput by 6 to 10 % compared to that of the current autonomous rate control (ARC) scheme. It also lowers transmission delays and ensures more fairness in delay outages than the round robin scheduler and ARC scheme.

1 Introduction

In 3G wireless networks, link adaptation and channel scheduling are key techniques used for dynamic resource management. In a cdma2000 1xEV-DO type downlink, adaptive modulation and coding (AMC) as well as proportional fairness (PF) scheduling have improved the overall performance of high data rate services [1]. The modified largest weight delay first (MLWDF) [2] and exponential (EXP) [3] algorithms have been proposed for real-time services. An adaptive EXP/PF algorithm has also been proposed for applications in multiple QoS service traffic experienced in a 1xEX-DO downlink [4].

The proposed uplink system should offer QoS to multimedia traffic as well as control multiple access interference, as not to overload the system. A cdma2000 1x (IS-2000) like uplink control scheme called autonomous data rate control (ARC) is employed in 1xEV uplink [5]. This distributed control scheme can neither support the various QoS requirements nor hold RoT constraint, which is associated with system loading, and used in highly traffic loaded environments. The weighted proportional fairness (WPF) algorithm was proposed to schedule the best effort services for the uplink [6]. However, it did not yet consider delay

I. Niemegeers and S. Heemstra de Groot (Eds.): PWC 2004, LNCS 3260, pp. 272–285, 2004.

requirements necessary for efficient QoS guaranteed services. In this paper, we take into consideration a scheduling algorithm for applications in QoS guaranteed services in cdma2000 type uplink, and propose a modified weighted g-fair (MWGF) algorithm for by using the notion of the WPF [6], MLWDF [2] and G -fair [7]. The total performance of the proposed algorithm is evaluated with simulations and compared with those of the ARC scheme as well as round robin scheduling algorithm.

The remainder of the paper is organized as follows. In section 2, the conventional uplink MAC algorithm for cdma2000 is presented. In section 3, proposed uplink scheduling schemes are presented. The total performance of the schemes is studied in section 4 by simulation. Finally, conclusions are made in Section 5.

2 Uplink MAC Algorithm for CDMA Cellular Networks

The uplink MAC is used to control the data rate at which the access terminals transmit. The access networks control the data rates of the mobile users by using two mechanisms: Reserve rate limit, and Reverse activity bit (RAB) and the transition probabilities.

2.1 Reverse Rate Limit

In autonomous data rate control (ARC) scheme, the user can send data anytime regardless of transmitting of the uplink traffic channel request message. The user initially starts to send his or her message at the lowest data rate. Then, if the user receives an *idle* RAB from the base station (BS), it will increase the data rate to the next higher level or transmission power level. If the user receives a *busy* RAB, it will decrease the data rate to the next lower level or transmission power level. However, when the data rate reaches its minimum, maximum, or the required transmit power of a mobile user exceeds the maximum available transmit power, the current data rate is maintained.

2.2 Reverse Activity Bit and Transition Probabilities

The RoT, defined as the total received power to thermal noise ratio occurring in a specific BS, is the parameter used to determine the network's capacity. The BS measures the received RoT given in (1).

$$RoT = \frac{\sum_i^{\text{active in cell}} P_{TX,i} L_i + \sum_k^{\text{other cells}} I_k + N_0 W}{N_0 W} \tag{1}$$

where $P_{TX,i}$ is the transmit power of the i-th user, L_i is the radio path loss, I_k is the other cell interference from the k-th cell, and $N_0 W$ is the thermal noise. The RAB is determined by equation (2) and broadcast to all users in the cell, in a regular manner.

$$RAB = \begin{cases} idle, & \text{if } RoT < RoT_{setpoint} \\ busy, & \text{if } RoT > RoT_{setpoint} \end{cases} \tag{2}$$

where $RoT_{setpoint}$ is the threshold of the RAB operation. This threshold is generally set smaller than RoT_{limit} which is the pre-determined receive interference level of BS. Fig. 1 shows the autonomous data rate transition process along with transition probabilities. Two rate transition probabilities can be defined for use in rate transition. If the RAB is *idle*, the user can transit to the higher data rate with the probability, p as shown in Fig. 1 (a). If the RAB is *busy*, the user can transit to the lower data rate with the probability, q as presented in Fig. 1 (b).

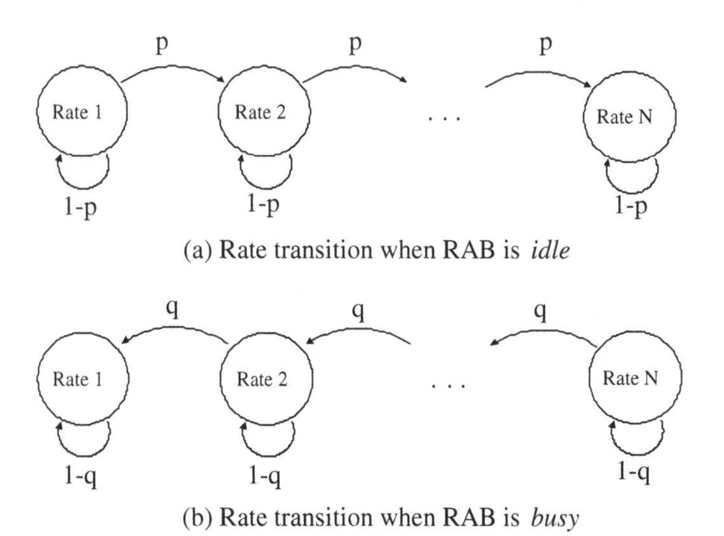

Fig. 1. Reverse rate transition process (N: the number of available data rates)

3 Uplink Scheduling for CDMA Cellular Networks

In a DS-CDMA uplink, there are two system constraints. One is the limited transmit power of each user and the other is the receive interference level of the base station (BS). Therefore, the uplink scheduler is used to determine how to allocate the available amount of power to each user without exceeding the pre-determined interference level as well as priority determination [6][8]. The channel scheduler of a cdma2000 uplink can serve some users simultaneously within the pre-determined RoT limit value, while the channel scheduler of a TDM type 1xEV-DO downlink allocates the full power of the BS to one user in each slot.

Since the BS does not identify queue status of each user's traffic, the scheduling and rate/power allocation should be determined at the BS by data rate requests and the queue status from users. Then, the uplink scheduling scheme has three stages as shown in Fig. 2. First, each user requests a desired transmit data rate and informs the queue status to BS. Secondly, the BS performs the scheduling process by executing two phases, i.e. scheduling and rate/power allocation. Since the cdma2000 uplink has two constrains on the available amount of transmit power as well as receive interference, which make joint determination of scheduling and simultaneous rate/power allocation very difficult, scheduling is first processed and subsequently followed by power/rate allocation. After scheduling and rate/power allocation processes are completed, the users can transmit their traffic by using contention-free dedicated resources. Since it required three stages to transmit data from initial demand of channel, uplink scheduling has at least two potential delay slots.

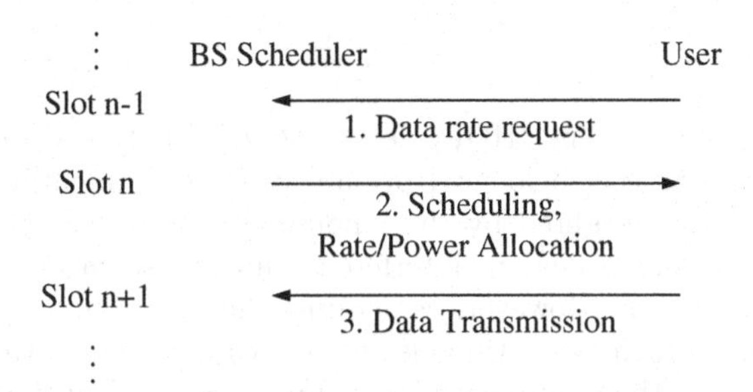

Fig. 2. Uplink scheduling schemes for a DS-CDMA system

Most distributed control schemes can neither support various QoS requirements nor hold the RoT constraint which is associated with system loading, and designed for use in highly traffic-loaded environments. The WPF centralized scheduling scheme was proposed to schedule the best effort services for use in the uplink [6]. We take into consideration a scheduling algorithm for use with QoS guaranteed services, and propose innovative MWGF algorithms.

3.1 Weighted Proportional Fairness (WPF)

In [6], the WPF algorithm was proposed for the uplink in order to ensure best effort services and results demonstrated the trade-off between throughput and fairness. The scheduler determines user priorities by applying the following metrics:

$$P_i = w_i \cdot \left(\frac{R_i^{req}}{R_i}\right)^{1/\alpha} \tag{3}$$

where w_i can be corrected to existing channel conditions in order to give a higher priority to the mobile users with better channels. Hence, it introduces another degree of freedom in adjusting fairness. The expression R_i^{req}/R_i is a well-known PF factor [6].

3.2 Proposed Scheduling Algorithm: Modified Weighted G-Fair (MWGF)

We adopt the MWPF scheduling algorithm for used in QoS guaranteed services, whose priority metrics are given in (4):

$$P_i = C_i \cdot (Q_i)^a \cdot (W_i)^b \cdot (F_i)^c \tag{4}$$

where C_i is the service grade factor of the i-th user and can be determined by the user's subscribing class grade such as *gold*, *silver* or *bronze*. The symbol Q_i represents the QoS factor and is given as (5):

$$Q_i = \frac{D_i}{D_i^{req}} \tag{5}$$

where D_i is the head of line (HOL) delay and D_i^{req} is the required delay QoS class in which the i-th user belongs. Normalization factor identified by D_i^{req} can have a priority value weighted by the amount of each user's delay sensitivity. In [6], the forward link signal to interference and noise ratio (FL SINR) was used as a channel factor. However, we assume that the uplink pilot channel in a cdma2000 1xEV provides continuous channel estimation, so that the channel factor could be set to the uplink path loss value given in (6), instead of FL SINR:

$$W_i = \text{UL Path Loss}_i \tag{6}$$

We also consider the normalized uplink path loss presented in (7), as another candidate of the channel factor:

$$W_i = \frac{\text{UL Path Loss}_i}{\text{UL Average Path Loss}_i} \tag{7}$$

Finally, F_i is the fairness factor that can be set to the G-Fair factor given in (8).

$$F_i = \frac{R_i^{req}}{R_i} \cdot \frac{h_i(\overline{R_i^{req}})}{\overline{R_i^{req}}} \tag{8}$$

where R_i^{req}/R_i is a well-known PF factor. The uplink rate request R_i^{req} is determined by the user's buffer status and does not include the channel condition as defined in (9).

$$R_i^{req} = min(\text{traffic rate}, \text{regional rate}, \text{service grade rate}) \tag{9}$$

where traffic rate represents the data rate required to transmit the remaining data volume. When the regional rate is the user's maximum available transmission rate predetermined by user's distance from the BS, and the service grade rate is the maximum available transmission rate associated with the user's service grade [5]. The symbol $h(x)$ represents a user-specific function that specifies the fairness behavior. If $h(x)$ is constant, the scheduler attempts to provide the same average throughput to all users. If $h(x) = x$, the g-fair scheduler becomes the proportional fair scheduler that we currently use [7].

3.3 Proposed Rise over Thermal Filling

Since cdma2000 uplink has two constrains concerning the limited transmit power and receive interference, which make joint determination of scheduling and simultaneous power/rate allocation very difficult. Therefore, scheduling is first processed and subsequently followed by power/rate allocation. Once user requests are prioritized, then the transmit power, $P_{TX,i}$ is calculated at the BS as follows,

$$\left(\frac{E_b}{N_0}\right)_{i,req} = \frac{W_i}{R_i^{req}} \frac{P_{TX,i} L_i}{I_{total}}, \tag{10}$$

where E_b/N_0 is the bit energy per noise ratio, W is the signal bandwidth, L is the channel path loss, and I_{total} is the total interference [9]. Once the required power levels are calculated, the scheduler starts allocating power to users in a decent order of priorities while computing the RoT value given in (1), and the BS can only allocate power levels within the available RoT limit value. This RoT filling technique is applied to the scheduling schemes. If the RoT value exceeds the RoT limit, while computing RoT, the data rate of the last filled user is decreased to the nearest level so that the computed RoT value does not exceed the RoT limit. Then, all users start transmission with these given rates and powers.

4 Simulations: QoS Guaranteed Service Assumption

4.1 Simulation Environments

Six video telephone users are assumed to be uniformly distributed within a cell. They have the minimum required rate of 64.3 kbps and very strict slot delay constraints (around 10 to 30 msec). These 19 circular cells with a 1.44 km cell radius are assumed, but simulation results have been extracted from the center cell only. The uplink reuse fraction is set at 0.55. Other simulation parameters are listed as follows:

- Signal BW: 1.2288 MHz
- Target Eb/No : 3.10 dB (for all users)
- Tx power limit: 23.0 dBm (200 mW)
- Slot size: 1.67 msec (1xEV-DO)
- Propagation model: $28.6 + 35log_{10}(d_{meter})$ dB
- Shadowing: Log normal with a standard deviation of 8.9 dB
- Rayleigh fading with Doppler frequency, 10.0 Hz
- Max propagation loss: 146.0 dB
- BS antenna gain: 5.0 dB
- Thermal noise density: -174.0 dBm/Hz
- RoT limit: 10.0 dB

The available transmission data rate of users can be selected according to

Table 1. User's average geometry and regional rate

User	Average distance from BS [m]	Regional rate [kbps]
1	72.01	1,024
2	996.86	614.4
3	724.11	1,024
4	129.58	1,024
5	903.46	614.4
6	541.41	1,024

Table 2. Investigated MWGF algorithms applied in various simulation scenarios

Factor	Algorithm 1	Algorithm 2	Algorithm 3
1. Service grade	No use	No use	No use
2. QoS	Normalized delay	Normalized delay	Normalized delay
3. Channel	UL path loss	Normalized path loss	Normalized path loss
4. Fairness	Conventional PF	Conventional PF	G-Fair ($h(x) = 1$)

1xEV data rates: 0(NULL), 9.6, 19.2, 38.4, 76.8, 153.6, 307.2, 614.4, and 1,024 kbps [5]. The regional rate differentiation is also applied to our simulation. The distances from the BS to the user as well as regional rate constraints of users are shown as Table 1. We have investigated three schemes: ARC, MRR, and the MWGF scheduler:

(Scheme 1) ARC scheme: simulations were conducted for the ARC of 9.0 dB RoT set point and 8.0 dB RoT set point respectively. The rate transition probabilities were set to $p = q = 0.5$.
(Scheme 2) Modified round robin (MRR) scheduler: a priority rotation method is used, in which the priorities were rotated among users. For example, if the priority orders were (User 1, User 2, User 3, User 4, User 5, User 6) in the n-th slot, the priority orders were (User 2, User 3, User 4, User 5, User 6, User 1) in (n+1)-th slot, and the priority orders will be (User 3, User 4, User 5, User 6, User 1, User 2) in the next slot. The priorities for the subsequent slots will be rotated in every slot.
(Scheme 3) Proposed MWGF scheduler: the proposed scheduling algorithms were investigated for three different factor choices. The discrepancy calculated three versions is shown in Table 2. The weighting indexes are set to $a = 2.0 \sim 6.0$, $b = 0.5$, $c = 1.0$. These values were chosen through trials, experimentations and simulations in order to find nearly optimal values.

4.2 Throughput and Delay Performance

Fig. 3 shows the total throughput of the investigated schemes. The scheduling schemes, i.e. MRR and MWGF, produce more throughput when compared to those of the ARC schemes by 6 to 10 %. These improvements originate from the multi-user diversity gain of the centralized scheduling schemes [1]. Fig. 4 shows the user throughput of the best geometry user (User 1) and that of the worst geometry user (User 2). Since User 2 has excessively poor channel conditions, User 2 can not be served with the 64.3 kbps minimal rate used in ARC schemes. Therefore, User 2 needs to be allocated more channel resources.

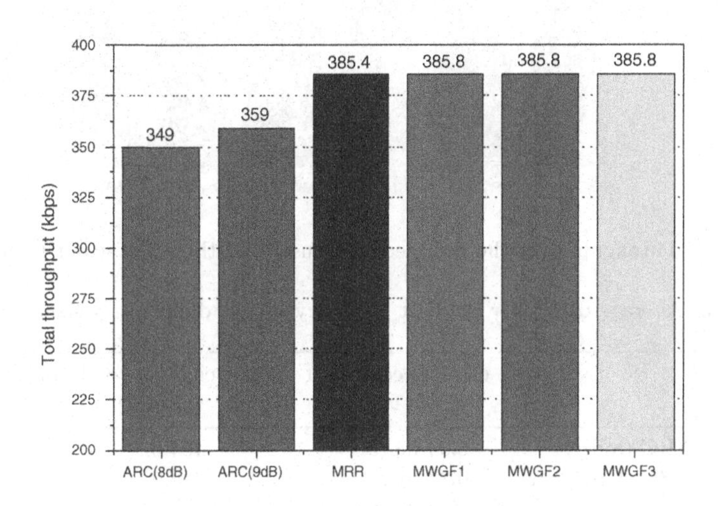

Fig. 3. Total throughput for the investigated schemes

Table 3 shows the average transmission delay of the best geometry user (User 1) and that of the worst geometry user (User 2). In the MRR scheme, it is shown that throughput of User 2 can be served with a transmission speed around the minimal rate, but User 2 suffers very a large delay compared to the MWGF scheme. This indicates that the round robin can allocate channel resources to users by fairly with reference time, but cannot allocate the channel resources in an adequate channel scheduling time.

4.3 Delay Outage Probability Performance

Fig. 5 displays the delay outage probability of users in the MRR algorithm. It is shown that user 2 suffers a very large delay outage probability. Fig. 6, 7 and 8 show the outage delay probability of the proposed algorithms: Algorithm 1, Algorithm 2, and Algorithm 3. In Algorithm 2 and Algorithm 3, the difference between user with the best delay performance and the worst one is smaller than in Algorithm 1. The figures indicate that the 10^{-2} outage of Algorithm 1 can be achieved within around 35 msec, and those of Algorithm 2 and Algorithm 3 can be obtained within 20 msec. Algorithm 2 and Algorithm 3 produce increased

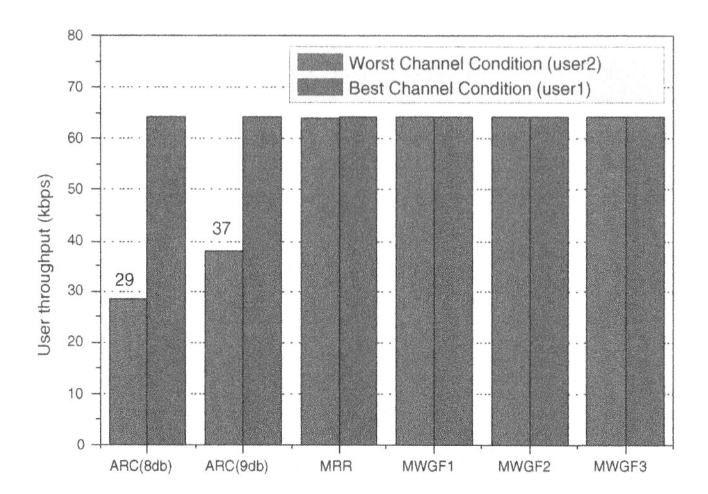

Fig. 4. Throughput for the best geometry user and the worst geometry user

Table 3. Average delays for the best geometry user and the worst geometry user

	Best channel condition (User 1)	Worst channel condition (User 2)
ARC (8 dB)	45.94 msec	464.0 sec
ARC (9 dB)	30.27 msec	376.0 sec
MRR	7.03 msec	4.53 sec
MWGF 1	9.03 msec	16.88 msec
MWGF 2	10.15 msec	9.93 msec
MWGF 3	10.32 msec	9.59 msec

fairness when compared to Algorithm 1 and the MRR with regard to delay outage.

4.4 Effect of Weighting Index on MWGF Scheduling Scheme

Fig. 9 shows the average transmission delay of the proposed schemes according to their weighting indexes. In this figure, we set the QoS weighting index a to 2.0 to 6.0. Symbols b and c are fixed at (b = 0.5, c = 1.0). Therefore, as QoS weight a increases, average transmission delays decrease. Since the channel factor of Algorithm 1 is not normalized, the effect of the increase of a is the largest among the proposed Algorithms. Fig. 10 shows average received RoT level of the proposed schemes, calculated according to their weighting indexes. It is shown that as a increases, the RoT level of Algorithm 1 decreases from 8.6 dB to 8.3 dB, and those of Algorithm 2 and Algorithm 3 increase 8.25 dB to 8.4 dB respectively.

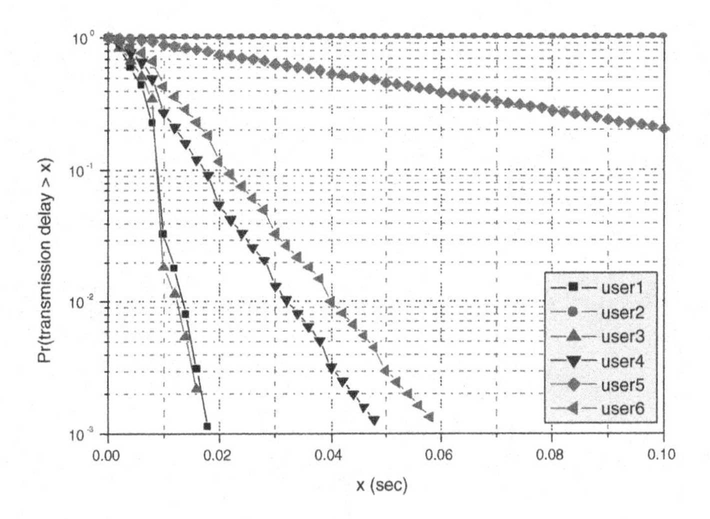

Fig. 5. Delay outage probability of the modified round robin scheduling scheme

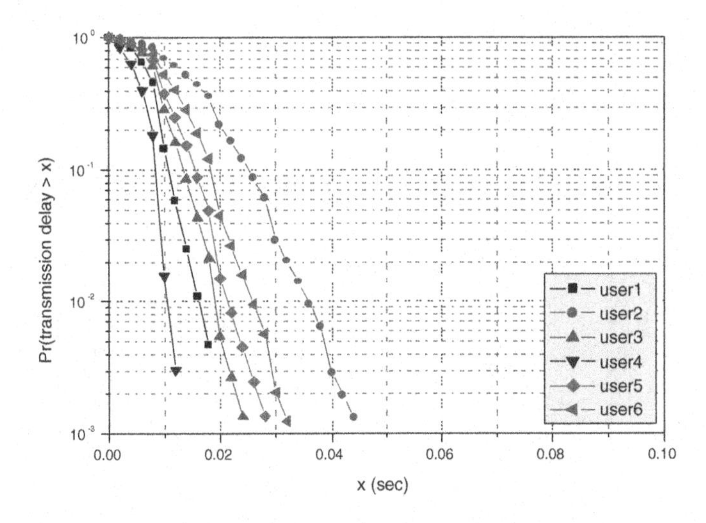

Fig. 6. Delay outage probability of the proposed MWGF Algorithm 1 (weighting index: $a = 4.0$, $b = 0.5$, $c = 1.0$)

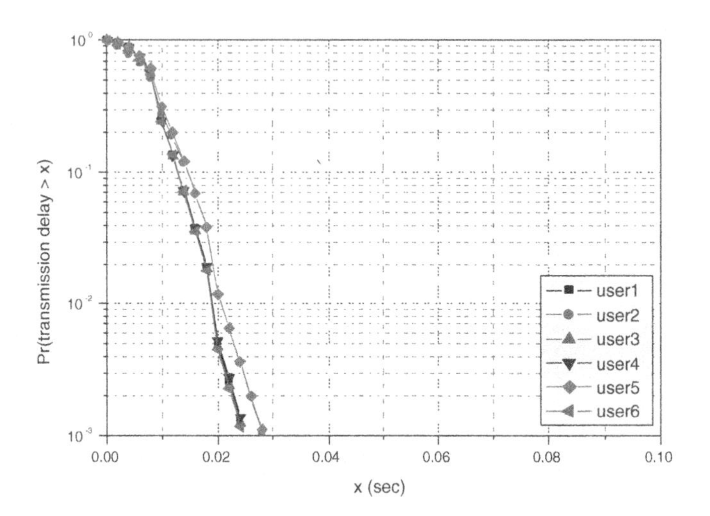

Fig. 7. Delay outage probability of the proposed MWGF Algorithm 2 (weighting index: $a = 4.0$, $b = 0.5$, $c = 1.0$)

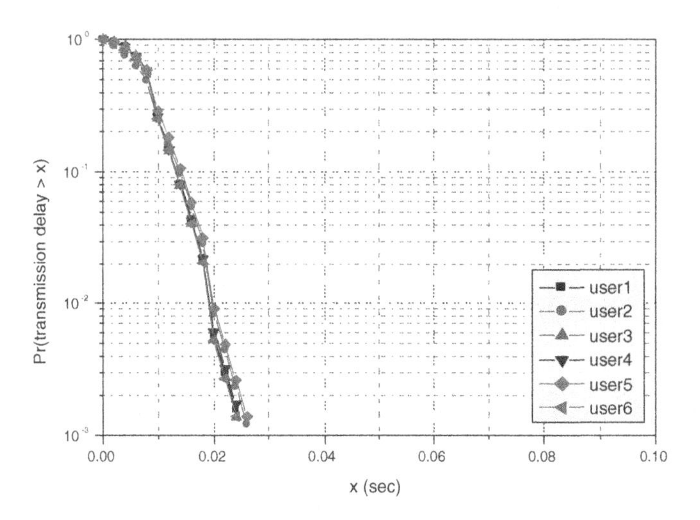

Fig. 8. Delay outage probability of the proposed MWGF Algorithm 3 (weighting index: $a = 4.0$, $b = 0.5$, $c = 1.0$)

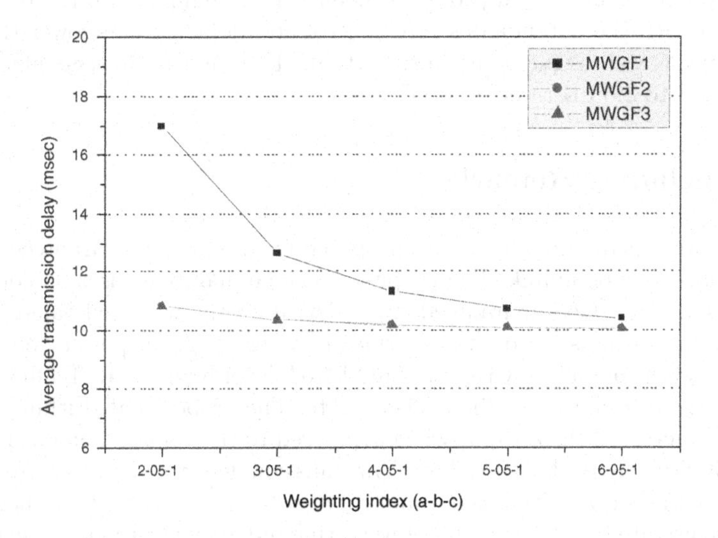

Fig. 9. Average transmission delay of the proposed schemes versus weighting indexes (weighting index: $a = 2.0 \sim 6.0$, $b = 0.5$, $c = 1.0$)

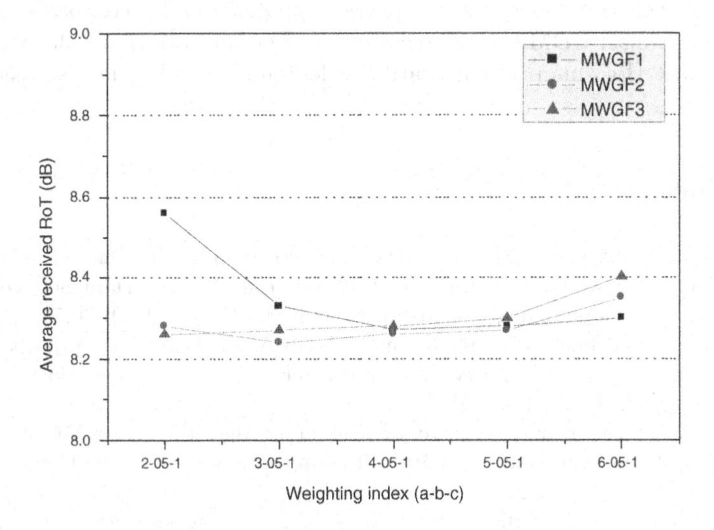

Fig. 10. Average received RoT level of the proposed schemes versus weighting indexes (weighting index: $a = 2.0 \sim 6.0$, $b = 0.5$, $c = 1.0$)

The proposed algorithm produces increased throughput of 6 to 10 %, when compared with the autonomous rate control and demonstrates lower transmission delays as well as increased fairness than the round robin scheduler. It also displays a 8 to 8.5 dB received RoT level.

5 Concluding Remarks

The MWGF scheduling scheme was proposed for use in QoS guaranteed services in a cdma2000 type uplink. The proposed scheme mainly focuses on computing priorities used for QoS guaranteed traffic based on the weighted values of QoS, channel, and fairness factor. Once priorities were assigned, power and rate for each user were computed within a limited RoT level by using RoT filling, which adjusted the overall user traffic transmit rate. The proposed algorithm produces 6 to 10 % increased throughput when compared to that of autonomous rate control (ARC) scheme and ensures lower transmission delays as well as more fairness in delay outages than either the round robin scheduler or the ARC scheme. It is also demonstrated that the choice of weighting index by the proposed scheduling algorithm allows trading off received RoT level and transmission delays.

Acknowledgment. This work was supported by grant No. (R01-2002-000-00531-0) from the Basic Research Program of the Korea Science & Engineering Foundation. Also, the work was supported by Qualcomm Incorporated through Qualcomm Yonsei CDMA Research Center. The authors thank Jelena Damnjanovic, Jack Holtzman and Edward Tiedemann for valuable discussions and comments.

References

1. Bender P., Black P., Grob M., Padovani M., Sindhushayana N., Viterbi A.: CDMA/HDR: A Bandwidth-Efficient High-Speed Wireless Data Service for Nomadic Users. Communications Magazine, Vol. 38. IEEE. (2000) 70-77
2. Andrews M., Kumaran K., Ramanan K., Stolyar A., Whiting P., Vijayakumar R,: Providing quality of service over a shared wireless link. Communications Magazine, Vol. 39. IEEE. (2001) 150-154
3. Shakkottai S., Stolyar A.: A study of Scheduling Algorithms for a Mixture of Real- and Non-Real-Time Data in HDR. 17th International Teletraffic Congress (ITC-17), (2001)
4. Rhee J., Kim D.: Scheduling of Real/Non-real Time Services in an AMC/TDM Systems: EXP/PF Algorithm. Lecture Notes in Computer Science, Vol. 2524. Springer-Verlag, Berlin Heidelberg New York (2003) 506-513
5. Koo C., Bae B., Jung J.: An Efficient Reverse Link Control Scheme for 3G Mobile Communication Systems. IST Mobile & Wireless Telecommunications Summit (2002)
6. Damnjanovic J., Jain A., Chen T., Sarkar S.: Scheduling the cdma2000 Reverse Link. IEEE 56th Vehicular Technology Conference (2002)

7. Application Note: G-fair Scheduler. 80-H0551-1 Rev. B. Qualcomm Incorporated. (2002)
8. Padovani R.: The Application of Spread Spectrum to PCS has Become Reality: Reverse Link Performance of IS-95 Based Cellular Systems. Personal Communications, Vol. 1, IEEE. (1994) 28-34
9. Dimou K., Godlewski P.: MAC Scheduling for Uplink Transmission in UMTS WCDMA. IEEE 54th Vehicular Technology Conference (2001)
10. Application Note: PLMAC Algorithm for IS-856 (1xEV). CL93-V3762-1 X1. Qualcomm Incorporated. (2001)
11. Shakkottai S., Rappaport T. S.: Cross-Layer Design for Wireless Networks. Communications Magazine, Vol. 41. IEEE. (2003) 74-80
12. TIA/EIA/IS-856. cdma2000 High Rate Packet Data Air Interface Specification, (2001)
13. cdma2000 release C. C.S000X-C, (2002)
14. 3GPP release 5. TS-25.211-25.214, (2002)
15. S. Keshav: An Engineering Approach to Computer Networking: ATM Networks, the Internet, and the Telephone Network. Addison Wesley. (1997)
16. Recommendation ITU-R M.1225, Guideline for Evaluation of Radio Transmission Technologies for IMT-2000, (1997)

Receiver-Based Flow Control Mechanism with Interlayer Collaboration for Real-Time Communication Quality in W-CDMA Networks*

Hiroyuki Koga[1], Katsuyoshi Iida[1], and Yuji Oie[2]

[1] Graduate School of Information Science,
Nara Institute of Science and Technology, Japan
{koga,katsu}@is.naist.jp
[2] Department of Computer Science and Electronics,
Kyushu Institute of Technology, Japan
oie@cse.kyutech.ac.jp

Abstract. Mobile networks are becoming increasingly prevalent, and this has led to an increase in the bandwidth available over wireless links in IMT-2000. Both non-real-time forms of communication, such as e-mail and web browsing, and real-time forms of communication, such as audio and video applications, are well suited to wireless networks. However, wireless networks are subject to relatively long transmission delays because of the need to recover lost packets caused by high bit error rates. This degrades the quality of real-time communications. Therefore, in the present paper, we propose a receiver-based flow control mechanism employing an interlayer collaboration concept to improve the quality of real-time communications without adversely affecting the performance of non-real-time communications on IMT-2000 networks. In addition, simulations are performed in order to evaluate the performance of the proposed mechanism and demonstrate its effectiveness.

1 Introduction

With the growth of both wireless networks and the Internet, wireless networks are becoming increasingly attractive as an option for data transmission services such as e-mail and web browsing. A next-generation mobile system, International Mobile Telecommunications-2000 (IMT-2000) [1], has been standardized by the International Telecommunication Union (ITU) in order to provide faster data transmission service as well as worldwide roaming capability. Wideband-Code Division Multiple Access (W-CDMA) technology, prescribed by the 3rd Generation Partnership Project (3GPP) [2] for wireless networks, is one of the technologies employed by the IMT-2000 system. The current mobile system provides a transmission rate of 9.6 Kb/s, whereas the W-CDMA system can provide

* This work was supported in part by Grant-in-Aid for the 21st Century COE Program "Ubiquitous Networked Media Computing" and National Institute of Information and Communications Technology.

I. Niemegeers and S. Heemstra de Groot (Eds.): PWC 2004, LNCS 3260, pp. 286–300, 2004.

a maximum of 384 Kb/s outdoors. In addition, the W-CDMA system uses robust error recovery technologies to minimize degradation due to bit errors in wireless channels.

In such W-CDMA networks, Koga et al. investigated how the performance of TCP [3], the transport protocol primarily used for non-real-time communications, is affected by packet losses caused by transmission errors and the ARQ mechanism of Layer 2 and proposed a TCP flow control mechanism to improve TCP performance [4,5]. The proposed mechanism is a receiver-based flow control mechanism that uses information on wireless link conditions provided by an interlayer collaboration concept on a mobile station. This mechanism focuses on TCP performance, in particular, the prevention of throughput degradation due to wireless link conditions, and the stability of the throughput performance.

Furthermore, by increasing the available bandwidth over wireless networks, both real-time communications, such as audio and video applications, and non-real-time communications become well suited to wireless networks. The coexistence of different types of communication over wireless networks is problematic. In non-real-time communications, importance is attached to efficiency and reliability of data transmission. TCP senders therefore execute flow control so as to use as much of the available bandwidth as possible and to retransmit lost packets [6]. In contrast, real-time communications are more tolerant of packet losses than non-real-time communications but are greatly affected by transmission delays and jitter, i.e., fluctuations in the interarrival times of packets. Therefore, real-time communications generate Constant Bit Rate (CBR) traffic transmitted by UDP datagrams, which do not retransmit lost packets.

Several studies have examined the effect of TCP traffic sharing a link with UDP traffic in wired networks. For example, TCP traffic is known to adversely affect the performance of UDP traffic [7]. Thus, various mechanisms, such as Class Based Queuing (CBQ) [8] and Random Early Drop with In & Out (RIO) [9], and Resource ReSerVation Protocol (RSVP) [10] have been proposed in order to provide Quality of Service (QoS) assurance for real-time communications. In wireless networks, QoS assurance over wireless networks as well as wired networks must be provided because the available bandwidth of these networks is very different. Although QoS mechanisms in wireless networks have been investigated [11,12,13], very few reports have examined QoS mechanisms for the IMT-2000 system.

In order to provide QoS assurance, particularly in wireless access networks, in which the available bandwidth is generally less stable than in wired networks, the concept of the interlayer collaboration, as discussed in [14,15], appears very useful. The interlayer collaboration concept allows an end host, i.e., a mobile station, to control its flow in an intelligent manner without any support by the network. In the present paper, we propose a receiver-based flow control mechanism employing an interlayer collaboration concept to improve the quality of real-time communications without adversely affecting the performance of non-real-time communications. The basic concept of this mechanism is to limit the bandwidth available for TCP by setting aside adequate bandwidth for UDP traffic. The information needed for managing the available bandwidth can be obtained from applications and from the data link layer through an interlayer collaboration concept. Moreover, except for a slight modification to the mobile station, this mechanism does not require any special support at intermediate nodes, including the base station. In addition, modification of the corresponding host is not necessary. Therefore, developing and deploying this mechanism appears very useful.

In the present study, we perform simulations focusing on the characteristics of loss probability and transmission delay time for UDP traffic and the throughput performance for TCP traffic in order to evaluate the effectiveness of the proposed mechanism. We demonstrate that the proposed mechanism achieves performance that is equivalent to the use of priority scheduling at the base station.

2 Quality of Real-Time Communications in Wireless Networks

The IMT-2000 system executes retransmissions on Layer 2 in order to minimize degradation due to bit errors. This affects the quality of communications over wireless links. In this section, we describe the W-CDMA system, which is one of the systems employed by the IMT-2000 system, and the major metrics that affect the quality of real-time communications examined herein.

2.1 W-CDMA

In the W-CDMA environment, the Radio Link Control (RLC) protocol [16] is standardized by 3GPP as a Layer 2 protocol. The RLC protocol divides a Service Data Unit (SDU), corresponding to an IP datagram, received from the upper layer into several Protocol Data Units (PDUs) of fixed size and transmits them. The RLC protocol employs Selective-Repeat ARQ for recovering lost PDUs. If ARQ is performed repeatedly, the delay becomes infinite. Therefore, SDU discard schemes are defined in order to prevent the delay required for recovering an SDU from reaching infinity. The defined schemes are timer-based discard and number-of-retransmissions-based discard [16], the latter of which is used in the present research [17,18].

In addition, 3GPP specifies a method for preserving the order of packet delivery [19]. We use the scheme whereby some Layer 2 PDUs are kept waiting in Layer 2 of the mobile station until a timeout timer expires. In the present research, an IP datagram is sent to the upper layer if all of the preceding PDUs that were transmitted arrive at the mobile station within the Waiting Timeout (WT) period [5].

2.2 Metrics Affecting the Quality of Real-Time Communications

The major metrics that affect the quality of real-time communications, particularly voice communications, are the Signal/Noise (S/N) ratio, the transmission delay time, and packet losses. The S/N ratio is determined based on the properties of the codec used for communication, whereas the transmission delay time and packet losses are due primarily to network conditions. Voice communication is interactive and requires a short one-way transmission time between speaker and listener. The Telecommunication Standardization sector of ITU (ITU-T) states in G.114 that a one-way transmission time of less than 150 ms is acceptable for most user applications and that a one-way transmission time of less than 400 ms is acceptable for delay-tolerant applications [20]. For Internet communications, a receiver provides a jitter buffer to ensure smooth play. Packets that suffer a transmission delay time that is so large that the buffer cannot remove the effects of jitter will be treated as lost packets. When the loss probability for packets is less than 10% in networks with short transmission delays, such as wired networks, the quality of communications is good, although it is necessary to lower the loss probability as transmission delay time increases [21].

3 Proposed Mechanisms

In wireless networks, retransmission on Layer 2 is performed in order to minimize degradation due to bit errors. This results in a larger transmission delay than that in wired networks, which in turn increases the number of packets treated as lost due to late arrival. Therefore, in order to improve the quality of real-time communications in wireless networks, the transmission delay time must be decreased. In this section, we propose a mechanism by which to improve the quality of real-time communications. In Sect. 3.1, we describe a priority scheduling mechanism, which is commonly used to provide QoS assurance. This mechanism is used in the present study for performance comparison. In Sect. 3.2, we propose a receiver-based flow control mechanism with an interlayer collaboration concept using application layer information and link layer information.

3.1 Priority Scheduling Mechanism in Layer 2

One method by which to provide good quality of service is to forward real-time communications packets with higher priority, compared to non-real-time communications packets, at intermediate routers. A static-priority scheduling mechanism [22] has been proposed as a packet scheduling method for such requirements. In the present study, we use the static-priority scheduling mechanism in Layer 2 of the base station, which is used for performance comparison, although such mechanisms are generally used in Layer 3. This is because received SDUs corresponding to the IP datagram are divided into several PDUs of Layer 2 at the base station and retransmissions on Layer 2 are executed in W-CDMA networks, so that even SDUs having priority are not necessarily processed preferentially, even if static-priority scheduling is performed in Layer 3.

This mechanism distinguishes the flow of received packets and gives priority to UDP traffic. When an SDU is divided into several PDUs, the PDUs are stored in a high-priority buffer if the received packet is a UDP packet. TCP packets,

Table 1. Packet classification for each buffer in the priority scheduling

Priority	Packet type
1st	Retransmission for UDP
2nd	Transmission for UDP
3rd	Retransmission for TCP
4th	Transmission for TCP

however, are stored in a low-priority buffer. Thus, the PDUs for TCP traffic can be transmitted only when the buffer for UDP traffic is empty. The mobile station also has two buffers with different priorities and maintains the sequence integrity of packet delivery for each type of traffic. In addition, each buffer on the base station contains a transmission buffer with a low propriety and a retransmission buffer with a high priority. The former stores newly received PDUs, and the latter stores the PDUs that should be retransmitted due to transmission errors. That is, the base station has four buffers and the mobile station has two buffers. Table 1 summarizes the priority for each type of traffic.

3.2 Receiver-Based Flow Control Mechanism Employing an Interlayer Collaboration Concept on the Mobile Station

In this section, we describe the receiver-based flow control mechanism employing an interlayer collaboration concept on the mobile station. In this mechanism, the TCP receiver on the mobile station informs the sender of the bandwidth that is currently available by sending ACK with the advertised window size (*awnd*), rather than the usual available capacity of its buffer. Note that the *awnd* based on buffer capacity is used when it is smaller than the *awnd* calculated from the available bandwidth. Consequently, the TCP sender can maintain a transmission rate that is the minimum value of the congestion window size (*cwnd*) and an *awnd* within the limited available bandwidth. Therefore, this mechanism limits the bandwidth available for TCP in order to leave adequate bandwidth for UDP traffic. This mechanism can reduce loss probability and transmission delay time for UDP packets, even if each type of traffic is transmitted through different paths in the networks.

The *awnd* value can be calculated based on the capacity of the wireless link and the bandwidth currently used by UDP traffic. The capacity of the wireless link, i.e., the bandwidth-delay product of the wireless link, is obtained from Layer 2 on the mobile station through the interlayer collaboration concept, whereas the bandwidth currently used by UDP traffic is obtained from Layer 7 at the mobile station. On Layer 2, the retransmission of the lost PDUs is executed due to bit errors, so that the delay over the wireless link is equal to the worst time T required to transmit one TCP packet to the mobile station and can be expressed as (1) under the conditions that the transmission delay time of the

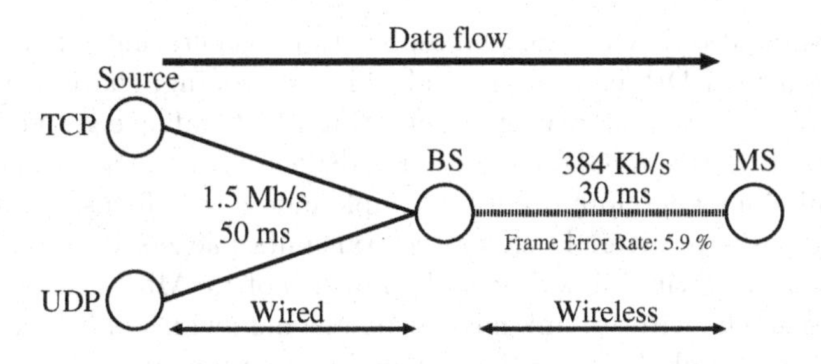

Fig. 1. Simulation model

ACK (NACK) on Layer 2 is ignored.

$$T = \left(\frac{\text{TCP packet size}}{\text{available bandwidth}} + 2 \times \text{propagation delay}\right) \times \text{the number of retransmissions.} \quad (1)$$

$$\text{available bandwidth} = \text{link bandwidth obtained from Layer 2} - \text{UDP rate obtained from Layer 7.} \quad (2)$$

Using this T, the ideal *awnd* can be obtained by

$$awnd = \frac{\text{available bandwidth} \times T}{\text{TCP packet size}}. \quad (3)$$

The priority scheduling mechanism mentioned in the previous subsection must be implemented on both the base station and the mobile station, which increases processing costs as the number of flows processed at the base station increases. In addition, the type of flow, i.e., TCP or UDP, in Layer 2 must be distinguished. On the other hand, except for a slight modification to the mobile station, the receiver-based flow control employing an interlayer collaboration concept proposed herein does not require any special support at the intermediate nodes, including the base station, and acts to increase the number of flows. In addition, the proposed mechanism does not require any modification to the corresponding host. Therefore, the proposed mechanism is more effective with respect to development and deployment.

4 Simulation Model

For the simulation of the present study, we consider a network model in which both real-time communications and non-real-time communications with different transmitting paths on wired links share an IMT-2000 link and focus our examination on the performance of QoS mechanisms. In this section, we describe the simulation model used in the present study. We used the VINT Network Simulator NS Version 2 [23] to which we added modules for Layer 2 ARQ schemes over wireless links.

In our simulation, real-time communications generating CBR traffic that is transmitted by UDP, e.g., audio and video streaming, share a wireless link with non-real-time communications generating TCP traffic, e.g., file transfer, as shown in Fig. 1. In the W-CDMA system, a dedicated channel is usually used for data transmission. Therefore, we use a simple model that focuses on one Mobile Station (MS) as a receiver. Each source transmits packets from wired links to wireless links. Both wired links have a bandwidth of 1.5 Mb/s and a propagation delay of 50 ms. The wireless link has a constant bandwidth of 384 Kb/s without interference and a link delay of 30 ms, including a propagation delay of 5 ms and processing delays of 25 ms in the Base Station (BS) and MS, so as to simplify the simulation model. The BS includes a Radio Network Controller (RNC). We assume that the wireless link is suffering from a burst error caused by Rayleigh fading [24]. More specifically, users are walking with an MS, which results in a two-path fading channel with a Doppler frequency of 5 Hz. The physical layer encodes Layer 2 PDUs in a Transmission Time Interval (TTI), during which bit errors occur. When the wireless link has the bandwidth of 384 Kb/s, the TTI is 20 ms and the transport block size is 7680 bits, according to the specification. The average error rate considered here is realistically 5.9% for the Frame Error Rate (FER) on Layer 2 with FEC, corresponding to a Bit Error Rate (BER) of 10^{-5}. We also assume that the ACK and Negative ACK (NACK) are error-free because they are much smaller than the data frame and their error rates approach zero. The BS stores PDUs in the buffer, while being managed on an SDU basis; i.e., an SDU cannot be partly stored in the buffer. In the following, the buffer will be 50 SDUs in length. The MS keeps some PDUs waiting for WT in order to preserve the sequence integrity of packet delivery, and the WT ranges from 0 to 500 ms.

In this simulation, we consider that TCP traffic is used for greedy file transfers, which continuously transmit infinity data from the sender. The TCP variant employed here is NewReno [25], which is used primarily by computers linked by the Internet. The TCP packet size is set to 1500 bytes, including the IP header [5]. The UDP packet size is set to 260 bytes, including the IP header, and its rates are set to 32, 64, and 128 Kb/s. The Layer 2 PDU size is set to 42 bytes; the header is 2 bytes and the payload is 40 bytes. In Layer 2, Selective-Repeat ARQ is employed, where the maximum number of allowable retransmissions is set to five for TCP traffic in order to achieve a large throughput [5] and ranges from zero to five for UDP traffic.

The simulation experiments are performed for 60 seconds, and the loss probability for UDP traffic and the average throughput performance of TCP traffic are discussed below. Two types of packet losses can occur in UDP traffic: (1) packets can actually be lost, i.e., a buffer overflow occurs in the BS and the Layer 2 ARQ cannot recover the lost PDUs, and (2) packet losses can occur in MS applications, i.e., packet delivery occurs out-of-order and the transmission delay time exceeds 300 ms. These losses are referred to as transport layer loss and application layer loss, respectively.

Table 2. Loss probability of UDP and average throughput of TCP with optimum parameters for TCP

UDP rate [Kb/s]	Loss probability of UDP [%]			TCP throughput [Kb/s]
	Transport	Application	Total	
32	0.35	93.22	93.57	293.29
64	0.37	91.72	92.09	258.98
128	1.32	87.16	88.48	194.75

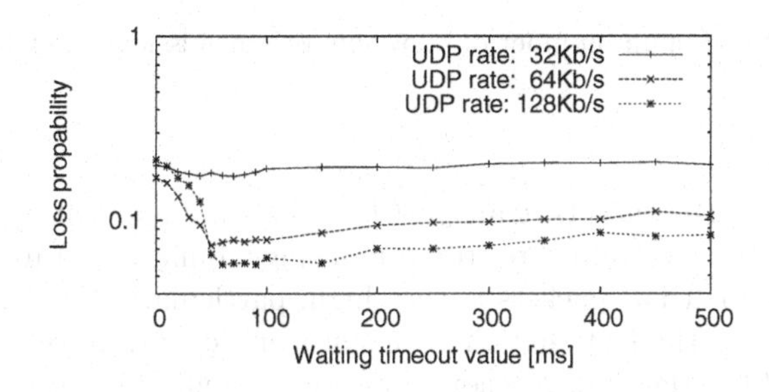

Fig. 2. Effect of WT: no QoS mechanisms

5 Simulation Results

In this section, we show the loss probability for UDP packets and the throughput performance of TCP in wireless networks based on our simulation results. We first examine the characteristics of current networks that do not provide QoS assurance for UDP traffic. Then, we show the characteristics when the priority scheduling mechanism and the receiver-based flow control mechanism are employed to improve the loss probability for UDP packets. In addition, we investigate the optimum values of tunable parameters, including the maximum number of allowable retransmissions in Layer 2 ARQ and WT.

5.1 Case Without QoS Mechanisms

Before investigating the characteristics for the case in which the QoS mechanisms are employed, we present the results for the case without QoS mechanisms for later comparison. First, we estimate the loss probability for UDP packets when the tunable parameters are set to the optimum values for TCP traffic reported in previous studies [4,5]. In order to improve TCP throughput performance, the maximum number of allowable retransmissions in Layer 2 ARQ should be set to high values and the WT should be high enough to preserve sequence integrity, even when a large number of retransmissions occurs over a wireless link. Table 2 shows the loss probability for UDP packets and the average throughput of TCP when the maximum number of allowable retransmissions in Layer 2 ARQ is set to five and WT is set to 500 ms. The table shows the two types of loss

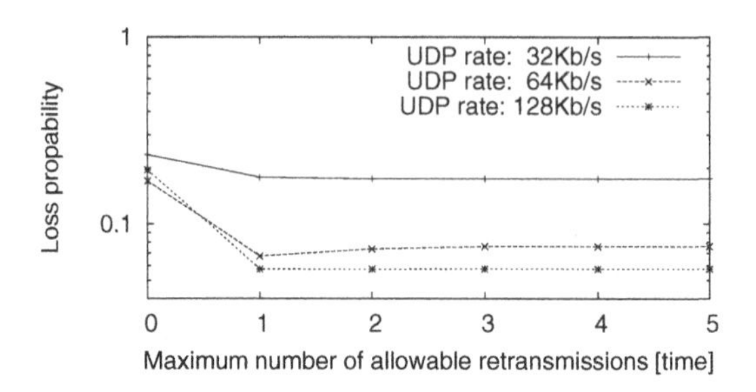

Fig. 3. Effect of maximum number of allowable retransmissions: no QoS mechanisms

probabilities for UDP packets, transport layer loss and application layer loss. All packets of each type of traffic are stored in a single buffer. As Table 2 shows, the loss probability for UDP packets is very high, particularly for application layer loss. This is due to the fact that the transmission delay time very often exceeds 300 ms. Therefore, the transmission delay time should be reduced in order to improve the quality of real-time communications.

Next, we investigate the optimum parameters for decreasing the loss probability for UDP packets. The loss probabilities for UDP packets for the range of WT from 0 to 500 ms are shown in Fig. 2. The buffer size is set at 10 packets that is optimum value in this scenario. As shown in the figure, the loss probability for UDP is minimal when WT is set to approximately 60 ms. A WT of less than 60 ms is not sufficient for maintaining the sequence integrity of UDP packets, so that out-of-order packet delivery causes a large loss probability. On the other hand, a WT of greater than 60 ms can increase the transmission delay time, while maintaining the sequence integrity with higher probability, even if out-of-order delivery occurs in Layer 2. Therefore, a WT of approximately 60 ms minimizes the loss probability as a result of a tradeoff relationship between the transmission delay time and the out-of-order packet delivery when the UDP rate is 64 or 128 Kb/s. In addition, we found that the WT does not affect loss probability when the UDP rate is 32 Kb/s. The reason for this is discussed later in this section.

Finally, Fig. 3 shows the loss probabilities for UDP packets for the maximum number of allowable retransmissions for UDP traffic in Layer 2 ARQ ranging from zero to five. The maximum number of allowable retransmissions for TCP traffic is five and WT is set to 60 ms. As shown in Fig. 3, Layer 2 ARQ can improve the loss probability for UDP packets.

The above results indicate that the optimum parameters for UDP are a maximum number of allowable retransmissions of one and a WT of 60 ms. Table 3 summarizes the loss probability of UDP packets and the average throughput of TCP when the parameters are set to these optimum values. Comparison of Tables 2 and 3 reveals that setting parameters to optimum values for UDP traffic greatly reduces application layer losses. However, high-quality real-time commu-

Table 3. Loss probability of UDP and average throughput of TCP with optimum parameters for UDP

UDP rate [Kb/s]	Loss probability of UDP [%]			TCP throughput [Kb/s]
	Transport	Application	Total	
32	3.14	14.72	17.86	277.53
64	3.49	3.24	6.73	241.89
128	5.38	0.36	5.74	185.18

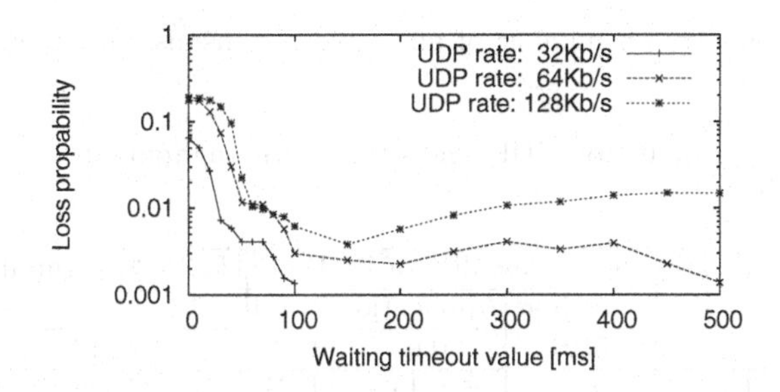

Fig. 4. Effect of WT: priority scheduling mechanism

nications are not yet achieved because of a large loss probability. In addition, TCP throughput performance is degraded and the loss probability for UDP increases as the UDP rate decreases because a lower UDP rate results in an increase in the number of TCP packets occupying the BS buffer. Consequently, the queuing delay for UDP packets increases, with the result that the transmission delay time exceeds 300 ms and application layer losses occur, particularly when the UDP rate is 32 Kb/s.

5.2 Case Employing a Priority Scheduling Mechanism in Layer 2

In this section, we employ a priority mechanism in Layer 2 in order to decrease the transmission delay time of UDP traffic. As described in Sect. 3.1, the BS has four buffers, and the MS has two buffers with different priorities and maintains the sequence integrity of packet delivery for each type of traffic in Layer 2. The sizes of the Layer 2 buffer and the WT for each type of traffic are assumed to be the same.

Figure 4 shows the loss probabilities for UDP packets for WT ranging from 0 to 500 ms. The maximum number of allowable retransmissions in Layer 2 ARQ is set to five. This is the optimum value for TCP traffic. As the figure shows, the loss probability is greatly reduced when the WT is 150 ms, which is a result of a tradeoff relationship between the transmission delay time and the out-of-order packet delivery, as mentioned in the previous subsection. In this case, WT is larger than that for a case without QoS mechanisms because the transmission delay time of high-priority UDP traffic when using a priority

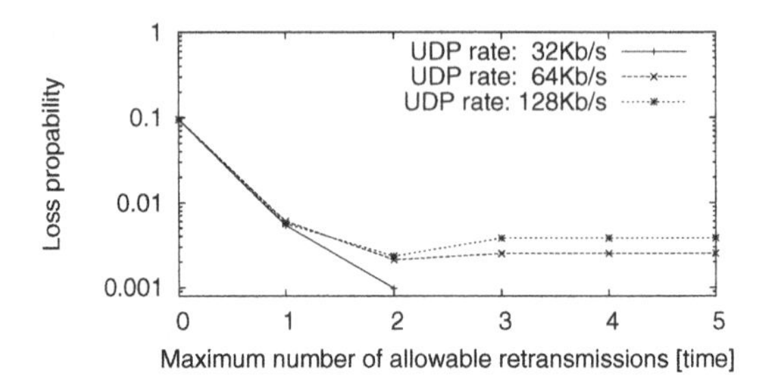

Fig. 5. Effect of maximum number of allowable retransmissions: priority scheduling mechanism

Table 4. Loss probability of UDP and average throughput of TCP with priority scheduling mechanism

UDP rate [Kb/s]	Loss probability of UDP [%]			TCP throughput [Kb/s]
	Transport	Application	Total	
32	0.10	0.00	0.10	294.13
64	0.07	0.14	0.21	261.86
128	0.05	0.18	0.23	196.55

scheduling mechanism becomes smaller, and this allows the WT to be larger in order to maintain the sequence integrity of packet delivery. Next, Fig. 5 shows the loss probability for UDP packets for a WT of 150 ms and the maximum number of allowable retransmissions in Layer 2 ARQ ranging from zero to five. The figure shows that setting the maximum number of allowable retransmissions to two minimizes the loss probability.

The above results suggest that the optimum parameters for the UDP are a maximum of two allowable retransmissions and a WT of 150 ms. Table 4 summarizes the loss probabilities for UDP packets and the average TCP throughput for the optimum parameters. Using this priority scheduling mechanism in Layer 2 drastically improves the loss probability for UDP packets, providing QoS assurance for real-time communications. Furthermore, TCP throughput is satisfactory.

5.3 Case Employing a Receiver-Based Flow Control Mechanism

The priority scheduling mechanism must be implemented in both the BS and the MS, as mentioned above. Since this is highly complex, we employ a receiver-based flow control mechanism using an interlayer collaboration concept, which does not require any special support in the intermediate nodes including the BS, except for a slight modification to the MS, as an alternative mechanism in this section. In this mechanism, instead of the usual available capacity of its buffer, the TCP receiver at the MS informs the sender of the currently available

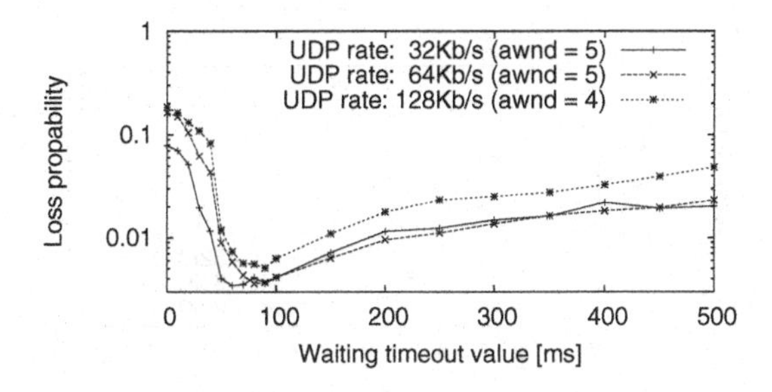

Fig. 6. Effect of WT: receiver-based flow control mechanism

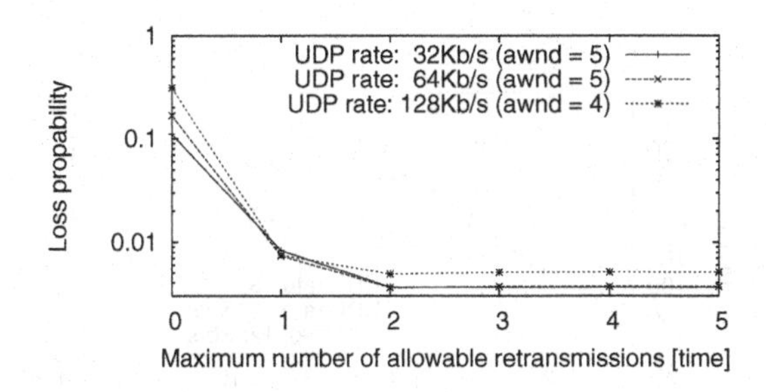

Fig. 7. Effect of maximum number of allowable retransmissions: receiver-based flow control mechanism

bandwidth by sending ACK with an *awnd* value, in order to limit the bandwidth available for TCP traffic so as to leave adequate bandwidth for UDP traffic, as described in Sect. 3.2. The *awnd* value can be calculated by (3). When UDP rates are 32, 64, and 128 Kb/s, the available bandwidths are 352, 320, and 256 Kb/s, respectively, as can be calculated using the information of the UDP rate obtained from Layer 7 and the bandwidth of the wireless link obtained from Layer 2. In addition, two retransmissions are actually executed in most of the simulation cases because of the assumption in the present study that the FER is 5.9%. Note that the number of retransmissions is obtained from Layer 2 at the MS. Thus, the ideal *awnd* values become 5.52, 5.20, and 4.56 packets for UDP rates of 32, 64, and 128 Kb/s, respectively.

Figure 6 shows the loss probability for UDP packets when the *awnd* is set to five packets for 32 Kb/s and 64 Kb/s and four packets for 128 Kb/s, for WT ranging from 0 to 500 ms. The maximum number of retransmissions allowed in Layer 2 ARQ is five. From this figure, we can see that the loss probability is reduced when the WT is approximately 90 ms, which is the tradeoff point mentioned above. Figure 7 shows the loss probability for UDP packets when WT is set to 90 ms and the maximum number of retransmissions allowed in Layer 2 ranges from zero to five. As this figure shows, the loss probability of UDP traffic reaches a minimum when the maximum number of allowable retransmissions is larger than one.

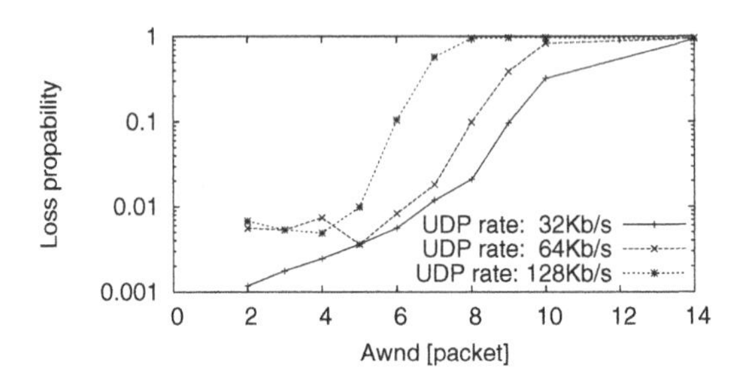

(a) Loss probability of UDP

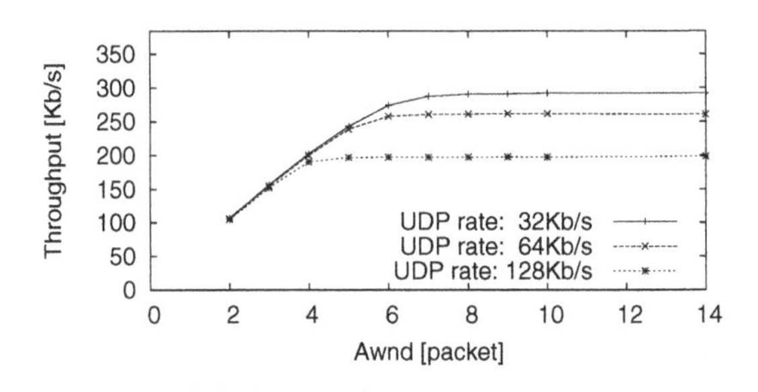

(b) Average throughput of TCP

Fig. 8. Effect of *awnd*: receiver-based flow control mechanism

Finally, we verify that the calculated *awnd* value agrees with the measured value. The loss probability for UDP packets and the average throughput of TCP for *awnd* ranging from 2 to 14 packets are shown in Figs. 8 (a) and (b), respectively. As these figures show, if we consider both loss probability and throughput performance, the optimum *awnd* values are six packets for 32 Kb/s, five packets for 64 Kb/s, and four packets for 128 Kb/s, which are approximately equal to the calculated values.

The above results indicate that the optimum parameters for UDP are a maximum of two retransmissions allowed, a WT of 90 ms, and *awnd* values of six, five, and four packets for UDP rates of 32, 64, and 128 Kb/s, respectively. Table 5 summarizes the loss probability for UDP packets and the average TCP throughput when the parameters are set to these optimum values. These mechanisms reduce the loss probability, providing QoS assurance for real-time communications, and maintaining good TCP throughput performance.

Table 5. Loss probability of UDP and average throughput of TCP with receiver-based flow control mechanism

UDP rate [Kb/s]	Loss probability of UDP [%]			TCP throughput [Kb/s]
	Transport	Application	Total	
32	0.02	0.54	0.56	273.98
64	0.07	0.29	0.36	238.74
128	0.11	0.38	0.49	190.29

6 Conclusions

In the present study, we have investigated how the quality of real-time communications is affected by non-real-time communications and have determined optimum parameters by which to improve the quality of real-time communications in W-CDMA networks. Through simulations, we found that the loss probability is too large to ensure high-quality real-time communications, even if the parameters associated with W-CDMA systems are set to optimum values for UDP traffic. Therefore, we proposed a receiver-based flow control mechanism employing an interlayer collaboration concept for improving the quality of real-time communications over IMT-2000 wireless networks. In addition, we introduced a priority scheduling mechanism in Layer 2 for comparing the performance of QoS mechanisms. The receiver-based flow control mechanism does not require any special support in the intermediate nodes, including the base station, except for a slight modification to the mobile station. The priority scheduling mechanism, however, must be implemented in both the base station and the mobile station, which increases the processing costs as the number of flows processed at the base station increases. Our simulations indicate that the receiver-based flow control mechanism can achieve comparable improvement in loss probability to that obtained by using a priority scheduling mechanism in Layer 2 of the base station; i.e., the receiver-based flow control mechanism can improve the quality of real-time communications. In addition, the mechanism does not adversely affect TCP throughput performance. Therefore, the receiver-based flow control mechanism employing an interlayer collaboration concept is effective in wireless networks.

References

1. International Mobile Telecommunications-2000, *http://www.imt-2000.org/*.
2. 3rd Generation Partnership Project, *http://www.3gpp.org/*.
3. Stevens, W.R.: TCP/IP Illustrated, Volume 1: The Protocols. Addison-Wesley, Readings, Massachusetts (1994)
4. Koga, H., Kawahara, K., Oie, Y.: TCP flow control using link layer information in mobile networks. In: Proc. SPIE Conference of Internet Performance and Control of Network System III. Volume 4865. (2002) 305–315

5. Koga, H., Ikenaga, T., Hori, Y., Oie, Y.: Out-of-sequence in packet arrivals due to layer 2 ARQ and its impact on TCP performance in W-CDMA networks. In: Proc. IEEE 2003 Symposium on Applications and the Internet (SAINT 2003). (2003) 398–401
6. Jacobson, V., Karels, M.J.: Congestion avoidance and control. In: Proc. ACM SIGCOMM'88. (1988) 314–329
7. Hori, Y., Sawashima, H., Sunahara, H., Oie, Y.: Performance evaluation of UDP traffic affected by TCP flows. IEICE Transactions on Communications **E81-B** (1998) 1616–1623
8. Floyd, S., Jacobson, V.: Link-sharing and resource management models for packet networks. IEEE/ACM Transactions on Networking **3** (1995) 365–386
9. Xiao, X., Ni, L.M.: Internet QoS: A big picture. IEEE Network **13** (1999) 8–18
10. Braden, R., Zhang, L., Berson, S., Herzog, S., Jamin, S.: Resource ReSerVation protocol (RSVP). RFC2205 (1997)
11. Vandalore, B., Jain, R., Fahmy, S., Dixit, S.: AQuaFWiN: Adaptive QoS framework for multimedia in wireless networks and its comparison with QoS frameworks. In: Proc. the 24th IEEE Conference on Local Computer Networks. (1999) 88–97
12. Maniatis, S.I., Nikolouzou, E.G., Venieris, I.S.: Convergence of UMTS and Internet services for end-to-end quality of service support. In: Proc. European Wireless 2002. (2002)
13. Marques, V., Aguiar, R.L., Pacyna, P., Fozdecki, J., Beaujean, C., Chaher, N., Gercia, C., Moreno, J.I., Einsiedler, H.: An architecture supporting end-to-end QoS with user mobility for systems beyond 3rd generation. In: Proc. IST Mobile & Wireless Telecommunications Summit 2002. (2002)
14. Chinta, M., Helal, A., Lee, C.: ILC-TCP: An interlayer collaboration protocol for tcp performance improvement in mobile and wireless environments. In: Proc. the Third IEEE Wireless Communications and Networking Conference (WCNC), New Orleans, Louisiana (2003)
15. Wu, G., Bai, Y., Lai, J., Ogielski, A.: Interactions between TCP and RLP in wireless Internet. In: Proc. IEEE Globecom'99. (1999) 661–666
16. 3GPP TS 25.322 V5.2.0: RLC protocol specification (2002)
17. Tynjala, T., Leppanen, S., Luukkala, V.: Verifying reliable data transmission over UMTS radio interface with high level petri nets. In: Proc. IFIP WG 6.1 International Conference on Formal Techniques for Networked and Distributed Systems (FORTE2002), Lecture Notes in Computer Science. Volume 2529. (2002) 178–193
18. Li, J.: A performance evaluation of the radio link control protocol in 3G UMTS. Diploma thesis, School of Mathematics and Statistics, Carleton University (2003)
19. 3GPP TS 25.301 V4.2.0: Radio interface protocol architecture (2001)
20. ITU-T Rec. G. 114: One-way transmission time (2002)
21. Kostas, T.J., Borella, M.S., Sidhu, I., Schuster, G.M., Grabiec, J., Mahler, J.: Real-time voice over packet-switched networks. IEEE Network **12** (1998) 18–27
22. Siegel, E.D.: Designing Quality of Service. John Wiley & Sons, Inc. (2000)
23. The Network Simulator ns-2, *http://www.isi.edu/nsnam/ns/*.
24. Adachi, F., Sawahashi, M., Suda, H.: Wideband DS-CDMS for next generation mobile communications systems. IEEE Communications Magazine **36** (1998) 56–69
25. Floyd, S., Henderson, T.: The NewReno modification to TCP's fast recovery algorithm. RFC2582 (1999)

Network-Level Simulation Results of Fair Channel-Dependent Scheduling in Enhanced UMTS

Irene de Bruin

Twente Institute for Wireless and Mobile Communications (WMC),
Institutenweg 30, 7521 PK Enschede, the Netherlands
irene.de.bruin@ti-wmc.nl

Abstract. In this paper, previously performed UMTS studies have been extended to enhanced UMTS, and combined with several scheduling algorithms (ref. [1,2]). Scheduling algorithms for W-CDMA systems ranging from C/I based scheduling to Round-Robin scheduling are presented and discussed. We also include a trade-off between these two extreme methods of scheduling: Fair channel-dependent scheduling. All these scheduling algorithms have been implemented in a network simulator in ns-2, employing input trace files of SNRs, modeled from link-level simulation results. The network-level simulation results indeed display that the advantages of both extremes in scheduling algorithms have been combined in the new algorithm: a good fairness, comparable to that of Round-Robin scheduling, together with an efficient use of resources, as seen in C/I based scheduling.

1 Introduction

While UMTS is still in its initial deployment phase in Europe, researchers are investigating improvements to the system's performance. UMTS and other third generation cellular systems provide global access to voice and data services. One of the most important features which distinguishes UMTS from many previous generation cellular network technologies is the underlying CDMA principle that is being used. The same frequency resources are shared by many users at the same time. As a result, ongoing connections cause interference to each other which affects the capacity of the corresponding cell.

A key improvement of UMTS is realized in HSDPA (High-Speed Downlink Packet Access), which is defined for release 5 of the 3GPP (3rd Generation Partnership Project) UMTS standard. HSDPA aims at increasing the systems capacity and the users' peak throughput from 2 Mbps to over 10 Mbps. Its transmission time interval (TTI) is smaller than that of regular UMTS and the scheduling functionality is no longer in the RNC (Radio Network Controller) but in the Node B (Base Station). As a result, the fluctuations in the radio channel fading characteristics can be better tracked. Scheduling of data flows is a key mechanism to both provide QoS, and to

I. Niemegeers and S. Heemstra de Groot (Eds.): PWC 2004, LNCS 3260, pp. 301–315, 2004.

optimise resource efficiency. Whereas the former objective is clearly the most important to the user, the latter is essential for the operator.

The network-level simulator [3] in ns-2, developed for this type of network layer studies employs SNR (signal to noise ratio) link-level results. It implements a complete end-to-end connection from the external IP network through the Core Network and UTRAN to the UE (User Equipment). The simulator focuses on MAC (Medium Access Control) and RLC (Radio Link Control), where versions of these protocols are implemented for HSDPA according to the 3GPP standard (release 5). The network simulator also requires a model that mimics the main characteristics of the radio channel based on physical layer simulation results. This is done by means of a link-level simulator which implements all physical layer aspects of Enhanced UMTS (release 5) as specified by 3GPP. The enhancements include HSDPA (with both QPSK and 16-QAM modulations [4]). Results of this link-level simulator are used in the network-level simulator by means of input trace files containing SNR levels fluctuating in time.

For the HS-DSCH (High-Speed Downlink Shared Channel, the wideband equivalent of the downlink shared channel in UMTS) a fast link adaptation method is applied. This involves a selection of the modulation and coding scheme. So, a bad channel condition does not result in a higher transmit power (as is the case for the downlink shared channel in UMTS), but instead prescribes another coding and modulation scheme. The focus of HSDPA is best effort traffic; mainly users in favourable channel conditions can benefit from the higher data rate achievable through the use of HS-DSCH.

Since the purpose of this study is the evaluation of different scheduling algorithms, we mainly consider UDP data (User Datagram Protocol). The Transmission Control Protocol (TCP) might blur the analysis as it introduces several parameters that should be tuned, which may affect the end-to-end delay experienced by the user. We focus on the 5-user case, in order to have the possibility of visually displaying the results for all users simultaneously, and for some aspects we simulate and display the results of the 20-user case as well. This enables us to show how the efficient use of resources, such as is the case for C/I scheduling, can be exploited when the number of users increases. In the rest of this paper, Chapter 2 is used to describe the scheduling algorithms, focusing on the FCDS algorithm. In Chapter 3, we consider the way the link-level results are post-processed (incorporating a model for the propagation loss), describe the error model, and we include a specification of the most critical parameters. Results of the pre-processing phase are presented in Chapter 4 and followed by the network simulator results in Chapter 5. Finally the conclusions are given in Chapter 6.

2 Scheduling Algorithm

In this chapter we discuss some methods that can be used to schedule data traffic over shared resources. Because of the expected asymmetry in the traffic, we focus on the data transfer in the downlink, *i.e.*, from the Node B to the user. Usually, the scheduling is based on either the current channel conditions (C/I based scheduling) or on simple Round-Robin mechanisms. While the latter is based on the principle of fairness, channel-dependent scheduling mechanisms tend to be unfair because receiving nodes which are close to the sending node are served more often than others. Since the C/I and FCDS scheduling base the user selection on the channel condition, it is of extreme importance that these algorithms are incorporated in the simulator as realistic as possible. All UEs signal their experienced channel condition over an uplink control channel by means of a so-called channel quality indicator (CQI). In the real system, only this integer value is known at the Scheduler. Furthermore, this signaling takes time, which is modeled by a time delay between the SNR and CQI level. In Table 1 we summarize the scheduling methods studied in this paper and define the corresponding acronyms.

Table 1. Scheduling methods used in this paper

Optimization	Corresponding technique
Absolute signal	C/I based scheduling
Relative signal	FCDS scheduling
Round Robin	Fair scheduling

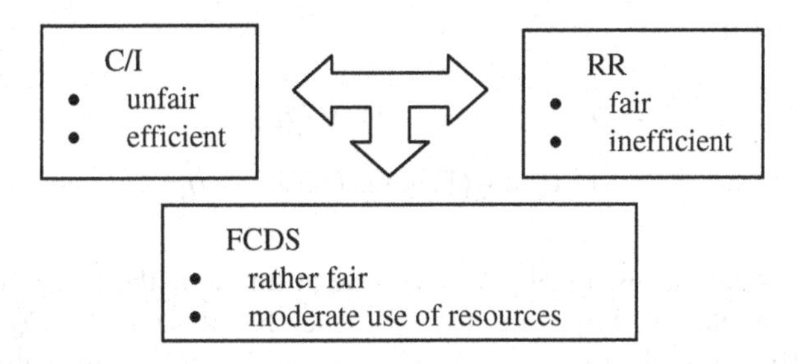

Fig. 1. The FCDS scheduling is a trade-off between C/I and RR scheduling

The Fair Channel-Dependent Scheduling (FCDS) technique introduced in this paper forms a trade-off between the two extremes: optimum capacity (system's perspective) and fairness (user's perspective). This trade-off is depicted in Figure 1. The C/I scheduling always picks the user with the best channel conditions. The FCDS instead tries to 'ride the waves': pick a user when it the channel condition is optimal for that particular user.

In practise, the signal is fluctuating around a mean value that, on its turn, also displays slow trends. This underlying slow fluctuation accounts for the distance from the base station. The time scale of so-called fading variations in the signal itself, due to multi-path reception and/or shadow fading, is much smaller than that of the variations of this so-called *local mean*. The FCDS scheduling is based on the relative indicator, *i.e.* the instantaneous CQI value relative to its own recent history. Therefore, the instantaneous level of all mobile terminals is translated with respect to their local means, and subsequently normalized with their local standard deviations. This latter step of normalization was previously concluded to be less important, based on time-driven Matlab simulations [2]. The network simulator results displayed here are based on link-level input traces and resulting SNR values that are much more realistic. It turned out that the distance from the UE to the Node B does have an impact on the variation in channel conditions. As a result, also the normalization step is included in the present study. A transmission is scheduled to the UE that has the highest value of this so-called relative CQI.

The idea of a relative indicator, introduced above, needs the definition of a local mean, with *local* referring to the recent time history. Exponential smoothing weights past observations with exponentially decreasing values in order to update the local mean. It takes the local mean of the previous period and adjusts it up or down based on what actually occurred in that period. Through the correct choice of weighting factor, this procedure can be made sensitive to a small or gradual drift in the process. This method is simple and therefore has low data storage and processing requirements, since only the actual (instantaneous) value and the old local mean value are needed to update the new local mean value. Compared with, for example, a moving averaging, the low storage and higher weighting of more recent samples, are two properties in favour of the FCDS method.

The local mean, introduced above, is updated each time unit according to the following algorithm [5]:

$$\begin{aligned} \mu_t &= \alpha \cdot \mu_{t-1} + (1-\alpha) \cdot CQI_t, \\ v_t &= \alpha \cdot v_{t-1} + (1-\alpha) \cdot (CQI_t - \mu_t)^2. \end{aligned} \tag{1}$$

with CQI_t the (instantaneous) received CQI value at time t, μ_t the local mean of CQI_t based on the time interval [t_0,t], $_t$ the local variance based on μ_t and CQI_t, and the smoothing coefficient with respect to the local mean. In other words: the new local mean is a weighted average of the instantaneous contribution and the old mean. Note that the lower script t either refers to the physical time unit or the corresponding integer index.

The criterium that determines the optimal mobile node for the next downlink transmission at time t, is formulated as follows, where the superscript i is used to denote the situation at node i:

$$\max_i \left\{ \left(CQI_t^i - \mu_t^i \right) / \sqrt{v_t^i} \right\} \tag{2}$$

Here, the instantaneous data point, CQI_t, is translated with respect to the local mean, $\mu_{t,}$, and next normalized with the corresponding local standard deviation. This translated indicator is referred to as the relative indicator, CQI_{rel}. At each time t, the value of CQI_{rel} is compared for all i nodes and the optimal node, *i.e.*, the highest CQI value, is selected for the downlink transmission.

3 Processing of HSDPA Link-Level Results

As already mentioned previously, power control is typically used in UMTS, while HSDPA employs link adaptation instead. The variation in channel conditions is taken care of through the use of different orders in the modulation and different coding rates. On its turn, these define several Transport Block Sizes (TBS), alongside different UE types [4]. The UE reports the observed quality to the Node-B by means of the CQI. The Node-B decides, based on the CQI and additional information, such as the amount of available data to send, what TBS to use in its transmission. Large TBSs require several channelization codes of a cell to be allocated to the High Speed Downlink Shared Channel (HS-DSCH), *e.g.* a ratio of 15/16 of all codes [4].

To estimate the performance of each single TBS, a link-level simulator has been used. This simulator considers radio communication between the Node-B and the UE using the HS-DSCH, based on the 3GPP specifications (for a detailed description, see reference [6,7]). The link-level results provide an Eb/No versus BLER curve for all possible CQIs, i.e. the interval [0 30]. It should be mentioned that consecutive CQIs have a near constant offset of 1 dB at a BLER of 10% [6,8].

Inside the network-level simulator the UE indicates the CQI to the Node-B. The CQI represents the largest TBS resulting in a BLER of less than 0.1. The relation between CQI and SNR for a BLER of 0.1 is approximated through a linear function, based on the 3GPP standard [4]:

$$CQI = \begin{cases} 0 & SNR <= -16 \\ \left\lfloor \dfrac{SNR}{1.02} + 16.62 \right\rfloor & -16 < SNR < 14 \\ 30 & 14 <= SNR \end{cases} \tag{3}$$

The RMSE of this approximation is less than 0.05 dB, based on integer CQI values [6]. Note that a CQI equal to 0 indicates out of range and the maximum CQI equals 30. So the function truncates at these limits [4].

These models provide a relation between SNR, CQI and BLER. The network-level simulation requires a trace of SNR values to determine the CQI. During each TTI (Time Transmission Interval) the SNR at the UE of a received DL transmission, results in a CQI that the UE reports to the Node-B in the next TTI. Another TTI is required by the Node-B to receive and process the CQI. The Node-B contains an algorithm that combines CQI information alongside information like the amount of data that needs to be transmitted to determine the scheduling and selection of the TBS for the next TTI [4]. Overall, a three-TTI delay exists between the SNR condition on which the CQI is based and the actual transmission of the resulting transport block over the HS-DSCH. The network simulator implements this as a simple delay.

By definition, the floor function in equation (3) introduces an extra contribution to the SNR value of, on average, 0.5 dB. This implies that the resulting BLER is smaller than expected. Together with the fact that the BLER curves get steeper for higher CQI values [6,8], this results in a smaller BLER for higher CQI values. The three-TTI delay mentioned above will also influence the resulting BLER since presumably a non-optimum TBS is selected for transmission.

The traces generated initially had CQI values in the interval [-24,24]. Next, they have been truncated according to the standards [4]. On the lower side, typically 5 to 10 % has been replaced by 1. Since a CQI of 0 implies an out of range, it is replaced by 1 as well. Similarly, on the upper side, 1 to 2% has been replaced by 22, since the CQIs of 23 to 30 all have the TBS of CQI 22 (UE type 1-6; ref [4]).

The SNR values result from the overall path loss and interference. Path loss consists of distance loss, shadow fading and multi-path fading. Without compromising the accuracy of the network-level simulation results, each UE has its own fixed distance to the Node-B, resulting in a fixed distance loss. The shadow fading has a log-normal distribution, correlated in time. A correlation distance of 40 m and a standard deviation of 8 dB is chosen [2,9]. The shadow fading is added to the fixed distance loss (in dB).

In a flooded W-CDMA system, where cells are surrounded by loaded cells, the inter-cell interference level is more or less constant over the complete area, except for the effect of shadowing. As the wanted signal is modeled with shadowing it suffices for network-level simulations to model the inter-cell interference as a fixed power level.

Intra-cell interference is strongly correlated to the wanted signal. From a propagation point of view it has an identical behaviour. The power of the intra-cell interference can fluctuate over time. Capacity control functions in UTRAN will however attempt to utilize the available power as efficiently as possible. The power is balanced between dedicated channels and the HS-DSCH, allocating more to the HS-DSCH when less power is needed by the dedicated channels. As a result the transmission power of the Node-B is roughly constant for systems applying HS-DSCH.

The transmission of channels is orthogonal to all transmissions using the same scrambling code. Frequency selective fading will destroy this property to some extent. A fraction of all DL transmissions turns into interference. For the purpose of the network simulation model it suffices to assume this to be a fixed power level, that undergoes the same propagation characteristics as the wanted signal. The assumptions specified above may introduce small errors in the link-level results, but, as already mentioned in the introduction, this is only effecting the absolute behavior, and not the relative characteristics in time. As the simulator aims at studying characteristics of network-level functions and procedures, it mainly prescribes that the underlying physical model's *relative* behavior is realistic, which is the case in this study.

All results displayed in this paper are based on the Pedestrian A environment. As already mentioned in the introduction, for all users we assume the same distance to the Node B during the whole simulation: 500 meters. So, apart from *local* fast and shadow fading effects, on the long run there is no distinction between the users. All simulations have a duration of 200 seconds (the equivalent of 100,000 TTIs) and the results displayed are based on either assuming 5 or 20 active users, all moving with a velocity of 3 km/h.

Regarding the H-ARQ (Hybrid Automatic Repeat Request) we assume that a H-ARQ process consists of up to 3 transmissions, with a period of 6 TTIs to wait for a retransmission, following HSDPA standards [4]. Furthermore, we assume a MAC-d PDU size of 40 bytes, which is a trade-off value [1]. As a result, this MAC-d PDU does not fit in the transport block size (TBS) corresponding to CQI values 1-4. Furthermore, the parameters that regulate the rate in the UDP CBR (constant bit rate) of data flowing through the system should be synchronized in order to avoid empty buffers in C/I scheduling where one user may send high data rates in subsequent TTI slots. The round-trip time in the flow control, between RNC and the Node B, equals 30 ms. The maximum number of MAC-d PDUs that can be stored in the transmission buffer is set to 250. The CBR is set to 1000 kbps.

4 Results of the Pre-processing Phase

We begin with two figures indicating how the link-level results have been used as input for the network simulator. Figure 2 shows a sample of the CQI generation process for a 3 km/h Pedestrian A channel. The figure clearly shows the delay of three TTIs in the CQI generation, where we note that a HSDPA TTI lasts 2 ms. In addition, the figure shows that the CQI is truncated to integer values, see formula (3).

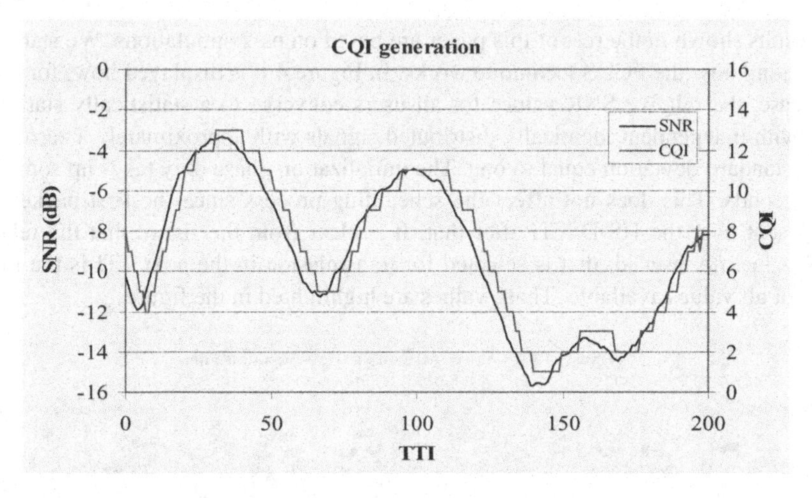

Fig. 2. Sample of CQI generation in a Pedestrian A environment

As a result of the delay in the CQI generation process, presumably a non-optimum TBS is selected for transmission. This effect can clearly be seen from Figure 3 which shows the BLER for each TTI. It is based on the actual SNR and the delayed CQI from Figure 2. While the CQI aims at a BLER of 0.1, quite extreme BLER variations can be seen, corresponding to the error in CQI as seen in Figure 2. Due to the floor function in formula (3), the resulting BLER is lower than the target value for most TTIs.

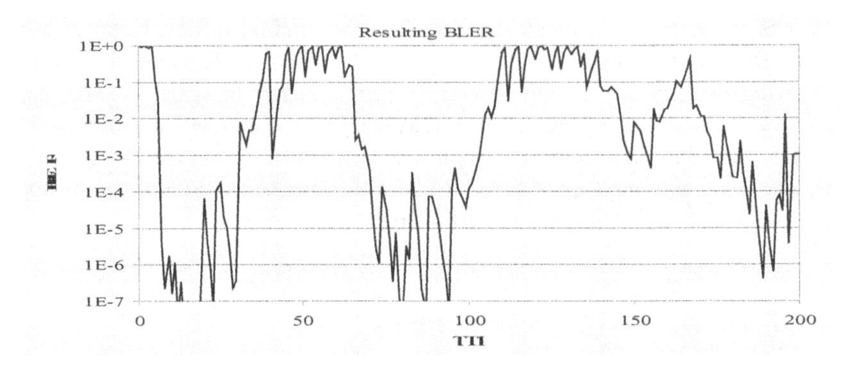

Fig. 3. Sample of BLER as a result of CQI and a 3 TTI delay (Pedestrian A)

5 Network Simulator Results

All results shown in the rest of this paper are based on ns-2 simulations. We start with illustrating how the FCDS technique works. In Figure 4 it is displayed how, for the 5-user case, the relative SNR values for all users converge to a statistically stationary state with independent identically distributed signals with approximately a zero mean and a standard deviation equal to one. The initialization phase only takes up some 100 milliseconds. This does not affect the scheduling process since the first packets are being sent over the HS-DSCH after that. It is clear from the figure that the relative SNR value, *i.e.* user id, that is selected for transmission in the next TTI is the maximum of all values available. These values are highlighted in the figure.

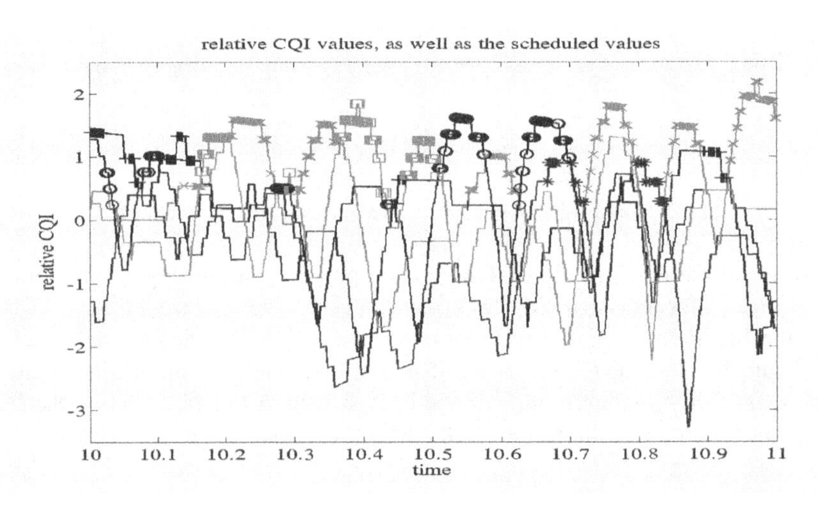

Fig. 4. Relative CQI values for the 5-user case. The values selected are highlighted

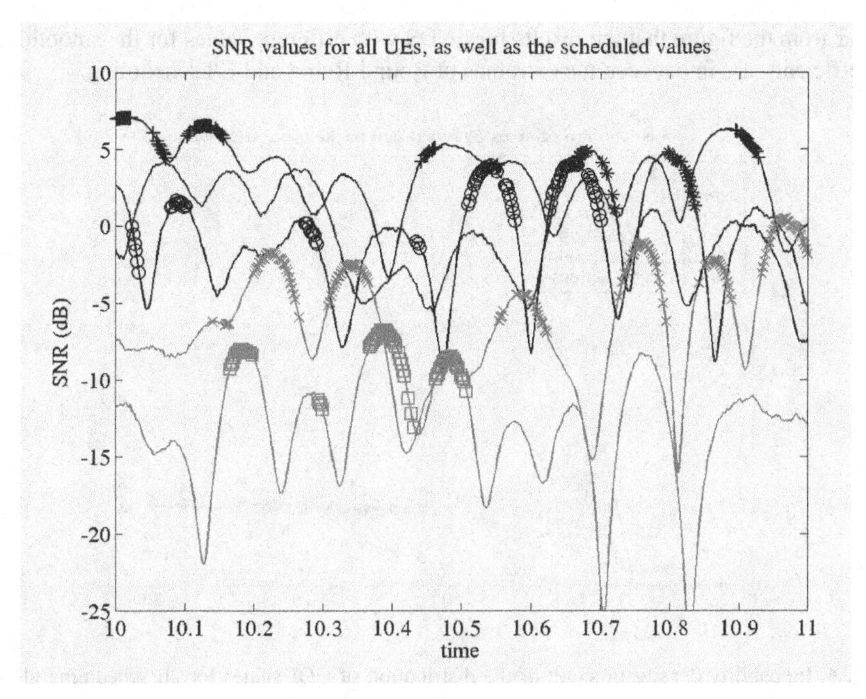

Fig. 5. SNR values for the 5-user case. The values selected for transmission (see Figure 4) are highlighted

In Figure 5 we see the impact of the FCDS scheduling algorithm with respect to the actual SNR values. It is clear that all five users are being scheduled regularly, and for equal amounts of time. The figure clearly illustrates how the FCDS scheduling technique is exploiting the most favourable transmission conditions, relative to their own history for all users.

Next we focus on the CQI in combination with the amount of data typically send during a TTI. In Figure 6 we have collected probability density functions of the distribution of CQI values for all scheduling algorithms considered. For comparison purposes we have also included the results of the input trace file. As expected, the round-robin (RR) result is close to that of the input trace file for most CQI values. A peak exists at a CQI value of 5, which can be explained from the way RR typically is implemented. A user with low channel condition quality (CQI 1-4) is not scheduled. As a result, its timer used in the Round-Robin scheduling is not reset again. This user is scheduled immediately when it comes out of the dip and reaches the CQI value of 5. This results in a higher contribution for the CQI=5 value for the RR scheduling. Next, the user has to wait for at most 10 ms until it is scheduled again. This explains why the next CQI values (CQI 6-8), which typically would occur as a next step in time, have a somewhat lower contribution.

The result for the other extreme in scheduling, C/I based scheduling, shows that mainly users at the highest CQI values are selected for transmission. We note that the C/I result for CQI 22 equals 0.75, and falls far beyond the range of the plot. It is also clear from the figure that the results for FCDS with different values for the smoothing coefficient are in between the extremes of Round-Robin and C/I scheduling.

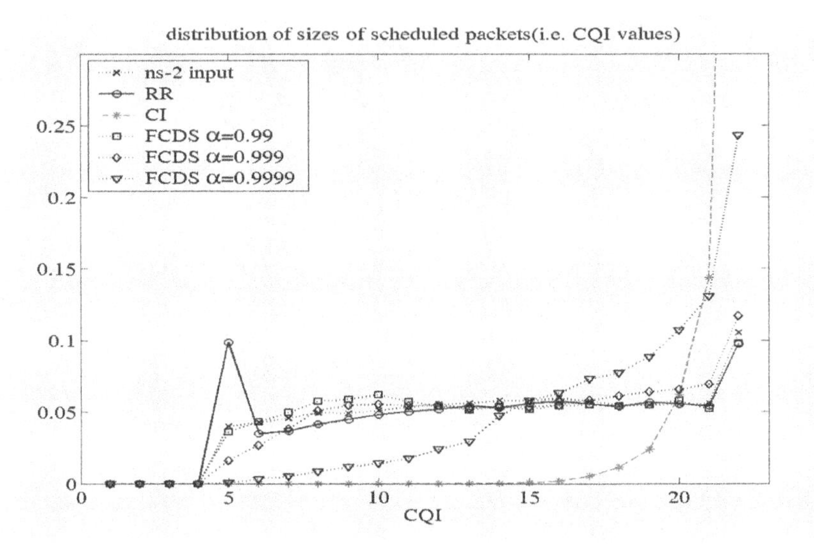

Fig. 6. Probability density function of the distribution of CQI values for all scheduling algorithms considered (20-user case). Also the CQI distribution of the input trace file is included. The results of all lines sum up to one

The CQI distribution is followed by the cumulative density function of the resulting transport block sizes (TBS) of the users selected for transmission. It should be stressed that this TBS is used even when the selected user could also suffice with a lower TBS. It ranges from 47 to 896 bytes, corresponding to a CQI value of 5 to 22. It is illustrated in Figure 7. Again, it is clear that the results for FCDS scheduling are in between the results for RR and C/I.

In Table 2 we have summarized the TBS for the 5-user and 20-user cases of all scheduling algorithms. We also included the theoretic limits based on the input trace file. These can only be reached if all users have their optimal channel condition in a successive way, which is not the case in practice, particularly not for few-user cases. Furthermore, the Transport Block Size remains the same throughout the whole HARQ process. So the HARQ retransmissions presumably do not transmit anymore at the most optimum transport block sizes. The C/I result increases when considering the higher number of users. And it is clear that the 20-user case is not far from the theoretical limit. Also the Round-Robin numbers can be compared with the theoretic value of 412 bytes. It is clear that the 5 users RR case is close to this value. And for the same reason as described in the analysis of Figure 6, it makes sense that the average TBS decreases when considering the larger user case, since there is an increasing probability of users moving out of a dip, thereby enforcing to be scheduled at a relatively low CQI value of 5. All FCDS results again fall in between the extremes of RR and C/I scheduling for both user settings.

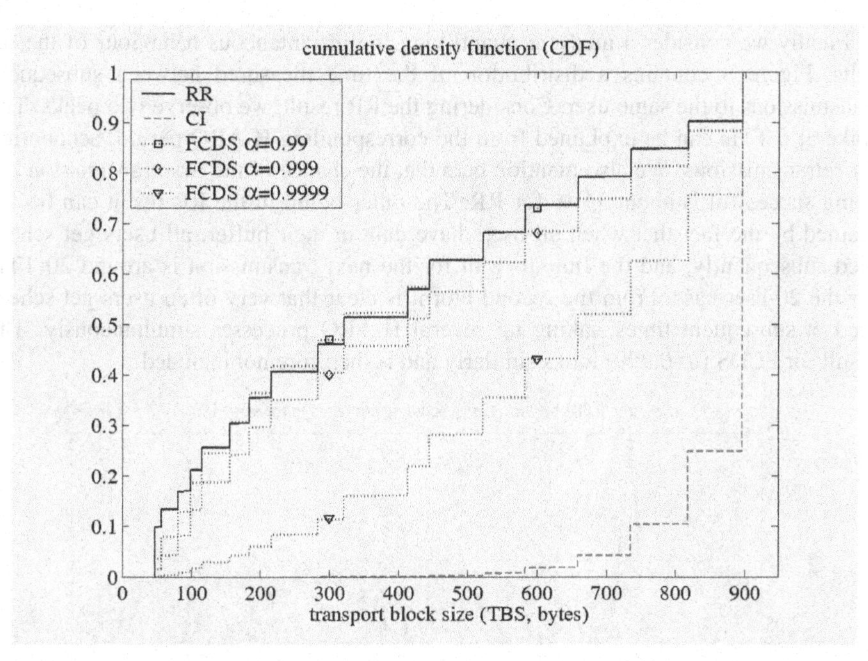

Fig. 7. Cumulative density function of the distribution of transport block sizes for all scheduling algorithms considered (20-user case). The symbols in the dotted lines of the FCDS results are only included as a marker of the line and do not refer to specific values of the TBS

In the table we have also included the BLER. The value for Round-Robin is the highest. This is mainly due to the fact that the SNR/BLER curves get steeper for higher CQI, as is the chance of success. Since the average TBS of RR is rather low, this explains the highest BLER. Depending on the parameter, the FCDS algorithm performs better or worse than the C/I scheduling. The increasing of the number of users has a negative effect on the BLER of RR. This can be explained with the decrease in the TBS. The increase of users has a positive effect on the BLER of all other scheduling algorithms. We come back to this later, when we consider this effect on a per CQI basis.

Table 2. Average transport block size (TBS, bytes) selected for transmission, and the resulting error rate

	TBS		BLER (%)	
	5 UEs	20 UEs	5 UEs	20 UEs
RR	414	398	16.1	16.8
FCDS α=0.99	451	399	9.1	5.2
FCDS α=0.999	474	444	11.8	10.3
FCDS α=0.9999	526	633	12.5	10.2
C/I	684	863	11.9	6.8
Theoretic limit	843	896		

Finally we consider a measure quantifying the instantaneous behaviour of the results. Figure 8 contains a distribution of the time measured between subsequent transmissions to the same user. Considering the RR result, we observe two peaks. The spike at 6 TTIs can be explained from the corresponding H-ARQ period, accounting for retransmissions. We also mention here that the chance of the first transmission not being successful is about 16 % for RR. The other bump in the RR result can be explained by the fact that when all users have data in their buffer, all users get scheduled subsequently, and the time to wait for the next transmission is around 20 TTIs for the 20-user case. From the second plot it is clear that very often users get scheduled at subsequent times, taking up several H-ARQ processes simultaneously. The result for FCDS (α=0.999) looks similarly and is therefore not included.

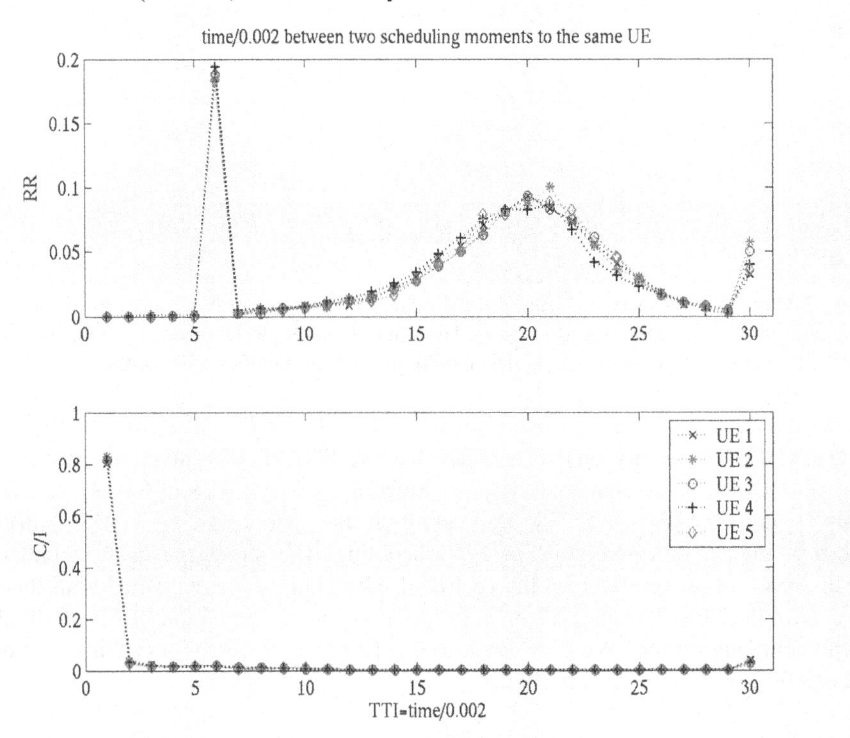

Fig. 8. Distribution of the time measured between two transmissions to the same user (20-user case). The results, still including the moments of H-ARQ retransmissions, contain the distribution for the RR and the C/I case.

Figure 9 compares the C/I and FCDS result at a different scale, zooming in on the middle part of the distribution. Also the separate contributions of all five users have been collected into a single line. Since the H-ARQ retransmissions are not taken into account in this plot, one can clearly distinguish the bumps at 40 and 80 TTIs in the FCDS results. This shows a correlation of 40 TTIs and supports the intuitive impression of the FCDS algorithm as 'riding the waves' on top of all signals, as already shown in Figure 5. Note that the zooming might blur the analysis: although the distribution also displays long waiting times, these occur rarely. However, in these scheduling scenarios, users may indeed experience waiting times of order seconds when the channel conditions are low.

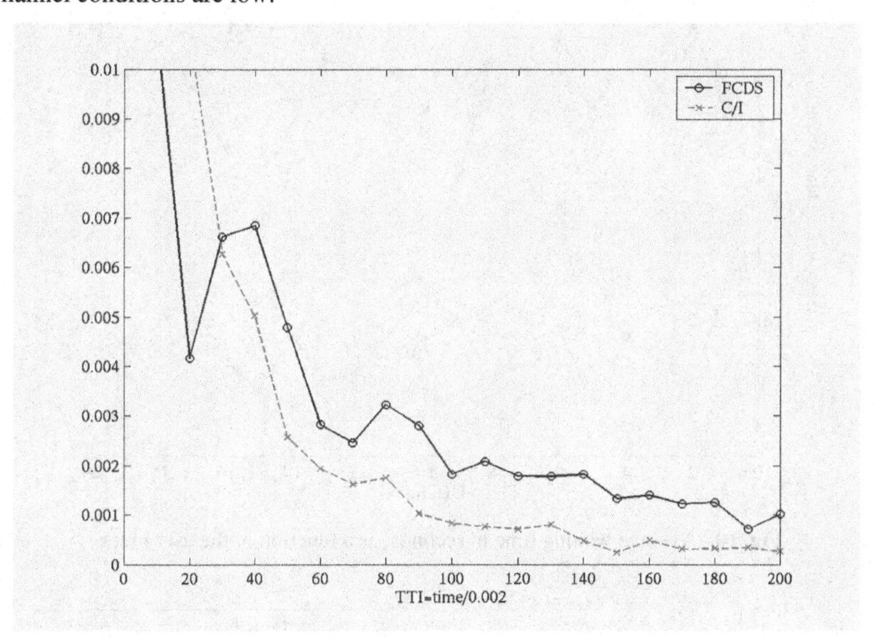

Fig. 9. Distribution of the time measured between two transmissions to the same user (20-user case) for the FCDS and C/I result. The moments of H-ARQ retransmissions are no longer taken into account.

Another measure describing the 'local fairness' is the average waiting time. In Figure 10, these results are collected per user and for all scheduling algorithms. It is clear from the figure that there is an order of magnitude in between the averaged waiting times of the three scheduling algorithms. Although all users on the very long term display similar statistics for the channel conditions, the values do change per user due to extreme local effects.

Finally we consider the Block Error Rate (BLER) from Table 2 on a per-CQI basis. In Figure 11 we have collected the results for all scheduling algorithms. We remark that the low CQI part of the C/I scheduling is based on a very low number of samples since the chance of selecting a UE with low CQI is rather small, and is therefore not included in the figure. It is clear from the figure that all curves decrease. This corresponds with the underlying BLER/SNR curve getting steeper for higher CQI values. The RR curve is the highest again, supporting the result from Table 2. It becomes clear from this plot that the FCDS scheduling performs best, considered per CQI. The overall C/I value becomes smaller than the tendency of the plot, since most of the times, a very high CQI value is scheduled.

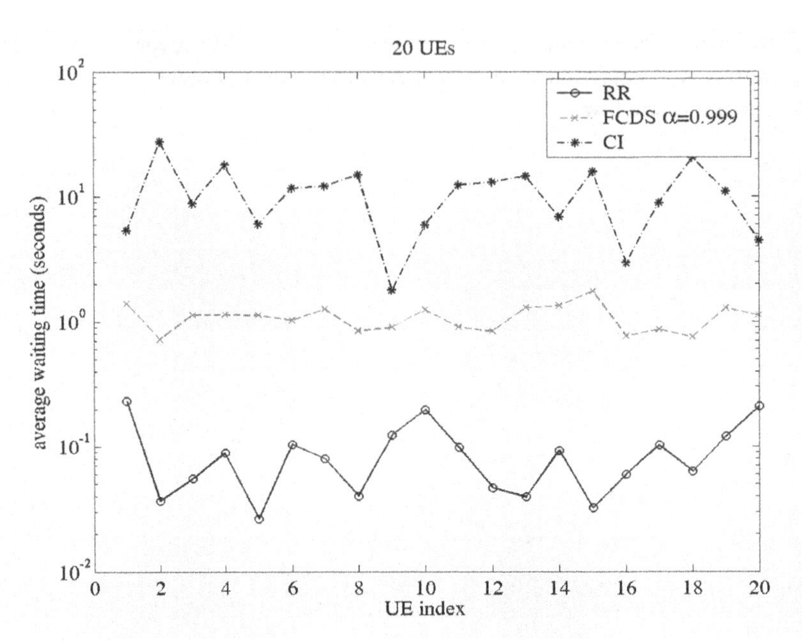

Fig. 10. Average waiting time in seconds, as a function of the user index

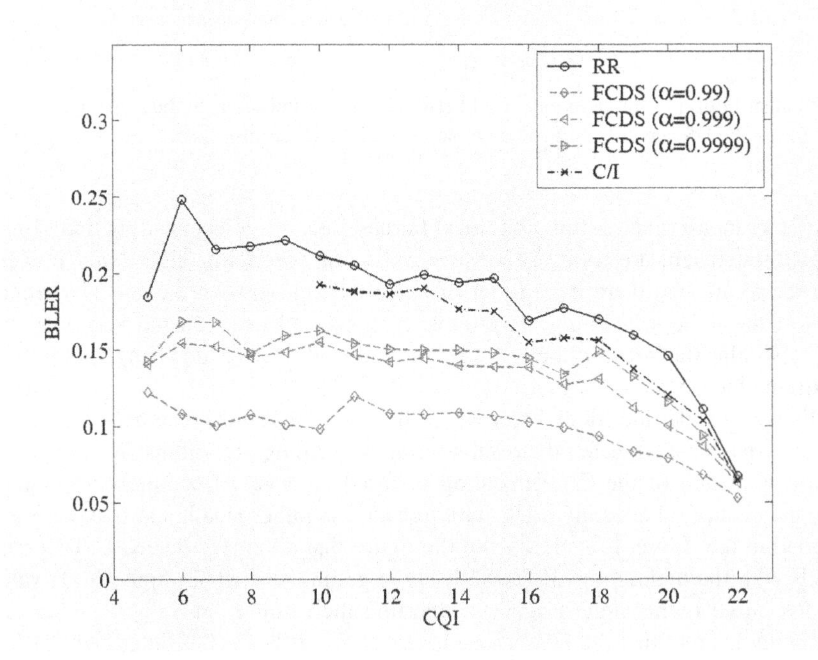

Fig. 11. Block Error Rate specified per CQI, for the 5-user case, for all scheduling algorithms

6 Conclusions

In this study we have evaluated the FCDS scheduling algorithm. It was introduced as a trade-off between the two extreme ways of scheduling: C/I based and Round-Robin scheduling. Based on the network simulator results (in ns-2), we can indeed conclude that the advantages of both these extremes have been combined in the new algorithm: a good fairness, comparable to that of the Round-Robin scheduling, together with a reasonable average transport block size and resulting throughput. Based on the type of service being considered, we have different demands on QoS and, for example, delay characteristics. This should be taken into account in the comparison of scheduling algorithms.

A next step of this study on scheduling techniques would be to consider more advanced protocols such as TCP (transmission control protocol). This also involves a careful tuning of the parameters that determine the TCP characteristics. The UDP study presented in this paper can next be considered as reference case in the study on the impact of TCP. Furthermore, a more diverse setting of different services, in particular with respect to the way data calls arrive and leave the system as well as distances between the UE and Node B should be considered.

References

1. Anthony Lo e.a., Performance of TCP over UMTS Common and Dedicated Channels, IST Mobile & Wireless Communications Summit 2003, pp. 128-142, Aveiro, Portugal, June 2003
2. I. de Bruin e.a., "Fair Channel-Dependent Scheduling in CDMA Systems", 12th IST Summit on Mobile and Wireless Communications, pp. 737-741, Aveiro, Portugal, June 2003.
3. EURANE (Enhanced UMTS Radio Access Networks Extensions for ns-2), http://www.ti-wmc.nl/eurane/
4. 3GPP TS 25.214 V5.5.0, "Physical layer procedures (FDD), Release 5"
5. S.W. Roberts, "Control chart test based on geometric moving averages", Technometrics, 1, 1959, 239-250
6. F. Brouwer e.a., "Usage of Link Level Performance indicators for HSDPA Network-Level Simulations in E-UMTS", accepted for presentation at ISSSTA, Sydney, 2004
7. I. De Bruin e.a., HSDPA Link Adaptation and Hybrid ARQ characteristics and its use in network-level simulations, submitted to IST Mobile Summit 2004
8. 3GPP TSG-RAN Working Group 4, R4-020612, "Revised HSDPA CQI Proposal", April 3-5, 2002.
9. M. Gudmundson, "Correlation Model for Shadow Fading in Mobile Radio Systems", Electronics Letters, Vol. 27, No. 23, pp. 2145-2146, 1991.

Downlink Scheduling and Resource Management for Best Effort Service in TDD-OFDMA Cellular Networks

Young Min Ki, Eun Sun Kim, and Dong Ku Kim

Yonsei University, Dept. of Electrical and Electronic Engineering
134 Shinchon Dong, Seodaemun Gu, Seoul, Korea
{mellow, esunkim, dkkim}@yonsei.ac.kr
http://mcl.yonsei.ac.kr

Abstract. Throughput performance and geographical service fairness of best effort service used for downlink of a 802.16e based TDD-OFDMA sectored cellular networks are evaluated in conjunction with different scheduling schemes and frequency reuse plans. The OFDM systems are based on two multiple access schemes, which are the OFDM-TDM and OFDMA, and considered scheduling schemes are round robin, max C/I, PF and G-fair schedulers with adaptive rate control. The 3-sectored 1 FA, 3-sectored 3 FA, and 6-sectored 3 FA plans are compared in terms of throughput, capacity, and geographical service fairness, which assist in determining the choice of a scheduling and frequency reuse plan.

1 Introduction

Link adaptation and channel scheduling are key techniques used in dynamic resource management for 3G wireless networks [1]. The single carrier 3G technologies, which are 1xEVolution data optimized (1xEV-DO) [2], 1xEVolution data and voice (1xEV-DV) [3] or high speed downlink packet access (HSDPA) [4], have proposed fast link adaptation as well as packet scheduling within the time domain. Multi-user time-domain schedulers have nearly doubled overall system capacity by handling fading channel time variation.

Emerging WLANs and future wireless mobile systems are expected to be based on a multi-carrier scheme (OFDM) with hundreds of carriers [1]. For a broadband channel, frequency variation as well as time variation should be handled more carefully in order to enhance overall performance. There are two strategies applied to combat frequency variations: "frequency diversity" and "frequency selectivity" [5]. With frequency diversity, the identical modulation and coding scheme is performed over all subchannels in order to prevent system performance from being dominated by a few deeply faded subcarries. In contrast, the frequency selective strategy can exploit frequency variations knowledge in order to provide link adaptation gains. The different signal quality values occurring at each subchannel can adopt different link adaptations and channel schedulers for use with each sub-band [5][6].

I. Niemegeers and S. Heemstra de Groot (Eds.): PWC 2004, LNCS 3260, pp. 316–329, 2004.

IEEE standards 802.16 [7] and 802.16a [8], which define the fixed broadband wireless access (FBWA) system, are based on OFDM/OFDMA transmission technologies. The IEEE 802.16 task group e (802.16 TGe) started in order ensure coexistence between fixed and mobile broadband wireless access (FMBWA) with fixed and mobile user terminals (UTs). The 802.16e amendment aims at the mobility support in 802.16 standard in 2 to 6 GHz licensed bands [9]. In this paper, the 802.16e is considered for use in OFDMA cellular networks and the overall performance of packet scheduling is evaluated for specific frequency reuse plans by performing simulations.

The remainder of the paper is organized as follows. In Section 2, the basic system model for 802.16e based OFDMA cellular networks is introduced. In Section 3, downlink scheduling schemes are presented. The overall performance of the schemes is evaluated in Section 4 by simulation. Finally, conclusions are made in Section 5.

2 OFDMA Cellular Networks

2.1 IEEE 802.16e PHY Frame Structure

The 802.16e PHY is based on OFDM/OFDMA and is highly aligned with the 802.16a. Furthermore, time division duplexing (TDD) is considered as a duplexing method. Fig. 1 shows the 802.16e PHY frame structure. The fixed length OFDMA frame consists of successive downlink (DL) OFDM symbols followed by successive (UL) OFDM symbols. The Tx/Rx transition gap (TTG) is defined as the transition gap between DL and UL, and the Rx/Tx transition gap (RTG) is defined as the transition gap between UL and DL. In the frequency domain, full RF bandwidth is divided into hundreds of subcarriers. The number of subcarriers is equal to the OFDM FFT size. Numbers of subcarriers are bunched into the subchannel, which can be handled as a minimal resource unit. The subchannel size is assumed to be larger than the coherence bandwidth [8][9].

2.2 OFDM-TDM Versus OFDMA

IEEE 802.16 specifies two categories of OFDM systems: one is OFDM-TDM and the other is OFDMA [8][9]. In the first scheme, all carriers are simultaneously assigned for transmission and downstream data is time-division multiplexed (TDM). The OFDMA represents the time and frequency division multiple access technique based on the OFDM. A number of a specific UT's traffic can be transmitted simultaneously by using different subchannels. Subchannels are allocated dynamically within 802.16e. There are two dynamic subchannel allocation methods for multiple access usage, which include the media access protocol (MAP) and frequency hopping (FH). In MAP based OFDMA, the data streams are assigned on subchannels through MAP messages sent via downstream transmission. In this case, MAP can be scheduled dynamically by the base station (BS). In FH-OFDMA, the data streams are assigned on subchannels by using

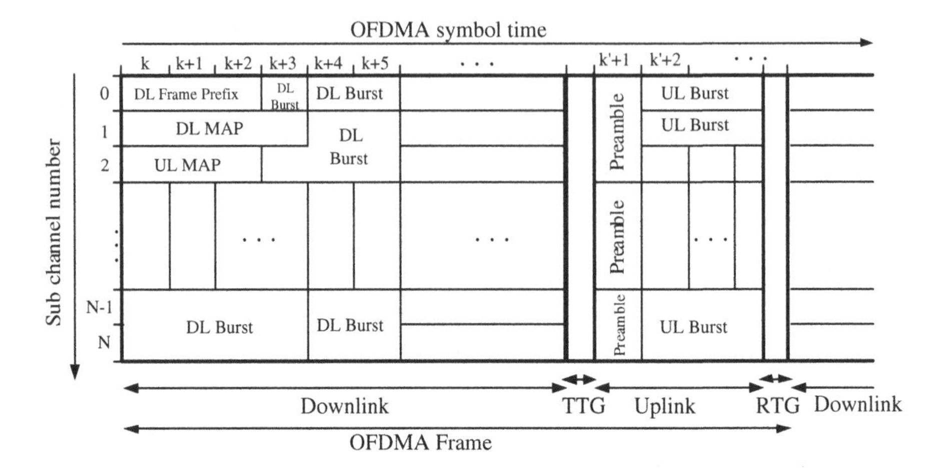

Fig. 1. IEEE 802.16e OFDMA frame structure

a hopping sequence, which is assigned to the UT during the call setup procedure. The FH-OFDMA technique is difficult to coordinate with dynamic resource management schemes such as packet scheduling or link adaptation.

2.3 Single Frequency Network and Multi Frequency Network

Two frequency reuse plans are considered: one is a single-frequency network (SFN) and the other is a multi-frequency network (MFN) [9]. Subsequently, multi-cell can be designed by single sectored or multi-sectored modes. The frequency reuse factor (FRF) is defined as the ratio between total number of cells (sectors) and the number of cells (sectors) that use the same frequency allocation (FA). Fig. 2 shows the examples of a frequency reuse plan: (a) shows SFN examples and (b) shows MFN examples. The "111" plan is a SFN with single sector cells and the "131" plan is a SFN with three-sector cells, in which the FRF equals 1. However, SFN experiences severe interference problems in the cell (sector) edge area. The "133" plan has 3-sectored cells in which each sector is operated at a different FA. In multi-frequency plans, edge area interference problems are improved, but frequency efficiency decreases.

3 Downlink Scheduling for TDD-OFDMA Networks

There are two types of scheduling: one is channel state independent scheduling and the other is channel state dependant scheduling. It is assumed that MAP is used to assign scheduled resources to UTs. These UTs measure the channel quality by using the received downlink pilot, and upload the proper modulation and coding scheme (MCS) levels to the BS through uplink feedback channels. Then, the BS schedules the downlink stream and transmits the UT's traffic by using the reported MCSs.

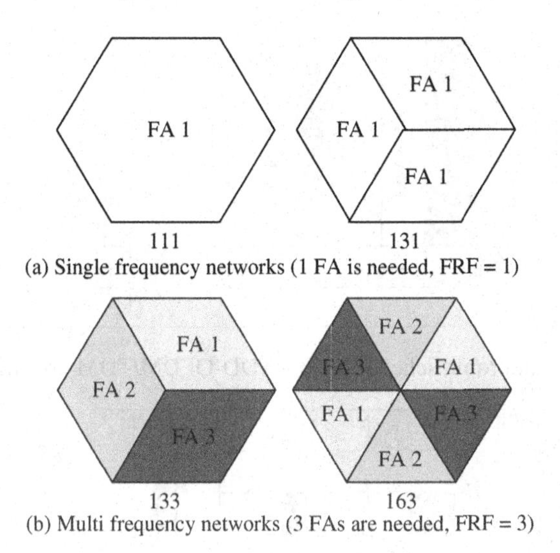

Fig. 2. Frequency reuse plan examples (surrounding cells are not shown, but are repeatedly present)

3.1 Round Robin (RR) Scheduling for TDD-OFDMA

A round robin (RR) represents channel state independent scheduling and does not guarantee throughput fairness in a wireless condition [1]. Fig. 3 shows round robin scheduling examples: (a) is the TDD-OFDM/TDM downlink and (b) is TDD-OFDMA downlink. Channel information is not needed for RR scheduling, but needed for adaptive rate control. In the OFDM/TDM system, scheduling for OFDMA symbol by only time division manner would be feasible. The RR scheduler of a specific OFDMA should be able to perform in the cluster unit which is defined as the group of contiguous subcarriers and OFDMA symbols. The minimal cluster will be one subchannel occurring during one OFDMA symbol.

3.2 Channel State Dependant Scheduling for TDD-OFDMA

Since a channel scheduler should use the available channel information, the scheduling time unit is determined to be the channel feedback period. In a frequency division duplexing (FDD) system such as 1xEV or HSDPA, continuous downstream and upstream channels coexist. Therefore, the scheduling period can be converged to fit the minimal transmission slot duration, which is generally used during the channel feedback period. The time division duplexing (TDD) system does not provide continuous downstream nor continuous upstream. The channel feedback period is determined to be the OFDMA frame duration. Therefore, scheduling can be performed during every frame. Fig. 4 (a) shows the channel scheduling example occurring in TDD-OFDM/TDM. Since channel feedback is reported every frame, only one user can be allocated for one DL frame. How-

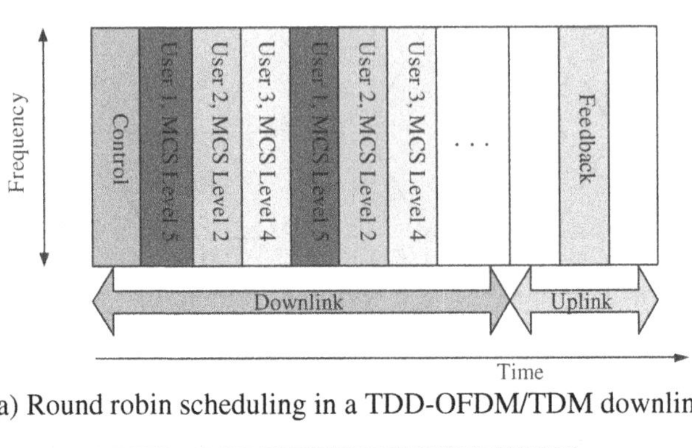

(a) Round robin scheduling in a TDD-OFDM/TDM downlink

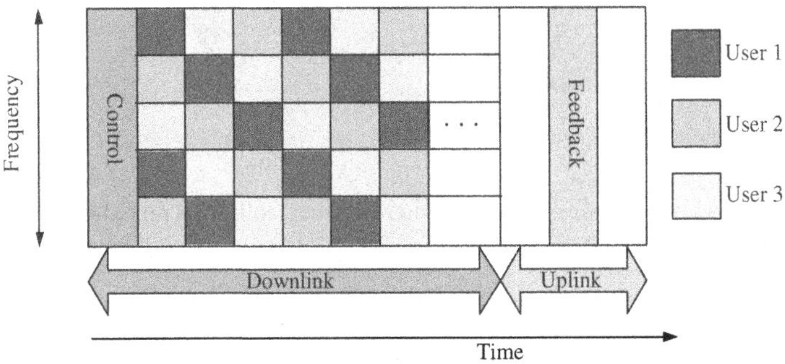

(b) Round robin scheduling in a TDD-OFDMA downlink

Fig. 3. Round Robin Scheduling Examples of TDD-OFDM/OFDMA Downlinks (adaptive rate control is performed, but power control is not performed)

ever, an OFDMA scheduler can allocate the frequency subchannels to different users as shown in Fig. 4 (b).

3.3 Conventional Channel Schedulers

One simple method is to serve the UT of index i_n^* at the n-th subchannel for every scheduling instance t, with respect to:

$$i_n^* = \arg\max_i MCS_{i,n}(t), \tag{1}$$

where $MCS_{i,n}(t)$ denotes the MCS level of the n-th subchannel of the i-th UT. This scheduling resembles the "max C/I" scheduler, and will maximize the total downlink throughput of a base station. However, this high throughput is achieved at the cost of unfairness among the various UTs. To remedy this problem, it was suggested that the selected UT, i_n^* should be denoted such that:

$$i_n^* = \arg\max_i MCS_{i,n}(t)/T_{i,n}(t), \tag{2}$$

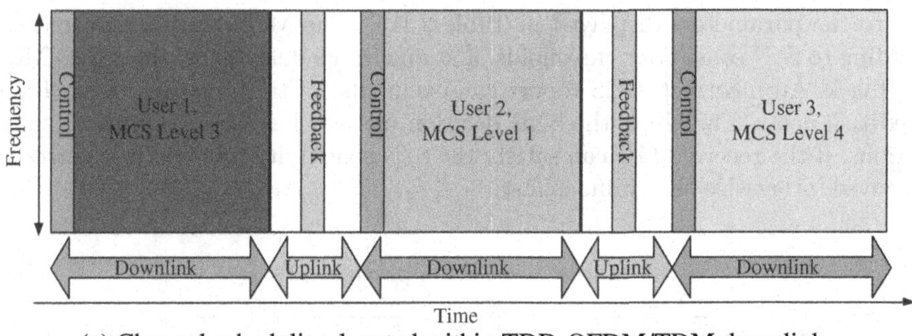

(a) Channel scheduling located within TDD-OFDM/TDM downlink

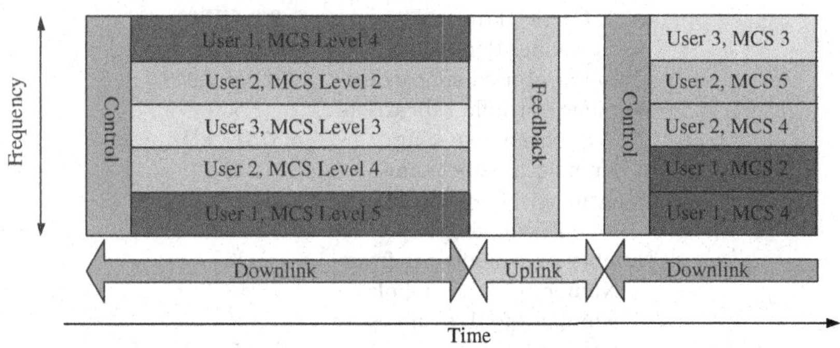

(b) Channel scheduling located within TDD-OFDMA downlink

Fig. 4. Priority-Based Channel Scheduling Examples used in TDD-OFDM/OFDMA Downlinks (adaptive rate control is performed, but power control is not performed)

where $T_{i,n}(t)$ denotes the average throughput of n-th subchannel of i-th UT. This scheme ensures proportional fairness (PF) scheduling algorithm [11]. In [12], a generalization of the proportional fair scheduler was shown, and it was suggested that the selected UT, i_n^* should be defined that:

$$i_n^* = \arg\max_i \frac{MCS_{i,n}(t)}{T_{i,n}(t)} \frac{h_i(M_{i,n}(t))}{M_{i,n}(t)}, \tag{3}$$

where $MCS_{i,n}(t)/T_{i,n}(t)$ is a well-known PF factor and $M_{i,n}(t)$ is the average MCS. This scheduling is defined as G-fair scheduling algorithm. The expression $h(x)$ is a user-specific function that specifies overall fairness behavior [12].

4 Simulations: Performance in a Multi-cell Environment

4.1 802.16e OFDMA Parameters

The 802.16e based TDD-OFDM/OFDMA system is considered [8][9]. Table 1 shows the OFDMA parameters, and the link adaptation table (MCS table) is set to the parameters displayed in Table 2 [15]. The MCS level is reported according to S/N sensitivity thresholds, and one frame delay is considered as MCS feedback. An incorrect MCS report can occur due to time-varying channel and feedback delay. Therefore, the SNR decision method has been adopted to simulations. If the received C/I can satisfy the S/N sensitivity, the traffic is correctly received, otherwise the traffic is lost.

Table 1. OFDMA parameters

Parameters	Value
Carrier Frequency	2.3 GHz
Channel Bandwidth	10 MHz
Number of used subcarriers	1,702 of 2,048
Number of traffic subcarriers	1,536
Subcarrier spacing	5.57617 kHz
Number of subchannels	32
Number of subcarriers	48
Frame length	5.0 msec
Number of symbols per frame	26
Number of DL symbols	18
Number of UL symbols	8
Sum of RTG and TTG	45.885 μsec
OFDMA symbol time	190.543 μsec
Guard interval	11.208 μsec
Cyclic prefix	1/16

Table 2. Modulation and coding (MCS) table [15]

Modulation	Code rate	Sensitivity threshold S/N	PHY bit/sec/Hz
QPSK	1/2	6.6 dB	0.807
16QAM	1/2	10.5 dB	1.613
64QAM	2/3	15.3 dB	3.227
64QAM	3/4	20.8 dB	3.63

4.2 Simulation Environments

It is assumed that there are 19 cells located within a 1 km radius. These 16 UTs are distributed in the center cell and the UT's location is generated more than 1,000 times. All UTs are assumed to be best-effort traffic with full-buffering. The channel C/I of the n-th subchannel of i-th UT is assumed to be:

$$(C/I)_{i,n} = \sum_{j=1}^{J} \|\gamma_{j,n}\|^2 \cdot \left(G_i^{-1} + \sum_{k=1}^{K} \|\psi_{k,n}\|^2 \right)^{-1}, \tag{4}$$

where G_i is the average geometry which is determined by path loss and shadowing and shown as

$$G_i = \frac{I_{or}}{I_{oc} + N_0 W} = \frac{1}{I_{oc}/I_{or} + 1/(I_{or}/N_0 W)}, \quad (5)$$

where I_{or} is the received serving-cell pilot strength, I_{oc} is the sum of the received other-cell pilot strength, and $N_0 W$ is the thermal noise power. The expression $\{\gamma_j\}$ represents the multi-path component within the guard interval, but $\{\psi_k\}$ is multi-path component which exceeds the guard interval. In simulations, $\sum_{j=1}^{J} \|\gamma_{j,n}\|^2$ is assumed to be Rayleigh fading and $\{\psi_k\}$ is ignored, since the cyclic prefix is assumed to be sufficiently larger than the overall delay spread. The path loss model is assumed to be a vehicular model $129.427 + 37.6 * log_{10}(d_{km})$ [14]. The standard deviation of Log-normal shadowing is assumed to be 10 dB. Short-term channel gains are assumed to be Rayleigh fading with a Doppler frequency of 6.4Hz (3km/Hr) and no correlation between subchannels is assumed. The BS transmit power is set to 20 W (43 dBm) and antenna gain is set to 14 dBi. Thermal noise density is assumed to be -174 dBm/Hz and max C/I limit is set to 30 dB.

Each Ring is defined as the area occupied by a 100 meter unit. For example, the n-th Ring is the area in that distance from the BS which are from (n-1) hundred meters to n hundred meters. Three frequency reuse plans are simulated: 131 (3 sector, 1 FA), 133 (3 sector, 3 FA), and 163 (6 sector 3 FA). Fig. 5 shows the average geometry distribution obtained by simulations. The 133 plan shows the best received geometry over almost all rings among the simulated frequency reuse plans, but it should be noted that the 133 plan has three FRF.

4.3 Throughput Performance

The four scheduling schemes, which include RR, max C/I, PF, and G-fair with $h(x) = 1$. Table 3 shows overall sector throughput. Sector throughput of the max C/I scheduler displays maximum sector capacity. In the OFDM/TDM system, throughput of the PF is close to that of the RR, but throughput of the PF is around 2 to 3 times higher than that of the RR for the OFDMA system. Furthermore, throughput of the G-fair is around half of that of the RR in OFDM/TDM, but throughput of the G-fair in the OFDMA is 1.5 to 2.5 times higher than that of the RR. Since the RR is performed during the symbol period while channel schedulers are working during every frame period, the TDM system produces lower multi-user scheduling gain. However, the OFDMA produces subsequently increased multi-user scheduling gain by utilizing frequency domain scheduling. For RR and max C/I, the throughput difference between TDM and OFDMA is negligible, but OFDMA throughput is around 2 to 3 times higher than that of the TDM in terms of PF and G-fair.

Table 4 shows normalized throughput determined as

$$T_{Normalized} = \frac{T_{Sector} \cdot N_{Sector}}{BW \cdot FRF}, \quad (6)$$

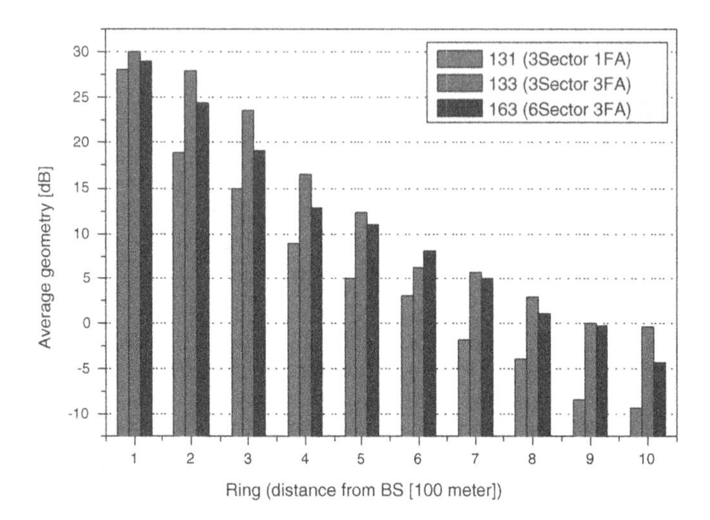

Fig. 5. Average received geometry for different frequency reuse plans

Table 3. Sector throughput [kbps/Sector]

	131 (3 sector 1FA)		133 (3 sector 3FA)		163 (6 sector 3FA)	
	TDM	OFDMA	TDM	OFDMA	TDM	OFDMA
RR	2,279.14	2,193.67	4,813.04	4,669.42	4,224.99	4,042.41
Max C/I	12,019.15	12,179.38	16,333.79	16,478.59	14,630.79	14,783.21
PF	2,301.14	7,000.82	4,781.71	9,361.92	4,143.76	8,452.39
G-Fair	1,112.00	5,343.32	2,184.87	6,743.79	1,986.67	5,804.66

where T_{Sector} is sector throughput, N_{Sector} is number of sectors per cell, and BW is the channel bandwidth for each FA. Table 4 shows that normalized throughput of plan 131 is the highest and the 133 plan produces the smallest throughput calculated among the investigated plans. In other words, the 133 plan can reduce other-cell interference and increase sector throughput but has lower frequency efficiency. The 131 plan displays the lowest sector throughput, but has the largest normalized throughput.

The throughput of UTs placed at around cell edge is an important key necessary to understand cell coverage. Fig. 6 shows the average user throughput of the first Ring (best channel condition) and Fig. 7 shows the average user throughput of the 10-th Ring (worst channel condition). PF and G-fair schedulers can provide higher throughput than max C/I in worst channel condition users. The PF and G-fair scheduler can provide more than 200 kbps to 10-th Ring users in the 133 plan, and can provide around 150 kbps in the 163 plan. Average user throughput of the 10-th Ring observed in the 131 plan is less than

Table 4. Normalized throughput [bps/Hz/Cell]

	131 (3 sector 1FA)		133 (3 sector 3FA)		163 (6 sector 3FA)	
	TDM	OFDMA	TDM	OFDMA	TDM	OFDMA
RR	0.68	0.66	0.48	0.47	0.84	0.81
Max C/I	3.61	3.65	1.63	1.65	2.93	2.96
PF	0.69	2.10	0.48	0.94	0.83	1.69
G-Fair	0.33	1.60	0.22	0.67	0.40	1.16

80 kbps. Therefore, though the 131 plan provides the highest channel capacity, it is not suitable for service in the cell edge area. Fig. 8 shows the average user throughput for each PF scheduling Ring for OFDMA schemes. It is shown that while the OFDMA 131 plan PF scheduler provides 70 kbps to 3.6 Mbps of user throughput, the 133 plan PF scheduler can provide 240 kbps to 2.4 Mbps and the 163 plan can provide 150 kbps to 2.1 Mbps.

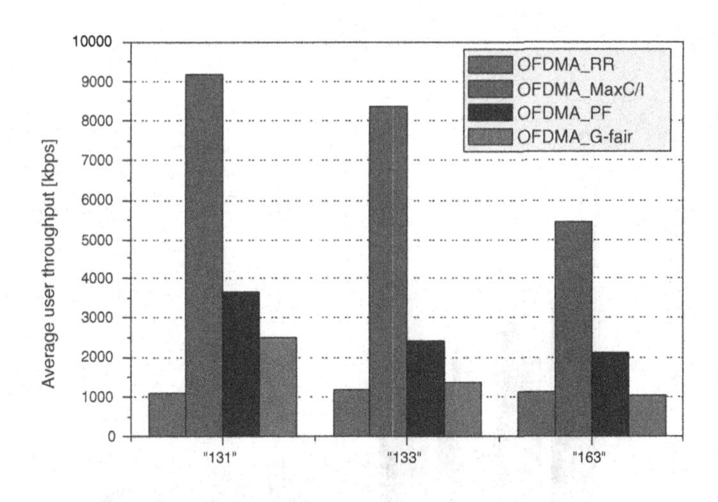

Fig. 6. Average user throughput in the first Ring (distance from BS: 0 to 100 meters) for the OFDMA system ("131": 3 sector 1 FA, "133": 3 sector 3 FA, "163": 6 sector 3 FA)

4.4 Fairness Performance

Figs 9 to 11 show the fairness CDF curves for use in OFDMA scheduling schemes for the 131, 133, and 163 plans. Fairness is measured by the throughput to average throughput ratio. The spike located at zero on the x-axis is a result of poor channel conditions or scheduling starvation. The max C/I scheduler

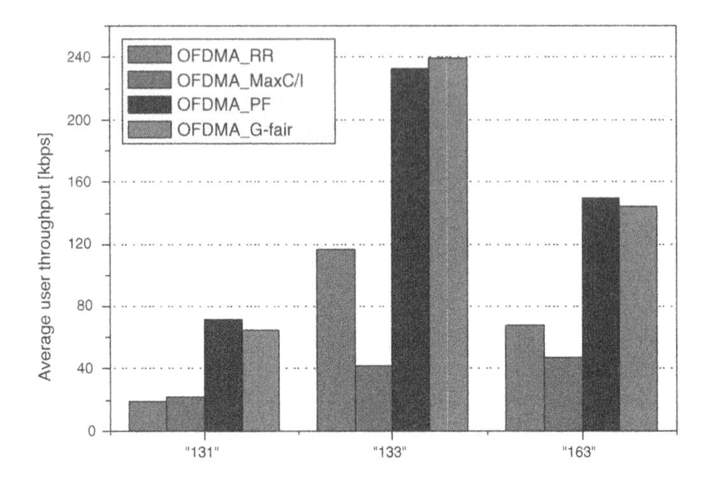

Fig. 7. Average user throughput in the 10-th Ring (distance from BS: 900 to 1,000 meters) for the OFDMA system ("131": 3 sector 1 FA, "133": 3 sector 3 FA, "163": 6 sector 3 FA)

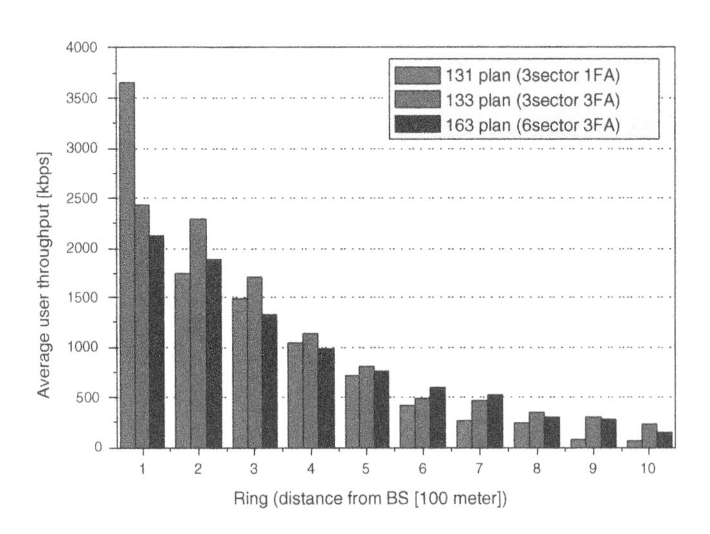

Fig. 8. Average user throughput of PF schedulers for use in an OFDMA system

accounts for approximately 90 % of the faded UTs that could not be served. Other schedulers account for around 65 %, 50 %, and 50 % fade for 131, 133, and 163 plans respectively. Since the max C/I serves only the best channel condition UT, it displays severe unfairness for all investigated plans. As expected, the G-fair scheduler provides the best fairness. The 133 and 163 plans also show more fairness than the 131 plan for RR, PF and G-fair schedulers.

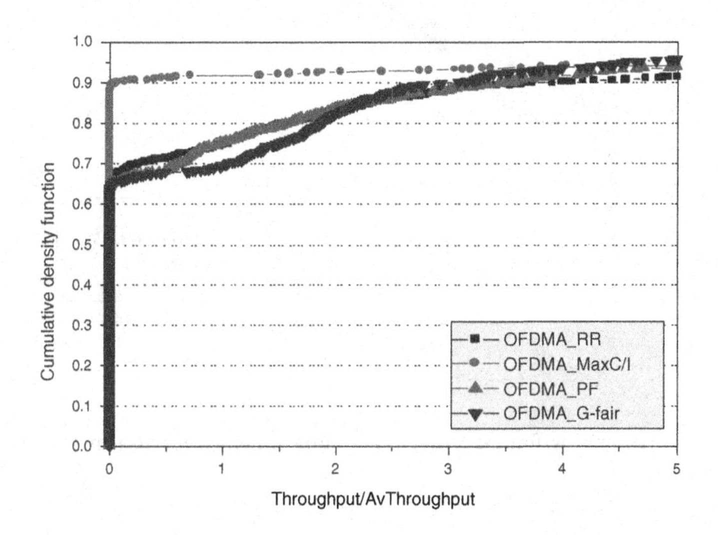

Fig. 9. Fairness curves for OFDMA scheduling schemes for use in the 131 plan (3 sector 1 FA)

5 Concluding Remarks

Throughput and its geographical service fairness of best effort service used for downlink of 802.16e based TDD-OFDMA sectored cellular networks were evaluated in conjunction with different scheduling schemes and frequency reuse plans. Both OFDM/TDM and OFDMA was considered and the 131 (3-sectored 1 FA), 133 (3-sectored 3 FA) and 163 (6-sectored 3 FA) plans were considered for use in a frequency reuse plan. The round robin, max C/I, PF and G-fair schedulers were evaluated along with adaptive rate control. The 133 plan shows the best throughput and geographical service fairness among the investigated plans, while it produces the worst frequency efficiency. Results indicate that the choice of a scheduling and frequency reuse plan should be determined with consideration in terms of the trade-off between system throughput and geographical service fairness.

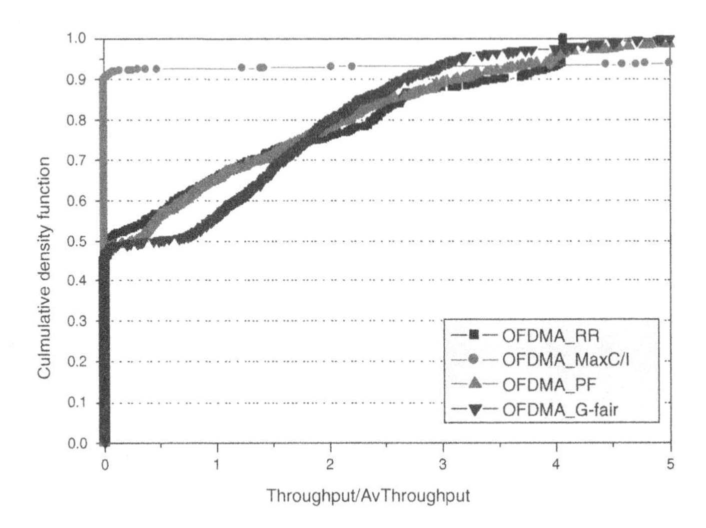

Fig. 10. Fairness curves for OFDMA scheduling schemes for use in the 133 plan (3 sector 3 FA)

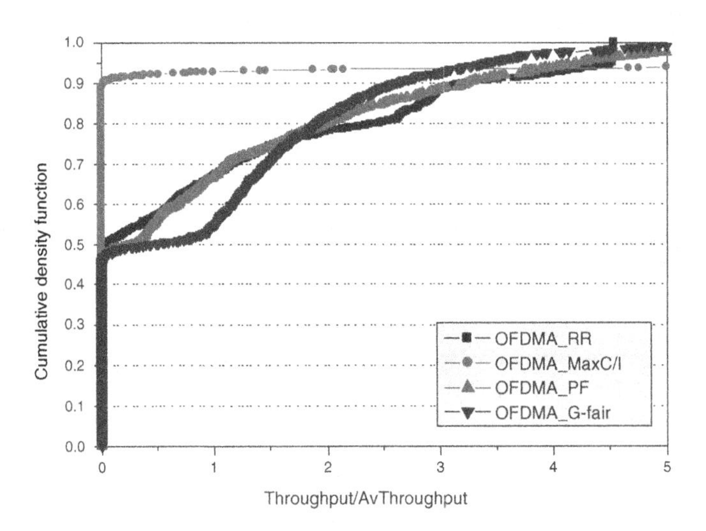

Fig. 11. Fairness curves for OFDMA scheduling schemes for use in the 163 plan (6 sector 3 FA)

Acknowledgments. This work was supported by grant No. (R01-2002-000-00531-0) from the Basic Research Program of the Korea Science & Engineering Foundation. Also, the work was supported by Qualcomm Incorporated through Qualcomm Yonsei CDMA Joint Research Center.

References

1. Shakkottai S., Rappaport T. S.: Cross-Layer Design for Wireless Networks. Communications Magazine, Vol. 41. IEEE. (2003) 74-80
2. TIA/EIA/IS-856. cdma2000 High Rate Packet Data Air Interface Specification, (2001)
3. cdma2000 release C. C.S000X-C, (2002)
4. 3GPP release 5. TS-25.211-25.214, (2002)
5. Classon B., Sartori P., Nangia V., Zhuang X., Baum K.: Multi-dimensional Adaptation and Multi-user Scheduling Techniques for Wireless OFDM Systems. IEEE International Conference on Communications (ICC '03), (2003)
6. Anchun W., Liang X., Shidong Z., Xibin X., Yan Y.: Dynamic Resource Management in the Forth Generation Wireless Systems. International Conference on Communication Technology (ICCT2003), (2003)
7. IEEE Standard 802.16, IEEE Standard for Local and Metropolitan Area Networks Part 16: Air Interface for Fixed Broadband Wireless Access Systems, (2001)
8. IEEE Standard 802.16a, IEEE Standard for Local and Metropolitan Area Networks Part 16: Air Interface for Fixed Broadband Wireless Access Systems Amendment2: Medium Access Control Modification and Additional Physical Layer Specifications for 2-11GHz, (2003)
9. http://www.ieee802.org/16/tge/
10. Nee R., Prasad R.: OFDM for Wireless Multimedia Communications. Artech House Publishers. (2000)
11. Bender P., Black P., Grob M., Padovani M., Sindhushayana N., Viterbi A.: CDMA/HDR: A Bandwidth-Efficient High-Speed Wireless Data Service for Nomadic Users. Communications Magazine, Vol. 38. IEEE. (2000) 70-77
12. Application Note: G-fair Scheduler. 80-H0551-1 Rev. B. Qualcomm Incorporated. (2002)
13. S. Keshav: An Engineering Approach to Computer Networking: ATM Networks, the Internet, and the Telephone Network. Addison Wesley. (1997)
14. Recommendation ITU-R M.1225, Guideline for Evaluation of Radio Transmission Technologies for IMT-2000, (1997)
15. IEEE C802.16e-03/22r1, Convergence simulations for OFDMA PHY mode, (2003)

On the Performance of TCP Vegas over UMTS/WCDMA Channels with Large Round-Trip Time Variations

Anthony Lo[1], Geert Heijenk[2], and Ignas Niemegeers[1]

[1] Delft University of Technology, P O Box 5031,
2600 GA Delft, The Netherlands
{A.Lo, I.Niemegeers}@ewi.tudelft.nl
http://www.wmc.ewi.tudelft.nl/

[2] University of Twente, P O Box 217,
7500 AE Enschede, The Netherlands
Geert.Heijenk@utwente.nl
http://wwwhome.cs.utwente.nl/~heijenk

Abstract. Universal Mobile Telecommunications System (UMTS) is a third-generation cellular network that enables high-speed mobile Internet access. This paper evaluates and compares the performance of two well-known versions of Transmission Control Protocol (TCP), namely, Vegas and Reno, in a UMTS environment. Bulk data transfer was considered in the simulation with varying radio channel conditions. We assume that data losses are only due to the radio channel. Simulation results show that the performance of Vegas is worse than Reno even though data losses incurred by the radio channel are completely recovered by the UMTS radio link control layer. This has led us to conduct a thorough investigation on the behavior of Vegas in order to identify the cause of performance degradation in Vegas. The poor performance of Vegas is attributed to the UMTS radio interface characteristics which resulted in large and highly variable TCP round-trip times. Vegas would interpret the round-trip time variation as a sign of congestion, and consequently, shrink its window size which reduces the transmission rate. Furthermore, a sudden increase in the instantaneous round-trip time can trigger spurious timeouts at the TCP sender using Vegas which performs unnecessary retransmissions. Spurious timeouts can lead to significant throughput reduction. Reno, on the other hand, does not show any abnormality and delivers the expected performance.

1 Introduction

Universal Mobile Telecommunications System (UMTS) [1,2] is a third-generation cellular mobile network where the radio interface is based on code division multiple access, known as Wideband Code Division Multiple Access (WCDMA). To date, a number of mobile operators in Europe and Asia have launched their UMTS commercial service and some plan to roll out their UMTS networks. In

I. Niemegeers and S. Heemstra de Groot (Eds.): PWC 2004, LNCS 3260, pp. 330–342, 2004.

addition to the legacy voice service, UMTS enables mobile users access to the Internet in a seamless fashion (i.e., always on) at data rates up to 2 Mb/s in indoor or small-cell environments, and wide-area coverage of up to 384 kb/s.

Today, the Internet is the most popular and widely used packet-switched network that supports applications like File Transfer Protocol (FTP), Email, etc. These Internet applications rely on two commonly used protocols, namely, Transmission Control Protocol and Internet Protocol (TCP/IP) [3], to reliably transport data across heterogeneous networks. IP is concerned with routing data from source to destination host through one or more networks connected by routers, while TCP provides reliable end-to-end data transfer and congestion control. TCP Reno is the most extensively used variant of TCP, while TCP Vegas [4] is a newer variant with improved congestion avoidance and retransmission mechanisms. Unlike Reno, Vegas constantly tries to detect congestion in the network before packet loss occurs and lower the rate linearly when sign of congestion is detected. On the contrary, Reno only reacts when packet losses are detected.

The performance of TCP over wireless networks has been extensively studied [5,6,7,8]. All these studies show that TCP performance is significantly degraded since TCP interprets packet losses due to the radio channel as signs of network congestion, which resulting in sender throttling and causes significant throughput reduction. Various solutions were proposed in the literature to combat TCP performance degradation, and in general, can be classified into three major categories: link-layer [7,8,9,10], split-connection [11] and proxy [6]. However, all these studies including the proposed solutions were purely targeted at Reno rather than Vegas. The solution employed by UMTS falls under the link-layer category. Presently, it is not clear how the link-layer solution used by UMTS would adversely affect the performance of Vegas.

The paper aims at evaluating the performance of TCP Vegas and compare its performance with Reno in a UMTS environment, in particular, how the performance of Vegas is adversely affected by the UMTS radio interface. We employ a simulation-based approach to analyze the performance of TCP Vegas and Reno over UMTS for FTP traffic with varying channel conditions.

2 Universal Mobile Telecommunications System (UMTS)

2.1 System Architecture

Fig. 1 shows a simplified architecture of UMTS for packet-switched operation [2,12], which consists of one or several User Equipments (UEs), the UMTS Terrestrial Radio Access Network (UTRAN) and the core network. The UTRAN is composed of several Node Bs connected to a Radio Network Controller (RNC). The core network, which is the backbone of UMTS, comprises the Serving GPRS Support Node (SGSN) and the Gateway GPRS Support Node (GGSN). The SGSNs route packets to and from UTRAN, while GGSNs interface with external IP networks. UE, which is a mobile station, is connected to Node B over the UMTS radio interface.

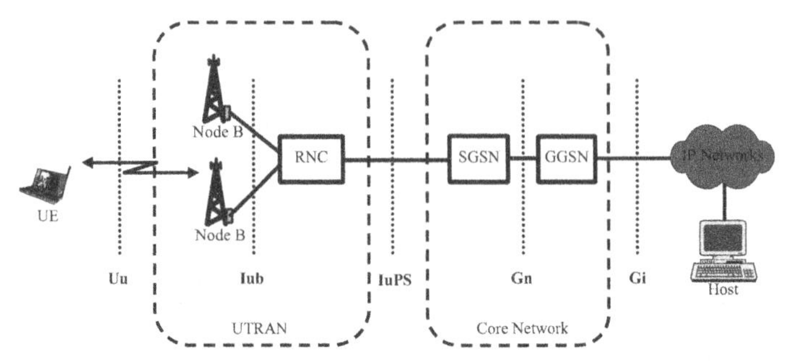

Fig. 1. UMTS Network Architecture

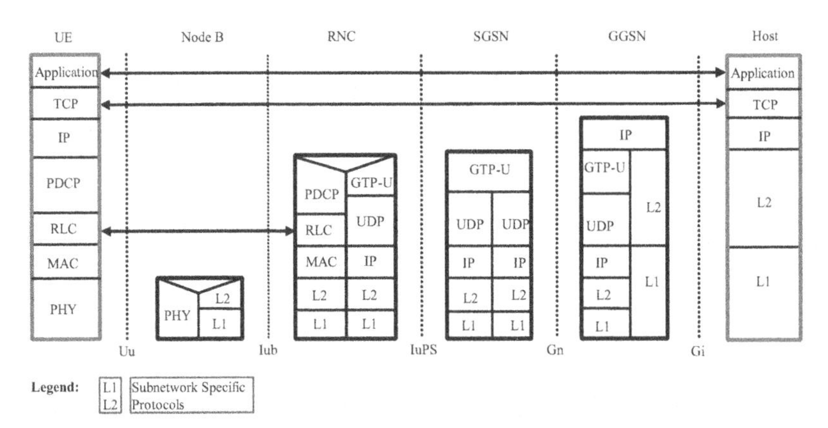

Fig. 2. UMTS Protocol Architecture for the User Plane

2.2 Protocol Architecture

Fig. 2 depicts the UMTS protocol architecture for the transmission of user data which is generated by TCP-based applications. The applications as well as the TCP/IP protocol suite are located at the end-nodes, namely, the UE and a host.

The Packet Data Convergence Protocol (PDCP) provides header compression functionality. The Radio Link Control (RLC) layer can operate in three different modes: acknowledged, unacknowledged and transparent. The acknowledged mode provides reliable data transfer over the error-prone radio interface. Both the unacknowledged and transparent modes do not guarantee data delivery.

The Medium Access Control (MAC) layer can operate in either dedicated or common mode. In the dedicated mode, dedicated physical channels are allocated and used exclusively by one user (or UE), whereas in the common mode, users share common physical channels for transmitting and receiving data.

The Physical (PHY) layer contains, besides all radio frequency functionality, spreading, and the signal processing including RAKE receiver, power control, forward error-correction, interleaving and rate matching.

3 TCP Vegas

Both TCP Reno and Vegas perceive packet losses as a sign of network congestion. Reno only reacts to network congestion when packet losses are detected via timeout or three duplicate acknowledgements. On the other hand, Vegas continuously monitors the state of the network and increment or decrement the current window size in order to prevent packet drops due to buffer overflowing in the intermediate routers. Vegas detects signs of incipient congestion by comparing the expected throughput to the measured throughput, which is given as follows [4]:

$$\Delta = (expected - actual) \times RTT_{base} \tag{1}$$

where $expected = windowSize/RTT_{base}$ and $actual = windowSize/RTT$.

$windowSize$ is the current window size which is the number of segments in transit; RTT_{base} is the minimum of all the instantaneous Round-Trip Times (RTTs); and RTT is the average round-trip time measured for each individual segment transmitted in the $windowSize$. Round-trip time is defined as the total time required by the TCP sender to transmit a segment through a network and receive an acknowledgement that the segment was received correctly.

Vegas defines two thresholds α and β, which are normally set to 1 and 3, respectively. When $\Delta < \alpha$, Vegas increases the congestion window linearly in the next round-trip time; and when $\Delta > \beta$, Vegas decreases the congestion window linearly in the next round-trip time. The congestion window is unchanged when $\alpha < \Delta < \beta$.

In the case of Reno, packet losses are detected via the receipt of three duplicate acknowledgements or retransmission timeout expiration. The latter resets the congestion window size to one segment, while the former reduces the congestion window by one half of the current window size.

4 Simulation Models

In order to analyze the performance of TCP Vegas and Reno over UMTS, network-level simulations were carried out using *ns-2* [13], which is an event-driven simulator. Several extensions were made to this simulator for modeling UMTS. The extensions were developed within the framework of the IST SEACORN project [14]. With the extensions, instances of UMTS nodes, viz., UE, Node B and RNC can be created.

The model used for simulation analysis is illustrated in Fig. 3. The model is based on the system architecture discussed in the previous section (see Fig. 1). UE, Node B, RNC and host are modeled according to the aforementioned protocol stack illustrated in Fig 2. The TCP/IP protocol stack of *ns-2* was used.

Since the primary aim of the simulation was to investigate the impact of the radio interface on end-to-end TCP performance, we assume that no packet losses, errors or congestion on either the Internet or the UMTS core network. Hence, the TCP performance is solely attributed to the radio interface. The links between two nodes are labeled with their bit rate (in bits per second) and

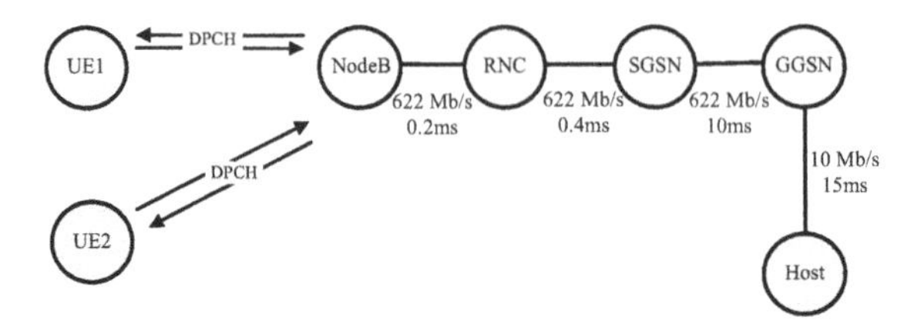

Fig. 3. Top Level Simulation Model

delay (in seconds). Each link capacity was chosen so that the radio channel was the connection bottleneck. Hence, the TCP performance is solely attributed to the UMTS radio interface. Consequently, the functionality of SGSN and GGSN was abstracted out and modeled as traditional *ns* nodes since generally they are wired nodes and, in many ways, mimic the behavior of IP router. Currently, no header compression technique is supported in the PDCP layer. In the following subsections, the UMTS model is described in detail.

4.1 RLC Model

The RLC model supports both the acknowledged and unacknowledged modes. For TCP-based applications, the acknowledged mode was used in the simulation since the acknowledged mode was designed to hide losses due to radio channels from TCP. The retransmission strategy adopted by the acknowledged mode is the Selective-Repeat ARQ (Automatic Repeat reQuest) scheme. With Selective-Repeat ARQ, the only RLC blocks retransmitted are those that receive a negative acknowledgement. An RLC block consists of a header and a payload which carries higher layer data.

A status message is used by the receiver for notifying loss or corruption of an RLC block. The status message is in bitmap format. That is, bit_j indicates whether the jth RLC block has been correctly received or not. The frequency of sending status messages is not specified in the standard [15]. However, several mechanisms are defined, which can trigger a status message. Either the sender or the receiver can trigger the status message. Table 1 and Table 2 list the triggering mechanisms for sender and receiver, respectively. It is important to note that not all the triggering mechanisms are needed for the Selective-Repeat ARQ to operate. However, a combination of triggering mechanisms, which deliver optimum performance, is sought.

The advantage of receiver-initiated mechanisms is that the receiver has direct information about missing blocks. For the sender-initiated mechanisms, the sender has first to request a status message by enabling the poll flag in the RLC block and wait for a reply, which has longer turn around time. Therefore, receiver-initiated mechanisms are preferred. Nevertheless, sender-initiated mechanisms are required to prevent deadlocks and stall conditions. Periodic mechanisms might be more robust compared to others but may result in too frequent status message. In addition, a timer is required at the sender and receiver for

Table 1. Sender-Initiated Mechanisms

Trigger	Explanation
Last Block in buffer or retransmission buffer	*status report is requested by enabling the poll flag in the RLC header*
Every m blocks	poll flag is enabled for every blocks
Every n service data units	poll flag is enabled for every n service data units
Utilization of Send Window	*poll flag is enabled when the Send Window is x% full*

Table 2. Receiver-Initiated Mechanisms

Trigger	Explanation
Detection of missing blocks	*status message is generated once a gap is detected in the RLC sequence number*
Estimated block counter	status message is generated if not all the retransmitted blocks are received within an estimated period

proper operation of the triggering mechanisms. At the sender, the timer is called poll timer, which is started when a request for status messages is sent to the receiver. If the status message from the receiver does not arrive before the timer expires, the sender repeats the same procedure again. The receiver is equipped with a timer called status prohibit timer, which controls the time interval between status messages if triggered consecutively. If the interval is too short, then bandwidth is wasted. On the other hand, if the interval is too long, bandwidth is preserved, but delay increases. The selected triggering mechanisms for the RLC model are the rows written in *italics*.

4.2 MAC Model

The MAC model implemented the dedicated mode. It requests the number of blocks buffered at the RLC layer, which are ready for transmission, and submits to the PHY layer as transport blocks. In this case, each transport block corresponds to an RLC block since no MAC header is required in the dedicated mode as depicted in Fig. 4. The frequency in which the PHY layer can accept transport blocks from MAC is defined by the Transmission Time Interval (TTI). In the UMTS standard, the values of TTI are 10 ms, 20 ms, 40 ms, and 80 ms, where TTI = 10 ms corresponds to the duration of one radio frame.

4.3 PHY Model

The PHY model is responsible for transmitting transport blocks over the physical channels using one or multiple radio frames. For the MAC dedicated mode, the transport blocks are sent over the Dedicated Physical Channel (DPCH). DPCH is a bi-directional channel dedicated to a single user only. The bit rates and

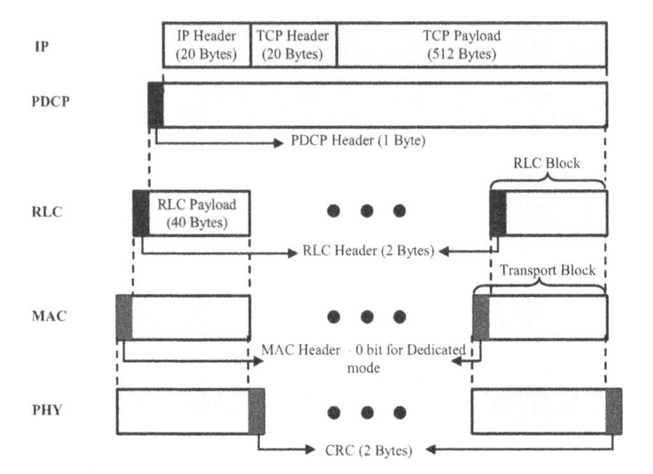

Fig. 4. IP Packet Data Transfer

the TTI associated with the DPCH channel used in the simulation are shown in Table 3. Note that the bit rates exclude RLC headers. Since the PHY layer passes the transport block to the MAC layer together with the error indication from the Cyclic Redundancy Check (CRC), the output of the PHY layer can be characterized by the overall probability of transport block error – also called Transport Block Error Rate (TBLER) in this paper. Thus, an error model based on uniform distribution of transport block errors, was used in the simulation. It is valid to assume that the erroneous transport blocks perceived by the RLC is independent and uniformly distributed as a result of interleaving and forward error-correction mechanisms used by the PHY layer. The TBLER, in the range from 0 to 30%, was considered in the simulation.

The transmission of an IP packet over the radio interface is illustrated in Fig. 4. The RLC entity receives a PDCP packet which comprises an IP packet of 552 bytes or an acknowledgement of 40 bytes, and additionally the PDCP header of 1 byte. This PDCP packet is segmented into multiple RLC blocks of fixed sizes. Each of these blocks fits into a transport block in which a CRC is attached at the PHY layer. In the simulation, the RLC header and the payload size was set to 2 bytes and 40 bytes, respectively. For this RLC payload size and a bit rate of 384 kb/s, twelve transport blocks can be transmitted within one TTI of 10 ms. The other simulation parameters are summarized in Table 3.

5 Simulation Results

5.1 TCP Throughput

End-to-end TCP *throughput* is used as the performance measure. The throughput (in bits per second) is defined as the amount of successfully received TCP segments by the receiver within the simulation duration. The TCP throughput

Table 3. Simulation Parameters

Application	File Transfer Protocol (FTP)		
TCP	TCP Variants	Vegas and Reno	
	Window Size (Segments)	64	
	Maximum Segment Size (Bytes)	512	
	TCP Header Size (Bytes)	20	
IP	IP Header Size (Bytes)	20	
	IP Packet Loss Rate in the Internet	0%	
PDCP	TCP/IP Header compression	No	
RLC	RLC Mode	Acknowledged Mode with In-sequence delivery	
	Window Size (Blocks)	4096	
	Payload Size (Bits)	320	
	RLC Header (Bits)	16	
	Max Bit Rate (kb/s)	Uplink	Downlink
		64	384
MAC	MAC Header (Bits)	0	
	MAC Multiplexing	Not required for DPCH	
PHY	Physical Channel Type	DPCH	
	Transport Block Size (Bits)	336	
	TTI (ms)	Uplink	Downlink
		20	10
	Transport BLER	0 – 30%	
	Error Model	Uniform Distribution	

was obtained using a single FTP session between a UE and a host. Data is transferred from the host to the UE. That means, the only higher layer data going in the opposite direction (or uplink channel) are TCP acknowledgements, which justifies for using lower bit rate in the uplink channel.

The FTP session was run for 1000 s, which is equivalent to 100,000 radio frames. Firstly, the simulation was run using TCP Vegas, and then, the same set of simulation was repeated for TCP Reno. Fig. 5(a) depicts the plots of TCP throughput as a function of TBLER for Vegas and Reno. For both Reno and Vegas, the throughput is normalized to the maximum downlink channel bit rate, i.e., 384 kb/s.

Under ideal radio channel condition (i.e., 0% TBLER), both Reno and Vegas attain the maximum throughput of approximately 96% of the radio channel capacity, which is the maximum achievable throughput, after discounting the overhead of RLC control messages (namely, status messages), PDCP and IP headers. As observed in Fig. 5(a), the performance of Vegas is rapidly deteriorating when the TBLER increases even though the simulation traces showed that the RLC acknowledged mode successfully delivered every transmitted TCP segment to the receiver. When the TBLER is 30%, Vegas' throughput drops to 5% of the radio channel capacity, which is 60% lower than Reno. The cause of poor TCP Vegas performance is analyzed and explained in detail in the next section.

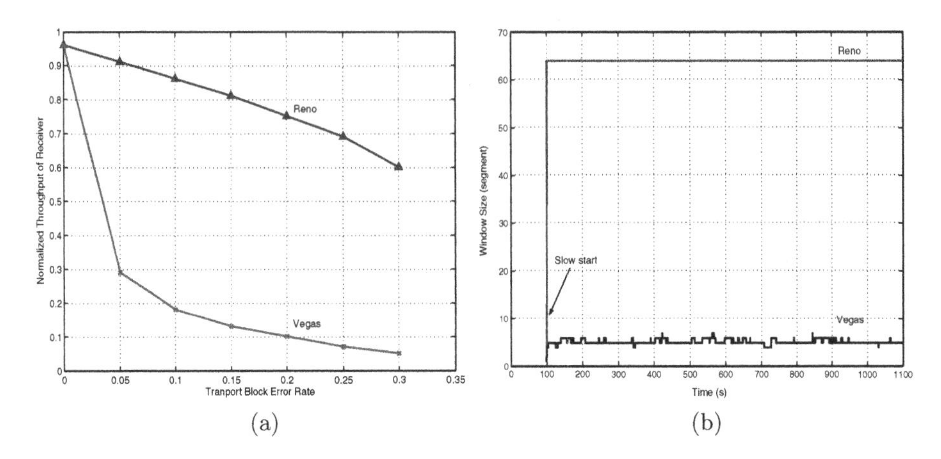

(a)

(b)

Fig. 5. (a) Throughput versus Transport Block Error Rate; (b) TCP Congestion Window versus Time for 5% TBLER

5.2 TCP Round-Trip Time Variability

Fig. 5(b) plots the Vegas and Reno congestion window sizes in segments over the simulation duration for 5% TBLER. The poor performance of Vegas is clearly evidenced by the small window size, while Reno's congestion window grows exponentially until the maximum window size is reached, and remains at this size throughout the simulation. The maximum window size of the TCP connection is 64 segments, which also corresponds to the bandwidth-delay product. In contrast, the congestion window size of Vegas only managed to reach a maximum window size of 7 segments for a short period (appeared as peaks in the Vegas' curve of Fig. 5(b)) and then dropped to 6 segments The first peak occurred at approximately 430 s. The Vegas' congestion window exhibits instability and oscillates about the mean window size of 5 segments. This mean window size is relatively small as compared with Reno, which explains for the low throughput achieved by Vegas. The peculiar behavior of Vegas is due to the congestion avoidance which uses the TCP round-trip times to adjust the window size. The round-trip time for each TCP segment was obtained from the simulation traces for 5% TBLER. In total, there were approximately 26,000 round-trip time samples produced from the simulation traces and plotted in Fig. 6(a). The y-axis shows the round-trip time expressed in seconds for each individual segment and the x-axis shows the time in seconds when each round-trip time was recorded. The round-trip time can be expressed as the sum of the end-to-end delay of the TCP segment and the corresponding acknowledgement,

$$RTT_{TCP} = t_{seg} + t_{ack} \tag{2}$$

where t_{seg} and t_{ack} are the end-to-end delay of the TCP segment and the acknowledgement, respectively. t_{seg} and t_{ack} consist of the delay components over

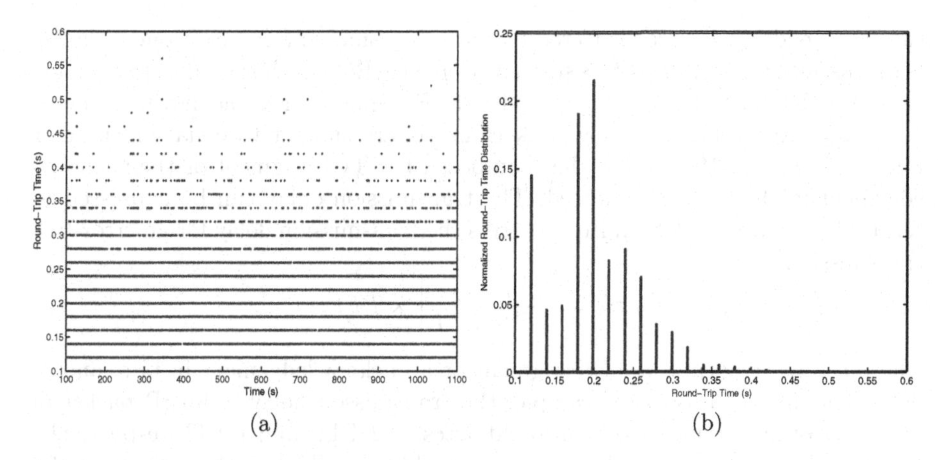

Fig. 6. (a) TCP Vegas Round-Trip Time for 5% TBLER; (b) TCP Vegas Round-Trip Time Distribution for 5% TBLER

the interfaces from U_u to G_i as shown in Fig. 1

$$t_{seg} \quad or \quad t_{ack} = t_{U_u} + t_{I_{ub}} + t_{I_{uPS}} + t_{G_n} + t_{G_i} \tag{3}$$

Each delay component in Eq. (3) is composed of the propagation delay and the transmission delay. The latter is the time required to transmit a TCP segment or an acknowledgement, that is the number of bits in a segment or acknowledgement divided by the channel bit rate in bits per second. The transmission delay over all the interfaces except U_u remains constant throughout the simulation since the TCP segment and the acknowledgement are of fixed length. The propagation delay is the time required by the signal to traverse the physical distance of the channel, which is limited by the speed of light in the tranmission medium. The propagation delay for all the interfaces except U_u is also fixed. The one-way propagation delay for each interface is marked in Fig. 3. The plot in Fig. 6(a) indicates that the TCP round-trip time exhibits relatively large variation between consecutive round-trip times with the maximum and the minimum values of 0.56 s and 0.12 s, respectively. The minimum round-trip time is the basic time to send a TCP segment and receive the corresponding acknowledgement in the absence of errors. Fig. 6(b) presents the round-trip time distribution of Fig. 6(a) in a histogram. The mean round-trip time is approximately 0.2 s.

Since the transmission delay and the propagation delay over the interfaces between G_i and I_{ub} are fixed in all the simulations, the delay component, t_{U_u} in Eq. (3), is the only contributor to the round-trip time variation. On the U_u interface, the propagation delay is negligible as compared with the transmission delay. Hence, the round-trip time variation is mainly due to the transmission delay on the U_u interface, which in turn, caused by the in-sequence delivery and retransmission strategy at the RLC layer. Without in-sequence delivery, the TCP receiver can generate spurious duplicate acknowledgements due to out of order segments. The duplicate acknowledgement triggers the fast retransmit of the TCP sender, which leads to redundant retransmission. As mentioned, the retransmission strategy of RLC is selective-repeat ARQ which is used to retransmit erroneous RLC blocks that composed a TCP segment or an acknowledgement. Note that, the uplink and downlink channels are subject to equal block error rates. Hence, the RLC blocks that composed the TCP segment or the acknowledgement might be retransmitted. The transmission delay can be expressed in terms of the number of TTI. Eq. (4) gives the transmission delay for an error-free transmission[1]

$$t_{U_u} = \left\lceil \frac{l}{C \times TTI} \right\rceil \times TTI \tag{4}$$

where l is the length of a TCP segment or an acknowledgement in bits; and C is the channel bit rate. For example, the transmisson delay of an IP packet of 552 bytes or an acknowledgement of 40 bytes, is 2 TTIs and 1 TTI, respectively. Retransmission of any RLC blocks requires additional TTIs with a minimum of 2 TTIs, i.e., one TTI for the status message to inform the sender of erroneous RLC blocks and the other for retransmitting the RLC blocks. The plot in Fig. 6(a) shows that the round-trip time values occurred at discrete levels, which is due to the fact that the transmission delay is characterized by the number of TTI. The interval between consecutive round-trip times is 20 ms which corresponds to two TTIs in the downlink or one TTI in the uplink. A large variation in round-trip time causes instability and can result in a destructive effect to Vegas. Vegas becomes unstable once the RTT_{base} in Eq. (1) is locked to the minimum round-trip time. For the 5% TBLER case, the RTT_{base} is 0.12 s. Therefore, if $windowSize$ is equal to 6 segments, and if the average round-trip time, $RTT \geq$ 0.3 s, the criterion $\Delta > \beta$ is satisfied, which linearly decreases the congestion window as observed in Fig. 5(a) For $windowSize = 5$ and RTT at the mean value of 0.2 s, the criterion, $\alpha < \Delta < \beta$, is satisfied and the congestion window is unchanged, which correlates to the mean window size of 5 segments.

A similar phenomenon was also observed for higher TBLERs. The round-trip time variation, however, is even greater as the TBLER increases, which means, the likelihood of Δ exceeding β is higher than small TBLERs. In addition, spurious timeouts were observed for TBLERs equal to or greater than 20%, which causes significant throughput degradation. Fig. 7(a) presents the Vegas and Reno's congestion window size over simulation duration for 30% TBLER. Note that, the behavior of Reno is similar to the 5% TBLER case, where the maximum window size is reached. As for Vegas, the congestion window size is relatively smaller than the 5% TBLER case. The congestion window opened up to 4 segments during slow start, and then dropped to 3 segments at around 103.6 s as depicted in Fig. 7(a), which is due to the congestion avoidance. Unlike Reno, Vegas' retransmission timeout is accurately reflecting the measured round-trip time of TCP segments. As a result, a sudden increase of the instantaneous round-trip delay beyond the sender's retransmission timeout value causes spurious timeouts. As shown in Fig. 7(a), several spurious timeouts occurred (marked by arrows) throughout the simulation, which prevents the congestion

[1] For a detailed derivation of an analytical model for TCP, see [16].

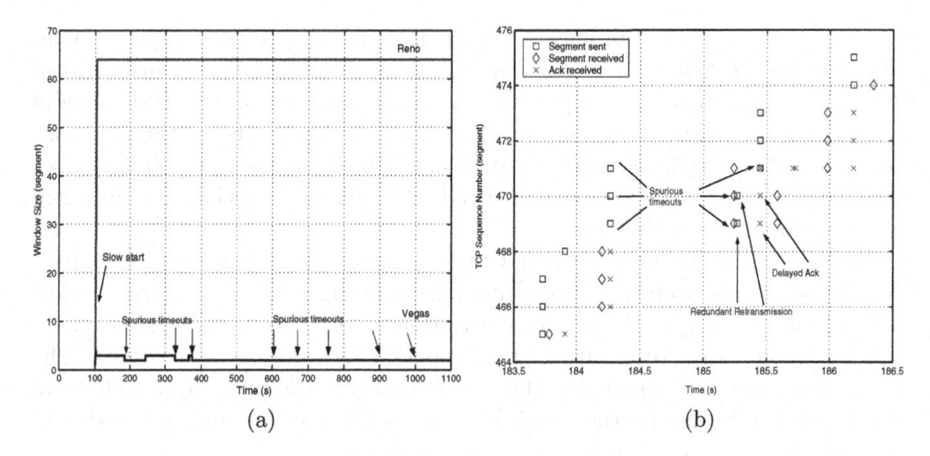

Fig. 7. (a) TCP Congestion Window versus Time for 30% TBLER; (b) TCP Vegas Trace shows Spurious Timeouts and Redundant Retransmissions for 30% TBLER

window from inflating. As a result, the congestion window size remained at 2 segments for more than fifty percent of the TCP connection time. Spurious timeouts can also cause redundant retransmissions which have contributed to the poor performance of Vegas. Fig. 7(b) shows the TCP trace for a spurious timeout event for 30% TBLER. The figure illustrates that the retransmission timeout value used by Vegas is vulnerable to the increased round-trip time and triggering unnecessary retransmissions.

In summary, Vegas would interpret an increase in round-trip delay as a sign of congestion in the network, and consequently, decrease the window size. However, this is not the desire behavior, but sender should perform the opposite.

6 Conclusion and Future Work

The paper has evaluated and compared the performance of TCP Vegas with TCP Reno over UMTS dedicated channels for bulk data transfers. Throughput was used as a performance measure. Throughput simulation results show that the performance of Vegas is worst than Reno under various transport block error rates. As a matter of fact, Vegas' throughput collapses for transport block error rates greater than 30% even though the RLC acknowledged mode successfully delivered every transmitted TCP segment. Conversely, Reno' behavior does not show any abnormality and it achieves the expected throughput over the different transport block error rates. The interaction between Vegas and the UMTS radio interface protocols, in particular, the radio link control layer was examined. The UMTS radio interface exhibits large and highly variable delay, which in turn, resulted in fluctuating TCP round-trip times. Vegas differs from Reno in the sense that it uses TCP round-trip times to detect congestion. Hence, the round-trip time variation is perceived as a sign of congestion and Vegas shrinks its window size, which has a detrimental effect on the performance. Moreover, a sudden increase in instantaneous round-trip time can trigger spurious timeouts at the sender running Vegas which performs unnecessary retransmissions. Spurious timeouts occur at high transport block error rates. The UMTS radio interface characteristics pose performance issues to Vegas due to the new congestion avoidance and retransmission features. On the other hand, Reno is less intelligent but its congestion avoidance and retransmission mechanisms are more robust as compared with Vegas.

In the near future, we anticipated that the use of TCP Vegas will be widely spread since it has been supported by some operating system, e.g. [17], in addition to Reno. Consequently, the poor performance of Vegas in wireless networks with large round-trip time variation such as UMTS needs to be improved. The proposed solutions, in particular, the split-connection and proxy approaches can be adapted for Vegas. Further work is required on adapting and to investigate the performance of Vegas using these two solutions.

References

1. 3GPP: available from `http://www.3gpp.org`
2. Kaaranen, H., et al.: UMTS Networks: Architecture, Mobility and Services, New York, John Wiley & Sons (2001)
3. Steven, W.: TCP/IP Illustrated,Vol. 1, Reading, Addison Wesley (1994)
4. Brakmo, L.S., Petersen, L.L.: TCP Vegas: End to End Congestion Avoidance on a Global Internet. IEEE Journal on Selected Areas in Communications, vol. 13, no. 8 (1995)
5. Caceres, R., Oftode, L.: Improving the Performance of Reliable Transport Protocols in Mobile Computing Environment. IEEE Journal on Selected Areas in Communications, vol. 13, no. 5 (1995)
6. Balakrishnan, H., et al.: Improving Reliable Transport and Handoff Performance in Cellular Wireless Networks. ACM Wireless Networks, vol. 1, no. 4 (1995)
7. Chaskar, N.M., et al.: TCP over Wireless with Link Level Error Control: Analysis and Design Methology. IEEE/ACM Transaction on Networking, vol. 7, no. 5 (1999)
8. Bai, Y., et al.: Interactions of TCP and Radio Link ARQ Protocol. Proceedings of the IEEE Vehicular Technology Conference (VTC'99 - Fall), Amsterdam, The Netherlands (1999)
9. Parsa, C., Garcia-Luna-Aceves, J.J.: Improving TCP Performance over Wireless Networks at the Link Layer. Mobile Networks and Applications, vol. 5, no. 1 (2000)
10. Ayanoglu, E., et al.: AIRMAIL: A Link-layer Protocol for Wireless Networks. Wireless Networks, vol. 1, no. 1 (1995)
11. Bakre, A., Badrinath, B.R.: Implementation and Performance Evaluation of Indirect-TCP. IEEE Transaction on Computers, vol. 46, no. 3 (1997)
12. 3GPP: Network Architecture. TS 23.002 (2000)
13. Fall, K., Varadhan, K.: The *ns* Manual. `http://www.isi.edu/nsnam/ns/ns-documentation.html`
14. SEACORN Home page: `http://seacorn.ptinovacao.pt/`
15. 3GPP: Radio Link Control (RLC) Protocol Specification. TS 25.322 (2002)
16. Peisa, J., Meyer, M.: Analytical Model for TCP File Transfer over UMTS. 3G Wireless 2001, San Francisca, CA (2001)
17. Cardwell, N.: A TCP Vegas Implementation for Linux. `http://flophouse.com/~neal/uw/linux-vegas`

On the Unsuitability of TCP RTO Estimation over Bursty Error Channels

Marta García, Ramón Agüero, and Luis Muñoz

Departamento de Ingeniería de Comunicaciones – ETSIIT
Universidad de Cantabria
Avda los Castros s/n, 39005 Santander (SPAIN)
{marta, ramon, luis}@tlmat.unican.es

Abstract. This work identifies the presence of long idle times as the main cause for the high performance degradation suffered by TCP over bursty error environments. After a comprehensive and fully experimental analysis, performed over an IEEE 802.11b real platform, it is derived that the traditional computation that TCP uses for the RTO estimation does not behave properly over channels prone to suffer from bursty errors. The authors propose a modification to that algorithm so as to avoid such an undesirable behavior.

1 Introduction and Objectives

INTERNET transport protocols were originally designed to work appropriately over traditional wired links, where losses of packets were mainly caused by the overwhelming of intermediate routers. By contrast, wireless channels are likely to damage some packets due to the hostile characteristics of the radio link. The original design and implementations of TCP were not targeted to behave well with these conditions, and several methods to overcome this drawback have been studied and proposed [1], [2]. Most of them focus on the TCP congestion control procedures and sometimes they have lead to new TCP versions, as the New Reno case [3], in which only a slight change was made to the Fast Recovery algorithm used on the Reno version. This work thoroughly analyzes the impact of wireless errors over TCP performance, by means of a completely experimental approach. In particular, the presence of long idle times at the sender side, originated by the consecutive expiration of the TCP retransmission timeout (RTO) when multiple data segments are lost within a single window, has been observed. These inactivity periods bring about a sharp decrease on the TCP throughput.

2 Influence of Error Bursts over TCP Performance

The results presented on this section have been obtained through an experimental measurement campaign over an IEEE 802.11b single-hop, comprising of two hosts

I. Niemegeers and S. Heemstra de Groot (Eds.): PWC 2004, LNCS 3260, pp. 343–348, 2004.

configured in ad-hoc mode. To generate TCP traffic, a 10 Mbytes file was transferred using an FTP application, resulting in around 7500 data segments, as a Maximum Segment Size (MSS) of 1448 bytes has been employed. Table I shows the measured throughput at a typical office environment with metallic obstacles and people moving around the radio channel in which the signal to noise ratio (SNR) was around 10 dB, as shown in Fig. 1.

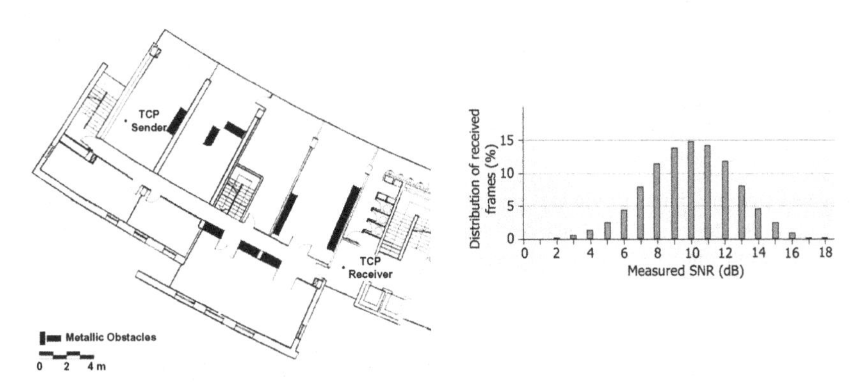

Fig. 1. Experimental environment and its SNR distribution

The IEEE 802.11 standard specifies the use of an idle repeat request (RQ) with implicit retransmission scheme at the MAC layer. In this sense, an erroneous frame might be retransmitted up to a certain number of times (three in our platform). Consequently, the IP loss, which can be defined as the number of datagrams not recovered by the IEEE 802.11 retransmission scheme, does not match the Frame Error Rate (FER). If this procedure is not enough to recover from errors, the TCP will trigger its retransmission mechanisms. All these retransmissions, triggered either by the MAC protocol or by the TCP itself may cause a performance degradation. The radio channel suffers from a extreme variability and, therefore, both erroneous frame and lost IP packet burst statistics obtained for different realizations of the experiment vary within a wide range.

Apart from the throughput, defined as the number of useful bytes received (the size of the file) over the transfer time, the following performance metrics have been collected, being reported on Table I

- Retransmissions: the total number of TCP data segments retransmitted by the TCP protocol sender entity.
- # Max Retx: maximum number of retransmissions for the same TCP data segment.
- RTT: Round Trip Time.
- Idle time: period of inactivity at the sender.
- FER: the ratio between the number of medium access control (MAC) frames received with cyclic redundancy check (CRC) error and the total number of received frames.
- IP Loss: percentage of datagrams that have not been recovered by the 802.11 retransmission scheme.

- Average Erroneous Frame/Lost IP Packet Burst: average length of erroneous frame/lost IP packet bursts.
- Maximum Erroneous Frame/Lost IP Packet Burst: maximum length of all the erroneous frame/lost IP packet bursts.

Table 1. TCP parameter report over an IEEE 802.11b link (11 Mbps) in a burst-error environment

#Test	Tput (Mbps)	FER	Avg EFB	Max EFB	IP loss	Avg LPB	Max LPB	TCP Rtx	Max Rtx	Avg RTT (ms)	StdDev RTT (ms)	MaxIdle Time (sec)	TotalIdle Time (sec)
1	2.91	0.135	1.8	29	1.20%	1.73	7	89	5	42.11	16.95	1.6	10.73
2	0.71	0.157	2.4	40	2.30%	2.27	10	178	8	41.83	35.6	57.4	102.23
3	4.49	0.058	1.1	4	0.03%	1.00	1	2	1	50.05	11.35	0.17	1.15
4	0.50	0.097	2.2	50	1.50%	2.54	12	112	6	45.86	44.74	120	152.83
5	4.59	0.004	1.4	4	0.04%	1.00	1	3	1	47.43	11.09	0.4	1.66

The achieved throughput can be roughly categorized into three different levels: poor, which is below 1 Mbps (tests #2, #4); medium, around 3 Mbps (test #1); and good, close to 5 Mbps (tests #3, #5), which the maximum throughput obtained over an error-free channel [4].

The dominant factor in the performance reduction for the tests belonging to the poor level is the presence of idle times. These are caused by the congestion control procedures implemented within TCP. It is worthwhile to perform a deeper analysis of these idle times, allowing to identify their specific causes and effects.

The measurement campaign was performed using Linux operating system (OS). The TCP implementation followed the well known Reno type, although some slight modifications were also used, as will be further explained [5]. Moreover, the Timestamp option was set during all the measurements, allowing us to track the variables used by TCP to manage its retransmission procedures. In addition, the selective acknowledgment option (SACK), which has proved to be the most effective way of dealing with multiple errors within the same window [6], was also activated during the experiments.

It is well-known that a TCP sender adapts its transmission rate to the state of the network; in this sense, it can only send a certain number of segments before receiving an acknowledgment, being this number indicated by the sender congestion window. Afterwards, it stops and if no confirmation is received within a period of time, it retransmits the first unacknowledged segment. A thorough analysis of the whole set of functions that are used to control the corresponding timers and variables is out of the scope of this text, but it is useful to give a brief introduction about how they are handled. The variable which is used to set the timeout value for a retransmission is the retransmission timeout (rto within the Linux OS), derived from both the mean value and standard deviation of the round trip time (srtt and mdev, respectively).

As the Timestamp option is activated, both srtt and mdev are updated each time new data is acknowledged. Furthermore, and following Nagle's algorithm [7], a TCP sender is not allowed to put any more new data on the channel as far as it has previously retransmitted some segments (by RTO expiration) and it still has more data to be confirmed by the receiver. On the other hand, and provided that the two aforementioned conditions are true (i.e. the TCP sender is in the loss state), the RTO is updated each time a new ACK is received, being multiplied by a backoff factor, whose value is doubled whenever a retransmission is triggered by timeout expiration. Fig. 2 shows the interchange of TCP segments between the two hosts, leading to an inactivity period on the TCP transmitter of 57 seconds (test #2 in Table I), being this the main reason for the throughput degradation within this particular measurement.

We can roughly follow the variation that the aforementioned variables have during this particular segment interchange, to see whether the 57 seconds inactivity period is due to the particularities of the TCP implementation. This is summarized on Table II, and a more detailed explanation follows. At the beginning of this particular connection chunk, the contention window managed by the TCP sender (snd_cwnd) was eight. After sending the 8th segment, the TCP sender stops and waits either for an ACK or the RTO expiration. On that moment, the RTO was at its minimum allowed value (200 ms for this Linux TCP implementation [5]) so, given that no ACK is received within 200 ms, the 1st segment is retransmitted. Furthermore, and following the slow start procedure, the value of snd_cwnd drops to one. However, as can be observed, the TCP sender does not receive confirmation that this segment arrived correctly at the receiver, so it has to be retransmitted again up to four times (each of them doubling the RTO, as specified in the backoff procedure).

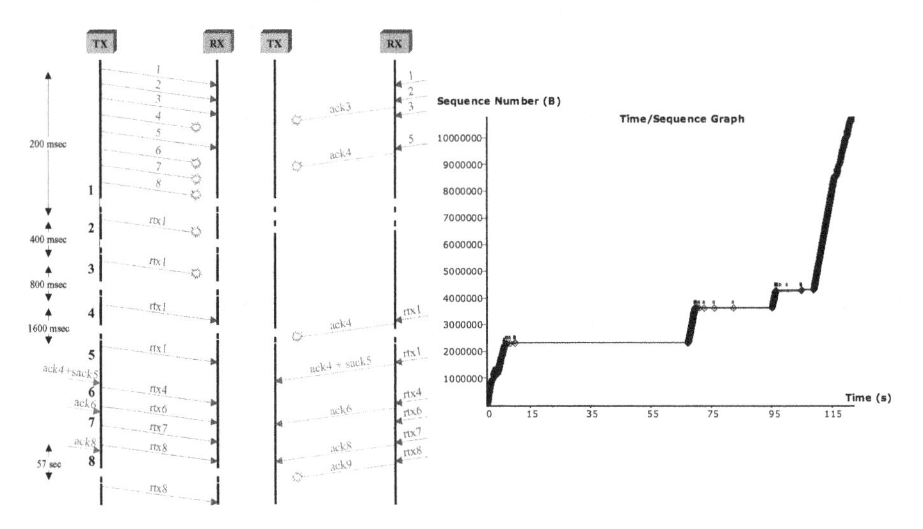

Fig. 2. Segment interchange and idle time of 57 seconds on a TCP connection

Table 2. Evolution of the variables in a TCP connection leading to a 57-second idle time

Situation	Event	srtt (ms)	mdev (ms)	snd_cwnd	Snd_ssthresh	retransmits	backoff	packets_out
1	Start	30	10	8	?	0	0	8
2	Rtx 1 (rto)	30	10	1	4	1	1	8
3	Rtx 1 (rto)	30	10	1	4	2	2	8
4	Rtx 1 (rto)	30	10	1	4	3	3	8
5	Rtx 1 (rto)	30	10	1	4	4	4	8
6	Rx ack(4)	400	750	2	4	4	4	5
7	Rx ack(6)	350	660	3	4	4	4	3
8	Rx ack(8)	310	580	4	4	4	4	1

After the reception of the acknowledgment of that 1st segment (situation #6), the TCP sender still has more data to be confirmed by the receiver, and therefore, the calculation of the RTO is done using the previous backoff value (which was 4), although the segment that caused it to increase has already been confirmed.

The retransmissions of the 4th, 6th, 7th and 8th segments don't affect the set of variables which are being analyzed, and are triggered upon the reception of a new ACK while being retransmitting. These new acknowledgments cause the snd_cwnd to increase up to 4.

When the acknowledgment for the 7th segment is received, RTO's value is updated to 57 seconds, as observed in Fig. 2. This value is derived from the following expression, applying the corresponding values (situation #8 in Table II) for all the variables:

$$rto = (srtt + 4mdev)\left(1 + \frac{1}{4} + \frac{1}{2^{snd\_cwnd-1}}\right)2^{backoff} \tag{1}$$

This expression is slightly different from the one which is specified in the standard (multiplied by a 5/4 and by a factor that depends on the contention window), so that it is more suitable for the Linux OS characteristics.

Unfortunately, the acknowledgment for this retransmission is lost (that is, four consecutive 802.11 frames arrived erroneously at the transmitter) and therefore the 8th segment is retransmitted again 57 seconds after its previous attempt. It is important to remark that the TCP sender still stays at the loss state so, despite not having filled the congestion window, it is not able to transmit any new segment. Moreover, it does not seem reasonable to maintain the backoff factor upon the reception of new acknowledgments, even though the segment that brought about this high value has already been confirmed. Hence, it is likely that modifying this approach, by cleaning the backoff variable upon the acknowledgment of the first lost segment, will help to alleviate the low throughput that has been observed.

3 Conclusion

In this work a deep analysis of the behavior of TCP over Wireless LAN has been performed, with special attention on the impact of wireless error bursts over its performance. It goes without saying that a lot of researching effort is being put on the enhancement of TCP wireless performance. However, most of the studies lack from an experimental approach and assume a TCP acknowledgment error free arrival, focusing on the improvement of the Fast Recovery algorithm. In this work we have shown that, on a real scenario, TCP acknowledgment losses can not be neglected, as they may lead to long idle times, causing a high performance degradation.

This study can be seen as a contribution towards the identification of new modifications to be done to the TCP RTO estimation procedure, adapting it to hostile wireless channels. In particular, resetting the backoff variable after the acknowledgement of the segment that caused it, might bring substantial improvements on the performance.

Although this work has been carried out over an IEEE 802.11b platform, its results can be easily extrapolated to other wireless infrastructures, such as IEEE 802.11g or Universal Mobile Telecommunications System (UMTS)

References

1. A. Chockalingam, M. Zorzi, and R. R. Rao, "Performance of TCP on Wireless Fading Links with Memory," in Proc. IEEE ICC´98, vol. 2. Atlanta, June 1998, pp. 595-600.
2. Yang, R. Wang, F. Wang, M. Y. Sanadidi, and M. Gerla, "TCPW with Bulk Repeat in Next Generation Wireless Networks," in Proc. ICC´03. Anchorage, Alaska, May 2003.
3. Floyd and T. Henderson, "The NewReno Modification to TCP´s Fast Recovery Algorithm, RFC 2582," April 1999.
4. Muñoz, M. García, J. Choque, R. Agüero, and P. Mähönen, "Optimizing Internet Flows over IEEE 802.11b Wireless Local Area Networks: A Performance Enhancing Proxy Based on Forward Error Correction," IEEE Communications Magazine, vol. 39, nº 12, pp. 60-67, December 2001.
5. Sarolahti and A. Kuznetsov, "Congestion Control in Linux TCP," in Proc Usenix 2002. Monterey, CA, USA, June 2002, pp. 49-62.
6. Fall and S. Floyd, "Simulation-based Comparisons of Tahoe, Reno, and SACK TCP," ACM Computer Communication Review, vol. 25 nº 3, pp. 5-21, July 1996.
7. Nagle, "Congestion Control in IP / TCP Internetworks, RFC 896," January 1998.

Performance Evaluation of Bit Loading, Subcarrier Allocation, and Power Control Algorithms for Cellular OFDMA Systems

Slawomir Pietrzyk[1] and Gerard J.M. Janssen[2]

[1] Access Network Unit, Polska Telefonia Cyfrowa Sp. z o.o., Warsaw, Poland,
S.Pietrzyk@ieee.org

[2] Wireless & Mobile Communications Group, Dept. of Electr. Eng., Math. and Comp. Sc. (EEMCS), Delft University of Technology, Delft, The Netherlands,
G.Janssen@ewi.tudelft.nl

Abstract. We consider the problem of radio resource allocation for QoS support in the downlink of a cellular OFDMA system. The major impairments present are co-channel interference (CCI) and frequency selective fading. The allocation problem involves assignment of base stations, subcarriers and bits, as well as power control, for multiple users. We evaluate the performance of a three-stage, low-complexity, heuristic algorithm, which allows to distribute radio resources among multiple users according to theirs individual QoS requirements, while at the same time maintaining the QoS of already established links in all co-channel cells. The evaluation includes checking system operation for various conditions described by different: a) delay spread, b) data rate required by a single user and c) path loss. It is shown that the proposed method is superior in terms of offered traffic and blocking probability to classical method based on FDMA with power control. Also, the performance of our scheme increases in highly frequency selective environment, disastrous for classical fixed schemes, since our method benefits from multiuser diversity.

1 Introduction

The cellular environment poses certain challenges to the resource allocation process, namely the needs: a) to handle co-channel interference (CCI) caused by the RF bandwidth reuse, b) to provide and maintain individual QoS profiles required by multiple users, and finally c) to assign radio resources efficiently. In case of cellular OFDMA system, resource allocation includes, besides OFDM subcarrier assignment, also assignment of modulation orders (bit loading), power levels and base stations (access points) serving the users, [1]. In order to meet these challenges, we apply a dynamic resource allocation approach, which requires the knowledge of channel conditions. In [2], it is claimed that such a design methodology results in higher system performance when compared to interference-averaging techniques (such as CDMA) and much higher performance when compared to fixed resource allocation methods (such as e.g. fixed frequency planning).

I. Niemegeers and S. Heemstra de Groot (Eds.): PWC 2004, LNCS 3260, pp. 349–363, 2004.

Previous work in the area of resource allocation for OFDMA systems that constitutes the base for this paper can be classified into the following groups of papers.

- Single-cell systems, [3], [4], [5]. These papers present various interesting resource allocation algorithms for OFDMA but they do not consider CCI.
- Cellular systems, [6], [7], [8]. Here, methods for OFDMA resource allocation in cellular environment are proposed. However, the considered algorithms do not allow for maintenance of QoS in the co-channel cells. CCI is limited rather by interference avoidance than control.
- Non-cellular systems with CCI control, [9], [10]. In these papers a mature consideration of a problem of bit, subcarrier and power allocation for OFDMA under CCI is given. However, the presented solutions exhibit quite high complexity and are suitable for point-to-point networks, such as xDSL or fixed wireless access, where allocation of transmitters to receivers is predetermined.

Our previous work in the area of resource allocation for OFDMA in cellular environment includes [11] and [12]. In [12] we proposed algorithms for appointing a serving base station, allocation of subcarriers, adaptation of modulation levels (bit loading), and finally control of transmit power to satisfy and maintain the users' individual QoS requirements, expressed in terms of bit rate and bit error rate, at the lowest possible cost of resource utilization in OFDMA system. In this paper, in conjunction with the ideas from the companion paper [12] we have focused on performance evaluation of the proposed algorithms under various settings of system parameters. This analysis allows us to better understand applicability of the proposed methods. Moreover, in this paper we propose a modification of the algorithms described in [12] to allow for performing adaptive cell selection (ACS) and we evaluate its performance by simulations. ACS was considered in an OFDMA scenario in [7] but its forms are known in contemporary systems (e.g. cell selection/reselection in GSM). Thanks to ACS, if a candidate cell is unable to serve a user, another cell is tried from the list of preferred cells. In this way, we increase the chances for user admission.

The paper is organized as follows. Section 2 presents the considered system model and formulates the allocation problem. Section 3 presents the proposed solution, while Section 4 verifies its performance by numerical experiments. Finally, conclusions are given in section 5.

2 System Model and Problem Formulation

The considered downlink of a cellular OFDMA system consists of K cells, each with one base station (BS) serving in total (i.e. in the entire system) U users, as shown in Figure 1 and in Figure 2. The total available bandwidth BW is partitioned into N narrowband OFDM subcarriers. In principle, the entire bandwidth BW is available in every cell (i.e. a reuse of one is applied) and the selection of a particular subcarrier is subject to local load and channel conditions. This

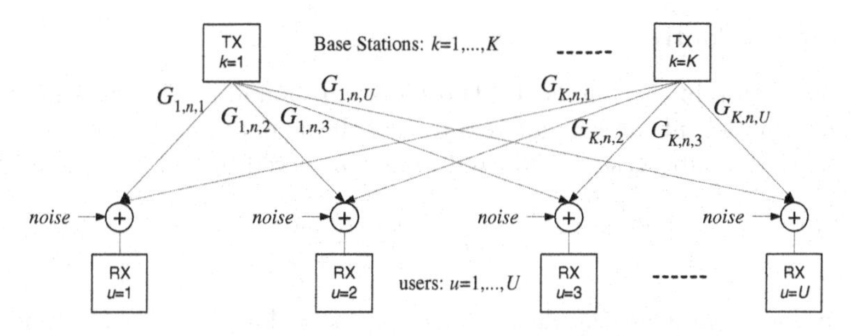

Fig. 1. Model of a downlink cellular OFDMA system for subcarrier n, where $n = 1,..,N$. Dotted lines indicate possible options for connecting transmitters (TX) with receivers (RX)

methodology is especially applicable to systems with a non-uniform spatial distribution of traffic and it does not require frequency planning, which is usually a complex task for the network operator. The bandwidth of each subcarrier is chosen to be sufficiently smaller than the coherence bandwidth of the channel in order to prevent inter-symbol interference (ISI). Then, each OFDM subcarrier n, belonging to a link between BS k and user u, is subject to flat fading, path loss and shadowing with channel gains $\boldsymbol{G} = \{G_{u,n,k}\}$. In addition, the signals suffer from AWGN noise, which is Gaussian distributed with zero-mean and variance σ^2 and the co-channel interference power $\boldsymbol{I} = \{I_{u,n,k}\}$, which is defined as

$$I_{u,n,k} = \sum_{i \neq k} G_{u,n,i} P_{v,n,i}, \; u = 1,..,U; \; v \neq u, \tag{1}$$

where: $G_{u,n,i}$ is the channel gain between the ith interfering BS and user u in cell k and $P_{v,n,i}$ is the transmit power of BS i on subcarrier n assigned to user $v \neq u$, which is allocated to BS i. $\boldsymbol{G} = \{G_{u,n,k}\}$, $\boldsymbol{I} = \{I_{u,n,k}\}$ and σ^2 are assumed to be known by the system. The allocation of subcarrier n to user u at BS k is expressed by the 3-dimensional allocation array $\boldsymbol{C} = \{C_{u,n,k}\}$, where $C_{u,n,k} = \{1, 0\}$ means that subcarrier n is allocated $\{1\}$ or is not allocated $\{0\}$ to user u served by BS k. Additionally, we use a user-to-cell allocation matrix $\boldsymbol{A} = \{A_{u,k}\}$, indicating that user u is allocated $\{1\}$ or is not allocated $\{0\}$ to cell k. Bit allocation is indicated by $\boldsymbol{b} = \{b_{u,n,k}\}$, where $b_{u,n,k}$ expresses the number of bits per symbol on subcarrier n allocated to user u served by BS k. Transmit power allocation is indicated by $\boldsymbol{P} = \{P_{u,n,k}\}$. Modulation levels are restricted, for practical reasons, to three M-QAM schemes with M = 4, 16, and 64 so the number of bits per symbol $b = \log_2 M$ is limited to $\{2, 4, 6\}$, i.e. $b_{min} = 2$ and $b_{max} = 6$. The user mobility is low (WLAN-like scenario) so that the Doppler spread can be neglected. This, together with the assumption of perfect time and frequency synchronization, gives a system free from inter-channel interference (ICI). Users are assumed to be uniformly distributed over the service area. Both BSs and user terminals are equipped with omni-directional antennas.

2.1 Power Control

The key relation in the cellular OFDMA system, bounding the transmit powers of the co-channel subcarriers, and expressing the signal to noise and interference ratio (SNIR) at subcarrier n allocated to user u served by cell k, is the following

$$SNIR_{u,n,k} = \frac{G_{u,n,k} P_{u,n,k}}{I_{u,n,k} + \sigma^2}. \tag{2}$$

In order to provide the data rate of $b_{u,n,k} = \log_2(M_{u,n,k})$ bits and error probability of Pe_u, $SNIR_{u,n,k} \geq \gamma_{u,n,k}$ should hold, where the threshold $\gamma_{u,n,k}$ is defined as, [13]

$$\gamma_{u,n,k} = \Gamma_{u,n,k}(2^{b_{u,n,k}} - 1). \tag{3}$$

The SNR gap $\Gamma_{u,n,k}$ for an uncoded system is defined using the well known Q-function as, [13]

$$\Gamma_{u,n,k} = \frac{1}{3}\left[Q^{-1}\left(\frac{Pe_u}{4}\right)\right]^2, \; n = 1,..,N; \; k = 1,..,K. \tag{4}$$

Rearranging $SNIR_{u,n,k} \geq \gamma_{u,n,k}$ using (1)-(3) leads to a set of linear inequalities, which should be fulfilled for all subcarriers in all the cells in order to guarantee the required level of $SNIR$: $\gamma_{u,n,k}$. This set of linear inequalities can be expressed for each subcarrier n in the matrix form as

$$\mathbf{H}(n)\,\mathbf{p}(n) \geq \sigma^2\gamma(n)\,, \; n = 1,..,N, \tag{5}$$

where

$$\mathbf{H}(n) = \begin{bmatrix} G_{1,n,1} & -\gamma_{1,n,2}G_{1,n,2} & \cdots & -\gamma_{1,n,\kappa}G_{1,n,\kappa} \\ -\gamma_{2,n,1}G_{2,n,1} & G_{2,n,2} & \ddots & \vdots \\ \vdots & \ddots & \ddots & \vdots \\ -\gamma_{\kappa,n,1}G_{\kappa,n,1} & \cdots & \cdots & G_{\kappa,n,\kappa} \end{bmatrix} \tag{6}$$

and

$$\gamma(n) = \left[\gamma_{1,n,1}\; \gamma_{2,n,2}\; \cdots\; \gamma_{\kappa,n,\kappa}\right]^T, \tag{7}$$

where κ is the number of cells using subcarrier n. The columns in (6) correspond to co-channel cells (at subcarrier n), while rows correspond to users allocated to these cells (one user per cell). Given $\mathbf{H}(n)$, $\gamma(n)$, σ^2 the goal is to find an all-positive BS transmit power vector $\mathbf{p}(n) = [p_1(n), p_2(n), \ldots, p_\kappa(n)]$ containing the transmit powers of each BS using the nth subcarrier. If such a solution exists it is called a feasible solution. As explained in [14], [15] in order to find a feasible solution we need to solve equation

$$\mathbf{H}(n)\,\mathbf{p}(n) = \sigma^2\gamma(n)\,, \; n = 1,..,N. \tag{8}$$

If the solution $\mathbf{p}(n)$ is all-positive then it also satisfies (5). Otherwise, no feasible solution exists, which means that we can not reuse subcarrier n in κ cells. Solving equation (8) can be based on the well-known Gaussian elimination process, [15]. Having $\mathbf{p}(n)$ and knowing allocations of users to cells $\boldsymbol{A}$, we can easily compose $\boldsymbol{P} = \{P_{u,n,k}\}$.

2.2 Resource Allocation Problem

The goal of the resource allocation algorithm is to find such allocation arrays $\boldsymbol{C} = \{C_{u,n,k}\}$, $\boldsymbol{b} = \{b_{u,n,k}\}$ and $\boldsymbol{P}_{min} = \min(\boldsymbol{P})$ so that the user's traffic requirements are met at the cost of minimum total transmit power. The *Allocation Problem* can be formulated as follows

$$\min \sum_{u=1}^{U} \sum_{n=1}^{N} \sum_{k=1}^{K} C_{u,n,k} P_{u,n,k} \tag{9}$$

subject to

$$R_u \leq oR_u = \sum_{n=1}^{N} \sum_{k=1}^{K} C_{u,n,k} b_{u,n,k};\ u = 1,..,U, \tag{10}$$

$$P^{max} \geq oP_k = \sum_{u=1}^{U} \sum_{n=1}^{N} C_{u,n,k} P_{u,n,k};\ k = 1,..,K, \tag{11}$$

$$\sum_{u=1}^{U} C_{u,n,k} \leq 1;\ n = 1,..,N; k = 1,..,K, \tag{12}$$

$$C_{u,n,k} \in \{0,1\},\ u = 1,..,U;\ n = 1,..,N;\ k = 1,..,K. \tag{13}$$

Constraint (10) expresses user data rate requirement R_u vs. offered data rate oR_u, constraint (11) is a limitation of a resulting transmit power oP_k, constraint (12) indicates that a subcarrier can be allocated to at most one user within a cell. Problem (9)-(13) is a 3-dimensional allocation problem, in which entries in the cost array $\boldsymbol{P} = \{P_{u,n,k}\}$ are mutually dependent due to CCI, as can be seen in relations (1)-(2), which is a major difficulty. This is also a non-linear, combinatorial optimization problem, since bits may take only integer values and the cost function is non-linear in one of the variables of interest ($\boldsymbol{P}$ depends non-linearly on the number of bits $\boldsymbol{b}$). Therefore, in order to solve the problem (9)-(13), we resort to a suboptimal heuristic algorithm.

3 Proposed Solution

The optimization problem stated in the previous section aims to minimize the sum transmit power of the entire system. Such general objective is fair regarding system-wide resource distribution, but the algorithm solving it closely would be difficult to implement in a cellular system. This is because a cellular system spanning usually over a service area of several (often more) cells has its dynamics caused by changes of the traffic and channel conditions due to new users arrival, users departure and users mobility in various places within a service area. These dynamics of a cellular wireless system are clearly more visible than that of fixed xDSL-like systems described in [9], [10]. Therefore, we take the following directions in designing the allocation algorithm:

- The (re)allocation of certain resources (not for entire-system) is triggered by the change of channel/traffic conditions of a single user.
- Not only the involved user has its resources allocated but also all the other users belonging to the candidate cell. This is done to exploit locally (within the candidate cell) multiuser diversity.
- Resources in all the other cells are not reallocated except the transmit power, which must be tuned to protect quality of the already existing and newly accessing links. Moreover, checking the existence of such power setting that satisfies all co-channel users, is a key part of admission control mechanism. A new user is admitted to the system only if its QoS requirement is fulfilled and the QoS of all the already existing connections is maintained at the acceptable cost.

In the following, we describe the suboptimal algorithm, which overcomes the main difficulty of the original allocation problem, namely the mutual dependency (due to CCI) of the entries in the cost matrix and which follows the above-presented design directions. The solution consists of three steps, where in each step we allocate different resources. We start to determine the list of preferred cells k_{pref} (sorted in descending order from the best one to the worst) for a considered user $\overline{u}$. For the first and not yet verified (for admission) cell l on the list, we check the *necessary condition* for admission that is if the sum of minimal number of subcarriers already allocated to cell l and required by a considered user to satisfy its rate requirement does not exceed the number of subcarriers available in the system, N. If it is not fulfilled then another cell on the k_{pref} list is tried (adaptive cell selection). If the *necessary condition* is fulfilled then, for cell l, we allocate subcarriers to users, including the new user and already existing users (within cell l). This *Step 2* is clearly more sensitive to local per-subcarrier channel gain levels. Finally, in *Step 3* we set the modulation level (bit loading) and the power level for each subcarrier allocated to a user. This last part requires verification of existence of a feasible power vector, (5)-(8). After *Step 3*, we check the *necessary condition* for admission i.e. whether the QoS objectives of all the users are met. If so, the new user is allocated to the system and new resource allocations are applied. Otherwise the next cell on the list k_{pref} is tried until we check all $k_{checked}^{max}$ cells. If the *necessary* and *sufficient* conditions are not met for any of the best $k_{checked}^{max}$ cells, user $\overline{u}$ is blocked.

The "master" algorithm for resource allocation and admission control of a new user is depicted below.

Resource allocations for existing users : $\boldsymbol{A}, \boldsymbol{C}, \boldsymbol{b}, \boldsymbol{P}_{min}$
Temporal resource allocations for new and existing users:
  $\boldsymbol{A}_{temp} = \boldsymbol{A}, \boldsymbol{C}_{temp} = \boldsymbol{C}, \boldsymbol{b}_{temp} = \boldsymbol{b}, \boldsymbol{P}_{min}^{temp} = \boldsymbol{P}_{min}$
$s_u^{min} = \lceil R_u / b_{max} \rceil$ is a min number of subcarriers required to satisfy R_u
$s_k^{min} = \sum_{u \in \text{cell}k} s_u^{min}$ is a min number of subcarriers allocated already to cell k
For each new user $\overline{u}$ with : $R_{\overline{u}}, Pe_{\overline{u}}, G_{\overline{u},n,k}, I_{\overline{u},n,k}$
  Satisfaction indicator: $satisf = 0$
  Number of tried cells: $k_{checked} = 0$
  Step 1: find a list of preferred cells k_{pref} for user $\overline{u}$
  Cell index: $i = 1$
  While $satisf = 0$ and $k_{checked} < k_{checked}^{max}$
    $l = k_{pref}(i)$
    // verification of the necessary condition for admission
    If $\left(s_{\overline{u}}^{min} + s_l^{min} \le N\right)$
      update $\boldsymbol{A}_{temp}$: $A_{\overline{u},l}^{temp} = 1$
      Step 2: find allocation of subcarriers to users in cell l, update $\boldsymbol{C}_{temp}$
      Step 3: per involved subcarrier find allocation of: bits in cell l
        and power levels in all cells; update $\boldsymbol{b}_{temp}$, $\boldsymbol{P}_{min}^{temp}$
      // verification of the sufficient condition for admission
      If $oR_u \ge R_u$ and $oPe_u \le Pe_u$ and $oP_k \le P_{max}, u = 1, ..., U$
        $satisf = 1$
      Else
        $satisf = 0; k_{checked} = k_{checked} + 1; i = i + 1$
      End
    Else
      $satisf = 0; k_{checked} = k_{checked} + 1; i = i + 1$
    End
    // final user admission or blocking
    If $satisf = 1$
      Allow user $\overline{u}$ to the system:
        $\boldsymbol{A} = \boldsymbol{A}_{temp}, \boldsymbol{C} = \boldsymbol{C}_{temp}, \boldsymbol{b} = \boldsymbol{b}_{temp}, \boldsymbol{P}_{min} = \boldsymbol{P}_{min}^{temp}$
    Else count user $\overline{u}$ as blocked
    End
  End
End

In the following, the algorithm's steps are discussed in more details.

3.1 Step 1: User to Cell (Base Station) Allocation

The basic idea behind "master" algorithm's *Step 1* is to produce a list of preferred cells k_{pref} based on the criterion of best average (over all subcarriers) normalized channel gain and interference level, which is particularly important in cellular environments and is usually not considered in classical cell selection schemes (e.g. based on received signal level or distance). We allow a user to be served by one base station only, in order to avoid prospective problems due to synchronization misalignments expected to arise if a user was connected to multiple base stations using different subcarriers. The average (over all subcarriers) normalized channel gain for a user u at cell k can be expressed by

$$\overline{T}_{u,k} = \frac{1}{N} \sum_{n=1}^{N} T_{u,n,k}, \tag{14}$$

where the normalized channel gain is defined as

$$T_{u,n,k} = \frac{G_{u,n,k}}{\left(I_{u,n,k} + \sigma^2\right) \Gamma_{u,n,k}}. \tag{15}$$

The list of preferred cells k_{pref} for user u is a vector of cell indices to the sorted (in descending order) normalized channel gains as shown below

$$k_{pref} = \arg\ \text{sort}_k \overline{T}_{u,k}. \tag{16}$$

3.2 Step 2: Subcarrier to User Allocation

Having temporarily allocated user $\overline{u}$ to a cell l the task is to allocate subcarriers for this user. In order to exploit multiuser diversity, the allocation of subcarriers for user $\overline{u}$ is done together with reallocating subcarriers of users already present in cell l. For this purpose we apply a slightly modified version of a two-phase algorithm proposed in [4].

In *phase A*, we determine the number of subcarriers each user would get (proportionally to its rate requirement) by verifying the relative reduction of cell transmit power after allocation of additional subcarrier. The modification of the original algorithm includes using $s_u \leq s_u^{max}$ as a stop-criterion in increasing the number of subcarriers for a user, since allocating more subcarriers than a user may operate leads to blocking these subcarriers for other users where they might be better exploited.

//*Phase A*: determine the number of subcarriers for each u
$s_u^{max} = \lceil R_u / b_{min} \rceil$; $s_u = s_u^{min} = \lceil R_u / b_{max} \rceil$ for each u
While $\sum_{u=1}^{U} s_u < N$ and $s_u \leq s_u^{max}$ do
  For $u \in$ cell l
    $\overline{P}_u = s_u \left(2^{R_u/s_u} - 1\right) / \overline{T}_{u,l}$ is the average power a user would require to transmit using s_u subcarriers in cell l
    $\overline{P}_u^{new} = (s_u + 1)\left(2^{R_u/(s_u+1)} - 1\right) \overline{T}_{u,l}$
    $\Delta P_u = \overline{P}_u - \overline{P}_u^{new}$
    $w = \arg\max_u \Delta P_u$
    $s_w = s_w + 1$
    $\overline{P}_w = \overline{P}_w + 1$
  End
End

The output of this phase is the number of subcarriers $\{s_u\}$ that each user in cell l should use. Knowing this, in *phase B* we allocate particular subcarriers for particular users within a given cell. The algorithm is based on a simple greedy routine distributing subcarriers among users based on best normalized channel gains. Such approach is motivated by a simple observation that users with good channels require less transmit power, thus causes less CCI and therefore have higher chances for admission. Various options of this routine have been described in many papers, including [5], [4] and [3]. An alternative strategy is to use the optimal Hungarian algorithm or to improve the greedy assignment by swapping subcarriers between users as in [5], [3]. These methods, however, are not considered here due to their increased computational complexity.

//*Phase B*: allocation of particular subcarriers to users
Initialize $C^{temp}_{u,n,l} = 0$ for all $\{u, n\}$, $u \in$ cell l, $n = 1, .., N$
While $\sum_{u \in \text{cell } l} s_u > 0$ do
  For $u \in$ cell l
    $m = \arg\max_{n \in \{\text{not assigned}\}} T_{u,n,l}$
    $s_u = s_u - 1$
    $C^{temp}_{u,m,l} = 1$
  End
End

3.3 Step 3: Bit and Power Allocation per Subcarrier

After having temporarily allocated users to cells according to $\boldsymbol{A}_{temp}$ and subcarriers to users according to $\boldsymbol{C}_{temp}$, the bit and power allocation boils down to single-user bit loading in the presence of background CCI. This algorithm step is similar to the one proposed in [9], [10] with the exception that it does not run over all subcarriers (and in case of a cellular system also over all base stations) since these allocations have been done already in previous steps. In effect, the computational load is reduced. Therefore, for each user u allocated in cell l, within a set of subcarriers allocated to user u, the algorithm increases the modulation level by one step (which corresponds to increasing the number of bits from $b_{u,n,l}$ to $b^{new}_{u,n,l}$) on a subcarrier, where it requires least transmit power increase. Subsequently, transmit power is set in a given cell and adjusted in all co-channel cells in order to maintain QoS of already existing connections. The procedure is repeated until user's rate requirement is fulfilled or the system is saturated for this user, which means that all subcarriers are not feasible (i.e. either maximal modulation level is already achieved or there is no feasible power setting over all co-channel cells). The aim of this routine is to determine the temporary number of bits for users within considered cell l and to set the temporary power levels within the cell l as well as within all the co-channel cells reusing the same subcarriers as cell l. Both bit and power allocations are parts of the all-system allocation arrays $\boldsymbol{b}_{temp}$ and $\boldsymbol{P}^{temp}_{min}$, in which only the involved entries are modified. The routine is outlined below.

For each $u \in$ cell l
  saturated $= 0$
  While $oR_u < R_u$ and saturated $= 0$
    For $n \in$ user u
      If $b_{u,n,l} = b_{max}$
        Mark entry (u, n, l) as not feasible
      Else
        Increase number of bits from $b_{u,n,l}$ to $b^{new}_{u,n,l}$
        Calculate $P^{current}_{u,n,l} = \sum \mathbf{p}(n, b_{u,n,l}) = \sum_{k=1}^{K} p_k(n, b_{u,n,l})$
          and $P^{new}_{u,n,l} = \sum \mathbf{p}(n, b^{new}_{u,n,l}) = \sum_{k=1}^{K} p_k(n, b^{new}_{u,n,l})$ using (3)-(8)
        If no feasible power vector $\mathbf{p}(n)$ i.e. $P^{new}_{u,n,l} = \emptyset$
          Mark entry (u, n, l) as not feasible
        End
      End
      If all considered (u, n, l) are not feasible
        Saturated$= 1$
      Else
        Calculate $\Delta P_{u,n,l} = P^{new}_{u,n,l} - P^{current}_{u,n,l}$
        $m = \arg\min_{n \in \text{user } u} \Delta P_{u,n,l}$
        Increase number of bits from $b_{u,m,l}$ to $b^{new}_{u,m,l}$
        Update power values in cell l according to $P^{new}_{u,m,l}$
          and over all CCI cells according to $\mathbf{p}(m)$ as in (8)
        Update $b^{new}_{u,m,l}$ in $\mathbf{b}_{temp}$ and $P^{new}_{u,m,l}$ with $\mathbf{p}(m)$ in $\mathbf{P}_{temp}$
      End
    End
  End
End

4 Numerical Experiments

The key point of this paper is the performance evaluation of the proposed algorithms, which is done by numerical experiments. We simulated the downlink of a cellular network consisting of $K = 19$ cells, each with one BS having one omni-directional antenna, as depicted in Figure 2 with users appearing one-by-one at random locations within the system service area. The following system parameters were fixed: system bandwidth $BW = 5$ MHz, mean Ricean K-factor: 4.9 dB, and variance of the Gaussian distributed power variations due to shadowing: 6 dB. Each user is assumed to require the same data rate $R_u = R$, and error probability $Pe_u = Pe$, though the model allows to set R_u and Pe_u per user individually. We have evaluated the influence of changing the following system parameters: a) mean rms delay spread $rds = \{50, 100, 200, 500, 1000, 1500\}$ ns, b) data rate required by a single user $R = \{0.5, 1, 2, 4, 8\}$ Mbit/s and c) path loss exponent $\alpha = \{1.5, 2, 2.5, 3, 3.5, 4\}$.

The following schemes have been taken for comparison:

- *proposed*: three modulation levels are applied, namely 4-QAM, 16-QAM and 64-QAM, adaptive cell selection (ACS) is allowed with $k^{max}_{checked} = 5$,

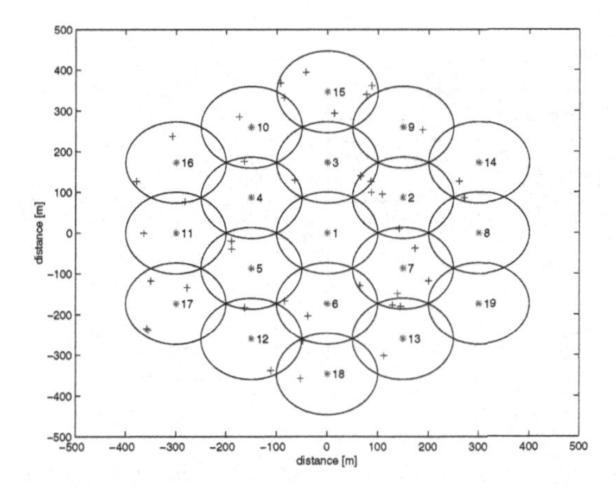

Fig. 2. Cellular network used in the simulations. Example distribution of $U = 40$ users (indicated by crosses) over $K = 19$ cells (base stations indicated by stars)

- *proposed no ACS*: three modulation levels are applied, namely 4-QAM, 16-QAM and 64-QAM, ACS is not allowed i.e. $k_{checked}^{max} = 1$,
- *proposed fixed no ACS*: modulation level is fixed to 4-QAM, ACS is not allowed i.e. $k_{checked}^{max} = 1$,
- *FDMA*: users are allocated to base stations as in *proposed fixed* scheme, subcarriers are allocated to users in a fixed classical FDMA way, according only to their data rate requirements; modulation level is 4-QAM. Power control is used as in proposed scheme in order to provide and maintain required QoS. ACS is not allowed i.e. $k_{checked}^{max} = 1$.

We have used the following metrics for performance comparison: (a) total data rate offered, which is the sum offered data rate over all users $\sum_u oR_u$ provided that the QoS requirement is met for all users (both new and existing) and (b) blocking probability, which is a probability that a new user will be blocked due to insufficient resources to support required QoS. The performance comparison metrics, averaged over 100 network realizations, are gathered at the reference load of 50 Mbit/s, which indicate the total data rate required in the network. In the simulations performed, no wrap-around technique was applied in order to reflect a small hot-spot network (WLAN-like), where the coverage is usually limited to a couple of base stations.

4.1 Influence of Delay Spread

The comparison of the considered schemes for various mean rms delay spread $rds = \{50, 100, 200, 500, 1000, 1500\}$ ns is depicted in Figure 3. It can be observed that the total data rate offered of the *proposed*, *proposed no ACS* and *proposed fixed no ACS* slightly degrades with the increased *rds* but are above the required reference data rate of 50 Mbit/s. This can be explained by the fact that the

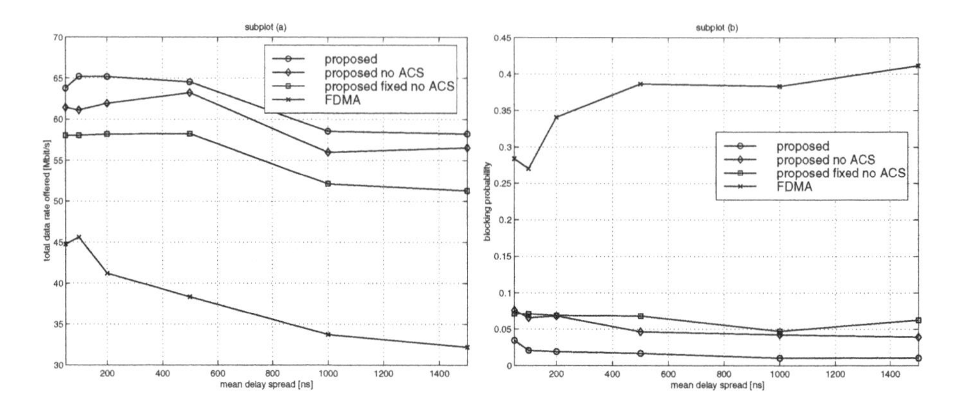

Fig. 3. Influence of various mean delay spread rms. The other system parameters are: $N = 16$ for $rds = \{50, 100, 200, 500\}$ ns and $N = 32$ for $rds = \{1000, 1500\}$ ns, $\alpha = 4$, $R = 2$ Mbit/s

offered data rates have a granularity determined by the number of subcarriers and the three modulation levels. For $rds = \{50, 100, 200, 500\}$, $N = 16$ thus granularity is lower than for $rds = \{1000, 1500\}$, where $N = 32$. For $N = 16$ users get higher data rates than the required 2 Mbit/s (per single user) because it is not possible to compose exactly 2 Mbit/s. For $N = 32$ we are closer to the required 2 Mbit/s.What is maybe more interesting is that with the increased rds (i.e. increased frequency selectivity), the blocking probability of the *proposed* and *proposed no ACS* actually improves, while in case of *FDMA* it degrades. This can be explained by the fact that the more frequency selectivity we have, the more we can gain from multiuser diversity. For fixed schemes (*FDMA*), on the other hand, the multiuser diversity is not exploited and thus a higher rds has a negative effect. It is interesting to note that blocking probability for the *proposed fixed no ACS* is relatively constant. This would mean that in this case blocking probability is improved rather by the application of adaptive modulation (as in *proposed* and *proposed no ACS*) than by adaptive subcarrier allocation applied to all users in the considered cell.

4.2 Influence of Data Rate Required by a Single User

The performance comparison of the considered schemes for various data rates R required by a single user, where $R = \{0.5, 1, 2, 4, 8\}$ [Mbit/s] is depicted in Figure 4. The observations in this case are similar to the previous observations on frequency selectivity. Also here, the total data rate offered of the *proposed*, *proposed no ACS* slightly degrades with the increased R but is above the required reference data rate of 50 Mbit/s. For *proposed fixed no ACS*, offered data rate drops quite rapidly, which indicates the importance of adaptive modulation (such drop is not observed in case of schemes employing adaptive modulation). When looking at the blocking probability, in case of *proposed fixed no ACS*, it

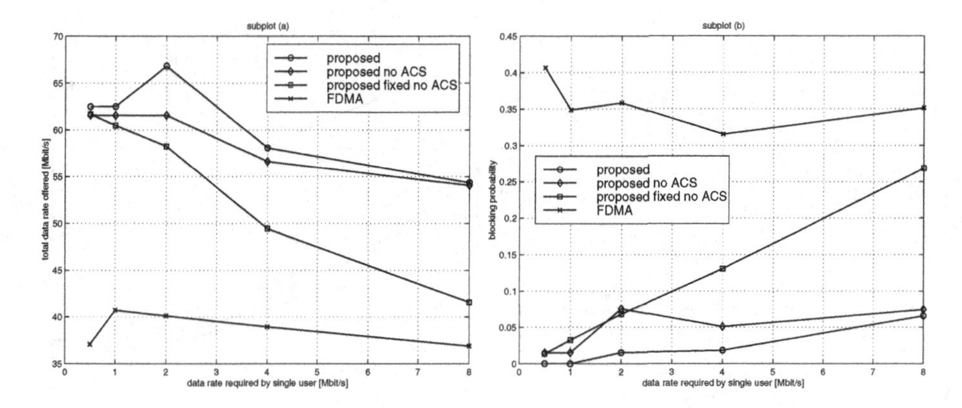

Fig. 4. Influence of various data rates required by a single user R. The other system parameters are: $N = 16$, $rds = 200$ ns, $\alpha = 4$

also degrades rapidly with the increased data rate per user. For *proposed* and *proposed no ACS* degradation is also observed but it is moderate. This can be explained by the fact that for *proposed fixed no ACS*, the better performance for low data rate per single user is due to multiuser diversity exploited by adaptive subcarrier allocation (and reallocation of users served by a considered cell), since this method can not exploit channel variability with the use of adaptive modulation. At a constant reference load of 50 Mbit/s, lower data rate requirement per single user means more users per cell. This obviously increases the gain from multiuser diversity, since we have more options (corresponding to users) to find good subcarriers. Heavy users (such as 8 Mbit/s in this case) result in low number of users per cell and thus lower multiuser diversity.

4.3 Influence of Path Loss

The performance comparison of the considered schemes for various path loss conditions $\alpha = \{1.5, 2, 2.5, 3, 3.5, 4\}$ is depicted in Figure 5. It can be observed that path loss exponent influences all the considered schemes in the same way: low α means poor shielding from CCI and thus low offered data rates and high blocking probability, while high α means good shieling from CCI thus increased offered data rates and decreased blocking probabilities.

5 Summary

In this paper we addressed a 3-dimensional problem of allocating users, base stations, subcarriers, bits and transmit power in a cellular multi-user OFDMA system. We have proposed a modification of the algorithms described in [12] to solve this allocation problem. The modification includes possibility to perform adaptive cell selection in order to find suitable cell to serve a user. In addition, we

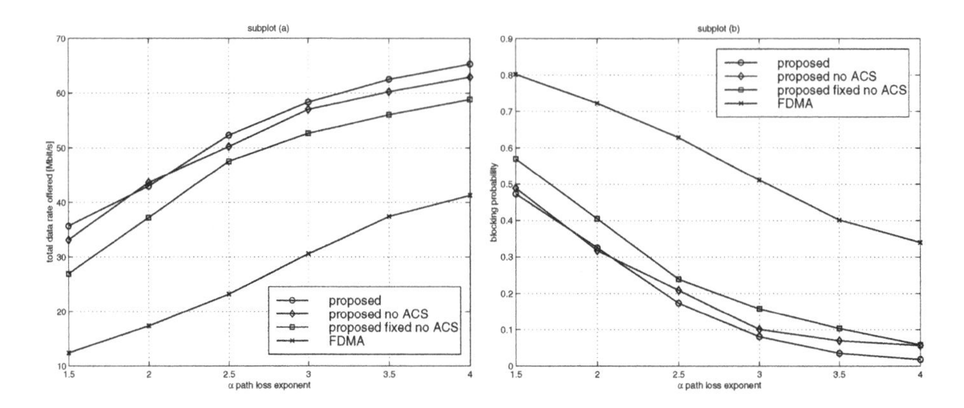

Fig. 5. Influence of various path loss exponent α. The other system parameters are: $N = 16$, $rds = 200$ ns, $R = 2$ Mbit/s

have evaluated the performance of the various options of the proposed algorithm in different scenario settings.

In all the considered cases, the best performance (offered data rate and blocking probability) is achieved with the *proposed* algorithm, which utilizes both adaptive cell selection and adaptive modulation. The worst performance has been observed with *FDMA* scheme. We have observed that the *proposed* algorithm improves its performance with increased frequency selectivity, which is disastrous for fixed *FDMA*. Moreover, we have observed that the more users per cell we have, the more we can exploit multiuser diversity with our adaptive techniques. This indicates the trade-off between allowed range of traffic profiles and offered system capacity. Another indication regarding the allowed traffic profiles is that it is important to match the required data rates to granularity offered by a systems in order to offer just-as-required data rates.

Potential implementations of the proposed schemes include mainly low-mobility, frequency selective, cellular radio access systems, such as OFDM-based WLAN (e.g. 802.11a, HiperLAN/2) or Broadband Wireless Access (e.g. 802.16).

References

1. J. Zander, "Radio Resource Management in Future Wireless Networks: Requirements and Limitations", IEEE Communications Magazie, August 1997.
2. G.J. Pottie, "System Design Choices in Personal Communications", IEEE Personal Communications, October 1995.
3. C.Y. Wong et al, "A Real-time Sub-carrier Allocation Scheme for Multiple Access Downlink OFDM Transmission", IEEE VTC'99.
4. D. Kivanc, H. Liu, "Subcarrier Allocation and Power Control for OFDMA", 34th Asilomar Conference on Signals, Systems and Computers, November 2000.
5. S. Pietrzyk, G.J.M. Janssen, "Multiuser Subcarrier Allocation for QoS Provision in the OFDMA Systems", IEEE VTC'02 Fall, September 2002, Vancouver, Canada.

6. J.C.I. Chuang and N.R. Sollenberger, "Spectrum Resource Allocation for Wireless Packet Access with Application to Advance Cellular Internet Service", IEEE JSAC, August 1998.
7. Y. Zhang and K.B. Letaief, "Multiuser Subcarrier and Bit Allocation along with Adaptive Cell Selection for OFDM Transmission", IEEE ICC, 2002.
8. J. Li, H. Kim, Y. Lee and Y. Kim, "A Novel Broadband Wireless OFDMA Scheme for Downlink in Cellular Communications", IEEE WCNC, 2003.
9. J. Lee, R.V. Sonalkar, J.M. Cioffi, "Multi-User Discrete Bit-Loading for DMT-based DSL Systems", IEEE Globecom 2002.
10. G. Kulkarni, M. Srivastava, "Subcarrier and Bit Allocation Strategies for OFDMA based Wireless Ad Hoc Networks", IEEE Globecom 2002.
11. S. Pietrzyk, G.J.M. Janssen, "Subcarrier Allocation and Power Control for QoS Provision in the Presence of CCI for the Downlink of Cellular OFDMA Systems", IEEE VTC'03 Spring, April 2003, Jeju, Korea.
12. S. Pietrzyk, G.J.M. Janssen, "Radio Resource Allocation for Cellular Networks Based on OFDMA with QoS Guarantees", (paper accepted), IEEE Globecom, December 2004.
13. J.M. Cioffi, "A Multicarrier Primer", Nov. 1991.
14. N. Bambos, G.J. Pottie, "Power Control Based Admission Policies in Cellular Radio Networks", IEEE Globecom 1992.
15. S.C. Chen, N. Bambos, G.J. Pottie, "Admission Control Schemes for Wireless Communication Networks with Adjustable Transmitter Powers", IEEE Infocom 1994, Toronto.

Multicode Multirate Compact Assignment of OVSF Codes for QoS Differentiated Terminals*

Yang Yang[1] and Tak-Shing Peter Yum[2]

[1] Department of Electronic and Computer Engineering,
Brunel University, Uxbridge, London, UB8 3PH, United Kingdom
[2] Department of Information Engineering,
The Chinese University of Hong Kong, Shatin, Hong Kong

Abstract. Orthogonal Variable Spreading Factor (OVSF) codes are used in both UTRA-FDD and UTRA-TDD of the third-generation (3G) mobile communication systems. They can support multirate transmissions for mobile terminals with multicode transmission capabilities. In this paper, a new OVSF code assignment scheme, namely "Multicode Multirate Compact Assignment" (MMCA), is proposed and analyzed. The design of MMCA is based on the concept of "compact index" and takes into consideration mobile terminals with different multicode transmission capabilities and Quality of Service (QoS) requirements. Specifically, priority differentiation between multirate realtime traffic and best-effort data traffic is supported in MMCA. Analytical and simulation results show that MMCA is efficient and fair.

1 Introduction

Orthogonal Variable Spreading Factor (OVSF) codes [1] are adopted in UTRA-FDD and TDD (Universal Terrestrial Radio Access – Frequency Division Duplex and Time Division Duplex) of the third-generation (3G) mobile communication systems to identify traffic channels for different users. According to the technical specifications [2,3], multiple parallel code (channel) transmissions are possible for a single user to support multirate multimedia applications. Although single-code transmission is simpler, multicode transmission has the advantages of finer granularity in bandwidth assignment, more flexible code assignment solutions, and therefore higher bandwidth efficiency.

From a user's perspective, traffic can be classified as either realtime calls or best-effort data packets. Realtime calls require realtime transmission with a fixed bandwidth or, in other words, at a fixed data rate. This traffic class includes audio and video telephonies, on-line TV/movie watching and so on. Best-effort data packets are those generated from the Internet and audio and video file transfers. Realtime calls have priority over data packets in code assignment. On

* This work was supported in part by the Hong Kong Research Grants Council under Grant CUHK 4325/02E and the Short Term Research Fellowship programme of British Telecommunications (BT).

I. Niemegeers and S. Heemstra de Groot (Eds.): PWC 2004, LNCS 3260, pp. 364–378, 2004.

the other hand, from the system's perspective, users are heterogeneous. First, they have different Quality of Service (QoS) requirements, e.g. realtime or best-effort transmission, fixed or variable bandwidth assignment, and fixed or variable packet size. Second, mobile terminals have different capabilities in supporting multicode transmission.

Code assignment schemes can be of the *non-rearrangeable* and *rearrangeable* type. Specifically, rearrangeable code assignment schemes allow the OVSF codes to be rearranged so that they have better performance at the expense of higher computational complexity. Many single-code rearrangeable code assignment schemes were proposed in literature [4,5,6,7,8,9,10,11,12,13,14]. Among them, the priority issue between realtime and best-effort traffic was considered in [4,7,9]. Several single-code non-rearrangeable code assignment schemes were proposed in [8,10,11,12,13]. Specifically, the algorithm in [8] is based on the first-fit scheme for the bin-packing problem. In [10], a fixed code configuration, which specifies the number of OVSF codes for each service class, is used for maximizing the average throughput. Tseng and Chao compare the performance of *random*, *leftmost* and *crowded-first* schemes in [11]. The concept of *crowded-first* is extended in [12] and a new code selection scheme based on the "weights" of candidate codes is proposed. In [13], a new measure called "compact index" is defined as the criterion for code assignment. By using this criterion, the proposed *Compact Assignment* (CA) scheme can offer comparable performance to rearrangeable schemes. Multicode rearrangeable code assignment schemes were proposed in [15,16] for uniform mobile terminals having exactly the same capability in supporting multicode transmission, and in [17,18] for different multicode capable terminals. All these multicode schemes consider only multirate realtime traffic class.

In this paper, based on the concept of "compact index", we design and analyze a non-rearrangeable multicode code assignment scheme, namely "Multicode Multirate Compact Assignment" (MMCA), for accommodating both multirate realtime and best-effort traffic. The design considers the coexistence of mobile terminals with different multicode transmission capabilities and QoS requirements. When multicode transmission is introduced, many slack capacities in the code tree can be taken up by the second and third codes and renders code rearrangement not essential. Also, when data packets are introduced, they can absorb these "wasted" capacity (usable but not used by realtime traffic).

The rest of this paper is organized as follows. In Section 2, the tree structure and some basic concepts of OVSF codes are reviewed. Based on that, the code assignment problem for accommodating mobile terminals with different QoS requirements and different multicode transmission capabilities is formulated. The algorithm of *Multicode Multirate Compact Assignment* (MMCA) is proposed and discussed in Section 3. In Section 4, the performance of MMCA is studied. Both the analytical and simulation results are given and compared.

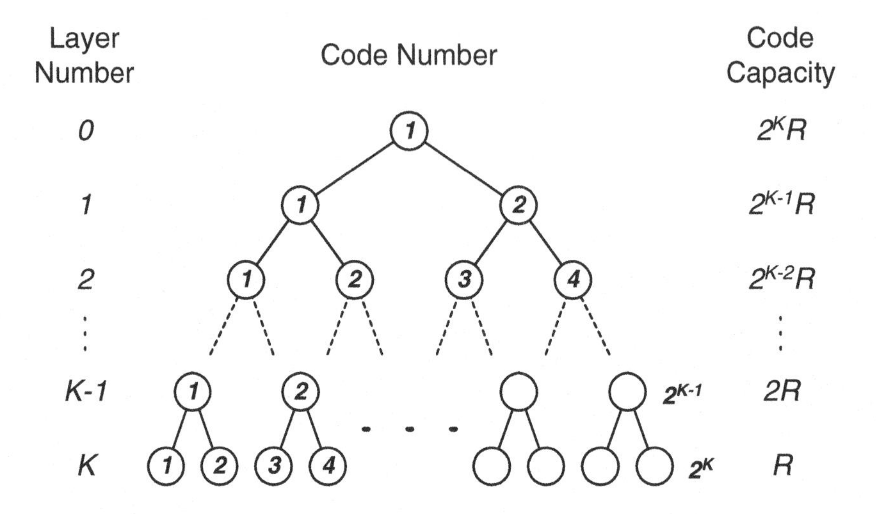

Fig. 1. A K-layer code tree.

2 The Code Assignment Problem

All OVSF codes in the system can be represented by the nodes in a binary tree [1]. Fig. 1 shows a K-layer code tree. Each layer corresponds to a particular spreading factor, so all codes in the same layer can support the same data rate. The data rate a code can support is called its capacity. Let the capacity of the leaf codes (in layer K) be R. Then, the capacity of the codes in layer k is $2^{K-k}R$, as shown in Fig. 1.

Layer k has 2^k codes and they are sequentially labeled from left to right, starting from one. The mth code in layer k is referred to as code (k, m). The total capacity of all the codes in each layer is $2^K R$, irrespective of the layer number. For a typical code (k, m), its ancestor code set, denoted by $S_A^{(k,m)}$, contains all the codes on the path from (k, m) to the root code $(0, 1)$. Its descendant code set, denoted by $S_D^{(k,m)}$, contains all the codes in the branch under (k, m).

Codes in the same layer that are connected by an i-layer sub-tree are defined as the ith-layer neighbours. Let $S_i^{(k,m)}$ denote the set of ith-layer neighbours of code (k, m). Then,

$$S_i^{(k,m)} = \{(k, m - p + q) \,|\, p = (m-1) \; mod \; 2^i, 0 \leq q \leq 2^i - 1\} \; . \tag{1}$$

Take code $(K, 3)$ in Fig. 1 as an example, the sets of 1st- and 2nd-layer neighbours are $S_1^{(K,3)} = \{(K, 3), (K, 4)\}$ and $S_2^{(K,3)} = \{(K, 1), (K, 2), (K, 3), (K, 4)\}$, respectively. The positional relationship between (k, m) and other layer-k codes are represented by the k different sets $S_i^{(k,m)}$ $(1 \leq i \leq k)$.

2.1 Assignable Codes

Consider code (k, m). When it is assigned to carry a realtime call, we stipulate that code (k, m) and all its ancestor and descendant codes are *non-preemptable.* When code (k, m) is assigned to carry a best-effort data packet, we stipulate that code (k, m) and all its descendant codes are *preemptable.* In this case, an ancestor code of (k, m) is also *preemptable* if it is not *non-preemptable.* Therefore, an ancestor code is *non-preemptable* if it has both *non-preemptable* and *preemptable* descendant codes. Preemptable codes can be only assigned and reassigned to realtime calls by suspending some ongoing packet transmissions. Besides *non-preemptable* and *preemptable* codes, all remaining codes in the tree are *assignable.* They can be freely assigned to carry either realtime calls or data packets. More importantly, assignable codes have the following properties.

Property 1: If code (k, m) (where $k \leq K - 1$) is assignable, so are all its descendant codes [13].

Property 2: If all the leaf descendant codes of code (k, m) (where $k \leq K - 1$) are assignable, so is code (k, m).

These assignable codes, preemptable codes and non-preemptable codes form a partition of the code tree. They can be characterized by the status index $I^{(k,m)}$, defined as

$$I^{(k,m)} = \begin{cases} 0, & \text{code } (k,m) \text{ is non-preemptable ;} \\ 1, & \text{code } (k,m) \text{ is preemptable ;} \\ 2, & \text{code } (k,m) \text{ is assignable .} \end{cases} \quad (2)$$

As an example, Fig. 2 shows the status index values of all codes in a 4-layer code tree.

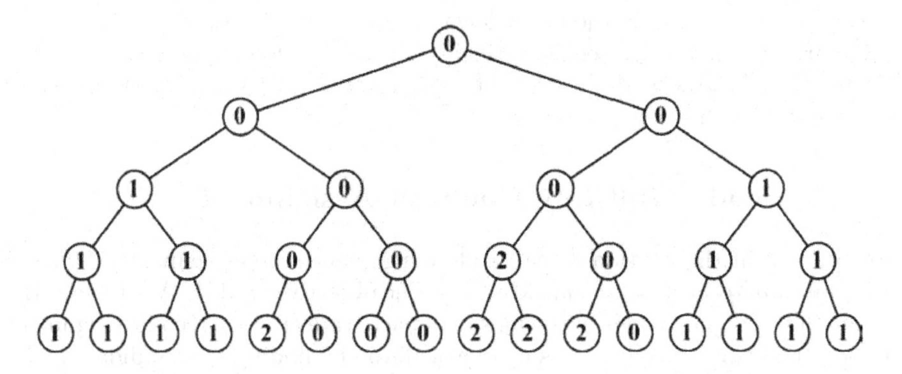

Fig. 2. Status index in a 4-layer code tree. Codes $\{(3,4),(4,6),(4,12)\}$ are carrying realtime calls, and codes $\{(2,1),(2,4)\}$ are carrying data packets.

Upon receiving a new transmission request (realtime call or data packet), the base station needs to first identify all candidate codes suitable for assignment.

Let S denote the set of all candidate codes in the tree and let S_K denote the set of leaf candidate codes in layer K. For data packets, S and S_K consist of assignable codes only. But for realtime calls, preemptable codes are also included in S and S_K since realtime calls have priority over data packets in code assignment. In other words, S_K is given by

$$S_K = \begin{cases} \{(K,m) \mid I^{(K,m)} = 2,\ 1 \le m \le 2^K\}, & \text{for data packets ;} \\ \{(K,m) \mid I^{(K,m)} \ge 1,\ 1 \le m \le 2^K\}, & \text{for realtime calls .} \end{cases} \tag{3}$$

According to *Property 2*, candidate code set S can be derived from S_K [1].

2.2 Compact Index $g^{(k,m)}$

To expand the capability of the code tree in supporting different data rates, new code assignments should be packed as tightly as possible into the existing busy codes, i.e. the candidate codes in the most congested positions are used to carry the new calls/packets. In this way, the code tree is kept as compact (and hence flexible for accommodating multiple data rates) as possible after each code assignment.

The candidate codes in the most congested positions can be identified by their *compact index* $g^{(k,m)}$, which is defined as the total number of candidate codes in the k different neighbourhoods of code (k,m) [13].

$$g^{(k,m)} = \sum_{i=1}^{k} |S_i^{(k,m)} \cap S| \ , \tag{4}$$

where $|x|$ denotes the size of set x. Given layer k, a smaller value of $g^{(k,m)}$ implies that candidate code (k,m) is surrounded by less number of other candidate codes in the same layer and is, therefore, located in a more congested position. For a newly arrived data packet seeing the code tree shown in Fig. 2, we have $g^{(4,5)} = 7$, $g^{(4,9)} = g^{(4,10)} = 12$ and $g^{(4,11)} = 11$, which implies code $(4,5)$ is in the most congested position.

3 Multicode Multirate Compact Assignment

We propose in this section a multicode assignment scheme, namely *Multicode Multirate Compact Assignment* (MMCA). The objective of MMCA is to keep the remaining candidate codes in the most compact state after each code assignment without rearranging codes. This can be achieved by finding the candidate codes in the most congested positions for the new calls/packets. In summary, MMCA is a natural extension of *Compact Assignment* (CA) [13] with the following features.

1. MMCA does not perform code rearrangement and is therefore simple.

[1] The mapping from S_K to S is a bijection.

2. MMCA provides priority differentiation between realtime calls and data packets.
3. MMCA supports mobile terminals with different multicode transmission capabilities.
4. MMCA balances transmission quality among the multiple codes assigned to the same user.
5. MMCA supports multirate realtime calls and keeps the code tree as flexible as possible in accepting new multirate calls.

3.1 Multicode Solutions

For a mobile terminal requiring bandwidth (or data rate) $j \cdot R$ and can transmit n codes, several code assignment solutions may be available for use. A solution, denoted by $(d_0, d_1, \cdots, d_K)$, consists of $K+1$ integers with d_k representing the number of candidate codes needed in layer k. The set of all solutions can be obtained by enumerating all integer-combinations under the constraints of bandwidth requirement and multicode transmission capability, i.e. $\sum_{k=0}^{K} d_k \cdot 2^{K-k} = j$ and $\sum_{k=0}^{K} d_k \leq n$.

We propose for use a more efficient algorithm called *Multicode Solution Generator*. It starts from the solution $(0, 0, \cdots, 0, j)$, which requires j leaf candidate codes. The next solution can be obtained by replacing two leaf codes by one $(K-1)$-layer code in the first solution, i.e. $(0, 0, \cdots, 1, j-2)$. Continuing this way, all possible multicode solutions satisfying the bandwidth requirement can be obtained. Next, we use multicode transmission capability to screen out solutions requiring more than n codes [2].

As an example, consider a 4-layer code tree with each solution represented by five integers. Table 1 lists all multicode solutions for different combinations of bandwidth requirement (from $j = 1$ to $j = 16$) and multicode transmission capability (from $n = 1$ to $n = 6$) [3].

As seen in Table 1, for some combinations of j and n, e.g. $j = 15$ and $n = 2$, no solution exists. These cases are marked by symbol "–" in the table. On the other hand, given j, mobile terminal with a larger n may have more choices in multicode assignment. As an example, for the case $j = 6$ and $n = 3$, there are three multicode solutions: $(0,0,1,1,0)$, $(0,0,0,3,0)$ and $(0,0,1,0,2)$. Intuitively, the solution requiring the least number of candidate codes (with large code capacity), i.e. $(0,0,1,1,0)$, is appealing. However, we found that the solution requiring a larger number of codes (with small code capacity) is more "system-friendly". Here are the reasons:

Reason 1: It is usually much easier to find small-capacity candidate codes for assignment, especially when the system is heavy loaded.

[2] Another approach using dynamic programming technique is given in [18].

[3] For simplicity, $(d_0, d_1, d_2, d_3, d_4)$ is represented by "$d_0d_1d_2d_3d_4$" in this table.

Table 1. Multicode solutions.

Data Rate	Multicode Transmission Capability					
	1	2	3	4	5	6
1R	00001					
2R	00010	00002				
3R	–	00011	00003			
4R	00100	00020	00012	00004		
5R	–	00101	00021	00013	00005	
6R	–	00110	00030 00102	00022	00014	00006
7R	–	–	00111	00031 00103	00023	00015
8R	01000	00200	00120	00040 00112	00032 00104	00024
9R	–	01001	00201	00121	00041 00113	00033 00105
10R	–	01010	00210 01002	00130 00202	00050 00122	00042 00114
11R	–	–	01011	00211 01003	00131 00203	00051 00123
12R	–	01100	00300 01020	00220 01012	00140 00212 01004	00060 00132 00204
13R	–	–	01101	00301 01021	00221 01013	00141 00213 01005
14R	–	–	01110	00310 01030 01102	00230 00302 01022	00150 00222 01014
15R	–	–	–	01111	00311 01031 01103	00231 00303 01023
16R	10000	02000	01200	00400 01120	00320 01040 01112	00240 00312 01032 01104

Reason 2: The use of small-capacity codes can offer better transmission quality because they have larger spreading factors.

Reason 3: There are more small-capacity candidate codes in the congested positions. They should be used first so as to keep the resulting code tree as flexible as possible in supporting different data rates.

When there are multiple choices for assigning the same number of candidate codes, we choose the solution with the minimum variance in code capacity (or spreading factor) so as to balance the transmission quality of these codes. Thus for the above example with $j = 6$ and $n = 3$, two suitable solutions both requiring three candidate codes, i.e. $(0, 0, 0, 3, 0)$ and $(0, 0, 1, 0, 2)$, are identified. Multicode solution $(0, 0, 0, 3, 0)$ (which requires three candidate codes with capacity $2R$) is chosen as it has smaller capacity variance.

3.2 Code Assignment

Upon receiving the transmission request from a particular mobile terminal, the base station first calculates the system assignable capacity according to traffic class. In unit of R, assignable capacity r is defined as

$$r = |S_K| = \begin{cases} \sum_{m=1}^{2^K} \left\lfloor \frac{I^{(K,m)}}{2} \right\rfloor, & \text{for data packets ;} \\ \sum_{m=1}^{2^K} \left\lceil \frac{I^{(K,m)}}{2} \right\rceil, & \text{for realtime calls .} \end{cases} \qquad (5)$$

where $\lfloor x \rfloor$ and $\lceil x \rceil$ denote the floor and the ceiling functions, respectively. For the code tree shown in Fig. 2, assignable capacity r is equal to 4 for data packets and 12 for realtime calls.

For transmitting a data packet, the mobile terminal does not need to specify the bandwidth requirement. When assignable capacity r is non-zero, it is used as "Bandwidth Requirement" (i.e. let $j = r$) in the code assignment algorithm. In other words, we apply the "greedy" policy and try to use all assignable capacity for data packet transmission so as to achieve full system utilization. Under the constraint n of multicode transmission capability, all multicode solutions can be pre-computed by the *Multicode Solution Generator* and stored in a table such as Table 1. If no solution exists (indicated by symbol "–" in the table) for some combinations of j and n, the code assignment algorithm will *reduce* the value of j gradually until the first solution is identified. On the other hand, when assignable capacity is zero, the data packet transmission request is put into a queue at the base station if the queue size limit is not exceeded, or blocked otherwise.

Now consider a realtime call request from a mobile terminal with bandwidth requirement j and multicode transmission capability n. When $r \geq j$, the base station performs code assignment assuming the absence of data packet traffic. Some data packets may need to reduce their transmission data rates, or even totally suspend their transmissions, to make codes available for the newly arrived realtime call. For some combinations of j and n, no solution exists and symbol "–" is observed in the table of multicode solutions. The code assignment algorithm will then *increase* j gradually until the first solution is identified. After a particular solution is chosen according to the criteria given in Section 3.1, compact index $g^{(k,m)}$ is used for code selection and assignment in each layer.

A realtime call request will be blocked if the system cannot meet the bandwidth or multicode requirements. Specifically, there are three blocking conditions.

Condition 1: The required bandwidth is larger than the assignable capacity, i.e. $j > r$.

Condition 2: $j \leq r$, but the summed bandwidth of every multicode solution is larger than the assignable capacity, i.e. $\sum_{k=0}^{K} d_k \cdot 2^{K-k} > r$.

Condition 3: $j \leq r$ and $\sum_{k=0}^{K} d_k \cdot 2^{K-k} \leq r$, but the candidate codes found in some layers are not sufficient, i.e. the number is less than d_k.

To illustrate, consider the code tree shown in Fig. 2. For realtime calls, the assignable capacity $r = 12$. However, a new call request with $j = 10$ and $n = 1$ will be blocked due to *Condition 2* (the identified solution $(1, 0, 0, 0, 0)$ has summed bandwidth of $16R$). Another request with $j = 12$ and $n = 3$ will be blocked due to *Condition 3* (all multicode solutions, namely $(0, 1, 1, 0, 0)$, $(0, 0, 3, 0, 0)$ and $(0, 1, 0, 2, 0)$, cannot be supported by the code tree).

The blockings due to *Condition 1* are unavoidable. The blockings due to *Condition 2* can be avoided only by improving the mobile terminal's multicode transmission capability. The blockings due to *Condition 3* can be avoided by either rearranging codes or improving multicode capability. For example, in Fig. 2, if the realtime call on code $(4, 12)$ is reassigned to code $(4, 5)$, a realtime call request with $j = 12$ and $n = 3$ can then be carried in the code tree by suspending all ongoing data packet transmissions. As seen in Table 1, when mobile terminals are multicode capable, a number of multicode solutions are usually available. *Condition 3* of blocking is therefore much less likely to occur, compared to the single-code transmission scenario.

3.3 Data Rate Reduction and Transmission Resumption

As data packet transmissions can be preempted by realtime calls, some mobile terminals have to reduce their transmission data rates to make codes available for realtime calls. For the mobile terminals that totally suspend the data transmissions, their identifications and the corresponding break points are recorded at the base station. When some occupied codes are released, they will be shared by these suspended terminals as fairly as possible.

4 Performance Analysis

Let there be N types of mobile terminals in the system where the type-n ($1 \leq n \leq N$) terminals can support the simultaneous transmission of n codes. Let p_n

be the fraction of type-n terminals. Further, let there be J classes of realtime calls where the class-j ($1 \le j \le J$) calls are characterized by *(i)* Poisson arrivals with rate λ_j; *(ii)* bandwidth requirements equal to $j \cdot R$; and *(iii)* exponentially distributed call holding time with mean μ_j^{-1}. Let $G_j = \lambda_j/\mu_j$ ($1 \le j \le J$) denote the offered traffic of class-j realtime calls. The total offered traffic G_R of realtime calls is simply the sum of G_j. For simplicity, we assume terminal type and service class are independent. Let λ_D and μ_D^{-1} denote the arrival rate and average packet length of data packets, respectively. The offered traffic of data packets is therefore given by $G_D = \lambda_D/\mu_D$.

Without loss of generality, a six-layer code tree ($K = 6$) and eight classes realtime calls ($J = 8$) with equal offered traffic ($G_1 = G_2 = \cdots = G_8$) are considered in the computer simulation. The arrival of data packets is assumed to be a Poisson process and the packet length is chosen from four exponential random variables with means R, $2R$, $4R$ and $8R$ with equal probabilities. Let there be four types of mobile terminals ($N = 4$) and let their combinations take on the following four cases.

Case 1: $p_1 : p_2 : p_3 : p_4 = 100 : 0 : 0 : 0$.
Case 2: $p_1 : p_2 : p_3 : p_4 = 40 : 30 : 20 : 10$.
Case 3: $p_1 : p_2 : p_3 : p_4 = 25 : 25 : 25 : 25$.
Case 4: $p_1 : p_2 : p_3 : p_4 = 10 : 20 : 30 : 40$.

Note that multicode transmission is not supported in *Case 1*. In the following figures, all simulation results are shown in dashed lines with markers. For each simulation experiment, the simulation time is increased until the 95% confidence interval is comparable to the marker size shown.

4.1 Blocking Probability of Realtime Calls

Blocking probability is the most important measure of QoS for realtime calls. Since the realtime calls have preemptive priority over data packets, as far as blocking performance is concerned, data packets are completely transparent to realtime calls. Consider the ideal case where all mobile terminals can use as many codes as required, i.e. $n = J$. Then, call blockings due to *Conditions 2* and *3* (section 3.2) can be completely avoided. The blocking probability in this case is the same as that under the "complete sharing policy" in shared resource environment [19]. This blocking result is therefore a lower bound (see Bound A in Fig. 3) for the restrictive multicode cases studied here. Due to the length limit, the derivation details of this lower bound are not shown in this paper.

Fig. 3 shows the overall blocking probability as a function of realtime offered traffic G_R. The solid lines are the analytical lower bounds. The blocking probabilities of the four cases discussed in Section 4 are obtained by computer simulation. As seen, the overall blocking probability can be significantly reduced with the use of multicode. As an example, at $G_R = 5.6$ (Erlang), the blocking probabilities for the four simulation cases and the analytical lower bound (marked as Bound A) are 2.21%, 1.14%, 0.92%, 0.58% and 0.45%, respectively. This lower

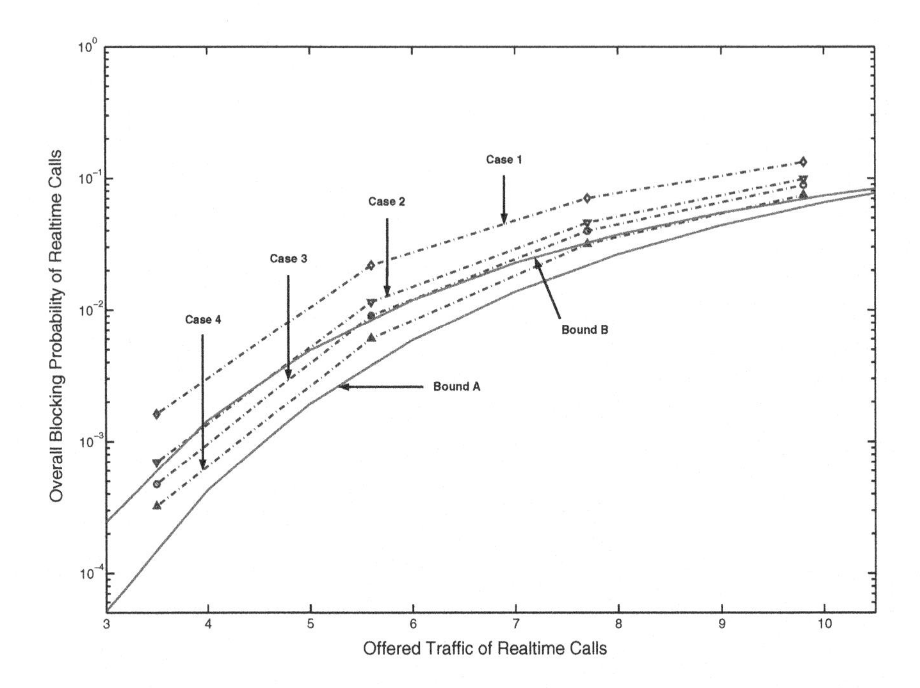

Fig. 3. Blocking probability of realtime calls.

bound can be achieved by letting all mobile terminals capable of transmitting any number of codes. This result indicates that the use of multicode can be an effective alternate to the rearrangeable single-code method in [13]. For comparison purpose, the lower bound for rearrangeable single-code system is also shown (marked as Bound B).

4.2 Throughput and Wasted Capacity of Realtime Calls

The throughput of realtime calls, denoted by T, is given by

$$T = \sum_{j=1}^{J} (1 - P_j) \cdot L_j \ , \tag{6}$$

where P_j and $L_j \triangleq j \cdot G_j$ denote the blocking probability and the offered load of class-j realtime calls, respectively. The total offered load L of realtime calls is simply the summation of L_j. The throughput in (6) gives the time-averaged required bandwidth from successful realtime terminals. The total assigned capacity for realtime calls may be larger. For example, to accommodate a type-1 realtime terminal with bandwidth requirement $6R$, the base station needs to assign a layer-$(K-3)$ code (with code capacity $8R$) to the terminal. The gap between these two values is called "wasted capacity".

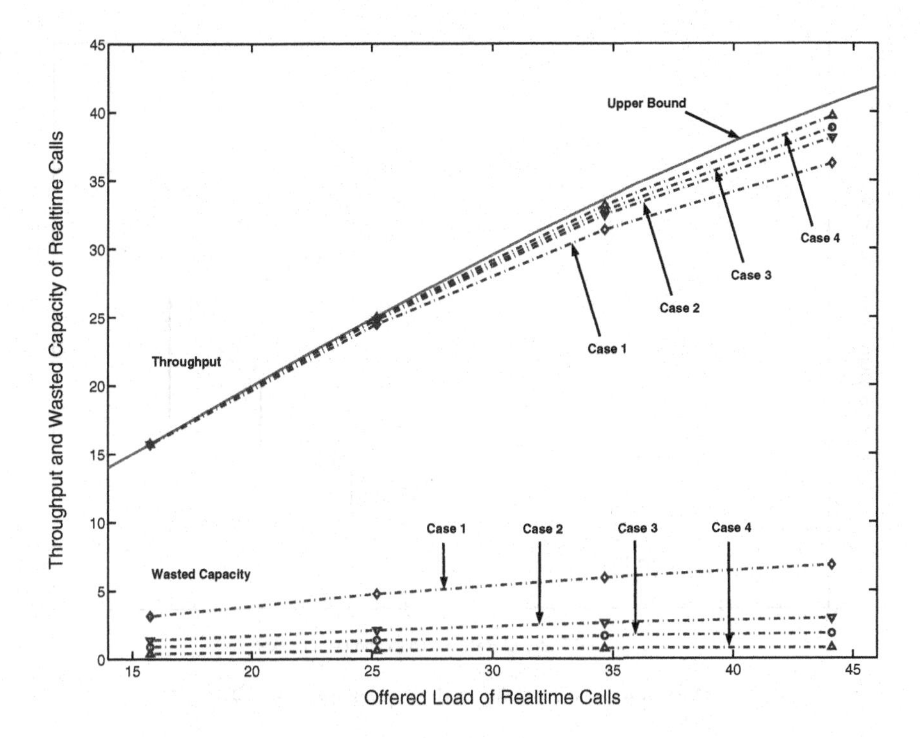

Fig. 4. Throughput and wasted capacity of realtime calls.

Fig. 4 shows the throughput and wasted capacity of realtime calls as a function of offered load L. The solid line is the analytical upper bound, or (6), on throughput. As seen, this upper bound can be approached by introducing more multicode-capable terminals. The same action can also reduce the amount of wasted capacity. In the limiting case where all terminals are capable of transmitting any number of codes, the wasted capacity is zero. Specifically, at offered load $L = 34.65$, the wasted capacity values in four simulation cases are 5.85, 2.65, 1.69 and 0.76, respectively.

4.3 Sojourn Time of Data Packets

For data packets, the average sojourn time is the most important QoS measure. It is defined as the time between a transmission request and the successful transmission of the whole packet. Fig. 5 shows the average sojourn time as a function of total offered traffic $G_R + G_D$. We assume in this case the ratio between realtime and data traffic is fixed at $G_R : G_D = 7 : 3$. The performance of the three multicode cases are similar and are all about 30% better than the single-code case, *Case 1*. This indicates that the sojourn time cannot be effectively reduced by manipulating the multicode capability mixes. As an example, at offered traffic $G_R + G_D = 11$, the average sojourn time values for the four cases are 1.71, 1.24, 1.20 and 1.16, respectively. By Little's formula, the same conclusion can be drawn on queue length.

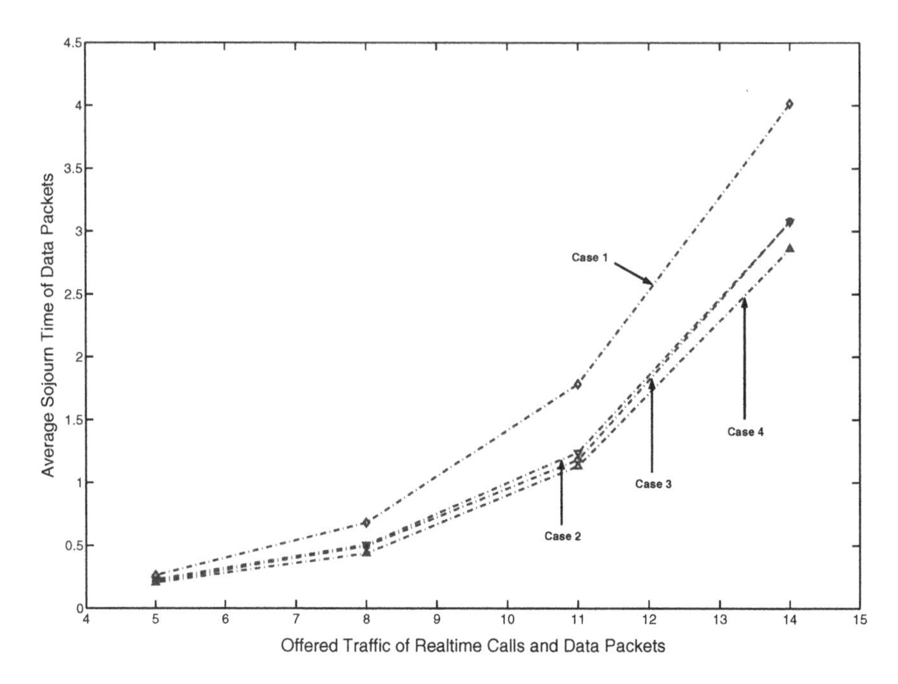

Fig. 5. Average sojourn time of data packets, $G_R : G_D = 7 : 3$.

4.4 Fairness Comparison

The major fairness concern is the chance of accessing system resource for different types of terminals with different bandwidth requirements. As an example, the fairness index for the terminals with different bandwidth requirements, denoted by F_R, is defined as [13]

$$F_R = \frac{\left[\sum_{j=1}^{J} (1 - P_j)\right]^2}{J \sum_{j=1}^{J} (1 - P_j)^2} \ . \tag{7}$$

In the ideal case where the terminals with different bandwidth requirements have the same opportunity of being served (i.e. the P_j values are equal), F_R achieves the maximum value of one.

Fig. 6 shows fairness index F_R as a function of offered traffic of realtime calls. As seen, even under heavy traffic, there is no substantial difference in the fair access among the realtime terminals with different bandwidth requirements. Although not shown, our results show that the same is true for the terminals with different multicode transmission capabilities.

5 Conclusions

Based on the concept of compact index, a new OVSF code assignment scheme, namely "Multicode Multirate Compact Assignment" (MMCA), is proposed for

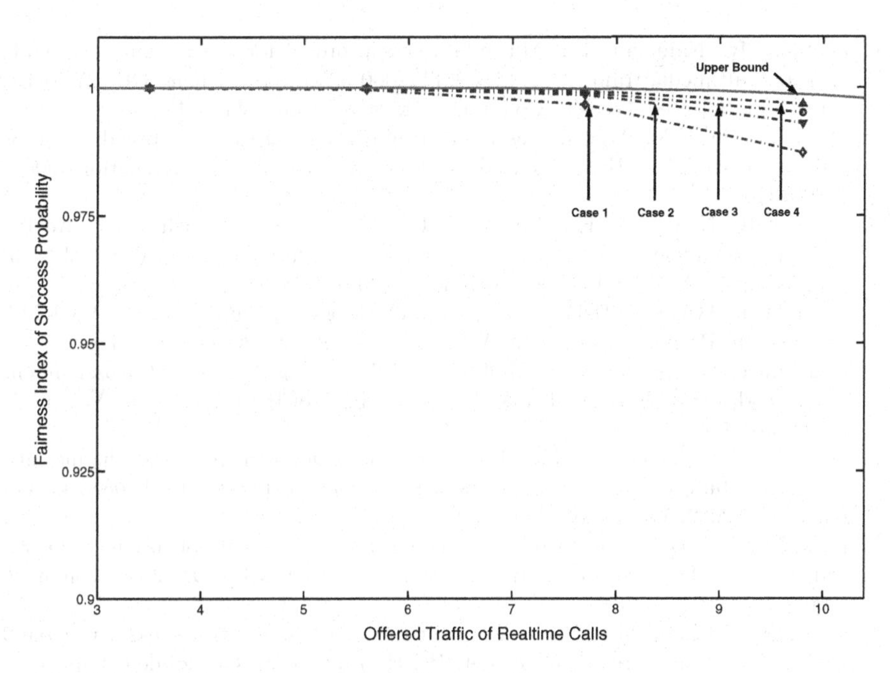

Fig. 6. Fairness index for realtime calls with different bandwidth requirements.

accommodating QoS differentiated mobile terminals. These terminals have different multicode transmission capabilities. They can also support different traffic types (realtime calls and data packets) with different priority and bandwidth requirements. When more mobile terminals have multicode transmission capability, the bandwidth granularity in code assignment becomes smaller and the system is more flexible in supporting multirate multimedia traffic classes. As a result, higher bandwidth efficiency is observed in MMCA. This is demonstrated by both analysis and simulation. In addition, MMCA is also shown to be a fair code assignment scheme for different service classes.

References

1. Adachi, F., Sawahashi, M., Okawa, K.: Tree-structured generation of orthogonal spreading codes with different lengths for forward link of DS-CDMA mobile radio. Electronics Letters**33** (1997) 27–28
2. 3GPP TS 25.213 (V6.0.0): Spreading and modulation (FDD). Technical Specification (Release 6), Technical Specification Group Radio Access Network, 3GPP (2003)
3. 3GPP TS 25.223 (V6.0.0): Spreading and modulation (TDD). Technical Specification (Release 6), Technical Specification Group Radio Access Network, 3GPP (2003)

4. Fantacci, R., Nannicini, S.: Multiple access protocol for integration of variable bit rate multimedia traffic in UMTS/IMT-2000 based on wideband CDMA. IEEE Journal on Selected Areas in Communications**18** (2000) 1441–1454
5. Minn, T., Siu, K.Y.: Dynamic assignment of orthogonal variable-spreading-factor codes in W-CDMA. IEEE Journal on Selected Areas in Communications(2000) 1429–1440
6. Assarut, R., Kawanishi, K., Yamamoto, U., Onozato, Y., Matsushita, M.: Region division assignment of orthogonal variable-spreading-factor codes in W-CDMA. In: Proceedings of IEEE VTC 2001. Volume 3. (2001) 1884–1888
7. Chen, W.T., Wu, Y.P., Hsiao, H.C.: A novel code assignment scheme for W-CDMA systems. In: Proceedings of IEEE VTC2001. Volume 2. (2001) 1182–1186
8. Dell'Amico, M., Merani, M.L., Maffioli, F.: Efficient algorithms for the assignment of ovsf codes in wideband CDMA. In: Proceedings of IEEE ICC 2002. Volume 5. (2002) 3055–3060
9. Fossa Jr., C.E., Davis IV, N.J.: Dynamic code assignment improves channel utilization for bursty traffic in third-generation wireless networks. In: Proceedings of IEEE ICC 2002. Volume 5. (2002) 3061–3065
10. Park, J.S., Lee, D.C.: On static and dynamic code assignment policies in the OVSF code tree for CDMA networks. In: Proceedings of IEEE MILCOM 2002. Volume 2. (2002) 785–789
11. Tseng, Y.C., Chao, C.M.: Code placement and replacement strategies for wideband CDMA OVSF code tree management. IEEE Transactions on Mobile Computing**1** (2002) 293–302
12. Rouskas, A.N., Skoutas, D.N.: OVSF codes assignment and reassignment at the forward link of W-CDMA 3G systems. In: Proceedings of IEEE PIMRC 2002. Volume 5. (2002) 2404–2408
13. Yang, Y., Yum, T.S.P.: Maximally flexible assignment of orthogonal variable spreading factor codes for multirate traffic. IEEE Transactions on Wireless Communications**3** (2004) 781–792
14. Sekine, Y., Kawanishi, K., Yamamoto, U., Onozato, Y.: Hybrid OVSF code assignment scheme in W-CDMA. In: Proceedings of 2003 IEEE Pacific Rim Conference on Communications, Computers and Signal Processing. Volume 1. (2003) 384–387
15. Cheng, R.G., Lin, P.: OVSF code channel assignment for IMT-2000. In: Proceedings of IEEE VTC2000. Volume 3. (2000) 2188–2192
16. Chao, C.M., Tseng, Y.C., Wang, L.C.: Reducing internal and external fragmentations of OVSF codes in WCDMA systems with multiple codes. In: Proceedings of IEEE WCNC 2003. Volume 1. (2003) 693–698
17. Shueh, F., Chen, W.S.E.: Code assignment for IMT-2000 on forward radio link. In: Proceedings of IEEE VTC 2001. Volume 2. (2001) 906–910
18. Yen, L.H., Tsou, M.C.: An OVSF code assignment scheme utilizing multiple rake combiners for W-CDMA. In: Proceedings of IEEE ICC 2003. Volume 5. (2003) 3312–3316
19. Kaufman, J.S.: Blocking in a shared resource environment. IEEE Transactions on Communications**COM-29** (1981) 1474–1481

Performance Analysis and Receiver Design for SDMA-Based Wireless Networks in Impulsive Noise

Anxin Li, Chao Zhang, Youzheng Wang, Weiyu Xu, and Zucheng Zhou

Department of Electronic Engineering, Tsinghua University,
Beijing, People's Republic of China,
lax00@mails.tsinghua.edu.cn

Abstract. In this paper, performance analysis and receiver design in the uplink of SDMA-based wireless systems are performed in the presence of impulsive noise modeled as a symmetric alpha-stable process. The optimal receiver and several suboptimal receivers are proposed and the symbol-error-rates (SER) or upper bound of SER of the receivers are derived. Simulation results show the proposed receivers can achieve significant performance gain compared with conventional detectors in SDMA-based wireless systems.

1 Introduction

Space Division Multiple Access (SDMA) has been proposed recently as a promising technique to satisfy the growing demand for system capacity and spectral efficiency in wireless Networks, such as wireless LANs, GSM, third-generation (3G) networks and beyond [1]. Most of the analyses of SDMA-based systems so far have been based on the ideal Gaussian noise model [1]-[3]. However,many sources, such as automobile ignitions, electric-devices, radiation from power lines and multiple access interference, will cause the noise in many actual channels to be impulsive [4]-[6].

Among the many impulsive noise models suggested so far, such as Laplace distribution, Generalized Gaussian Distribution and student-t distribution, the family of alpha stable distribution is especially attractive [4]-[8]. This is mainly due to the Generalized Central Limit Theorem (GCLT), which indicates that the stable distribution arises in the same way as the Gaussian distribution does and can describe the noise resulting from a large number of impulsive effects. The alpha stable distribution can describe the impulsive noise and actually includes the Gaussian distribution as a special case.

In this paper, we consider the performance and receiver design in the uplink of SDMA-based systems in alpha stable impulsive noise. On the one hand, many detectors developed for Multi-Input Multi-Output (MIMO) systems and CDMA systems can be extended to SDMA-based systems, such as maximum likelihood (ML) detector and VBLAST detector [3][9]. But almost all of them are based on Gaussian noise assumption, so their performances are greatly degraded when

I. Niemegeers and S. Heemstra de Groot (Eds.): PWC 2004, LNCS 3260, pp. 379–388, 2004.

impulsive noise appears. On the other hand, some enhanced receivers are developed according to the statistical characteristic of the alpha stable noise in Single-Input Single-Output (SISO) systems [4][5][8]. But they cannot be applied to SDMA-based systems because they do not take into account the multiple access interference (MAI). In this paper, we develop receivers of SDMA-based systems on the basis of impulsive noise assumption and analyze their performance.

2 System Model

2.1 Model of the Uplink of SDMA-Based Systems

Consider the uplink of a SDMA-based wireless system. The system has M terminals and the base station (BS) has N receive antennas. The channels between terminals and receive antennas are assumed to be independent Rayleigh flat fading with the symmetric alpha-stable ($S\alpha S$) impulsive noise. The channels' transfer coefficients are assumed to be known by channel estimation at the BS. The system can be described as follows:

$$\mathbf{r}(l) = \mathbf{H}(l)\mathbf{x}(l) + \mathbf{n}(l) \tag{1}$$

where $\mathbf{x}(l) \in \mathcal{C}^{M\times 1}$ has entries $x_m(l), m = 1, \ldots, M$, being the signal transmitted from terminal m at time l; $\mathbf{H}(l) \in \mathcal{C}^{N\times M}$ has entries $h_{nm}(l), n = 1, \ldots, N, m = 1, \ldots, M$, being the complex channel transfer coefficient from terminal m to receiver antenna n; $\mathbf{r}(l) \in \mathcal{C}^{N\times 1}$ has entries $r_n(l), n = 1, \ldots, N$, being the signal received from receive antenna n; and $\mathbf{n}(l) \in \mathcal{C}^{N\times 1}$ has the entries $n_n(l), n = 1, \ldots, N$, being the $S\alpha S$ impulsive noise observed at receive antenna n.

2.2 Model of the Alpha Stable Impulsive Noise

The elements of $\mathbf{n}(l)$ are modeled as independently and identically distributed complex $S\alpha S$ random variables, that is $\forall n_i(l) = \Re(n_i) + j\Im(n_i), 1 \le i \le N$, $\Re(n_i)$ and $\Im(n_i)$ obey the bivariate joint $S\alpha S$ distribution. The probability density function (pdf) of $n_i(l)$ can be written as

$$\begin{aligned} &f_{\alpha,\gamma}\left(\Re(n_i), \Im(n_i)\right) \\ &= \frac{1}{4\pi^2} \int_{-\infty}^{\infty}\int_{-\infty}^{\infty} \exp(-\gamma \left|\omega_1^2 + \omega_2^2\right|^{\frac{\alpha}{2}}) \exp(-j\omega_1\Re(n_i)) \exp(-j\omega_2\Im(n_i)) d\omega_1 d\omega_2 \end{aligned} \tag{2}$$

where $\alpha \in (0, 2]$ is *characteristic exponent*, which implies the impulsiveness of the respective $S\alpha S$ noise. The larger the value of α is, the less impulsive the $S\alpha S$ noise is, and when $\alpha = 2$, the $S\alpha S$ impulsive noise is reduced to the Gaussian noise. The parameter $\gamma \in (0, \infty)$ is *dispersion*, which plays an analogous role to the variance of Gaussian distribution. For simplicity, we define special operator $\|\cdot\|^2_{F(\alpha,\gamma)}$ to denote (2)

$$f_{\alpha,\gamma}\left(\Re(n_i), \Im(n_i)\right) = \|n_i(l)\|^2_{F(\alpha,\gamma)} = \|\Re(n_i) + j\Im(n_i)\|^2_{F(\alpha,\gamma)} \tag{3}$$

Unfortunately, there are no closed form expressions for general complex $S\alpha S$ random variables except for the Gaussian ($\alpha = 2$) distribution and the Cauchy ($\alpha = 1$) distribution

$$Gaussian: \|\Re(n_i) + j\Im(n_i)\|^2_{F(2,\gamma)} = \frac{1}{4\pi\gamma}\exp\left(-\frac{\Re(n_i)^2 + \Im(n_i)^2}{4\gamma}\right) \quad (4)$$

$$Cauchy: \|\Re(n_i) + j\Im(n_i)\|^2_{F(1,\gamma)} = \frac{\gamma}{2\pi}\left(\sqrt{\Re(n_i)^2 + \Im(n_i)^2 + \gamma^2}\right)^{-3} \quad (5)$$

This leads to difficulties in performance analysis and receiver design in the alpha stable noise.

3 Receiver Design and Performance Analysis

In this section several receivers are proposed on the basis of $S\alpha S$ noise model and the SER or the upper bound of the SER of the receivers are derived.

3.1 The Zero-Forcing Receiver (ZF)

The zero-forcing receiver decorrelates the received signal vector to cancel the multiple access interference (MAI) and carries out the hard decisions according to the statistical characteristics of the noise, which can be formulated as

$$\tilde{\mathbf{x}}(l) = (\mathbf{H}(l)^H\mathbf{H}(l))^{-1}\mathbf{H}(l)^H \cdot \mathbf{r}(l) = \mathbf{x}(l) + (\mathbf{H}(l)^H\mathbf{H}(l))^{-1}\mathbf{H}(l)^H \cdot \mathbf{n}(l) \quad (6)$$

where $(\cdot)^H$ denotes the conjugate transpose. Due to the *stable property*, the elements of the vector $(\mathbf{H}(l)^H\mathbf{H}(l))^{-1}\mathbf{H}(l)^H \cdot \mathbf{n}(l)$ obey the stable distribution with parameters α and γ_{eq} ([10], pp.35). So the ZF receiver can be obtained as

$$\hat{x}_m(l)_{ZF} = \arg\max_{s_q \in \mathcal{C}^Q}\left(\|\tilde{x}_m(l) - s_q\|^2_{F(\alpha,\gamma_{eq})}\right), 1 \le m \le M \quad (7)$$

where $\hat{x}_m(l)_{ZF}$ denotes the (hard) estimate of the symbol transmitted from the mth terminal; $\tilde{x}_m(l)$ denotes the mth element of $\tilde{\mathbf{x}}(l)$; and $\left\{s_q \in \mathcal{C}^Q \middle| q = 1, \ldots, Q\right\}$ denotes the transmitted symbol taken from a discrete constellation.

If QPSK is adopted at terminals and the transmit power of terminals is E_s. The SER for ZF receiver can be calculated as

$$\begin{aligned} SER_{ZF} &= \frac{1}{M}\sum_{m=1}^{M} P\left(\hat{x}_m(l) \ne x_m(l)\right) = \frac{1}{M}\sum_{m=1}^{M}\left(1 - P\left(\hat{x}_m(l) = x_m(l)\right)\right) \\ &= \frac{1}{M}\sum_{m=1}^{M}\left(1 - \sum_{q=1}^{4} P\left(s_q\right) P\left(\hat{x}_m(l) = s_q \left| s_q\ sent\right.\right)\right) \end{aligned} \quad (8)$$

If the transmitted symbols are equal probability, namely $P\left(s_q\right) = 1/4$, we have

$$SER_{ZF} = \frac{1}{M}\sum_{m=1}^{M}\left(1 - P\left(\hat{x}_m(l) = s_q \left| s_q\ sent\right.\right)\right) \quad (9)$$

where

$$P\left(\hat{x}_m(l) = s_q \left| s_q \; sent \right.\right)$$
$$= P\left(\arg\max_{s_i}\left(\|\tilde{x}_m(l) - s_i\|^2_{F(\alpha,\gamma_{eq})}\right) = s_q \left| s_q \; sent \right.\right)$$
$$= \int_0^\infty \int_0^\infty \left(\|\tilde{x}_m(l) - s_q\|^2_{F(\alpha,\gamma_{eq})}\right) d\Re\left(\tilde{x}_m(l) - s_q\right) d\Im\left(\tilde{x}_m(l) - s_q\right) \quad (10)$$

Since the alpha stable distribution usually has no closed form pdf, (10) usually has no closed form expression. Nevertheless numerical methods can be adopted to evaluate the performance of the ZF receiver by (8)-(10).

3.2 The Optimal Receiver in the Alpha Stable Noise (ML-Alpha)

It is known that the optimal ML receiver performs a computation as follows:

$$\hat{\mathbf{x}}(l)_{ML} = \arg\max_{\mathbf{x}(l)_q \in \mathcal{C}^{Q^M}} \left(p\left(\mathbf{r}(l)|\mathbf{x}(l)_q\right)\right) \quad (11)$$

where $p\left(\mathbf{r}(l)|\mathbf{x}(l)_q\right)$ is the conditional probability density function of the received vector $\mathbf{r}(l)$, given that $\mathbf{x}(l)_q$ is transmitted, and $\mathbf{x}(l)_q$ is taken from the set of all possible transmitted vectors with a size of Q^M, where Q is the size of the constellation. The ML receiver searches all possible transmitted vectors and selects the one which gives the maximum value of conditional pdf.

For a specific channel $\mathbf{H}(l)$ and a given $\mathbf{x}(l)_q$, it is easy to see the received vector $\mathbf{r}(l)$ follows the same distribution as $\mathbf{n}(l)$ (with different parameters). Since the impulsive noise observed at different receive antennas is assumed to be independent, the joint pdf of the impulsive noise can be written as

$$f_N\left(\mathbf{n}(l)\right) = f_N\left(n_1(l), n_2(l), \ldots, n_N(l)\right) = \prod_{n=1}^{N} \|n_n(l)\|^2_{F(\alpha,\gamma)} \quad (12)$$

Use logarithmic likelihood, then the ML detection rule in (11) becomes

$$\hat{\mathbf{x}}(l)_{ML-Alpha} = \arg\max_{\mathbf{x}(l)_q \in \mathcal{C}^{Q^M}} \log\left(f_N\left(\mathbf{r}(l) - \mathbf{H}(l)\cdot\mathbf{x}(l)_q\right)\right)$$
$$= \arg\max_{\mathbf{x}(l)_q \in \mathcal{C}^{Q^M}} \left(\sum_{n=1}^{N} \log\left(\|r_n(l) - \mathbf{H}_n(l)\cdot\mathbf{x}(l)_q\|^2_{F(\alpha,\gamma)}\right)\right) \quad (13)$$

where $\mathbf{H}_n(l)$ denotes the nth row of channel matrix $\mathbf{H}(l)$.

An upper bound on SER for ML-Alpha receiver can be obtained by assuming all possible code words have the same distance. From this point, and after some manipulation, the upper bound of SER can be obtained as

$$SER_{ML-Alpha} \le \frac{Q^M - 1}{M}\sum_{m=1}^{M}\left(1 - \sum_{q=1}^{4} P\left(s_q\right) P\left(\hat{x}_m(l) = s_q \left| s_q \; sent \right.\right)\right)$$
$$= \frac{Q^M - 1}{M}\sum_{m=1}^{M}\left(1 - \sum_{q=1}^{4} P\left(s_q\right)\int_0^\infty\int_0^\infty\left(\|\tilde{x}_m(l) - s_q\|^2_{F(\alpha,\gamma)}\right) d\Re(\tilde{x}_m(l) - s_q)\, d\Im(\tilde{x}_m(l) - s_q)\right) \quad (14)$$

3.3 The Optimal Receiver in the Gaussian Noise (ML-G)

Since Gaussian noise is a special case of alpha stable noise, the ML-G receiver is a special case of ML-Alpha receiver. Take $\alpha = 2$ in (13) and combine with (4), the ML receiver in the Gaussian noise can be obtained as

$$\begin{aligned}\hat{\mathbf{x}}(l)_{ML-G} &= \arg\min_{\mathbf{x}(l)_q \in C^{Q^M}} \sum_{n=1}^{N} \left(\Re^2(r_n(l) - \mathbf{H}_n(l)\cdot \mathbf{x}(l)_q) + \Im^2(r_n(l) - \mathbf{H}_n(l)\cdot \mathbf{x}(l)_q)\right) \\ &= \arg\min_{\mathbf{x}(l)_q \in C^{Q^M}} \left(\left\|\mathbf{r}(l) - \mathbf{H}(l)\cdot \mathbf{x}(l)_q\right\|^2\right) \end{aligned} \tag{15}$$

Formula (15) is in accordance with the receiver developed in [3]. As is shown here it is only a special receiver in the alpha stable noise. It is the optimal receiver in the alpha stable noise when $\alpha = 2$. To derive the upper bound of SER of this receiver, use (4) and for QPSK: $\|s_q\|^2 = E_S, \forall q, 1 \le q \le 4$, thus

$$\begin{aligned} & P\left(\hat{x}_m(l) = s_q \,|\, s_q \ sent\right) \\ &= \frac{1}{4\pi\gamma}\int_0^\infty\int_0^\infty \exp\left(-\frac{\Re^2\left(\tilde{x}_m(l) - s_q\right) + \Im^2\left(\tilde{x}_m(l) - s_q\right)}{4\gamma}\right) \\ &\quad \cdot d\Re\left(\tilde{x}_m(l) - s_q\right) d\Im\left(\tilde{x}_m(l) - s_q\right) \\ &= \left(1 - \frac{1}{2}erfc\left(\frac{\Re(s_q)}{2\sqrt{\gamma}}\right)\right)\left(1 - \frac{1}{2}erfc\left(\frac{\Im(s_q)}{2\sqrt{\gamma}}\right)\right) = \left(1 - \frac{1}{2}erfc\left(\frac{1}{2}\sqrt{\frac{E_s}{2\gamma}}\right)\right)^2 \end{aligned} \tag{16}$$

where $erfc(x) \triangleq \frac{2}{\sqrt{\pi}}\int_x^\infty \exp\left(-t^2\right)dt$.

Therefore the upper bound of ML-G receiver can be got by substituting (16) into (14), and after some simplification

$$SER_{ML-G} \le \left(Q^M - 1\right)\cdot erfc\left(\frac{1}{2}\sqrt{\frac{E_s}{2\gamma}}\right)\cdot\left(1 - \frac{1}{4}erfc\left(\frac{1}{2}\sqrt{\frac{E_s}{2\gamma}}\right)\right) \tag{17}$$

3.4 Cauchy Receiver

The Cauchy receiver is the optimal receiver when Cauchy noise appears. Take $\alpha = 1$ in (13) and combine with (5), the Cauchy receiver can be developed as

$$\hat{\mathbf{x}}(l)_{Cauchy} = \arg\min_{\mathbf{x}(l)_q \in C^{Q^M}} \left(\sum_{n=1}^{N} \log\left(\left\|r_n(l) - \mathbf{H}_n(l)\cdot \mathbf{x}(l)_q\right\|^2 + \gamma^2\right)\right) \tag{18}$$

Comparing (15) with (18), we can see that the Cauchy receiver depends on the *dispersion* of the alpha noise, while ML-G receiver dose not, which is the main reason why their performances are strikingly different in the impulsive noise, as is shown in the simulation results.

For Cauchy receiver, following the same steps as for ML-G, we can get the upper bound as follows:

$$SER_{Cauchy} \leq \frac{Q^M - 1}{M} \sum_{m=1}^{M} \left(1 - \sum_{q=1}^{4} P(s_q) P(\hat{x}_m(l) = s_q | s_q \ sent) \right) \quad (19)$$

where

$$P(\hat{x}_m(l) = s_q | s_q \ sent) = \frac{\gamma}{2\pi} \int_0^\infty \int_0^\infty \left(\Re^2 (\tilde{x}_m(l) - s_q) + \Im^2 (\tilde{x}_m(l) - s_q) + \gamma^2 \right)^{-\frac{3}{2}} \cdot d\Re(\tilde{x}_m(l) - s_q) \, d\Im(\tilde{x}_m(l) - s_q) \quad (20)$$

3.5 VBLAST Receiver

The VBLAST receiver is one of the well-known space-time signal processing algorithm adopting a cancelling and nulling precessing [9]. We include it here for comparison with other receivers.

4 Comparisons of the Receivers

4.1 Performance

The performance of the receivers is mainly determined by two factors. One is the diversity order and the other is to what degree the receivers take into account the statistical characteristic of the noise. (1) ML-Alpha, ML-G and Cauchy Receiver have the same diversity order of N, but due to the different degrees they take into account the statistical characteristic of the alpha stable noise, their performances are different. ML-Alpha considers the pdf of the alpha stable noise, so its performance is the best of the three. The performances of the other two receivers depend on the value of α. (2) When α is away from 2, the Cauchy receiver performs better than ML-G and when α is close to 2, ML-G performs better than Cauchy receiver. (3) The ZF performs the worst in all cases due to the smallest diversity order $N - M + 1$ it has.

Table 1. Performances of Receivers

$\alpha = 1$	ML-Alpha = Cauchy Receiver > ML-G > VBLAST > ZF
$\alpha = 2$	ML-Alpha = ML-G > Cauchy Receiver > VBLAST > ZF
$0 < \alpha < 2$ and $\alpha \neq 1$	ML-Alpha > Cauchy Receiver, ML-G > VBLAST > ZF

Table 2. Complexity of Receivers

	ZF	ML-Alpha	ML-G	Cauchy Receiver	VBLAST
Size of candidate signal set	Q	Q^M	Q^M	Q^M	Q

4.2 Complexity

The complexity of the receivers is shown in Table.2. (1)ML-Alpha is the most complex one, which needs to search a vector set of size Q^M and requires lots of numerical integrals for each candidate vector. Despite the optimal performance ML-Alpha has, the computational complexity prevents it from practical applications. (2) For ML-G and Cauchy Receiver, they have the almost the same computational complexities, because the candidate sets they need to search are of the same size, and the their computational complexities for each candidate vector are approximate the same.

5 Simulation Results

An SDMA-based system with 2 terminals and 2 receiver antennas at BS is simulated. Both terminals adopt the QPSK modulation. The simulation results are plotted as BER vs. Signal-to-Noise-Dispersion Ratio (SNDR) rather than common SNR, since the variance of the impulsive noise does not exist for $\alpha < 2$. The SNDR is defined as $SNDR = \frac{1}{N} \sum_{i=1}^{N} SNDR^i$, where $SNDR^i = \frac{M \cdot Es}{2\gamma}$ is the ratio of received signal power from all M terminals to the dispersion of alpha stable noise at the ith receive antenna. When $\alpha = 2$, the $SNDR$ is identical to the common SNR definition in Gaussian noise.

5.1 Performances of VBLAST and ML-G

Fig.1 shows the performances of VBLAST and ML-G in the alpha stable noise. It can be seen that:(1) Their performances in the impulsive noise are much worse than in the pure Gaussian noise. The more impulsive the noise is, the worse they perform. For example, when $\alpha = 1.5$, namely middle impulsive noise, at $BER = 3 \times 10^{-3}$, there are about 9dB and 11dB performance losses in VBLAST and ML-G respectively. When $\alpha = 0.5$, the performances of both systems degrade to unacceptably low levels. (2)The more impulsive the noise is, the less performance gain ML-G can attain over VBLAST. For example, when $\alpha = 2.0$, at $BER = 3 \times 10^{-3}$, there is about 5dB performance gain, but when $\alpha = 0.5$, their performances are almost the same. This is because both systems are based on Gaussian noise, so when the impulsive noise appears, their performances are mainly determined by the impulsive noise. As a result, the performance gain of ML-G in the Gaussian noise is lost.

5.2 ZF Versus ML-G

Fig.2 shows the performances of ZF and ML-G in the alpha stable noise. It can be seen that: ZF performs badly in the impulsive noise. This is because ZF pays much attention to cancel the MAI rather than the noise, so the noise is enlarged during the decorrelation process.

5.3 Cauchy Receiver Versus ML-G

Fig.3 shows the performances of Cauchy Receiver and ML-G in the alpha stable noise. It can be seen that:(1) Cauchy receiver can achieve a significant performance gain over ML-G in the impulsive noise. Even in middle impulsive noise, e.g. $\alpha = 1.5$, at $BER = 3 \times 10^{-3}$, it has about 5dB performance gain over ML-G. (2) Cauchy receiver seems quite robust in the impulsive noise although it is based on the Cauchy noise ($\alpha = 1.0$). Even in the Gaussian noise, its performance is only a little worse than ML-G, the optimal receiver in this occasion.

5.4 ML-Alpha, Cauchy Receiver Versus ML-G

Fig.4 shows the performances of ML-Alpha, ML-G and Cauchy receiver. It can be seen that: (1) The receivers designed by taking into account the statistical characteristic of the impulsive noise will gain a lot performance gain compared with the receiver designed on the basis of Gaussian noise. For example, ML-Alpha can attain about 6dB performance gain over ML-G at $BER = 3 \times 10^{-3}$ when $\alpha = 1.5$, and 9dB at $BER = 1 \times 10^{-2}$ when $\alpha = 1.0$. (2) Cauchy receiver is very robust compared to the optimal receiver ML-Alpha. This conclusion accords with the case in the SISO system, where Cauchy receiver is found to perform almost as well as the optimal receiver for a wide range of α [8].

Combining the analysis in section 4 and the simulation results, we deduce that: (1) The performances of conventional receivers of SDMA-based systems designed on the basis of the Gaussian noise are greatly degraded by the impulsive noise. When high impulsive noise appears, the performances of these receivers are degraded to unacceptably low levels. (2) The optimal receiver, ML-Alpha, has the optimal performance in the alpha stable noise, but it is not suitable for practical applications due to its complexity. (3) The ZF receiver can attain little performance gain in impulsive noise. (4) Cauchy receiver has good performance with reasonable computational complexity and is very robust in the $S\alpha S$ noise. So it is a very attractive scheme for SDMA-based system in impulsive noise.

6 Conclusions

In this paper the performance and receiver design in the uplink of SDMA-based wireless systems in impulsive noise are analyzed and discussed. The impulsive noise is modeled as a complex symmetric alpha stable process, which is an extension of Gaussian process and includes the Gaussian process and Cauchy process as the special cases. The optimal ML receiver and several suboptimal receivers,

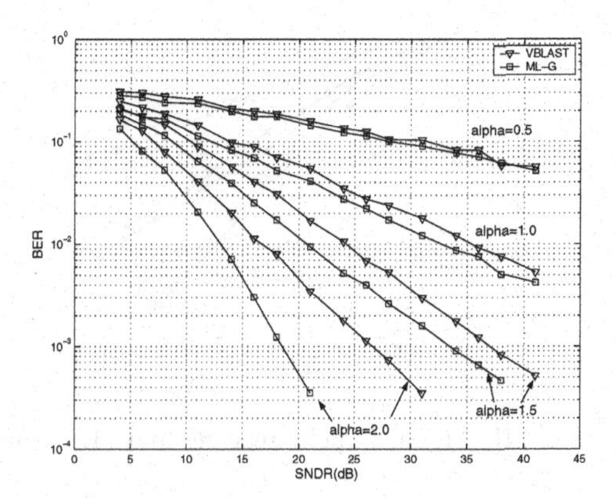

Fig. 1. Performances of VBLAST and ML-G in the impulsive noise

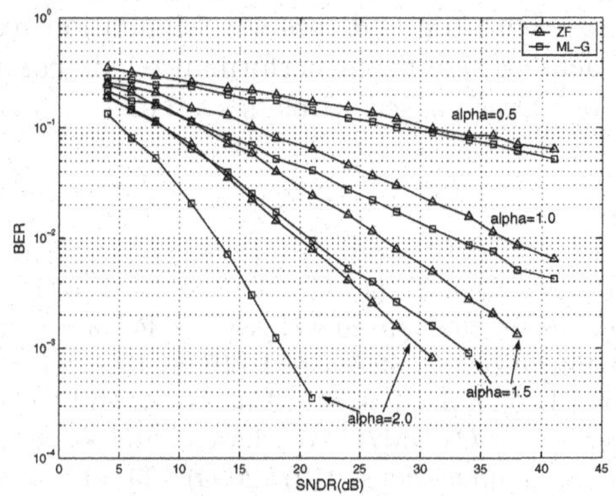

Fig. 2. Performances of ZF and ML-G in the impulsive noise

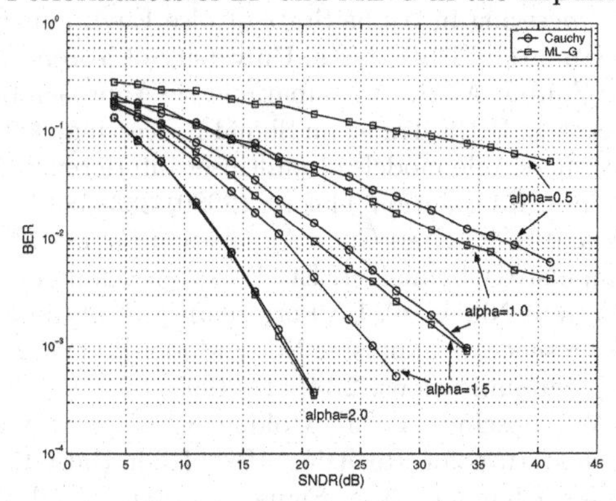

Fig. 3. Performances of Cauchy receiver and ML-G in the impulsive noise

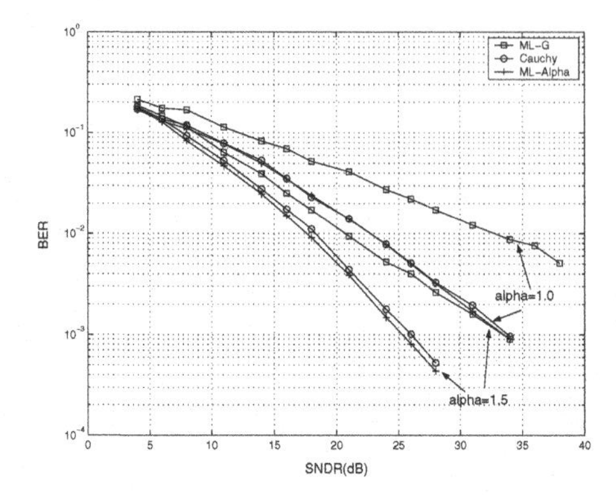

Fig. 4. Performances of ML-Alpha, Cauchy receiver and ML-G in the impulsive noise

such as ZF, ML-G and Cauchy receiver, are proposed. The SER or upper bound of SER is derived for each proposed receiver. Simulation results show the proposed receivers can achieve significant performance gain compared with the conventional detectors of SDMA-based wireless systems.

References

1. Vandenameele, P., Perre, L. V. D., Gyselinckx, B., Engels, M. and Man, H. D.: An SDMA algorithm for high-speed WLAN. Performance and complexity. Proc. Global Telecommunications Conference. **1** (1998) 189–194
2. Thoen, S., Deneire, L. Van der Perre, L. Engels, M. and De Man, H.: Constrained least squares detector for OFDM/SDMA-based wireless networks. IEEE Transactions on Wireless Communications. **1**(2) (2003) 129–140
3. Nee, R. V., Zelst, A. V. and Awater, G.: Maximum likelihood decoding in a space division multiplexing system. Proc. Vehicular Technology Conference. **1** (2000)
4. Ilow, J.: Signal Processing in Alpha-Stable Noise Environments: Noise Modeling, Detection and Estimation. PhD thesis, University of Toronto, Canada. (1995)
5. Brown, C. L. and Zoubir, A. M.: A Nonparametric Approach to Signal Detection in Impulsive Interference. IEEE Trans. on Signal Processing. **48**(9) (2000) 2665–2669
6. Hughes, B. L.: Alpha-stable models of multiuser interference. Proc. IEEE International Symposium on Information Theory. (2000) 383–383
7. S. Yoon, I. Song and S. Y. Kim: Code Acquisition for DS/SS Communications in Non-Gaussian Impulsive Channels. IEEE Trans. on Communications. **2**(52) (2004)
8. Tsihrintzis, G. A. and Nikias, C. L.: Performance of Optimum and Suboptimum Receivers in the Presence of Impulsive Noise Modeled as an Alpha-Stable Process. IEEE Trans. on Communications. **234**(43) (1995) 904–914
9. Wolniansky, P. W., Foschini, G. J., Golden, G. D. and Valenzuela, R. A.: V-BLAST: An Architecture for Realizing Very High Data Rates Over the Rich-Scattering Wireless Channel. Proc. Signals, Systems, and Electronics,URSI International Symposium on. (1998) 295–30
10. Tsakalides, P.: Array Signal Processing with Alpha-Stable Distributions. PhD thesis, University of southern California, USA. (1995)

Performance Analysis and Allocation Strategy for the WCDMA Second Scrambling Code

Frank Brouwer

Twente Institute Wireless & Mobile Communications, Institutenweg 30, 7521 PK Enschede, The Netherlands
Frank.Brouwer@ti-wmc.nl

Abstract. The WCDMA downlink is limited by Node-B cell power and the number of channelization codes. With upcoming capacity enhancement techniques, like MIMO and HSDPA, the potential limitation in channelization codes becomes visible more frequently. The WCDMA standard allows for allocation of traffic on a secondary scrambling code, which is non-orthogonal to the primary scrambling code. This paper analyzes the capacity impact of using the secondary scrambling code. In addition this paper introduces an allocation strategy for traffic to the secondary scrambling code. The analysis shows a potential gain in secondary scrambling code usage for this strategy.

1 Introduction

The air-interface is a very expensive resource in cellular systems. In order to utilize this resource to its full extend, WCDMA based systems will be enhanced with techniques like MIMO and HSDPA that improve the capacity of the system. When increasing the capacity of a WCDMA based system, an important limit to consider is the number of channelization codes in the downlink, which create the individual channels. The Orthogonal Variable Spreading Factor (OVSF) technique is used for generating channelization codes. The OVSF technique use of the Hadamard matrix,

$$H_{2n} = \begin{bmatrix} H_n & H_n \\ H_n & -H_n \end{bmatrix} \quad \text{and} \quad H_1 = [1], \tag{1}$$

where each row of H represents a single OVSF code word, and n is the Spreading Factor (SF). Figure 1 shows the way this process works. Both the matrix notation and the figure show that the number of OVSF codes is equal to the SF.

When the capacity increases the risk that the required number of codes exceeds the number of available codes is realistic. The reason for such limitation can be various:

- Capacity enhancement techniques aim at increasing the overall capacity, which typically result in more codes to be allocated.
- Mobile systems are used indoor much. Such environments have little time dispersion, causing little intra-cell interference, and therefore provide very high capacity.
- The code tree can have a scattered allocation. A specific code is only available for allocation when its children are available. (E.g. code $C_4(1)$ is a child of $C_2(1)$.)

I. Niemegeers and S. Heemstra de Groot (Eds.): PWC 2004, LNCS 3260, pp. 389–398, 2004.

- Some techniques, like SSTD, allocate more codes that they actually use.
- High capacity area Node-Bs will use the HS-DSCH, which occupies semi-statically between 5/16 and 15/16 [7] of a complete code tree. When little HS-DSCH traffic is carried the allocated codes are under-utilised.

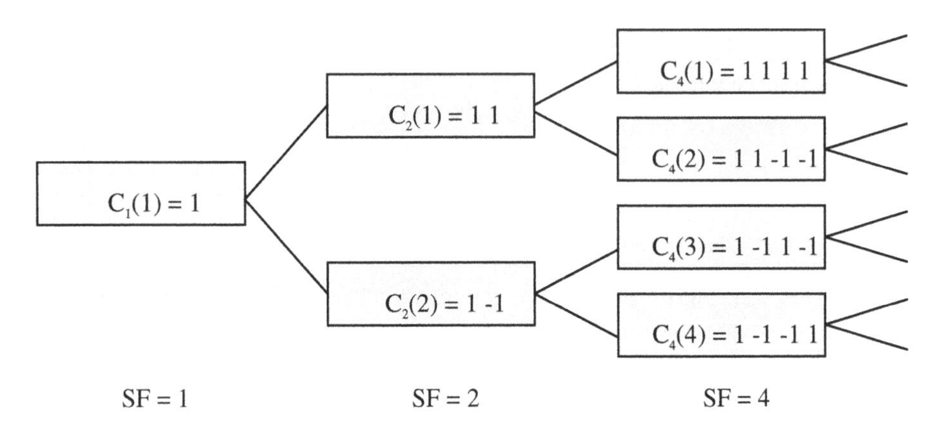

Fig. 1. OVSF channelization code tree [1], [2]

When more channelization codes are required than fit into a single OVSF code tree, WCDMA offers the possibility to allocate a secondary code tree. Surplus traffic is allocated to a secondary scrambling code. This secondary code has its own OVSF code tree. Doing such, the number of channelization codes is doubled. As scrambling codes are non-orthogonal, adding this secondary scrambling code will increase the intra-cell interference. This paper analyzes the capacity impact of adding this secondary scrambling code, and analyzes the performance of a proposed allocation strategy.

2 Performance Analysis

The number of available channelization codes and the available power determine the downlink capacity of the system. To find the downlink capacity of a W-CDMA system requires the combined analysis of both.

2.1 Available Channelization Codes

Common and shared channels occupy a part of the first code tree. The downlink channelization code capacity for dedicated channels is the part that is not allocated to the common and shared channels. For a service with spreading factor SF this capacity in terms of OVSF channelization codes can be calculated as

$$M_{SF} = SF \cdot \left(k - \frac{1}{SF_{common}} - \frac{1}{SF_{shared}} \right), \tag{2}$$

M_{SF}	DL OVSF code capacity for DCHs with spreading factor SF
SF	Spreading factor
SF_{common}	SF for common channels
SF_{shared}	SF for shared channels
k	Number of code trees.

2.2 Available Node-B Transmit Power

The Node-B transmit power is a shared resource. A fixed portion of the power is assigned to common channels. The power assignment of shared channels can be fixed or variable. In the latter case a strategy is to allocate the remainder of the Node-B transmit-power to the shared channel, with a pre-determined minimum and maximum.

2.2.1 General Expressions for Downlink Power

The Node-B dedicated channel transmit power is power-controlled. The UE compares the quality of the received signal with a service dependent target SIR, and requests an increase or decrease of the power to remain at the target as close as possible. Many studies report E_b/N_o instead of SIR. The SIR is the ratio of E_b/N_o and processing gain (PG). The PG is the ratio of the chip rate and bit rate including signalling overhead.

$$\gamma = \frac{E_b / N_o}{PG} = \frac{E_b / N_o}{R_{chip} / R_{DCH}}, \tag{3}$$

γ	SIR
E_b/N_o	Energy per bit over noise density
PG	Processing Gain
R_{chip}	Chip rate
R_{DCH}	DCH bit rate (including signalling overhead).

The received SIR results from the transmit power and the interference. This interference consists of received non-orthogonal power from the own and other Node-Bs.

Signals transmitted with the same scrambling code are in principle orthogonal. The propagation channel between Node-B and UE introduces time dispersion, reducing orthogonality between transmissions using the same scrambling code. This effect is environment dependent, and can vary over a cell. For ease of analysis it is assumed that an average orthogonality factor can be derived in a cell.

The synchronization channel (SCH) is a special case. The SCH is not modulated with any scrambling code, and is therefore by definition non-orthogonal. The SCH power determines the quality of the synchronization, and is therefore of large importance as system parameter. Only extensive analysis can provide a good estimate for its setting. Setting the SCH power to 1% of the maximum power of a Node-B is generally assumed to be a realistic assumption, and is used in this analysis.

The SIR is the ratio between the wanted and the unwanted signals. The wanted signal is the transmitted signal times the path-gain. The unwanted signal is the sum of all non-orthogonal power and thermal noise. The SIR of a received signal at UE i from Node-B j can be expressed as

$$\gamma_{i,j} = \frac{g_{i,j} p_{i,j}}{g_{i,j}(\alpha_{i,j} p_{orth} + p_{non-orth}) + \sum_{j \neq k} g_{i,k} p_k + N}, \tag{4}$$

$\gamma_{i,j}$	SIR of signal at UE i of Node-B j
$g_{i,j}$	path-gain between UE i and Node-B j
$p_{i,j}$	power of the signal for UE i from Node-B j
p_{orth}	other power transmitted from the same scrambling code
$p_{non-orth}$	non-orthogonal transmit power (SCH and other scrambling code(s))
p_k	total transmit power of Node-B k
$\alpha_{i,j}$	non-orthogonality between UE i and Node-B j.

When r_t is the received power of a signal that is transmitted orthogonal, and r_n is the received non-orthogonal power of the signal, α is the ratio of r_n over r_t. α depends on the environment and the receiver technology. It does not depend on the signal.

Loaded cells surround the cell under consideration. For a constant transmit power of the surrounding Node-Bs the received power will vary between 1 dB for free space and 5 dB for dense urban, disregarding effects of shadowing. Being small in relation to the variation in received power from the own Node-B, it is of secondary importance to the analysis in this paper. The total interference is modelled as a non-orthogonal cell transmitting at full power p_{max} at distance β times the site-to-site distance. β may vary per environment between 0.38 and 0.55. In this analysis is fixed to 0.4.

Thermal noise is small compared to the inter-cell interference in most cases. It is therefore neglected here. In addition the orthogonality is assumed constant over the cell. Taking the above assumptions the formula for SIR reduces to

$$\gamma_{i,j} = \frac{g_{i,j} p_{i,j}}{g_{i,j}(\alpha p_{orth} + p_{non-orth}) + g_\beta p_{\max}} = \frac{p_{i,j}}{\alpha p_{orth} + p_{non-orth} + p_{\max} \, g_\beta / g_{i,j}}. \tag{5}$$

The SIR $\gamma_{i,j}$ is input to power control, which will keep the effective SIR as close as possible to a target. Instead of having a SIR per connection, the analysis uses the target SIR γ as input and the Node-B transmit power as result. The SIR target is set equal for all DCHs. The power at which the Node-B transmits the signal for UE i equals

$$p_{i,j} = \gamma(\alpha \cdot p_{orth} + p_{non-orth} + p_{\max} \, g_\beta / g_{i,j}) \tag{6}$$

The quality of reception at the cell boundary determines the power for common and shared channels. This implies that $g_{i,j}$ equals the gain at the cell edge, being at 0.5 times the site-to-site distance. Common and shared channels transmit at a different bit rate. Assuming that the same E_b/N_0 applies to these channels the SIR γ for the DCH can be corrected with the ratio of the bit rates. As a result the combined common and shared channel transmit power is expressed as

$$p_c = R_c \gamma(\alpha \cdot p_{orth} + p_{non-orth} + p_{\max} \cdot k_c) \text{ with } R_c = \frac{R_{common} + R_{shared}}{R_{DCH}} \text{ and } k_c = g_\beta / g_{0.5}, \tag{7}$$

R_{common} combined bit rate on all common channels (including overhead)
R_{shared} combined bit rate on all shared channels (including overhead)
R_{DCH} bit rate on a DCH (including overhead).

The path-gain for the DCH varies over the cell. With a uniform user distribution over the cell area, it is possible to exchange this varying path-gain by an average path-gain, where the average must conserve the sum of transmit powers of all connections. This is valid using the following calculation:

$$p_{\max} \cdot k_{DCH} = p_{\max} \, {}^{g_\beta}\!/_{g_{avg}} = \frac{1}{cellarea} \int_{cell} p_{\max} \, {}^{g_\beta}\!/_{g_{pos}} \, dpos \;\Rightarrow\; k_{DCH} = \frac{1}{cellarea} \int_{cell} {}^{g_\beta}\!/_{g_{pos}} \, dpos \,. \quad (8)$$

So the average transmit-power per DCH equals

$$p_{DCH} = \gamma(\alpha \cdot p_{orth} + p_{non-orth} + p_{\max} \cdot k_{DCH}) \,. \quad (9)$$

The split in orthogonal and non-orthogonal power for the DCH depends on its allocation to primary or secondary scrambling code, and the amount of DCHs on each code. For N_1 DCHs on the primary scrambling code and N_2 DCHs on the secondary scrambling code the sum of powers for all DCHs on primary and secondary scrambling code and the power for the common and shared channel respectively are

$$\begin{aligned} p_1 &= N_1 \cdot \gamma(\alpha \cdot (p_c + p_1) + p_{SCH} + p_2 + p_{\max} \cdot k_{DCH}) \\ p_2 &= N_2 \cdot \gamma(\alpha \cdot p_2 + p_{SCH} + p_c + p_1 + p_{\max} \cdot k_{DCH}) \\ p_c &= R_c \cdot \gamma(\alpha \cdot (p_c + p_1) + p_{SCH} + p_2 + p_{\max} \cdot k_c) \,. \end{aligned} \quad (10)$$

The maximum transmit power of a Node-B p_{max} and the power for common and shared channels p_c are fixed. So dedicated channels can be hosted as long as

$$p_1 + p_2 + p_c + p_{SCH} \le p_{\max} \,. \quad (11)$$

As all values are scaled in p_{max} the analysis simplifies by expressing all powers as a fraction f of p_{max}, resulting in the following set of expressions.

$$\begin{aligned} f_1 &= N_1 \cdot \gamma(\alpha \cdot (f_c + f_1) + f_{SCH} + f_2 + k_{DCH}) \\ f_2 &= N_2 \cdot \gamma(\alpha \cdot f_2 + f_{SCH} + f_c + f_1 + k_{DCH}) \\ f_c &= R_c \cdot \gamma(\alpha \cdot (f_c + f_1) + f_{SCH} + f_2 + k_c) \\ & f_1 + f_2 + f_c + f_{SCH} \le 1 \,. \end{aligned} \quad (12)$$

2.2.2 Capacity for Only the Primary Scrambling Code

When the power related downlink capacity is less than the number of channelization codes for DCHs M_{SF} (or when disregarding this limit) all DCHs are allocated to the primary scrambling code. In the above formulas N_2 and effectively f_2 equal zero. The capacity than equals the maximum number N_1 that fulfils the set of equations:

$$\begin{aligned} f_1 &= N_1 \cdot \gamma(\alpha \cdot (f_c + f_1) + f_{SCH} + k_{DCH}) \\ f_c &= R_c \cdot \gamma(\alpha \cdot (f_c + f_1) + f_{SCH} + k_c) \\ & f_1 + f_c + f_{SCH} \le 1 \,, \end{aligned} \quad (13)$$

which can be shown to equal

$$N_1 = \left\lfloor \frac{1 - f_c - f_{SCH}}{\gamma(\alpha \cdot (1 - f_{SCH}) + f_{SCH} + k_{DCH})} \right\rfloor \text{ with } f_c = R_c \cdot \gamma(\alpha \cdot (1 - f_{SCH}) + f_{SCH} + k_c) \,. \quad (14)$$

2.2.3 Capacity for Primary and Secondary Scrambling Code

When the power related capacity limit exceeds the capacity limit for the primary scrambling code, only additional traffic is allocated to the secondary scrambling code. The number of codes on the primary scrambling code N_1 equals the capacity limit M_{SF}. Working out the set of equations 12 leads to a capacity limit N_2 of

$$N_2 = \left\lfloor \frac{1}{\gamma(\frac{1 + k_{DCH}}{f_2} + \alpha - 1)} \right\rfloor \text{ with } f_2 = 1 - f_{SCH} - \frac{M_{SF} \cdot (1 + k_{DCH}) + R_c \cdot (1 + k_c)}{1/\gamma + (1 - \alpha) \cdot (M_{SF} + R_c)} \,. \quad (15)$$

2.3 Example Capacity Limits with Secondary Scrambling Code

Numerical examples in this section provide an interpretation of the formulas. Figure 2 shows the impact of the secondary scrambling code. The figure shows a virtual capacity. This assumes that all DCHs are allocated to the primary code tree, ignoring limit M_{SF}. The true capacity, taking the allocation to the secondary code tree into consideration is significantly less. This shows an impact of the mutual non-orthogonality between scrambling codes.

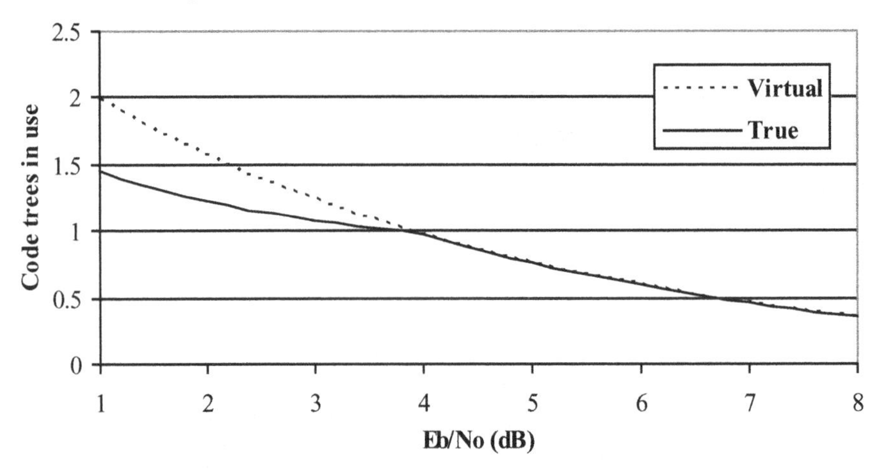

Fig. 2. Capacity impact of secondary scrambling code

This result shows that recovering codes on the primary code tree is very beneficial. There are two main options for recovering codes.

1. Unassigned codes may be unavailable due to scattering of code allocation. A code can only be assigned when all its children codes are unassigned. The SF is service dependent. Different services will be present in the same cell, so a mix of small

and large SF is assigned. Large SF codes, scattered in the code tree, may be unassigned, but not leaving any small SF code assignable. Concatenating free codes by reassigning large SF services to other codes, enables small SF code allocation. Various proposals are provided in literature. [4], [5], [6]

2. The HS-DSCH is of great importance in increasing the performance of W-CDMA. This channel can be assigned semi-statically up to 94% of the code tree. The usage of the HS-DSCH depends on the amount of relevant traffic in the cell. As this traffic can fluctuate heavily, the HS-DSCH may be underutilized during some intervals. At the same time traffic not suitable for HS-DSCH may peak. Freeing codes during a period of low utilization of the HS-DSCH therefore is very important. So far no studies were found in literature.

For a more detailed analysis of the impact of the non-orthogonality between the scrambling codes the true capacity exceeding the primary scrambling code is expressed as a fraction U of the virtual capacity exceeding one code tree:

$$U = \frac{\left\lfloor \dfrac{1}{\gamma((1+k_{DCH})/f_2+\alpha-1)} \right\rfloor}{\left\lfloor \dfrac{1-f_c-f_{SCH}}{\gamma(\alpha\cdot(1-f_{SCH})+f_{SCH}+k_{DCH})} \right\rfloor - M_{SF}} \quad \text{with} \tag{16}$$

$$f_2 = 1 - f_{SCH} - \frac{M_{SF}\cdot(1+k_{DCH})+R_c\cdot(1+k_c)}{1/\gamma+(1-\alpha)\cdot(M_{SF}+R_c)} \quad \text{and}$$

$$f_c = R_c\cdot\gamma(\alpha\cdot(1-f_{SCH})+f_{SCH}+k_c).$$

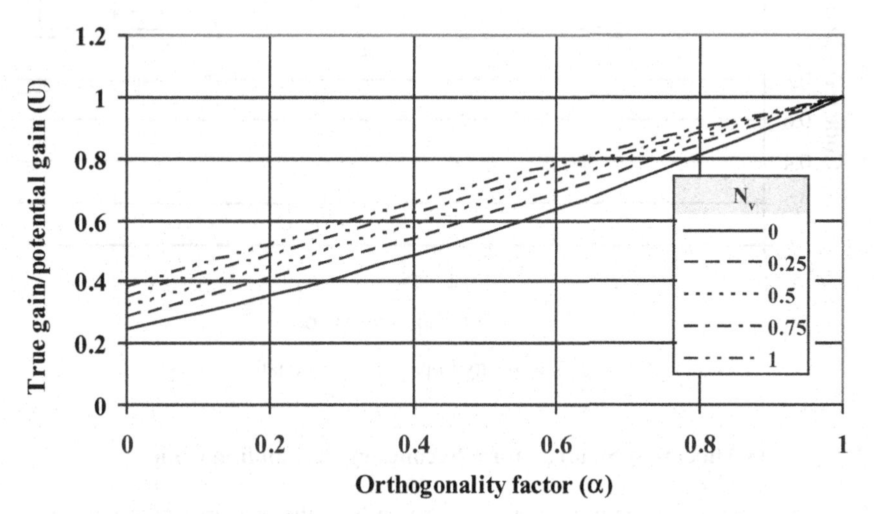

Fig. 3. Usability of the capacity gain on secondary scrambling code

Figure 3 shows the usability of a capacity gain when this exceeds the primary scrambling code. The gain is shown as function of the orthogonality factor and the potential loading of the secondary code tree N_v. N_v is varied between 0 (no traffic on the secondary code tree) and 1 (the secondary code tree could be used to its full extend when the non-orthogonality between code trees would not exist).

The figure clearly shows that the impact on capacity gain is much larger for the orthogonality than for the loading of the second scrambling code. α depends on the environment and indicates the amount of power that is observed as non-orthogonal. Without time dispersion in the received signal α is equal to 0. When α equals 1 all orthogonal power is received non-orthogonal. This corresponds to a hypothetical condition of infinite time dispersion. α depends on receiver technology and environment.

Practical values for α in W-CDMA range roughly from 0.1 in office environments to 0.5 in rural environments. In this range one third to half of the potential gain remains when traffic needs to be allocated to the secondary scrambling code. So reassigning codes from the secondary code tree to the primary code tree pays back. E.g. reassignment of one connection from secondary to primary scrambling code frees sufficient power to allocate one or two additional connections on the primary code tree.

The orthogonality also has a direct impact on the capacity. This effect strongly reduces when a significant amount of traffic is allocated to the secondary scrambling code. Figure 4 shows the capacity versus the orthogonality. As soon as the secondary scrambling code is used, the capacity gain of a smaller α diminishes. Comparing this result with the effect of a reduced SIR shows that the performance gain of reducing SIR and α are quite similar when only the primary scrambling code is in use. When also the secondary scrambling code needs to be used a better performance in SIR provides a much better gain than reducing α.

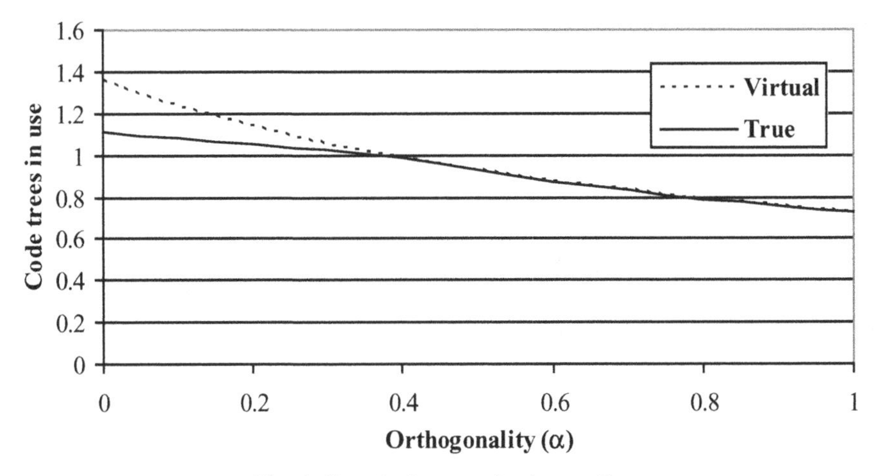

Fig. 4. Capacity impact of orthogonality

2.4 Code Allocation Strategy for a Secondary Scrambling Code

So far the implicit assumption is that traffic on both the primary and secondary code tree is uniformly distributed over the cell area. Traffic on the secondary code tree uses more power than traffic on the primary code tree. Traffic near the cell boundary uses more power than traffic near the cell centre. It may be expected that selecting traffic as near as possible to the cell centre for allocation on the secondary code tree is beneficial. For comparison also the opposite case where as far as possible traffic is selected for allocation on the secondary code.

For analysis of such algorithm the average path-gain is calculated separately for the two scrambling codes, splitting the cell area corresponding to the ratio between traffic using each scrambling code. This gain of such algorithm can be expressed as

$$gain = \frac{N_2(\text{with algorithm}) - N_2(\text{without algorithm})}{N_2(\text{without algorithm})} . \tag{17}$$

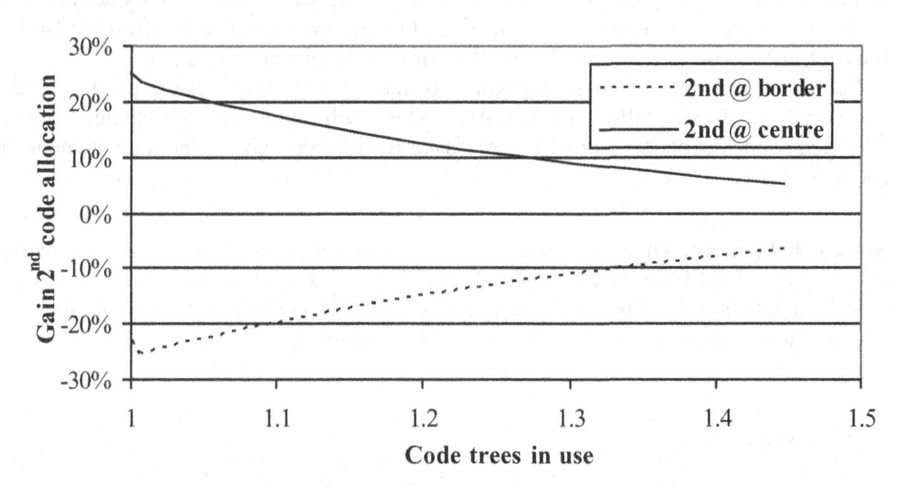

Fig. 5. Capacity gain for code allocation strategies

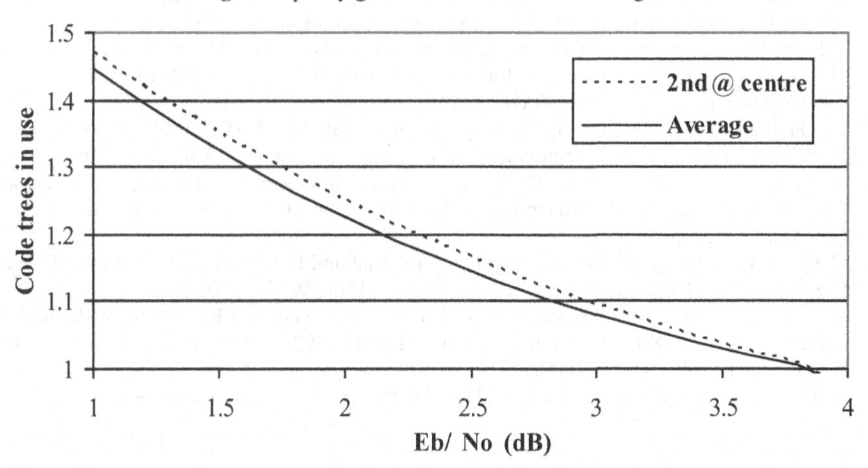

Fig. 6. Effective capacity for code allocation strategies

Figure 5 shows the gain of allocating traffic to the secondary code tree as close as possible to the cell centre or cell border. Clearly allocating near the cell centre is beneficial, while allocating near the cell border costs capacity. The effect however decreases with increasing use of the secondary scrambling code.

Figure 6 compares the resulting capacity for the strategy of allocating connections on the secondary scrambling code as close as possible to the cell centre with the uniform allocation. The capacity gain is apparent, though not dramatic.

3 Conclusions

The DL capacity limit of a Node-B is determined by the available power. The limited number of channelization codes per scrambling code does not directly limit the capacity, but does impact the capacity significantly. The cost for allocating traffic to the secondary scrambling code is two to three times as high as for the primary scrambling code. This cost can be reduced to some extend by preferring traffic far from the Node-B over traffic close to the Node-B for allocation to the primary code tree.

For maximum capacity it is important to manage the codes on the primary code tree very good. Code reallocation can free codes, such that as little traffic needs to be allocated to the secondary code tree. An issue for further study is the management of codes for the HS-DSCH.

Acknowledgement. This paper reflects part of the work performed in the framework of project TSIT1025 Beyond3G. The Beyond3G project investigates the enhancements required to 3G systems like WCDMA for integration in systems beyond the 3rd generation mobile communication systems.

References

1. H. Holma, A. Toskala (Editors), "WCDMA for UMTS - Radio Access for Third Generation Mobile Communications", John Wiley and Sons, Second Edition, 2002.
2. H.D. Schotten and Hafez Hadinejad-Mahram, "Analysis of a CDMA Downlink with Non-Orthogonal Spreading Sequences for Fading Channels", Proc. VTC2000-Spring, pages 1782-1786, Tokyo, Japan, May 2000.
3. E. Dahlman, B. Gudmundson, M. Nilsson, and J. Sköld, "UMTS/IMT- 2000 based on Wideband CDMA", IEEE Communication Magazine, vol. 36, pp. 70 80, Sep. 1998.
4. Y. Tseng, C. Chao and S. Wu, "Code Placement and Replacement Strategies for Wideband CDMA OVSF Code Tree Management", IEEE Trans. on Mobile Computing, Vol. 1, No. 4, Oct.-Dec. 2002, pp. 293-302.
5. C. Chao, Y. Tseng, and L. Wang, "Reducing Internal and External Fragmentations of OVSF Codes in WCDMA Systems with Multiple Codes", Proc. WCNC, 2003.
6. A.N. Rouskas, and D.N. Skoutas, "OVSF Codes Assignment and Reassignment at the Forward Link of W-CDMA 3G Systems", Proc. PIMRC 2002, Vol. 5, pp. 2404-2408, Lisbon, Portugal, September 15-18, 2002.
7. 3GPP, "Spreading and modulation (FDD)", TS 25.213 v5.4.0, http://www.3gpp.org

An Adaptive Localized Scheme for Energy-Efficient Broadcasting in Ad Hoc Networks with Directional Antennas*

Julien Cartigny[1], David Simplot-Ryl[1], and Ivan Stojmenović[2]

[1] IRCICA/LIFL, University of Lille 1, INRIA Futurs, France.
{cartigny,simplot}@lifl.fr.
[2] SITE at the University of Ottawa, Canada. ivan@site.uottawa.ca.

Abstract. Several solutions for energy-efficient broadcasting, mostly centralized, have been proposed with directional antennas. However, such globalized protocols are not suitable for ad hoc networks, because each node needs a full knowledge of the topology. Recently, a localized algorithm, called DRBOP, using one-to-one communication model has been proposed. It uses RNG graphs, which can be locally computed by each node. However, for energy consumption reasons, it can be useful to reach more than one neighbor at a time. In this paper, we propose an efficient protocol which uses both one-to-one and one-to-many communication models. First, we present a variant of DRBOP efficient for sparse networks, based on LMST graph which is a local adaptation of minimal spanning tree. Then, a one-to-many protocol efficient for dense networks is proposed. From these two algorithms we derive an adaptive protocol which is shown to be efficient for both sparse and dense networks.

1 Introduction

In mobile wireless ad hoc networks, each node participates in networking tasks by relaying messages, in order to provide a full coverage of the network. This implies a high energy consumption of the radio interface and thus limits the lifespan of the battery unit. The problem of energy consumption is very significant in this type of network, because nodes are energy limited. Many solutions have been proposed to decrease the energy consumption. An idea is to control the emitted transmission power by decreasing the range of the radio beam, and thus reducing the energy consumption [1]. Special devices, like directional antennas, offer better energy savings by reducing the angle of the beam, to only cover part of the neighborhood. However, theses adjustments have to be managed while maintaining the connectivity of the network.

There exist several communication models. In one-to-all model, nodes use omnidirectional antennas to cover all their neighbors. In one-to-many model, a

* This work was partially supported by a grant from Gemplus Research Labs., an ACI *Jeunes Chercheurs* "Objets Mobiles Communicants" (1049CDR1) from the Ministry of Education and Scientific Research, France, the CPER Nord-Pas-de-Calais TACT LOMC C21 and NSERC.

I. Niemegeers and S. Heemstra de Groot (Eds.): PWC 2004, LNCS 3260, pp. 399–413, 2004.

node can target a subset of its neighborhood with one transmission by choosing direction and width of the beam, using directional antennas. The one-to-one model is a particular case of the previous one, where nodes use a constant narrow beam to transmit toward a particular neighboring node. Since the communication area is a narrow beam with a small angle, the directional antennas provide more energy saving and interference reduction.

Several solutions have been proposed [2], [3] for energy-efficient broadcasting with directional antennas. However, they are globalized, meaning either a centralized entity has to gather knowledge of the full topology and diffuse information to organize the network, or each node has to know the total topology of the network to locally compute the broadcast spanning tree. This approach is not efficient in ad hoc networks, because of a high communication overhead. We are interested in localized protocols, which require only information about the neighborhood. Furthermore, we are looking to use directional antennas for the one-to-one and one-to-many models.

The problem of finding a broadcast spanning tree with minimal power, and the problem of broadcasting a message with a minimal number of retransmissions are well-known to be both NP-complete [4]. Consequently, several heuristics have been proposed. For instance, MST (Minimum Spanning Tree) [1] is a globalized algorithm which builds the spanning tree by choosing shortest links between nodes. The BIP (Broadcast Incremental Power) algorithm [1] (and the directional version DBIP [3]), proposed by Wieselthier *et al.* is a globalized greedy algorithm inspired by Prim's algorithm, known to be the most efficient existing broadcast protocol. Some localized solutions also exist. For instance, RBOP (RNG Broadcast Oriented Protocol) [5] (and the directional version DRBOP [6]) is a localized energy-efficient broadcast protocol, which only requires local information to construct a RNG subgraph (Relative Neighborhood Graph) [7] while keeping the network connected. Another recent solution is BLMST, (Broadcast with Local Minimum Spanning Tree) [8] (and a similar independently proposed solution in [9]) is a localized energy-efficient broadcast algorithm for omnidirectional antennas. It is based on LMST (Local Minimum Spanning Tree) [10], a MST algorithm applied on the local neighborhood.

In this paper, we propose DLBOP (Directed LMST Broadcast Oriented Protocol), an algorithm based on LMST and using directional antennas. This protocol is a straightforward variation of DRBOP [6]. A second algorithm, called OM-DLBOP (One-to-Many Directed LMST Broadcast Oriented Protocol), is used instead of DLBOP when energy cost of one-to-one transmissions is too high (*i.e.* when network is dense). These two algorithms are combined in a hybrid protocol, called A-DLBOP (Adaptive Directed LMST Broadcast Oriented Protocol), which adaptively decides which communication model to use among one-to-one and one-to-many models. This protocol is energy-efficient and applied on general energy model (proposed in [6]) to optimize the energy consumption. Each node requires only the knowledge of neighbor positions. This information can be measured by using signal strength or time delay combined with direction evaluation by smart antennas. The position of each node can be also extracted with positioning system (like GPS).

The paper is organized as follows. Some preliminaries are given in Section 2. We describe the existing works on energy-saving broadcast in Section 3. Protocols and their consumption estimations are given in Section 4. The experimental results and the comparison with theoretical bounds and other protocols are proposed in Section 5. Finally, Section 6 concludes the paper.

2 Preliminaries

In the unit graph model, two nodes can communicate if and only if their distance is $\leq R$, where R is the transmission radius, equal for all nodes. We denote by $N(u)$ the set of neighbors of u.

In one-to-all communication model (omnidirectional antenna), each node in the network can only change its transmission power. In one-to-one and one-to-many models, we will assume that all the nodes have directional antennas. They can hear messages from every neighbor and send messages to every neighbor in unicast communications, by aim the beam to the addressee.

We need to evaluate the energy consumption for each node. We use the following formula proposed in [6]. The cost of a transmission of range r with angle θ is calculated by:

$$e(\theta, r) = \begin{cases} \frac{\theta}{2\pi}(r^\alpha + C_1) + C_2 & \text{if } r \neq 0, \\ 0 & \text{otherwise.} \end{cases}$$

This model generalizes several energy consumption models proposed in the literature. The parameter α gives the power loss. The constant C_1 associates a energy cost for aiming the angle beam. The constant C_2 is a constant overhead for each sending, representing the minimum needed energy for signal processing and MAC control mechanism, and the power needed for neighboring nodes to receive the message. Although this power is spent by neighbors, we simplified the model by charging the node itself, assuming that each node eventually still cares about receiving one full copy of the packet, and can decide to switch off the receiver at the beginning of subsequent reception of the same message. For constants $\alpha = 2$ and $C_1 = C_2 = 0$, the model is a generalization of the one commonly used. We considered in [6] a directional version of a specific model proposed by Rodoplu and Meng in [11]. This model uses the following constants $\alpha = 4$, $C_1 = 8.10^7$ and $C_2 = 2.10^7$, thus charging high cost for each directional transmission. In one-to-one model, the beam angle is a constant value while it varies between a minimal angle β and 2π in one-to-many model.

3 Literature Review

Wieselthier *et al.* proposed two extensions of the BIP protocol [1] with directional antennas. The protocol BIP is an omnidirectional protocol which constructs a spanning tree with respect to the energy consumption. For each step, the algorithm decides if the best solution is to create a new transmission beam or to increase the range of an existing transmission. The directional version [3] of BIP

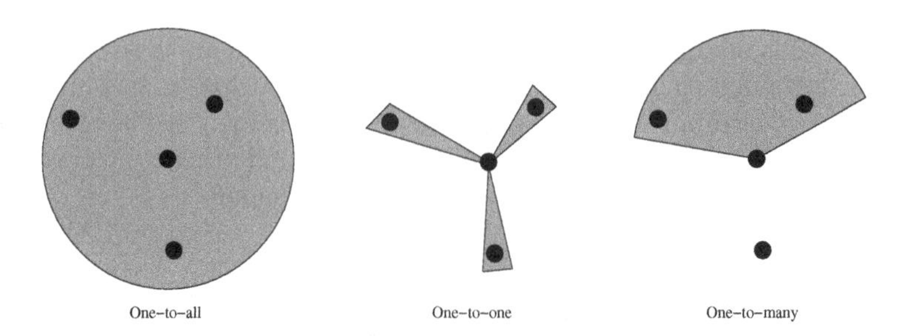

Fig. 1. One-to-all, one-to-one and one-to-many communication models.

proposes two protocols. The first protocol is called RB-BIP (Reduced Beam BIP) and uses one-to-one communication model (with minimal angle) to join neighbors in the BIP tree. Because of the tree construction, the BIP spanning tree is exactly the same as the tree built by the MST algorithm. The second protocol is D-BIP (Directional BIP). Each node can send only one message by broadcast, the protocol has to decide, at each step, if it is better to extend the beam and/or the range of a node, or to add a new communication beam. This decision is made with respect to the energy consumption. Hence, if the constants C_1 and C_2 are not null, the natural tendencies of D-BIP are to favor transmissions with large radii and beam angles, to avoid retransmissions by every node.

Cartigny *et al.* have proposed RBOP (RNG Broadcast Oriented Protocol) [5], a localized broadcast protocol for reducing energy consumption with omnidirectional antennas. Each node constructs an RNG (Relative Neighborhood Graph) [7] subgraph from its neighboring graph. Let $G = (V, E)$ be a graph. The RNG subgraph, denoted by $RNG(G) = (V, E_{rng})$, is defined by:

$$E_{rng}(G) = \{(u,v) \in E \mid \nexists w \in V \quad d(u,w) < d(u,v) \wedge d(v,w) < d(u,v)\} \ .$$

RNG has several advantages: each node needs to know only its neighbors and the distance between them. Furthermore, the required information can be gathered in a localized manner. The RNG transformation removes some edges from the set E. In obtained graph, the average degree of nodes is approximately 2.6, and connected neighbors are the closest neighbors of the node. RNG preserves connectivity. The protocol RBOP consists of a Neighbor Elimination Scheme (NES) [12], [13] limited to RNG neighbors where a transmitting node adjusts its communication range to reach all non-covered RNG neighbors. In a NES protocol, each node eliminates from the list of neighboring nodes for retransmission those nodes that are supposed to receive the same packet received by given node one or more times in previous retransmissions.

A directional version of this protocol, called DRBOP (Directional RNG Broadcast Oriented Protocol), has been proposed in [6]. This algorithm proposes that each node sends a separate unicast message to each of its non-covered RNG neighbors. The protocol is efficient and gives results reasonably close to the cen-

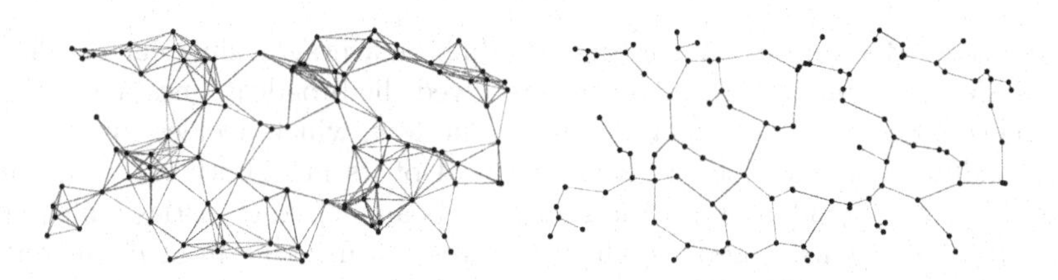

Fig. 2. Sample of network with its LMST subgraph (average degree 10).

tralized MST protocol because DBIP protocol has an average neighbor degree of 1.99 (which is average degree of a minimal spanning tree), and DRBOP has average degree about 2.6, as found by experiments.

Li *et al.* [10] have proposed LMST (Local Minimum Spanning Tree) which offers a better graph reduction than RNG, with a degree of approximately 2.04. The LMST method is simple: each node applies MST algorithm on its local topology (the list of neighbors and links between them), and keeps only links that are present in LMST of both endpoints. The LMST algorithm is localized and offers a lower subgraph degree than RNG (in fact, LMST is a subgraph of RNG). Both LMST and RNG require 2-hops informations to be computed. Experimentally, the LMST degree is approximately 2.04, which is closer to the BIP degree (1.99) than RNG (2.6). The authors proposed an omnidirectional protocol for topology control using LMST and proved the correctness of the algorithm for preserving the connectivity. They also proposed BLMST, an energy-saving broadcast protocol using LMST in [8] (a similar protocol is also proposed in [9]). An example of a graph and its LMST subgraph are presented in Figure 2.

4 One-to-One and One-to-Many Protocols

First, we are going to present DLBOP, a variant of DRBOP based on LMST graph which is a local adaptation of minimal spanning tree. This protocol is shown to be efficient for sparse networks. Then, we present OM-DLBOP, a protocol using one-to-many communications. It is close to the omnidirectional case and shown to be efficient for dense networks. Finally, we present A-DLBOP, an efficient adaptive protocol which uses both one-to-one and one-to-many communication models which is shown to be efficient for both sparse and dense networks.

4.1 Directed LMST Broadcast Oriented Protocol (DLBOP)

The directed LMST broadcast oriented protocol (DLBOP) is a variant of DRBOP where the RNG graph is replaced by the LMST set. DLBOP uses one-to-one communication model. Hence, each node u sends to its LMST neighbors v an unicast message, with a beam of angle β and range $d(u,v)$. This scheme is more efficient with LMST than RNG, because the average degree of LMST (2.04) is lower than the one of RNG (2.6). Thus, each node has to send approximately one message (RNG needs one message and half on average), since each node only covers LMST neighbors which have not received the broadcast message. In order to achieve NES with directional antenna, a node u, which decides to retransmit the broadcasted message to a given subset A of its neighborhood with one or several beams, includes its position and the beams characteristics which cover nodes from A. Hence a node v which receives the message from u can remove nodes of A from its NES list.

The energy consumption of the DLBOP protocol with directional antennas can be derived by summing the power expenditures of one-to-one messages from each node. A node sends a unicast message to each of its LMST neighbors with a beam of minimal angle β. Since the degree of each node is approximately 2 on the LMST subgraph (including the local forwarder of the broadcast message) we can expect that each node will broadcast in average one unicast message. Let d_{lmst} denotes the average distance between LMST neighbors. The energy consumption of the DLBOP protocol is approximately:

$$E_{DLBOP} = n \times e(\beta, d_{lmst}) \ . \tag{1}$$

Note that this formula does not follow from our general model, and the approximation assumes that each node has degree two in LMST, which is not the exact distribution. However, we made simplification to have simple theoretical estimates, verified by experiments. To evaluate the energy consumption with this model, it is necessary to know the average distance between LMST neighbors, with respect to the area S and the number of nodes n. We propose to approximate LMST distance by the length of the edge of regular hexagonal mesh of n nodes covering the area S. In a regular hexagonal mesh, we observe that a node is located at the intersection of three hexagons, and thus two nodes are needed for each hexagon. The size of a hexagon side is then:

$$d_{lmst} \simeq r_{hex} = \sqrt{\frac{4S}{3\sqrt{3}n}} \ . \tag{2}$$

This assumption can be verified experimentally. The Figure 3 presents the graph of the average distance in LMST from the theoretical and experimental point of view ($S = 2000 \times 2000$m and $R = 250$m). Although if the experimental data are lower (because of the border effect), they have the same behavior

Finally, the energy consumption formula (1) can be rewritten, using the formula (2), as:

$$E_{DLBOP} = n\left(\frac{\beta}{2\pi}\left(\left(\frac{4S}{3\sqrt{3}n}\right)^{\frac{\alpha}{2}} + C_1\right) + C_2\right) \ . \tag{3}$$

4.2 One-to-Many Directed LMST Broadcast Oriented Protocol (OM-DLBOP)

The DLBOP algorithm has high energy consumption if $C_1 > 0$ and $C_2 > 0$. Hence, in this case it can be beneficial to reach several neighbors with the same

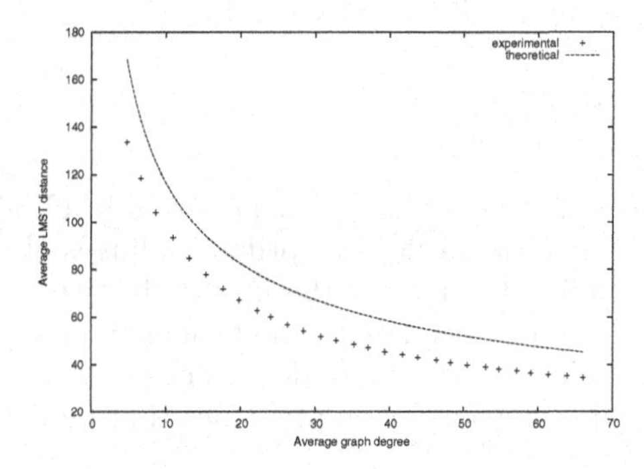

Fig. 3. Average distance between lmst neighbors, from the theoretical and experimental point of view.

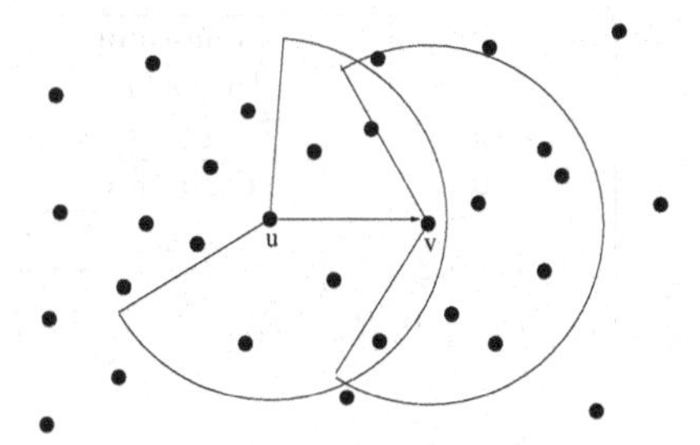

Fig. 4. Beam coverage with OM-DLBOP broadcast.

beam. The one-to-many variant of DLBOP, denoted by OM-DLBOP, consists of sending a single variable angle beam instead of several narrow beams. A node which decides to retransmit the message - because of LMST neighbor elimination scheme reason - uses a single beam with an appropriate angle which allows reaching non-covered LMST neighbors.

To increase energy savings, it can be useful to extend the range in order to avoid excessive retransmissions that can be expensive if constants C_1 or C_2 are not null. In fact, it is not always efficient to minimize transmission range because the reduction of transmission range implies a greater number of transmissions in the entire network. This property is also observed in one-to-all communication model [14]. For instance, let us consider the broadcasting by uniform beams with γ angle ($\gamma \in [\beta, 2\pi]$ where β is the minimal angle) and range r. For a given angle γ, we look for an optimal radius $R_{opt}(\gamma)$ which minimizes the total energy consumption.

To cover the circular area around itself, a node needs to send $2\pi/\gamma$ beams with the cost $e(\gamma, r)$ associated with each beam. Suppose that the full area S is ideally covered with such beams (of course, this is impossible but nevertheless leads to useful conclusions). The total energy consumption, denoted by $E_{area}(r)$

is then:

$$E_{area}(r) = \frac{S}{\pi r^2}\frac{2\pi}{\gamma}e(\gamma, r) = \frac{S}{\pi}\left(r^{\alpha-2} + C_1 r^{-2} + \frac{2\pi C_2 r^{-2}}{\gamma}\right) .$$

The behavior of the function $E_{area}(r)$ depends on α, C_1 and C_2 and is summarized in Table 1. The formula for the optimal radius is obtained by standard calculus method of finding the root of the first derivative over r. Interestingly, the optimal radius does not depend on the node density or average node degrees. However, it is valid only for reasonably dense networks since otherwise sparse networks may have large portions of empty zones that do not need coverage.

Table 1. Behavior of total energy consumption.

	$C_1 = C_2 = 0$	$C_1 \neq 0 \vee C_2 \neq 0$
$\alpha = 2$	constant No $R_{opt}(\gamma)$	monotone decreasing $R_{opt}(\gamma) = R$
$\alpha > 2$	monotone increasing $R_{opt}(\gamma) = 0$	minimal at $r(\gamma) = \sqrt[\alpha]{\frac{2C_1 + \frac{4\pi C_2}{\gamma}}{\alpha-2}}$ $R_{opt}(\gamma) = min(r(\gamma), R)$

Now, we give the complete OM-DLBOP algorithm. We choose to send beams with angle $4\pi/3$. This angle minimizes the overlap communication zone and provides a good coverage of the neighborhood as illustrated Figure 4 (in case of uniform transmission range). Note that the beam can be wider than $4\pi/3$ as explained below. The angle of $4\pi/3$ represents the ideal case and is used to determine the transmission range. The beam angle chosen by v is set symmetrically with respect to the line uv as shown in the same figure because, since the angle between any two LMST neighbors is at least $\pi/6$, the mentioned beam contains the remaining LMST neighbors of v. LMST neighbors that already received the same message can be determined from the position of sender, the positions of the neighbors and the transmission range and the beam direction of sender. After applying NES restricted to LMST neighbors, a node u which decides to retransmit computes its transmitting angle and range as follows:

- Let A be the set of uncovered neighbors and $B \subseteq A$ the set of uncovered LMST neighbors.
- The node u computes the set of nodes closer than $R_{opt}(4\pi/3)$:

$$A' = \{v \in A \mid d(u, v) \leq R_{opt}(4\pi/3)\} .$$

  The "goal" of u is to reach nodes of $C = A' \cup B$, *i.e.* nodes closer than $R_{opt}(4\pi/3)$, for optimization reason, and covering LMST neighbors, for coverage reason. If $C_1 = C_2 = 0$, the optimal radius cannot be evaluated or is null. In this case, we always consider $R_{opt}(4\pi/3) = 0$. This implies that $A' = \emptyset$ and the sender must cover only its non-covered LMST neighbors.

- The node calculates the angle θ needed to cover C and the distance d to the furthest node of C. If $\theta < \beta$ then set $\theta = \beta$. If C is empty, the retransmission is canceled.
- If $d > R_{opt}(\theta)$ then send θ-beam with d range. Otherwise, the node sets the range of θ-beam (without modifying the orientation) in order to reach all nodes of A closer than $R_{opt}(\theta)$ (thus the selected radius is generally somewhat lower than $R_{opt}(\theta)$).

An evaluation of energy consumption of OM-DLBOP can be obtained if we consider an ideal $4\pi/3$ beam tessellation. Let us consider an area S with N relaying nodes. Consider an approximation where relaying nodes are placed according to a honeycomb mesh. Suppose that the area S is divided into hexagons with side $R_{opt} = R_{opt}(4\pi/3)$. Then $N = 2S/A_{hex}$ where $A_{hex} = 3R_{opt}^2\sqrt{3}/2$. Therefore the energy consumption is then

$$\begin{aligned} E_{OM-DLBOP} &= N \times e(4\pi/3, R_{opt}) \\ &= \frac{8S}{9\sqrt{3}}(R_{opt})^{\alpha-2} + \frac{4S}{3\sqrt{3}}\left(\frac{2}{3}C_1 + C_2\right) {R_{opt}}^{-2} . \end{aligned} \tag{4}$$

A comparison between $E_{OM-DLBOP}$ and $E_{area}(r)$ gives the same behavior, as seen in Table 1:

- If $\alpha = 2$ and $C_1 = C_2 = 0$, the optimal radius does not matter, as the energy consumption of OM-DLBOP is $E_{OM-DLBOP} = \frac{8S}{9\sqrt{3}}$.
- With the case $\alpha = 2$, $C_1 \neq 0$ and $C_2 \neq 0$, it is shown in the Table 1 that the best solution is to maximize R_{opt}. This is confirmed by the value of $E_{OM-DLBOP}$, equals to $E_{OM-DLBOP} = \frac{8S}{9\sqrt{3}} + \frac{4S}{3\sqrt{3}}\left(\frac{2}{3}C_1 + C_2\right){R_{opt}}^{-2}$. And so, the energy consumption with $R_{opt} = R$ is $E_{OM-DLBOP} = \frac{8S}{9\sqrt{3}} + \frac{4S}{3\sqrt{3}}\left(\frac{2}{3}C_1 + C_2\right)R^{-2}$.
- If $\alpha > 2$ and $C_1 = C_2 = 0$, as presented by the Table 1, it is better to minimize the range. This is confirmed by the rewriting of $E_{OM-DLBOP} = \frac{8S}{9\sqrt{3}}{R_{opt}}^{\alpha-2}$. The best solution is to use the minimal range in the graph, so $R_{opt} = d_{lmst}$. Hence, the energy consumption can be write as $E_{OM-DLBOP} = \frac{8S}{9\sqrt{3}}d_{lmst}^{\alpha-2} = \frac{2n}{3}\left(\frac{4S}{3\sqrt{3}n}\right)^{\frac{\alpha}{2}}$.
- For the last case $\alpha > 2$, $C_1 \neq 0$ and $C_2 \neq 0$, the original equation (4) has no simplification.

4.3 Looking for a Threshold

It seems that the two approaches (one-to-one and one-to-many) are valid. The protocol DLBOP offers a subgraph with a minimal degree for each node, and OM-DLBOP covers large group of nodes to reduce the cost associated with each sending. We are now developing theoretically what will be the energy consumption, and showing which protocol is better in respect of the selected energy consumption model.

For the four energy models, we investigate now the cases when $E_{OM-DLBOP}$ is performing better than E_{DLBOP}. The inequality $E_{OM-DLBOP} < E_{DLBOP}$ is resolved by using the formula (4) and (3) with the values of α, C_1 and C_2 of the energy models.

- For $\alpha = 2$, $C_1 = 0$ and $C_2 = 0$, $E_{OM-DLBOP} < E_{DLBOP}$ if and only if $4\pi/3 < \beta$.
- With $\alpha = 2$ and $C_1 \neq 0$ or $C_2 \neq 0$, the inequality $E_{OM-DLBOP} < E_{DLBOP}$ is true if and only if:

$$d > \frac{4\pi\left(\left(\frac{2}{3} - \frac{\beta}{2\pi}\right) R^2 + \frac{2}{3}C_1 + C_2\right)}{3\sqrt{3}\left(\frac{\beta}{2\pi}C_1 + C_2\right)} , \qquad (5)$$

  where d is the density in number of nodes per communication area. For a density higher than this threshold, the best solution is to use the OM-DLBOP algorithm. Otherwise, it is better to use DLBOP.
- For $\alpha > 2$, $C_1 = 0$ and $C_2 = 0$, $E_{OM-DLBOP} < E_{DLBOP}$ if and only if $4\pi/3 < \beta$.
- With $\alpha > 2$ and $C_1 \neq 0$ or $C_2 \neq 0$, because $E_{OM-DLBOP}$ is a constant (it does not depend of n and β is fixed to $4\pi/3$), the inequality $E_{OM-DLBOP} < E_{DLBOP}$ can be reduced as:

$$n^{\left(1-\frac{\alpha}{2}\right)} \frac{\beta}{2\pi} \left(\frac{4S}{3\sqrt{3}}\right)^{\left(\frac{\alpha}{2}\right)} + n\left(\frac{\beta}{2\pi}C_1 + C_2\right) > E_{OM-DLBOP} . \qquad (6)$$

  The value $E_{OM-DLBOP}$ is a constant, whatever is the number of nodes. This is a polynomial, whose solutions will be described in the next section.

In DLBOP and OM-DLBOP, full network coverage is ensured by NES over LMST neighbor set. Regardless of MAC layer problems, the way of reaching theses LMST neighbors has no impact on coverage. It means that these exist, in the same broadcasting task, nodes applying DLBOP and nodes applying OM-DLBOP. The protocol we propose in next section is a solution where nodes locally decide independently between the two modes.

4.4 Adaptive Directed LMST Broadcast Oriented Protocol (A-DLBOP)

We are now in position to describe A-DLBOP algorithm which combines one-to-one and one-to-many communication models. The protocol A-DLBOP is a flooding protocol based on DLBOP and OM-DLBOP described above. Hence a node which receives the broadcasted messages starts a NES limited to LMST neighbors. At the end of the quiet period, the node has to choose between one-to-one or one-to-many communication models. For a given node u, the decision algorithm is the following one:

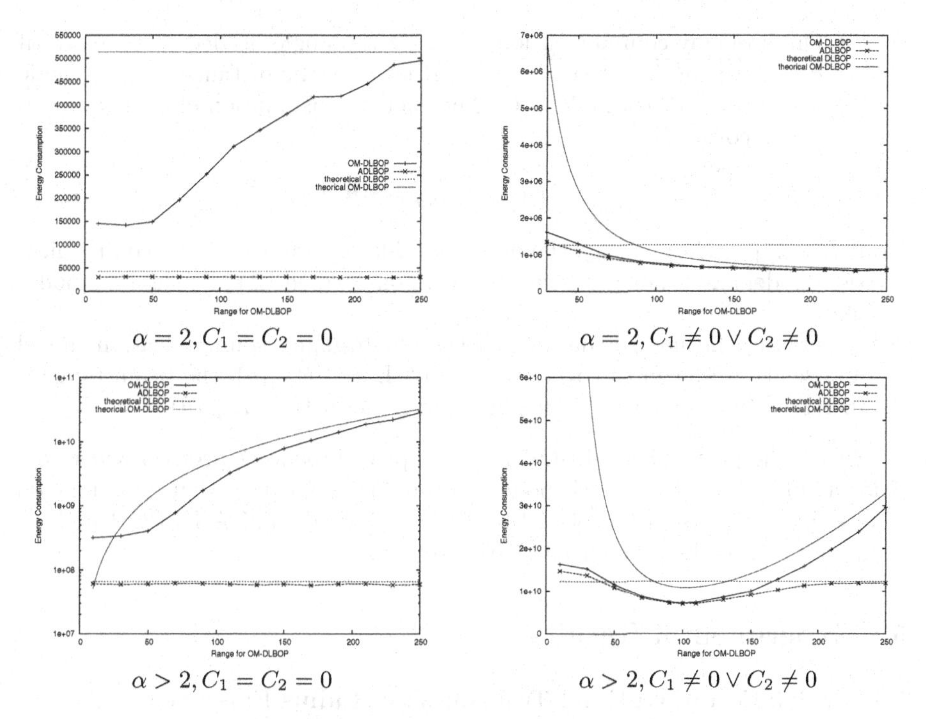

Fig. 5. Influence of R_{opt} on energy consumption.

- Let A be the non-covered neighbors set and $B \subseteq A$ the non-covered LMST neighbors set. We denote by A' the set of nodes belonging to A and closer than $R_{opt}(4\pi/3)$. As previously, if $C_1 = C_2 = 0$ then $R_{opt}(4\pi/3) = 0$. The "goal" of the node u is to cover the set $C = A' \cup B$. If the set C is empty, the retransmission is canceled.
- The communication model choice is made from a comparison of energy consumption needed to cover C:
  - <u>One-to-one communication:</u> while flooding over the subset C with one-to-one communication model, each node retransmits the message to its non-covered LMST neighbors. On average, each node has only one non-covered LMST neighbor. Hence the energy consumption with one-to-one communication model using β-beams can be evaluated by:

$$E_{1-to-1} = |C| \times e(\beta, d_{lmst}) \ .$$

The node u ignores the distance d_{lmst} which represents the average LMST edge length. It can simply estimate it with distance between itself and its LMST neighbors:

$$d_{lmst} \simeq \frac{1}{|B|} \sum_{v \in B} d(u, v) \ .$$

- One-to-many communication: let θ be the angle needed to cover C (if $\theta < \beta$ we consider that $\theta = \beta$) and let d be the distance between node u and the furthest node of C. The energy consumption of a single beam which covers C is:

$$E_{1-to-many} = e(\theta, d) \ .$$

- If $E_{1-to-1} < E_{1-to-many}$, the node u decides to use one-to-one communication model and sends a β-beam to each non-covered LMST neighbor (nodes of B).
- Otherwise, the node u decides to use one-to-many communication model and sends a θ-beam to cover nodes of C. If $d < R_{opt}(\theta)$, the beam range is increased in order to reach neighbors of A closer than $R_{opt}(\theta)$.

Hence, the protocol A-DLBOP is an adaptive broadcast protocol where decision is made locally by each node. We present in next section experimental results for our three protocols DLBOP, OM-DLBOP and A-DLBOP and we compare them to the centralized DBIP protocol.

5 Experimental Results

5.1 Validation of Optimal Transmission Radius Existence

The two protocols OM-DLBOP and A-DLBOP use an optimal transmission radius function R_{opt} which depends on energy consumption model (parameters α, C_1, C_2). This function R_{opt} has been theoretically studied in Section 4 and results are summarized in Table 1. The goal of the first experiments is to valid these results by replacing the function R_{opt} by a value we control.

We use randomly generated networks of 500 nodes (only connected graphs are considered). The other parameters are the size of the square grid $S = 1000 \times 1000$m (for a density around of 78 nodes by communication zone) and the minimal angle beam $\beta = \pi/9$. We evaluate theses instances for varying radius target R_{opt} from 10m to the maximal range (250m). NES timeout is randomly generated and the simulator uses an ideal MAC layer (with absence of collisions).

We compare the OM-DLBOP protocol and the A-DLBOP protocol, because they are the only ones which are influenced by the value of R_{opt}. We add to the graphs the values of E_{DLBOP} and $E_{OM-DLBOP}$ for the given configuration, to compare and justify the validity of the theoretical models.

The Figure 5 shows the impact of R_{opt} on energy consumption for different protocols. The graphs presents the EER (Expanded Energy Ratio: the results are normalized in function of the best energy economy, which is equal to 100) for each enery model. Concerning the case where $\alpha = 2$ and $C_1 = C_2 = 0$, although that theoretical study indicates that the energy consumption does not vary with transmission radius (Table 1), experimentally, the OM-DLBOP energy consumptions increases. This is probably due to extra retransmissions from NES. This fact validates the choice of $R_{opt} = 0$ in OM-DLBOP and A-DLBOP protocols for this energy model.

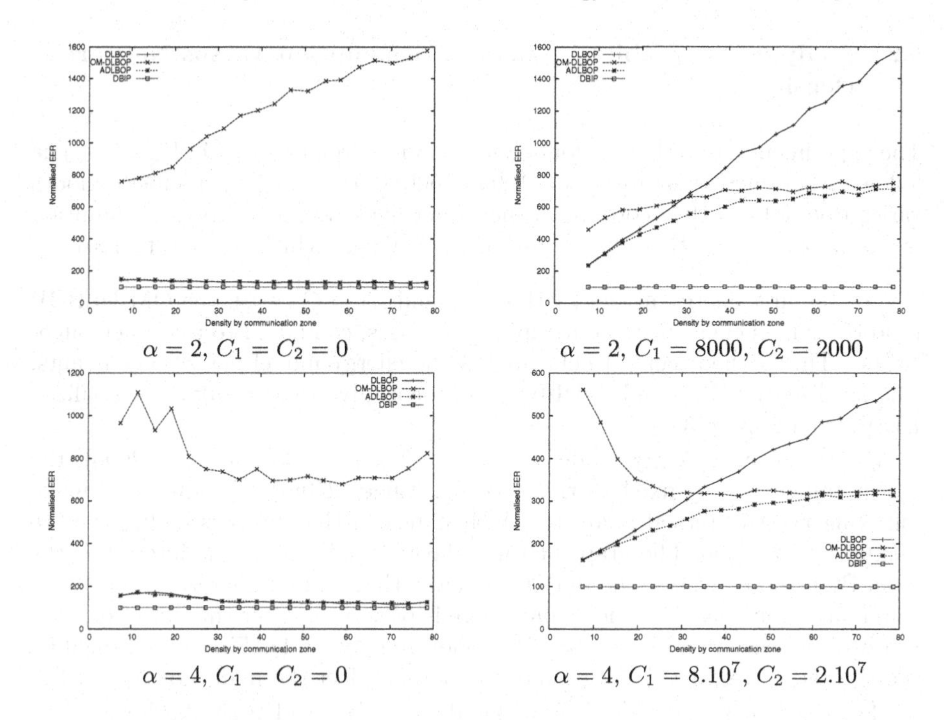

Fig. 6. Normalized Energy consumption for $S = 1000 \times 1000$ and $\beta = \pi/9$.

For the case $\alpha = 2$ and $C_1 \neq 0$ or $C_2 \neq 0$, we have demonstrated that the best solution is to maximize R_{opt}. Our simulations show that the values of the OM-DLBOP protocol and A-DLBOP are very close, and offer results identical to the theoretical value of the OM-DLBOP mode.

For the energy model $\alpha > 2$ and $C_1 = C_2 = 0$ (presented with a logarithmic scale for the ordinate), we have again confirmed the theoretical model validity, as the OM-DLBOP protocol follows the values given by $E_{OM-DLBOP}$. Again, the best energy saving is given by the DLBOP method. More generally, if the constants are null, then the best solution is to minimize the radius of each sending, and so the DLBOP protocol is used. Even if the A-DLBOP model gives better results, it is not necessary the same case for other values of β, n and S, as shown by the turning points between the theoretical models of DLBOP and OM-DLBOP.

For the last energy model, $\alpha > 2$ and $C_1 \neq 0$ or $C_2 \neq 0$, the OM-DLBOP protocol reaches minimal value for $R_{opt} = 99$ (compared with the theoretical value of $R_{opt} = 102.41$). The A-DLBOP protocol succeed to adapt its behavior as function of R_{opt}: it switches to the OM-DLBOP protocol when it becomes more interesting. Furthermore, once again, the experimental results follow the theoretical values.

5.2 Study of Adaptability: Energy Consumption Versus Network Density

The experimental data for the four energy models are presented in Figure 6. The following parameters are used: $S = 1000 \times 1000$m; $\beta = \pi/9$. The number of nodes varies from 50 to 500 according to the target density. The energy consumption is normalized as function of the best energy savings (which is equal to 100).

For the first energy model (with $\alpha = 2$ and $C_1 = C_2 = 0$), the OM-DLBOP protocol is inefficient, because it rapidly increases, compared to the others algorithms. This is the excepted behavior, as the energy model has zero constants. The localized DLBOP and A-DLBOP protocols give close results to globalized DBIP: from 26% to 33%.

For the second energy model (with $\alpha = 2$, $C_1 = 8000$ or $C_2 = 2000$), the radius target R_{opt} is fixed to the maximal range (250m). The choice is made following the theoretical analysis, which indicates that the best solution is to maximize the range. The graphs clearly show the advantage the interest of the A-DLBOP mode. When the density is lower than approximately 37 nodes by communication zone, the one-to-one mode is preferred. After this threshold, the protocol switches to a higher use of the one-to-many mode. This is confirmed by the theoretical result (eq. 5), despite the border effect (54 nodes for $\beta = \pi/9$).

The third energy model (*i.e.* $\alpha = 4$ and $C_1 = C_2 = 0$) has similar behavior to the first energy model: the DLBOP mode is always better. More generally, with null constants, it is better to use DLBOP solution. In this case, the OM-DLBOP mode is inefficient.

For the last energy model ($\alpha = 4$, $C_1 = 8.10^7$ and $C_2 = 2.10^7$), the OM-DLBOP algorithm is too energy-consuming for low densities. When the density increases, between 25 and 30 nodes by communication zone, the OM-DLBOP algorithm becomes better than DLBOP. The interesting fact is the A-DLBOP protocol takes advantages of the two communication models, by using them at the same time. Hence, A-DLBOP has better energy-saving than DLBOP and OM-DLBOP. The figure confirms the inequality (6), with an experimental threshold value of 30 for a theoretical value of 35. The other inequality (6) is also correct in respect of the figure.

6 Conclusion

We have proposed in this paper a new localized protocol called A-DLBOP, that combines one-to-one and one-to-many communication models. The first suffers from high overhead with high density, compared to DBIP. The second communication model corrects this problem by using variable angle to cover large number of nodes The protocol A-DLBOP uses either the first or the second model as function of a local evaluation of the energy consumption. We proposed a theoretical evaluation of the performance of the algorithm. These results are confirmed by the experimental data.

References

1. Wieselthier, J., Nguyen, G., Ephremides, A.: On the construction of energy-efficient broadcast and multicast trees in wireless networks. In: Proc. IEEE INFOCOM 2000, Tel Aviv, Israel (2000) 585–594
2. Spyropoulos, A., Raghavendra, C.: Energy efficient communications in ad hoc networks using directional antennas. In: Proc. IEEE INFOCOM 2002, New-York, USA (2002)
3. Wieselthier, J., Nguyen, G., Ephremides, A.: Energy-limited wireless networking with directional antennas: the case of session-based multicasting. In: Proc. IEEE INFOCOM 2002, New-York, USA (2002)
4. Kirousis, L., Kranakis, E., Krizanc, D., Pelc, A.: Power consumption in packet radio networks. In R.Reischuk, M.Morvan, eds.: Proc. 14th Symposium on Theoretical Computer Science (STACS'97). Volume 1200 of Lecture Notes in Computer Science., Hansestadt Lübeck, Germany, Springer-Verlag, Berlin (1997) 363–374
5. Cartigny, J., Simplot, D., Stojmenović, I.: Localized minimum-energy broadcasting in ad-hoc networks. In: Proc. IEEE INFOCOM 2003, San Fransisco, USA (2003)
6. Cartigny, J., Simplot, D., Stojmenović, I.: Localized energy efficient broadcast for wireless networks with directional antennas. In: Proc. IFIP Mediterranean Ad Hoc Networking Workshop (MED-HOC-NET'2002), Sardegna, Italy (2002)
7. Toussaint, G.: The relative neighborhood graph of finite planar set. Pattern Recognition **12** (1980) 261–268
8. Li, N., Hou, J.: BLMST: A scalable, power-efficient broadcast algorithm for wireless sensor networks. Submitted (2002)
9. Cartigny, J., Ingelrest, F., Simplot-Ryl, D., Stojmenović, I.: Localized LMST and RNG based minimum-energy broadcast protocols in ad hoc networks. Ad Hoc Networks (2004) to appear.
10. Li, N., Hou, J., Sha, L.: Design and analysis of an MST-based topology control algorithm. In: Proc. IEEE INFOCOM 2003, San Francisco, USA (2003)
11. Rodoplu, V., Meng, T.: Minimum energy mobile wireless networks. In: IEEE J. Selected Area in Comm. Volume 17. (1999) 1333–1344
12. Peng, W., Lu, X.: On the reduction of broadcast redundancy in mobile ad hoc networks. In: Proc. Annual Workshop on Mobile and Ad Hoc Networking and Computing (MobiHOC 2000), Boston, Massachusetts, USA (2000) 129–130
13. Stojmenović, I., Seddigh, M., Zunic, J.: Dominating sets and neighbor elimination based broadcasting algorithms in wireless networks. IEEE Transactions on Parallel and Distributed Systems **13** (2002) 14–25
14. Ingelrest, F., Simplot-Ryl, D., Stojmenovic, I.: Target radius over LMST for energy-efficient broadcast protocol in ad hoc networks. In: IEEE International Conference on Communications (ICC'2004), Paris, France (2004)

An Enhanced Power Save Mode for IEEE 802.11 Station in Ad Hoc Networks*

Yeonkwon Jeong[1], Jongchul Park[1], Joongsoo Ma[1], and Daeyoung Kim[2]

[1]School of Engineering, Information and Communications University,
Daejeon, 305-714, KOREA
{ykwjeong, jcpark, jsma}@icu.ac.kr
[2] Dept. of InfoCom Engineering, Chungnam National University,
Daejeon, 305-764, KOREA
dykim@cnu.ac.kr

Abstract. The IEEE 802.11 standard requires that a beacon station has to stay in awake after generating a beacon frame to serve probe request. This requirement make a reason to consume valuable power of beacon station even though there are no frame exchanges between stations or there exists only one station in the ad hoc network. This paper proposes an ePSM(Enhanced Power Save Mode) to reduce effectively the power consumption of beacon station in stand-by state. We also develop a power consumption model and show that the ePSM can save 73% energy at beacon station per a BI(Beacon Interval) with measurement results.

1 Introduction

Today wireless LANs based on the IEEE 802.11 standard are widely used at offices and homes. Stations can form a network using one of the two modes of operation specified in the standard: the infrastructure mode and the ad hoc mode.

The infrastructure mode is suitable for stations to connect to the enterprise network and the Internet. In this mode, one station must be designated as an AP(Access Point). The AP is the smart master and other stations follow the control of the AP. Stations must communicate via the AP. Wireless LANs(Local Area Network) use the infrastructure mode.

On the other hand, the ad hoc mode allows stations to form an ad hoc network in a peer-to-peer manner. Stations communicate to each other directly without requiring any central controller such as an AP. Therefore networking can take place anywhere the stations meet. The ad hoc mode is useful in connecting computers, peripheral devices and multimedia appliances at home or offices. It is particularly useful when notebook computers or PDAs are joined together to play network games. Since there is no AP in the middle to route packets between stations, the throughput can be much higher with a less delay. However these stations are often portable and rely on their battery-supplied power. Therefore they must save power, particularly during the stand-by periods in which they are not sending or receiving any packet.

* This research was supported by University IT Research Center Project.

I. Niemegeers and S. Heemstra de Groot (Eds.): PWC 2004, LNCS 3260, pp. 414–420, 2004.

So far the majority of the IEEE 802.11 market has been in the infrastructure mode, and it is not well known how effective the standard method of saving power in the ad hoc mode is. There are a few research activities to conserve power consumption by considering characteristics from MAC(Medium Access Control) to application layer. Their researches focus on measurement by experiments, and analyze those results with existing WaveLAN at infrastructure mode[2], or Lucent IEEE802.11 at ad hoc mode[3]. Other researches measure power consumption at application oriented view such as email[4] and web[5]. The last approaches use a coordinator which stay in awake continuously[6], or a soft-state timer for transition between active and sleep mode[7].

In this paper, we propose an ePSM(Enhanced Power Save Mode) to reduce effectively the power consumption of beacon station in stand-by state. We also develop a power consumption model of the IEEE 802.11 ad hoc mode operation. Our model is based on physical measurements. We place our emphasis in investigating the effectiveness of ad hoc mode power saving features during the stand-by periods. This environment is important because we expect that the IEEE802.11 ad hoc mode will be most popular in offices and homes, where a few stations are connected and left connected for a whole day, waiting for occasional communication works to occur.

The remainder of this paper is organized as follows. Section II reviews the IEEE 802.11 standard in view of PSM and ad hoc mode. Section III shows the observed problem of current standard and proposes new ePSM to resolve this issue. Section IV presents numerical and experimental results. Finally Section V concludes this paper.

2 Overview of the IEEE 802.11

We will review ad hoc network setup and maintenance procedures in the aspect of the beacon frame, scanning procedures, and PSM of IEEE 802.11.

2.1 Beacon Frame

In an ad hoc network, the generation of beacon frame is completely distributed among the mobile stations of the network. Each station will attempt a back-off for a random time to send a beacon frame every BI(Beacon Interval) . In the random time, if a station hears a beacon it cancels its transmission. A beacon frame contains ad hoc network configuration parameters. Those are not only timing information for synchronization, but also network identifier, a BI, ATIM(Announce Traffic Indication Message) window size, and the data rates that can be supported.

2.2 Scanning Procedures

Scanning procedures are necessary when station want to join an already existing an ad hoc network. If the scanning procedure does not find any network, the station may start with the creation of a new network. The scanning procedures can be either passive or active. In a passive scanning the station only listens to the channel for hearing a beacon frame. However, the station of active scanning generates probe request frames, and processes probe response frames received. Only the station that generated the last beacon frame will respond to a probe request, in order to avoid the waste of bandwidth with repetitive control frames.

2.3 Power Management

A station may be in one of two different power states: awake which is fully powered or doze which is not able to transmit or receive frames because of low power. Stations having frames destined to a power-conserving station are first announced ATIM frame to wake up one or more doze stations during ATIM window. A station operating in the PSM listens to these announcements and, based on them, decides whether it has to remain awake or not.

3 Observed Problem and Solution

3.1 Observed Problem

The IEEE 802.11 standard requires that the station which send a beacon frame shall remain in the awake state until the end of the next ATIM[1]. Basically, the reason of this restriction comes from an intention to use active scanning as a default mode. Only the station that generated the last beacon frame will respond to a probe request, in order to avoid the waste of bandwidth with repetitive control frames.

However, the standard prevent from conserving the power consumption of beacon station even though it is possible. The standard is missing the following cases when the network is ad hoc. One of possible examples is when a beacon station is being stand-by. Stand-by means that a station does not have any user traffics to transmit or receive frames. The other is that as a subset of previous case, an ad hoc network has only one member being stand-by. In both cases the station should always stay in awake to meet the IEEE 802.11 requirements without doing anything.

If the standard permits beacon station being stand-by to go doze state, the power consumption of beacon station can be saved as much as $\Delta(mW)$. The power gains is $\Delta(mW) = \int_{T_{doze}(t)} (p_{awake} - p_{doze})dt$. Where, $T_{doze}(t)$ is a time period over BI-ATIM window. p_{awake} and p_{doze} are the power required for station to be in awake and doze state.

3.2 Enhanced Power Save Mode

In this paper we propose new ePSM mechanism to improve power consumption of beacon station in stand-by state. ePSM is that

- Ad hoc network use the passive scanning instead of the active scanning of IEEE 802.11 standard.

- A beacon station may go to doze state if the station does not have transmit or receive frames. Like a Fig. 1, the beacon node of IEEE 802.11 shall remain awake during one BI irrespective of having transmitting or receiving frames. But we permit this node to go doze in that case.

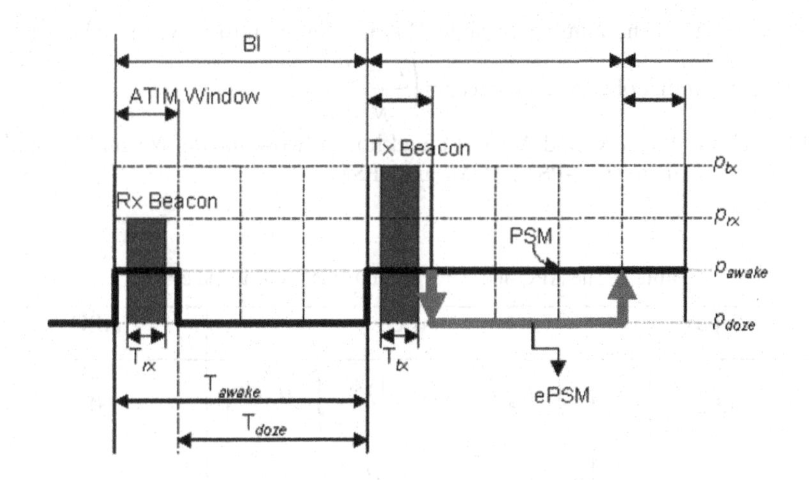

Fig. 1. The operation of PSM and ePSM

4 Numerical and Experimental Results

4.1 Numerical Results

We will use APC(Average Power Consumption) and APG(Average Power Gains) as the comparison parameters of different PSMs. APC is the averaged amount of power required to operate during one BI in an ad hoc network having N stand-by stations. The power consumption of any station is expressed by the sum of these values dependent on the state of the station:

$$\text{transmit: } \int_{T_{tx}(t)} p_{tx}dt, \text{ receive: } \int_{T_{rx}(t)} p_{rx}dt,$$

$$\text{awake: } \int_{T_{awake}(t)} p_{awake}dt, \text{ doze: } \int_{T_{doze}(t)} p_{doze}dt$$

In the non-PSM, an ad hoc network consists of a beacon transmit station and the beacon receive stations of N-1. All station of non-PSM remains always awake state and the APC of non-PSM is sum of transmit, receive and awake state power. In the PSM, the beacon station is awake and the other stations of ad hoc network are doze. The APC of PSM is calculated by reduced value from non-PSM as much as the number of doze state stations. In the ePSM, all station is doze state irrespective of beacon generation. So we can minimize the energy consumption among these three types.

APG is a relative value which is reducible amount from the power consumption of non-PSM. Hence the APG is zero for non-PSM. However APG for PSM is $\left(\frac{N-1}{N}\right)\Delta$, and ePSM is Δ. If the number of node *N* is increasing infinitely, the APG of PSM will converge on that of ePSM such as $\lim_{N\to\infty}\left(\frac{N-1}{N}\right)\Delta \simeq \Delta$.

Table 1 shows the APC and APG of an ad hoc network having *N* stand-by stations for 3 type mechanisms: non-PSM, PSM, and ePSM.

Table 1. The APC and APG for non-PSM, PSM, and ePSM

Types	APC	APG
non-PSM	$\dfrac{\left\{\begin{array}{l}\int_{T_{tx}(t)} p_{tx}dt+(N-1)*\int_{T_{rx}(t)} p_{rx}dt \\ +N*\int_{T_{awake}(t)} p_{awake}dt\end{array}\right\}}{N}$	0
PSM	$non\_PSM-(\frac{N-1}{N})\Delta$	$(\frac{N-1}{N})\Delta$
ePSM	$non\_PSM-\Delta$	Δ

Fig. 2 shows the transition of APG when Δ is normalized to 1 and the number of stations participated in ad hoc network are increasing. As you can see the results, ePSM is effective when the number of stations is small than any other mechanisms. This is a significant gain because there would be only one station when the ad hoc network is started, and there would be many applications where two or three stations are connected.

4.2 Experimental Results

We conduct measurement to get the practical power values of Table 1 such as p_{tx}, p_{rx}, p_{awake}, and p_{doze}. The circuit and methodology are closely based on the experiments reported in [3]. We are using SAMSUNG SENSE T10 laptop computers running Linux RedHat8.0 with Cisco Aironet 350 series client adapter and PCCextend 140A

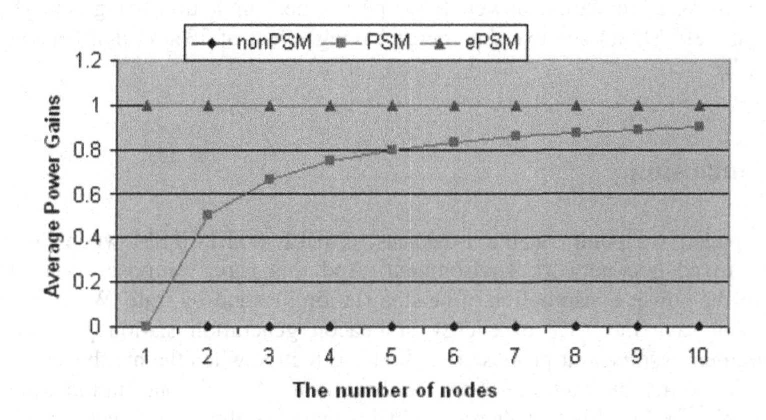

Fig. 2. The APG of non-PSM, PSM, and ePSM in an ad hoc network

card bus extender. The BI and ATIM window values applied to this measurement are 100msec and 5msec respectively. Transmit power is 100mW and input voltage is 4.8165.

Measurement results show that awake state station keeps going on idle phase for a BI continuously. There is also another interesting point on the doze state station. After ATIM window, dose state station goes through sleep phase, and reaches warm-up phase to prepare a transition from dose to awake state. Warm-up takes for 13.14msec. Table 2 is the summary of measured average currents and power consumptions in each phase.

Table 2. Aironet 350 NIC characteristics

Phase	Current(mA)	Power(mW)
Transmit	270.83	1,304.476
Receive	198.17	954.507
Idle	164.12	790.47
Sleep	35.07	168.91
Warm-up	105.75	509.35

Let's calculate the absolute power gains Δ(mW) using the measurement results. Our interest is the difference of power consumption between awake and doze state station during the interval of BI-ATIM window.

$$P_{awake}(mW) = 790.47 * (BI - ATIM\ \ Window)$$

$$P_{dose}(mW) = 168.91 * T_{Sleep} + 509.35 * T_{Warmup}$$

Therefore, the Δ (mW) is 54577.1186. A doze state station uses the small portion 27% of awake state station power. If we permit beacon station being stand-by to go doze state, ePSM can achieve the energy saving effect of 73% at that beacon station per one BI.

5 Conclusion

In this paper, we point out the weakness of IEEE 802.11 PSM when the PSM is applied to ad hoc network environment. And this paper proposes new ePSM to enhance the power consumption of beacon station in stand-by state. We show that the ePSM can save the 73% of energy at beacon generation station per a BI using measurement results. Our proposed method is effective when the number of stations is small. This is a significant gain because there would be only one station when the ad hoc network is started, and there would be many applications where two or three stations are connected in home or office environments.

References

1. IEEE Standard 802.11b, Part 11: Wireless LAN Medium Access Control(MAC) and Physical Layer(PHY) specifications : High Speed Physical Layer Extension in the 2.4GHz Band, 1999.
2. R. Kravets, and P. Krishnan, “Power Management Techniques for Mobile Communication”, Proc. of MOBICOM 1998, Oct. 1998.
3. L. M. Feeney, and M. Nilsson, “Investigating the Energy Consumption of a Wireless Network Interface in an Ad Hoc Networking Environment”, Proc. of INFOCOM 20001, Apr. 2001.
4. M. Stemm, and R. H. Katz, “Measuring and Reducing Energy Consumption of Network Interfaces in Handheld Devices”, IEICE Trans. on Fundamentals of Elec., Comm., and Computer Science, Aug. 1997
5. P. Gauthier, D. Harada, and M. Stemm, “Reducing Power Consumption for the Next Generation of PDAs: It’s in the Network Interface!”, Proc. of MoMuc 1996, Sep. 1996
6. B. Chen, K. Jamieson, H.Balakrishnan, and R. Morris, “Span : An energy-efficient coordination algorithm for topology maintenance in ad hoc wireless networks”, Proc. of MOBICOM 2001, Jul. 2001
7. R. Zheng, R. Kravets, “On-demand Power Management for Ad Hoc Networks”, Proc. of INFOCOM, 2003

Achieving "Always Best Connected" Through Extensive Profile Management

Lucian Suciu[1], Jean-Marie Bonnin[1], Karine Guillouard[2], and Bruno Stévant[1]

[1] École Nationale Supérieure des Télécommunications de Bretagne
BP 78 - 2, rue de la Châtaigneraie, 35512 Cesson-Sévigné, France
{lucian.suciu, jm.bonnin, bruno.stevant}@enst-bretagne.fr
[2] France Télécom R&D, DMR/DDH Laboratory
BP 59 - 4, rue de clos Courtel, 35512 Cesson-Sévigné, France
karine.guillouard@francetelecom.com

Abstract. The integration of various access networks into a ubiquitous, yet heterogeneous, wireless environment is on the way. This evolution of the mobile network will give the end-user a greater choice of access technologies, and, therefore, the decision to select the "best" interface and access network from many possible combinations has to be taken. The decision will depend on information such as: performances and capabilities of the available networks, requirements from applications, user preferences, or network operators' constraints. Our work focuses on an advanced middleware which deals with profile management to support the interface automatic configuration and selection. Furthermore, the proposed mechanism supports the dynamic (re)mapping of the application flows by taking into consideration multiple selection criteria.

1 Introduction

The main characteristics of the next-generation all-IP mobile architectures can be foreseen by carefully considering the current trends. First of all, we notice a greater choice of access networks and simultaneous multi-access of these networks, including IEEE 802.11a/b/g WLAN, IEEE 802.15 WPAN (embracing Bluetooth), IMT-2000, IEEE 802.20 MBWAN, and so on. Then, there is an increasing number of multimedia communicators and mobile terminals with outstanding performances, such as smartphones, PDAs, tablet PCs, and laptops. Finally, we reckon that there will be a great demand of advanced yet simple to use mobile services comprising mobile commerce, adaptive and self-configuring services, context aware applications, user profiling and personalisation, etc.

This heterogeneous communication environment has already opened up new research areas, e.g. the Mobile IP and its micro-mobility suite, the Quality of Service ([1], [2]), or the Authentication, Authorization and Accounting. However, we believe that two basic requirements have already clearly emerged. The first one states that users should be provided with seamless roaming amongst various access networks (e.g. [3]), including simultaneous or successive connections to several access technologies. The second one mandates that users should

I. Niemegeers and S. Heemstra de Groot (Eds.): PWC 2004, LNCS 3260, pp. 421–430, 2004.

be allowed to always stay connected through the "best" access network (e.g. [4], [5]).

To achieve these goals we propose, design and implement an add-on middleware which is adaptable and reconfigurable. It is adaptable because the terminal always considers the current context (e.g., users' preferences, terminal resources, networks' conditions and applications' needs) and it tries to continuously adjust to the context when communicating. It is reconfigurable because the user or the network operator can redefine their preferences, subscribe to new services, add new configurations for the network interface cards, and so on.

As for adaptive applications we provide a clearly defined API towards them; we have also worked out a solution which handles the legacy applications, i.e. the applications which are unaware of our add-on middleware.

The rest of the paper is organised as follows. We will first present the related work in Section 2. Then will we describe our proposed architecture in Section 3. In the next section we will show the Profile Manager (PM) in detail and will look into our Selection Decision Algorithm (SDA) in Section 5. Section 6 will present the implementation and the results obtained so far. We will reveal the future work and conclude with Section 7.

2 Related Work

There is a growing number of research and standardisation efforts related to profile definition and handling. For example, in [6], and references therein, the notion of Generic User Profile (GUP) is defined as being a collection of data stored and managed by different entities such as the user equipment, the home or visited networks, and which affects the way the end-user experiences the different services offered. Then, the WAP User Agent Profile (UAProf), defined in [7], is concerned with capturing classes of device capabilities and preferences. The Composite Capability/Preference Profiles (CC/PP) framework (see [8]) is yet another mechanism for handling the preferences associated to users and user agents accessing the World Wide Web.

The Information Society Technologies (IST) AQUILA project tries to provide dynamic control to DiffServ based traffic; an objective of this project is to define and manage application profiles which contain the concrete application descriptions. Moreover, the IST-TRUST project tries to understand the users' requirements related to reconfigurable radio systems. It also defines a layered architecture which contains a policies and profiles management component. These profiles are further refined within IST-SCOUT project.

On the other hand, in the recent years, the interface selection problem for a multi-interface terminal, communicating in heterogeneous wireless environment has gained importance (e.g., [9], [10], [11]). However, to our knowledge, less work has been done to integrate the profile management with the optimal interface selection issue.

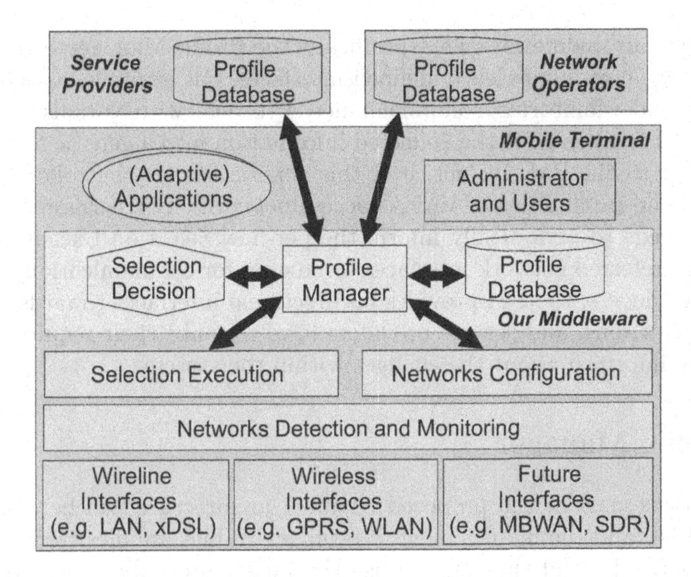

Fig. 1. Proposed architecture

3 The Proposed Middleware

Currently, after terminal start-up, the user surveys the communication and, if several access networks exist, the user decides which one to employ. To overcome this inconvenient, our approach relies on investigating the access networks capabilities and the applications limits and also interacting with the users in order to obtain their preferences.

We will gather in a structured way all these capabilities, requirements, and preferences in well-defined profiles. Then, the decision to use or not to use an access network will be based on these profiles. The final goal is to provide for each application flow the "best" access technology within multi-interface mobile terminals. Fig. 1 shows the envisaged mobile terminal architecture, its components and the possible interactions amongst them. Interfaces above our middleware collect the users, the operators and the service providers preferences and handle the applications' requirements. The layers below the middleware detect the available networks, provide real-time information about communication interfaces and access networks capabilities, or perform on-demand network interface configurations. Furthermore, a separate component handles the selection execution process, i.e., it actually maps the application flows on particular interfaces. Our middleware will control all these "low-level" layers by initiating network configurations and performing interface selection decisions.

The architecture is split in various functional bricks because this modular design facilitates implementation and testing and it also permits the gradual integration of better selection decision algorithms, of novel network detection and monitoring techniques (e.g. [12]), or of fine-grained selection execution modules (e.g. Per-Flow Movement).

Between our middleware, i.e., specifically the Profile Manager, and the external blocks we have used clearly defined interfaces. For example, as the Network Detection and Monitoring component needs to deal with various access technologies, it has to convert the collected information into a generic format which is then sent to the PM. In fact, it is this generic format that allows the comparison of the capabilities of various access networks. Furthermore, the Profile Manager needs to periodically inform the per-flow Selection Execution module about the preferred network interfaces to be used for communication. As for the "high-level" layers, the PM provides bi-directional interfaces towards them, i.e., users, applications, and service providers have to make their requirements and have to be informed about the changes within the system.

4 Profile Manager

An important part of the proposed terminal architecture will be dedicated to the definition and the management of profiles. Profiles are files stored in Profile Databases (PDB) and they summarise key information about the components of the system and its interactions with the environment, i.e., users, applications, access networks, or service providers. Specifically, the profile handling mechanism serves the following purposes:
– triggers and assists the SDA when it makes the choice of the "best" access;
– automatizes the selection of an access network by maintaining all the necessary information for proper interface configuration;
– sets forth a solution which works both for adaptive applications and for unaware applications.

We propose three kinds of profiles within the Profile Databases: generic, specific and active profiles. The generic profiles describe what information could be stored in the various types of profiles, i.e., they can be seen as patterns or schemas. We consider four generic profile types within our Profile Databases:

1. Preferences and Resources Profile (PRP): it has been noticed that one's preferences depend on the currently existing resources or the present situations. Thus, the generic PRP specifies how the system should behave based on the available resources or the current context. The preference parameters considered here are, e.g., selection mode, selection goal, and preferred and forbidden access networks. The system itself may provide the information about the current context, e.g., battery status, geographical location, or subscription type.
2. Flow Description Profile (FDP): the first part of this profile holds the application's QoS requirements (e.g., service class, minimum necessary bit rate, typical delay expected, maximum delay variation) and the second part contains the QoS monitored by the system (e.g., mean bit rate, bit error rate, average latency). We also propose default parameter values for some of the common applications.
3. Network Interface Profile (NIP): it comprises network interface card parameters that can be obtained from technical specifications (e.g., maximum

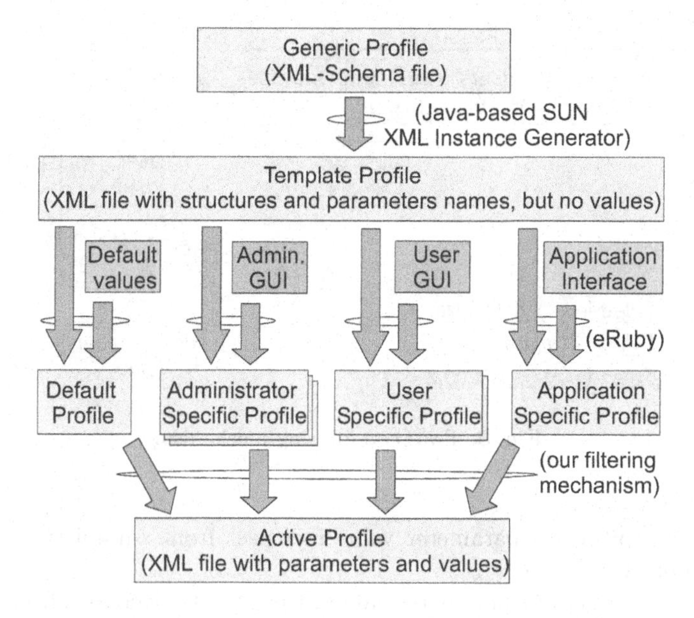

Fig. 2. Obtaining the active profile

theoretical throughput supported), or throughout offline statistical measurements (e.g., maximum real throughput).

4. Access Network Profile (ANP): it specifies all the necessary information, for both Layer 2 and Layer 3, required to successfully configure and use an access network. The Layer 2 part of the ANP contains the mandatory parameters required to associate with the network, the authentication and encryption protocols, the QoS settings and the billing and charging parameters. The Layer 3 part comprises the IPv6 address configuration if needed so, the Mobile IPv6 settings, and various tunnelling configurations. The accounting information from the ANP is intentionally simplified for the time being (i.e., a flat cost model is used), and it is globally defined for an access network, regardless of the traffic type (i.e., business or leisure).

Likewise, the administrators, the network operators, the users and the applications can instantiate a generic profile (i.e., automatically provided) to build specific profiles corresponding to specific cases (e.g. Jean/Player/802.11/Office).

Finally, the third kind of profile is the active profile which is obtained by filtering different specific profiles of the same type. The profile handling mechanism described above is depicted in Fig. 2. Because the administrator (or network operators), the users and the applications (or service provides) can have their own specific profiles, we need to filter somehow the various values defined for the same parameter in different specific profiles of the same type. The following rules apply to the parameter values when an active profile is inferred:
– all parameters have a value type, i.e., mandatory or proposed, and the mandatory value for a parameter has always priority over the proposed value; – the

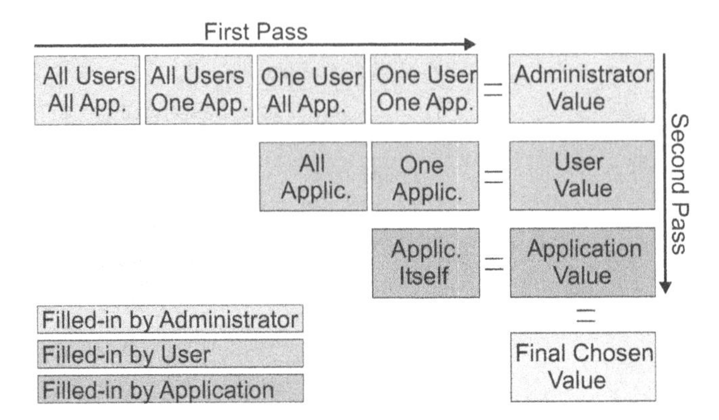

Fig. 3. Two-pass filtering mechanism

priority for a proposed parameter value increases from administrator to user, then to application;
– if there are mandatory parameter values, the priority decreases from administrator to user, then to application;
– some of the parameters may obey to the mutual exclusion rule, e.g., if the administrator sets a forbidden network for a user, the user can set the same access network to "preferred", but its preference is ignored in this case.

In Preferences and Resources Profile case we need to further extend the process to a 2-pass filtering mechanism. This is necessary as the administrator (or e.g., the network operator or the service provider) needs to designate its preferences on a more fine-grained basis: she or he can define a specific PRP for all users and all applications, for a user and all applications, for all users and an application, or for a user and an application. Furthermore, the users themselves can define their preferences as being common for all applications they use, or just for a particular application.

Thus, for each parameter within specific PRPs, we need to construct a kind of hierarchy when an active PRP needs to be obtained, as shown in Fig. 3. Then, we perform the first pass and construct the input for the second pass (i.e., the last column) by picking up the last proposed parameter value or the first mandatory parameter value on each line. Next, we execute the second pass and we chose as final value the last proposed parameter value or the first mandatory parameter value found in the last column.

The information stored within all profiles is managed in a uniform and extensible manner using the XML paradigm. Furthermore, as depicted in Fig. 1, the profiles may be distributed amongst different entities and thus, e.g., a SOAP or an XML-RPC protocol will be required to handle them. However, it can be pointed out that the Profile Manager acts as a dispatcher within our architecture: it interacts with all the entities which supply the profiles, it knows which information needs to store in the Profile Database, and it implements the filtering mechanism. Moreover, after updating the PDB, the PM determines if

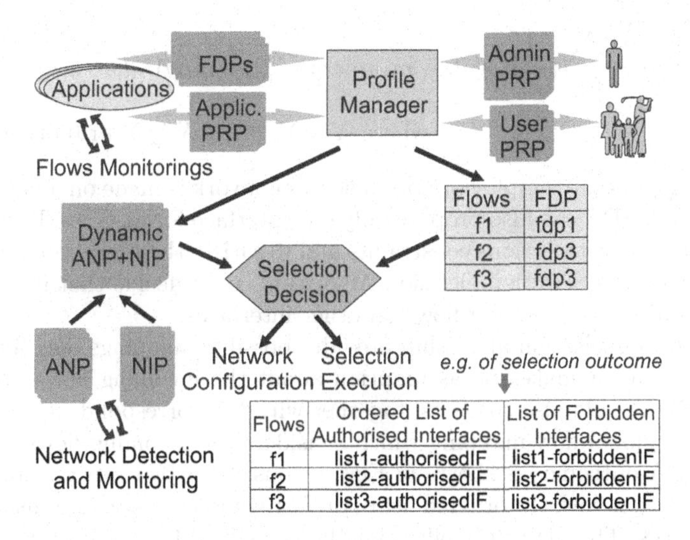

Fig. 4. Profile interdependencies

the SDA should be triggered or not. Fig. 4 shows the relations amongst various profiles defined within our framework and the selection algorithm outcome.

5 Selection Decision Algorithm

Most of the interface selection algorithms or the current handover algorithms take into consideration just one selection criterion, usually the Received Signal Strength (RSS) or the Signal-to-Noise Ratio (SNR). More recently the multiple selection criteria algorithms have emerged as a better alternative.

The Profile Manager determines if the Selection Decision Algorithm needs to be informed or not about the changes within the terminal. Thus, the SDA does not need to know how the parameters are collected within profiles or how the selection decisions are enforced.

The PM must restrain the plethora of various triggers, otherwise the SDA could be activated too often and it will exhaust the CPU or the battery. Based on the triggers received from the Profile Manager, the SDA interrogates the Profile Database and starts its computational procedure. Thus, various active profiles (i.e., the ones resulted after the filtering process) are used as input by the SDA in order to select the "best" interface for each application flow. The procedure used by the SDA is to define access network score functions and application utility functions which are both maximized. The score is calculated for each network i and the utility for each network-flow i, j 2-tuple. Then, to solve this multiple-goal problem, we use the weighting objectives method.

$$Score(i) = \sum_{i=1}^{N} (w_i \cdot \|monitoredValue_i\|); \text{ where } \|\| \text{ is the } ln \text{ or } tanh \text{ function}$$

$$Utility(i,j) = \sum_{i=1}^{N}(\|monVal_i - minNecVal_j\|) \cdot (monVal_i \otimes minNecVal_j);$$
$$\text{where } X \otimes Y = 1 \text{ if } X \geq Y \text{ or } 0 \text{ otherwise}$$

The suggestion to employ or not an access network is made on a flow-per-flow basis, but the SDA can also propose only one interface for all flows. Furthermore, in order to cope with various constraints and to satisfy the users and applications requirements, the Selection Decision Algorithm could decide that it is better to (re)map some of the existing flows on other interfaces.

The current selection algorithm provides possible mappings only for the outgoing flows and it makes no assumption about the incoming flows. Yet, a distributed SDA could also suggests, together with the correspondent nodes or the networks, a global flow mapping which considers the incoming flows as well.

The SDA outcome, as shown in Fig. 4, consists of two lists of interfaces for each application flow: the ordered list of preferred interfaces and the list of forbidden interfaces. The SDA only offers middle term (i.e., hundreds of milliseconds) handover decisions. Nevertheless, our framework supports an extensible interface with the per-flow Selection Execution component, which actually maps the flows and makes short-term (tens of ms) adaptations when needed. This happens, e.g., when RSS/SNR drops below the communication sustainable limit and the Selection Execution module immediately re-maps the flows on the next preferred interface from the list provided by the SDA.

6 Implementation and Results

To implement and test the proposed architecture we have chosen two terminals: IPaq 3970 with Familiar Linux, and Dell Latitude C610 running Debian Linux. To support the L3 mobility we installed the Mobile IPv6 for Linux distribution (i.e., MIPL). As wireless access technologies, we have 802.11b and Bluetooth access points and a commercially available GPRS network.

We prefer to store all the profiles within the local Profile Database for the time being. Nonetheless, we implement all four types of profiles using XML and XML Schemas. The handling and filtering of profiles within the Profile Manager and PDB, and the Selection Decision Algorithm are implemented using Ruby1.8, which is a portable, lightweight, object-oriented scripting language.

The inter-module communication within our middleware is done through Distributed Ruby, and with external modules through an XML-RPC-like protocol. To detect and configure the network interfaces we employ bash scripts.

We use the following selection initiation triggers: interface ready, interface not ready, preferences and resources changed, flow created, and flow deleted. As decision criteria, we use from active PRP the list of forbidden access networks for user and application, the monetary cost vs. QoS goal parameter, the required security level for application, and the battery status. From ANPs we take the monitored bit rate, the average bit error rate, the cost per byte and the security level of the access network. Finally, from NIPs we obtain the theoretical bit rate, and from FDPs we use the minimum necessary bit rate, the supported bit error rate and the maximum delay. Fig. 5 illustrates two of the most common use-cases, application launch and, respectively, interface status change (either the interface loses the connection with the network, or it gets associated).

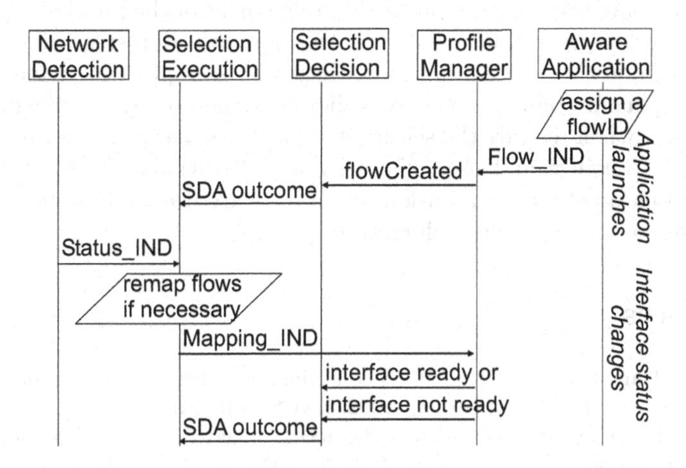

Fig. 5. Two common use-cases

The results obtained so far are encouraging. The selection latency (i.e., starting from selection initiation trigger till the phase when decision to (re)map a flow was taken) depends on trigger type, and it takes 2-3 seconds on the PDA and under a second on the laptop. Yet, we need to do a thoroughly evaluation of our architecture by taking into account, e.g., the selection cost gain (in general the monetary cost gain, but the aggregate bandwidth gain or security gain should not be overlooked), the number of flow re-mapped in similar conditions and during a well-defined period of time, the number of cut off flows, and so on.

However, as we already have a full functioning prototype, it is quite easy to update the profiles with new parameters and to test new selection algorithms.

7 Conclusions and Future Work

In this paper we have presented an adaptable and reconfigurable architecture for the mobile terminals supporting multiple access network interfaces. Our first ambition was to provide seamless access over heterogeneous networks, including simultaneous or successive connections to several access technologies. The second goal was to allow the user to always stay connected through the "best" access network. The proposed architecture includes adaptation mechanisms and relies on tight interactions amongst the different layers, from the application layer to the data link layer.

Thus, we argue that future multi-criteria handover algorithms will be more complex than nowadays, and they need to be divided into two parts: a contributory middle-term handover algorithm (i.e., the one which uses various profiles, as presented in this paper) and an essential short-term handover algorithm (i.e., the one traditionally based on RSS or SNR).

Our on-going work focuses on further refinement of the profiles, e.g., uniform monetary cost representation. In addition, more selection strategies need to be investigated and the most promising of them will be implemented and evaluated. Specifically, in our opinion, three areas should be thoroughly examined: initiation triggers (i.e., when to start the selection algorithm), decision criteria (i.e., which parameters to collect and use), and selection algorithms (i.e., how to combine the chosen parameters). In addition, more tests are needed in order to grasp all the benefits of our terminal architecture.

References

1. Fu, X., Karl, H., Kappler, C.: QoS-conditionalized handoff for Mobile IPv6. In E. Gregori et al (Eds.): Networking 2002, LNCS 2345, May 2002
2. Wisely D., et al.: Transparent IP radio access for next-generation mobile networks. IEEE Wireless Communications, August 2003
3. Zahariadis T.: Trends in the path to 4G. Communications Engineer, February 2003
4. Gustafsson E., Jonsson A.: Always best connected. IEEE Wireless Communications, February 2003
5. O'Droma M., et al.: "Always best connected" enabled 4G wireless world. IST Mobile and Wireless Communications Summit 2003, June 2003
6. 3GPP TS 24.241 V0.4.0: 3GPP generic user profile (GUP) common objects; Stage 3. 3GPP, August 2003
7. WAP 248-UAPROF V20: User agent profiling specification. WAP Forum, October 2001
8. W3C: Composite capabilities/preference profiles (CC/PP): structure and vocabularies. W3C Working Draft, March 2003
9. Park T., Dadej A.: Adaptive handover control in IP-based mobility networks. Proc. of 1st Workshop on the Internet, Telecommunications and Signal Processing (WITSP'02), December 2002
10. Zhang W., Jaehnert J., Dolzer K.: Design and evaluation of a handover decision strategy for 4th generation mobile networks. 57th Semiannual Vehicular Technology Conference VTC 2003-Spring, April 2003
11. F. André et al.: Optimised support of multiple wireless interfaces within an IPv6 end-terminal. Smart Objects Conference, May 2003
12. F. Adrangi (Ed.): Network discovery and selection within the EAP framework. Work in progress, Internet Engineering Task Force, draft-adrangi-eap-network-discovery-and-selection-00.txt, October 2003

Efficient Role Based Access Control Method in Wireless Environment

Song-hwa Chae[1], Wonil Kim[2], and Dong-kyoo Kim[3*]

[1] Graduate School of Information and Communication, Ajou University, Suwon, Korea
portula@ajou.ac.kr
[2] College of Electronics and Information Engineering, Sejong University, Seoul, Korea
wikim@sejong.ac.kr
[3] College of Information and Computer Engineering, Ajou University, Suwon, Korea
dkkim@ajou.ac.kr

Abstract. As resource sharing is common practice in distributed and wireless network, authentication and authorization become important issue in the security field. Many access control mechanisms including Role-Based Access Control (RBAC) are widely used for authorization. The common use of the wireless network imposes not only existing problems such as secure system management but also new problems such as limited storage and transaction size. In this paper, we propose an access control method that solves these problems for wireless environment. The proposed system uses bit pattern to represent access control information hence reduces transaction size and enhances security level. This system employs neural networks instead of access control tables, which reduces storages for role-permission tables and extra mutual exclusive data tables.

1 Introduction

Recently wireless networking has become more common, through which provides various types of contents to users. As technology advances, wireless terminals provide in various forms of services such as multi media contents, mailing and banking services. For the secure communications in between, it is essential for service providers to know user's information such as '*who connect*' and '*what is user's rights*'. It can only be supported by proper authentication and authorization methods.

Wireless networks have several problems that are different from those encountered in wired network. Consequently implementing security service is more difficult than in wired network. Wireless network is more vulnerable to unauthorized access. Today, a mobile phone is a common terminal for wireless service. The mobile phone has limited memory and the power that is less than personal computer. For that reason, it is difficult to have full fledged security service to mobile phone. It is a formidable task how to make secure data with these reduced transmissions.

Much progress has been made on access control mechanism in information security. Generally access control mechanism is categorized into three areas, such as

* Author for correspondence +82-2-3408-3795

I. Niemegeers and S. Heemstra de Groot (Eds.): PWC 2004, LNCS 3260, pp. 431–439, 2004.

mandatory access control (MAC), discretionary access control (DAC) and role-based access control (RBAC). MAC is suitable for military system, in which data and users have their own classification and clearance levels respectively. DAC is another access control method on the objects with user and group identifications. RBAC has emerged as a widely acceptable alternative to classical MAC and DAC [2][5]. It can be used in various computer systems. Since service provider should support different levels of service to users according to user's level, access control is important security service.

In this paper, we propose an efficient authorization method in wireless environment. In the proposed system, access control information is represented by bit patterns which are obtained via neural network. It reduces the size of access control information and enhances security level of wireless communication in which small data size for transaction size is preferable. We employ neural network for role-based access control mechanism instead of fixed tables, hence the proposed system uses bit pattern to represent access control information. In addition, it enables to eliminate the search time of relation tables and easily detects mutual exclusive roles.

This paper is organized as follows Chapter 2 explains the basic concept of RBAC and the neural network. Chapter 3 describes the role based access control method in wireless environment. Chapter 4 simulates the proposed system and Chapter 5 concludes with future works.

2 Background

2.1 RBAC

RBAC uses the concept of *role*. It does not allow users to be directly associated with permissions, instead each user can have several roles and each role can have multiple permissions. There are three components of RBAC: users, roles, and permissions. Each group can be represented as a set of user U, a set of role R, and a set of permission P.

- $U = \{u_1, u_2, \ldots, u_i\}$
- $R = \{r_1, r_2, \ldots, r_j\}$
- $P = \{p_1, p_2, \ldots, p_k\}$

Two different types of association must be managed by the system; one is the association between user and role, the other is the association between role and permission. It is characterized as user-role (UR) and role-permission (RP) relationship. Consequently, in order to have proper management, the system needs to maintain two separate association tables.

- $UR = \{u \in U \mid u \rightarrow 2^{|R|}\}$
- $RP = \{r \in R \mid r \rightarrow 2^{|P|}\}$

Conflicts of interest in a role-based system may arise as a result of a user gaining authorization for permissions associated with conflicting roles [4]. For example, if one role requests expenditures and another role approves them, the system must prohibit the same user from being assigned or active to both roles. In order to solve these conflict problems, there have been many researches since middle of 1990's.

Finally, National Institute of Standards and Technology (NIST) proposed two Separation of Duty (SOD) reference models in 2001; Static Separation of Duty (SSD) and Dynamic Separation of Duty (DSD). To implement these reference models, the system must have extra tables to protect against activating mutual exclusive roles. According to the NIST reference models, SSD should check the mutual exclusive role in every role assignment, and the role-permission in every user's session request. DSD should check all the three relations for every user's session request. SSD is rarely used in practice and DSD imposes a lot of overload in a system with many users due to excessive table accessing time. In fact, general processes do not always use mutual exclusive roles.

2.2 Neural Network

Artificial neural networks refer to computing systems whose central theme is borrowed form the analogy of biological neural networks [3]. It is composed of a large number of highly interconnected processing elements that are analogous to neurons and are tied together with weighted connections. These connection weights store the knowledge necessary to solve specific problems. Artificial neural networks are being applied to an increasing number of real world problems of considerable complexity, control problems, where the input variables are measurements used to drive an output actuator, and the network learns the control function. They are often good at solving problems that are too complex for conventional technologies and are often well suited to problems that people are good at solving, but for which traditional methods are not [4].

The proposed system employs backpropagation algorithm. It is one of method of neural network learning algorithm, which uses input data and desired output data for supervised learning. The weights between nodes are updated to reduce the difference of actual output and desired output by iterative learning process.

3 Role Based Access Control in Wireless Environment

Wireless network is popular way to use Internet. As technology advances, various services are possible in handheld terminals such as tablet PC, mobile phone and PDA. Most of wired service providers now offer wireless Internet services on banking and stock trading. These services should be processed in secure environment. Especially, authentication and authorization are important security service to achieve safe transaction. ID/password checking method is widely used authentication method whereas role-based access control is popular authorization method.

The resources of wireless network are limited hence a system developer must consider both the capability of the system and the handheld terminal. For that reason, implementing security service in wireless network is more difficult than in wire network. It is important issue that how to make smaller secure data and reduce the number of transactions.

In this paper, we propose an access control method to alleviate the mentioned problems. This system employs neural networks for authorization which represents access control information using bit patterns. This pattern reduces the size of access control information. In order to achieve efficient access control, we also propose a new implementation method for RBAC. It uses neural networks instead of access control tables. In addition, it enables to eliminate the search time of relation tables and easily detects mutual exclusive roles which was an inherent problem in RBAC.

3.1 System Architecture

When the users log into a system, they should be authenticated by that system. The system should use minimal authentication transactions for wireless environment. The ID/password method is a simple authentication method. After authenticating user, the system generates user's access control information from user information in the system. The user is given access control information that is represented by bit pattern. The user submits this bit pattern to use specific services from system. The system checks user's right of resources using neural network whenever the user wants to access a resource.

The proposed system represents all the access control information by bit patterns. Generally for a system employing RBAC, it needs at least 3 fixed tables for a session, such as user-role table, role-permission table and mutual excusive role table. It requires extra storage as well as checking time for both role-permission and mutual exclusive role tables. To cope with these problems, the proposed RBAC system uses neural networks instead of fixed RBAC relation tables. By employing neural network in RBAC, the system can check the user's permissions without using relation tables in each corresponding session. This method does not only reduce access time for authorization but also prevent a user from being activated with mutual exclusive permission.

It is assumed that roles, permissions and their associations are static information, whereas associations between user and role are dynamically changed. User can have several roles and role may have multiple permissions too. The process of this system consists of the following three phases: 1) neural network learning phase, 2) role assignment phase, 3) permission extraction phase. Depending on whether a user's role set contains mutual exclusive roles or not, two cases are considered in system processing.

3.2 Non-mutual Exclusive Role Case

This case is SSD in RBAC. Users do not contain mutual exclusive roles. Thus, the system checks only user's role and permission.

1. neural network learning phase
   In the learning phase, the training data is the role-permission relation provided by the system administrator. Each role and permission is represented by input and output respectively. The value of input (role) and output (permission) is either '1' (active/permit) or '0' (non-active/denial). Since the proposed system should be able to accommodate hierarchical RBAC, the high level role may contain the low level permissions too. After the neural network learns the relation, the system should be able to respond user's permissions for access control.

2. role assignment phase
   In this system, a user is defined as a human being and a role is a job function within the context of an organization [4]. Therefore, the system can assign multiple roles to the user according to the required job. The association of the user and the role can be changed dynamically. Our RBAC system can successfully respond to these changes. It produces the proper set of permissions dynamically even though the user-role association changes.
3. permission extraction phase
   When the user tries to access data, the system makes decisions such as permit or denial. The proposed system makes the decision using the user's permission set. This permission set is generated by the neural network in the user logging phase. This set has the user's whole permissions. Since all the permissions do not contain any mutual exclusive role, mutual exclusive resolution is not necessary.

3.3 Mutual Exclusive Role Case

This case uses DSD in RBAC. The user can have multiple roles, potentially including mutually exclusive roles, such as applying for expenses and approving those same expenses. These two permissions come from two different roles. The main point of this process is to reduce the role set, so that the reduced role set does not have any mutual exclusive permission. With this process, the neural network will be able to produce the reduced permission set according the reduced role set. The process of producing the reduced permission set is as follows.

1. neural network learning phase
   In this case, we represent permissions with three types of values; '0','1', and '0.5'. The '0' and '1' has the same meaning as defined in non-mutual exclusive role case. Especially, '0.5' means mutual exclusive permission. An example of this is shown in Table 2. If there is a high level role containing low level mutual exclusive permission, that node (permission) is also set to '0.5'. In this case, second neural network is employed to produce the reduced role set. It will lean the relation between permissions and roles. Each permission and role will be represented by input and output respectively. An example of this is shown in Table 3.
2. role assignment phase
   Same as in non-mutual exclusive case.
3. permission extraction phase
   The DSD system must decide permit or denial on the given request in every session. The proposed system makes the decision using the user's permission set. This permission set is generated by the neural network in the user logging phase. This set has the user's whole permissions including mutual exclusive permission. The process of making the whole permission set is similar to non-mutual exclusive case. When the user tries to access data defined as mutual exclusive permission, the system recognizes mutual exclusion case using permission's value, which is approximately '0.5'. In this case, the system should make a least user's permission set for the session. This permission set is a reduced set which does not contain mutual exclusive permission. Second neural network is employed to produce reduced role set. In order to limit user's permission, it changed a bit to '1' in user's permission set and others to '0'. This changed user's permission set is used to

produce the limited role set. The limited role set is used to produce the reduced permission set using the first neural network for this session. After this process, the user has reduced permission set which is not containing mutual exclusive permission. In any session, if the request permission is found to be the mutual exclusive permission, the system does the same process recursively. After the session, it will return to the previous permission set. This process protects from executing mutual exclusive permission in any given session.

4 Simulation

The proposed system was evaluated on a typical customer class hierarchy in Fig 1. It consists of 2 Internet services in a wireless terminal that performs stock trading and remote banking service. The customer class has 10 roles and 6 mutual exclusive roles. This system can be extended to accommodate unlimited number of users depending on services. The evaluation result showed that the proposed system was able to detect mutual exclusive role activation and produced the user's permission set using reduced role set. The role-permission relationship of Fig 1 is represented in Table 2. The wireless terminal does not need to contain user's whole access control information. It holds just small size of bit pattern which is user's whole role set. When user needs service, he/she submits this bit pattern to server which supports specific service.

For instance, if the system has 8 roles and a user assigned 2 roles, it can be represented as Table 1. In this case, User1's access control information is [0100000100].

Table 1. Example of roles

Role	FC1	GC1	GC3	VIP1	Admin	VIP2	GC4	GC2	Studt1	Customer
Bit	0	1	2	3	4	5	6	7	8	9
User	0	1	0	0	0	0	0	1	0	0

We assumed that VIP1 and VIP2 had mutual exclusive roles. GC1 and GC3, GC 2 and GC4 had mutual exclusive roles too. Therefore, the permissions of these roles are set to '0.5'.

We trained two neural networks using Table 2 and Table 3 respectively, with learning and the momentum rates are 0.001. The proposed neural network had one hidden layer (30 nodes) and used sigmoid functions for hidden and output nodes. The five sample users and their role set (representations) are given below;

- User1 : {GC1, GC2} : (0100000100)
- User2 : {GC1, GC4} : (0100001000)
- User3 : {GC1} : (0100000000)
- User4 : {VIP1, GC2} : (0001000100)
- User5 : {Admin} : (0000100000)

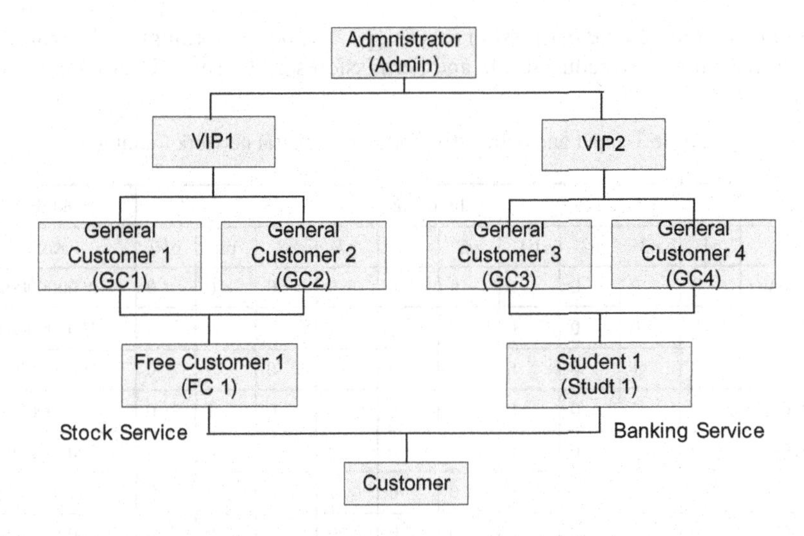

Fig. 1. Example of a role structure for wireless service

Table 2. Input and output data for first neural network training

	Input data	Out data									
	Role	p1	p2	p3	p4	p5	p6	p7	p8	p9	p10
Customer	0000000001	0	0	0	1	0	0	0	0	0	0
FC 1	1000000000	0	0	0	1	0	0	1	0	0	0
Studt1	0000000010	0	1	0	1	0	0	0	0	0	0
GC1	0100000000	0	0	0	1	0	0	1	0	1	0
GC2	0000000100	0	1	0	1	0	0	0	0	0	1
GC3	0010000000	1	0	0	1	0	0	1	0	0	0
GC4	0000001000	0	1	0	1	0	0	0	1	0	0
VIP1	0001000000	0.5	0	1	1	0	0	1	0	0.5	0
VIP2	0000010000	0	1	0	1	0	1	0	0.5	0	0.5
Admin	0000100000	0.5	1	0.5	1	1	0.5	1	0.5	0.5	0.5

The system produces user's whole permission set using first neural network. If the user tries to access any mutual exclusive permission, it produces reduced role set using second neural network. For example, user5 is assigned as Admin, the first neural network will produced the whole permission set of {p1, p2, p3, p4, p5, p6, p7, p8, p9, p10} with access values between 0.0 and 1.0. If Admin tries to access any mutual exclusive permission, such as p9, the system recognizes its mutual exclusive case and produces the reduced role GC1 using second neural network. As a result, the role of user5 which was originally defined as Admin, has reduced to GC1 in this particular session and the permission sets as {p4, p7, p9} accordingly. The whole role and permission set and reduced role and permission set of User1~ User5 are shown in Table 4.

Table 3. Input and output data for second neural network training

	Input data										Out data
	p1	p2	p3	p4	p5	p6	p7	p8	p9	p10	Role
Customer	0	0	0	1	0	0	0	0	0	0	0000000001
FC 1	0	0	0	1	0	0	1	0	0	0	1000000000
Studt1	0	1	0	1	0	0	0	0	0	0	0000000010
GC1	0	0	0	1	0	0	1	0	1	0	0100000000
GC2	0	1	0	1	0	0	0	0	0	1	0000000100
GC3	1	0	0	1	0	0	1	0	0	0	0010000000
GC4	0	1	0	1	0	0	0	1	0	0	0000001000
VIP1	0	0	1	1	0	0	1	0	0	0	0001000000
VIP2	0	1	0	1	0	1	0	0	0	0	0000010000
Admin	0	1	0	1	1	0	1	0	0	0	0000100000

Table 4. The whole role set and reduced role set

Case	Whole Role Set	Whole permission Set	Requested Job(permission)	Reduced Role Set	Reduce Permission Set
user1	GC1,GC2	{p2,p4,p7,p9,p10}	p9	GC1	{p4,p7,p9}
user2	GC1,GC4	{p2,p4,p7,p8,p9}	p8	GC4	{p2,p4,p8}
user3	GC1	{p4,p7,p9}	p9	GC1	{p4,p7,p9}
user4	VIP1,GC2	{p1,p2,p3,p4,p7,p9,p10}	p10	GC3	{p2,p4,p10}
user5	Admin	{p1,p2,p3,p4,p5,p6,p7,p8,p9,p10}	p9	GC1	{p4,p7,p9}

5 Conclusion and Future Works

Wireless network becomes common in these days as many people use Internet service in wireless environment. As wireless technology advances, users can access various types of contents such as multi media, bank account and stock trading. However, users' terminals such as mobile phone and PDA still have limited memory and computing power. Consequently they are equipped with limited security feature. Therefore, reducing data and transaction size is very important in wireless Internet. In addition, both authentication and authorization should be well defined and properly controlled for secure Internet service. Access control method such as RBAC is a popular method used to satisfy these security requirements.

In this paper, we proposed a novel access control method in wireless environment. In order to reduce access control information, we use bit pattern and process this pattern to neural networks. This reduced amount of information in wireless communication enhances security level. The proposed methods can be applied dynamically with user's role change. It has advantages of not using multiple storages for role-permission tables and extra mutual exclusive data tables. It also reduces access time by eliminating excessive table search for mutual exclusive roles. This method can be easily extended to various access control mechanisms and suitable for wireless network environment.

References

1. Rumelhart, D. E., Hinton, G. E., and Williams, R. J. Learning representations by back-propagating errors. Nature, 323, (1986) 533-536
2. E.H.Choun, A Model and administration of Role Based Privileges Enforcing Separation of Duty. Ph.D. Dissertation, Ajou University(1998)
3. K.Mehrotra, C.K.Mohan,S.Ranka, Elements of Artificial Neural Networks, MIT Press (1997)
4. D.F.Ferraiolo, R.Sandhu, E.Gavrila, D.R.Kuhn, R.Chandramouli, Proposed NIST Standard for Role-Based Access Control, ACM Transactions on Information and System Security, Vol4, No3 (2001) 224-274
5. G.Ahn, R.Sandhu, Role-Based Authorization Constraints Specification, ACM Transactions on Information and System Security, Vol3, No4,207-226 (2000)
6. D.FFerraiolo, J.F.Barkley, D.R. Kuhn, A Role-Based Access Control Model and Reference implementation Within a Corporate Intranet, ACM Transactions on Information and System Security, Vol2, No1 (1999) 34-64
7. S. Farrell, An Internet Attribute Certificate Profile for Authorization, RFC 3281,(2002)

Signaling Load of Hierarchical Mobile IPv6 Protocol in IPv6 Networks*

Ki-Sik Kong[1], Sung-Ju Roh[2], and Chong-Sun Hwang[1]

[1] Dept. of Computer Science and Engineering, Korea Univ.
1, 5-Ga, Anam-Dong, Sungbuk-Gu, Seoul 136-701, Korea
{kskong, hwang}@disys.korea.ac.kr
[2] Core Network Development Team
Technology R&D Center, LG Telecom Co., Korea.
sjroh@lgtel.co.kr

Abstract. Hierarchical Mobile IPv6 (HMIPv6) has been proposed to accommodate frequent mobility of the mobile nodes and reduce the signaling load in the Internet. Though it is being considered as an efficient local mobility management protocol, its performance may vary widely depending on the various mobility and traffic related parameters. Therefore, it is essential to investigate the effects of these parameters and conduct in-depth performance study of HMIPv6. For the analysis of HMIPv6, we present a new analytical method using the mobility model based on imbedded Markov chain and a simplistic hierarchical network model. Based on these models, we analytically derive the location update cost (i.e.,binding update cost plus binding renewal cost), packet tunnelling cost, and total signaling cost, respectively, in HMIPv6. In addition, we investigate the effects of various parameters such as the speed of a mobile node, binding lifetime, and packet arrival rate on the total signaling cost generated by a mobile node during its average MAP domain residence time. The analytical results demonstrate that the signaling load generated by HMIPv6 decreases as the speed of a mobile node and binding lifetime get larger, and its packet arrival rate gets smaller.

1 Introduction

Recently, the demand for wireless communications has grown tremendously. The demand for "anywhere, anytime" high-speed Internet access has been a driving force for the increasing growth and advances in wireless communication and portable devices. As a consequence, these trends have prompted research into mobility support in networking protocols.

Mobile IPv6 (MIPv6) [1] has been developed by the IETF with some new functionalities, which is based on the next generation Internet protocol. However, it still has some problems. That is, MIPv6 handles local mobility of a mobile node (MN) in the same way as it handles global mobility. As a result, an MN

* This work was supported by the Korea Research Foundation Grant (KRF-2003-041-D00403).

I. Niemegeers and S. Heemstra de Groot (Eds.): PWC 2004, LNCS 3260, pp. 440–450, 2004.

sends the binding update message to its Home Agent (HA) and its correspondent node (CN) each time it changes its point-of-attachment regardless of its locality. Such an approach may cause excessive signaling traffic, especially for the MNs with relatively high mobility or long distance to their HAs or CNs. In addition, this is not scalable since the generated signaling traffic can become quite overwhelming as the number of the MNs increases.

In order to overcome these drawbacks, Hierarchical Mobile IPv6 (HMIPv6) [2,3] has been proposed to accommodate frequent mobility of the MNs and reduce the signaling load in the Internet. HMIPv6 introduces a new entity, the Mobility Anchor Point (MAP) which works as a proxy for the HA in a foreign network. When an MN moves into a network controlled by a new MAP, it is assigned two new CoAs: a Regional CoA on the MAP's subnet (RCoA) and an on-link address (LCoA), which is the same as used for MIPv6. When an MN moves to a new subnet within the same MAP domain, only the MAP has to be informed. Note, however, that this does not imply any change to the periodic binding update message an MN has to send to the HA and the CNs, and now an MN additionally should send it to the MAP.

Generally, the performance of IP mobility protocol is highly dependent on various mobility and traffic related parameters. Therefore, it is essential to analyze and evaluate the IP mobility protocol under the various conditions, and more in-depth study needs to be performed. There have been several researches for the performance study on the IP mobility protocols.

In [3], they present a hierarchical mobility architecture that separates local mobility from global mobility to propose a mobility management scheme that is hierarchical, flexible and scalable. But, they mainly focus on evaluating the signaling bandwidth according to the binding update emission frequency. As already introduced and studied in location management for PCS networks [4], in order to evaluate the efficiency of IP mobility management protocol, the tradeoff relationship between the location update cost and the packet tunnelling cost also has to be taken into consideration in terms of total signaling cost [5]. Nevertheless, in [3], they do not consider the extra tunnelling cost for packet delivery. When the network administrator and network designer consider the deployment of HMIPv6, they should fully understand how various mobility and traffic related parameters may have an effect on the system performance. However, they just show few of the effects and relations of various mobility and traffic related parameters.

In [6], the author investigated the performance of MIPv4 regional registration. The performance measures used are registration delay and the CPU processing overheads loaded on the agents to handle mobility of the MNs. They also do not consider the signaling cost caused by the packet tunnelling.

In [7], the authors propose an analytic model for the performance analysis of HMIPv6 in IP-based cellular networks, which is based on the random walk mobility model. Based on this model, they formulate location update cost and packet delivery cost. Then, they analyze the impact of cell residence time on the location update cost and the impact of user population on the packet delivery cost. Though their analysis is well-defined, however, they do not take both the periodic binding update and the effect of binding lifetime into consideration, which may have much effect on the total signaling cost. In addition, their analysis about the packet delivery cost of HMIPv6 is not likely to be the pure extra signaling bandwidth consumption incurred by the packet tunnelling but the network bandwidth consumption including the data traffic as well as the signaling traffic. However, from the viewpoint of IP mobility management, the consideration of the extra signaling bandwidth consumption (not including the data traffic) occurred during the processes of the location update and the packet tunnelling should be taken into [5].

In contrast to the related literature mentioned above, we perform a detailed and in-depth study of HMIPv6 in terms of the total signaling cost. Moreover, while the previous analyses do not consider either the periodic binding update or the extra packet tunnelling, our work considers both of them for the analysis. Also, we present a new analytical method using mobility model based on imbedded Markov model and a simplistic hierarchical network model. Based on these models, we analytically derive the binding update cost, binding renewal cost, packet tunnelling cost, and total signaling cost, respectively, in HMIPv6. In addition, in terms of the total signaling cost generated by an MN during its average MAP domain residence time, we show performance evaluation by investigating the effects and relations of various parameters such as the speed of an MN, binding lifetime and its packet arrival rate.

The remainder of this paper is organized as follows. In Sect.2, we describe the user mobility model and network model for the performance analysis of HMIPv6, and then the signaling cost functions of HMIPv6 are analytically derived. In Sect.3, we investigate the results of Sect.2 by applying various numerical examples. Finally, conclusions and future works are given in Sect.4.

2 Analytical Modelling of Hierarchical Mobile IPv6

In this section, we first introduce our user mobility model and network model to evaluate the performance of HMIPv6. Then, for the analysis, we derive the location update cost, packet tunnelling cost, and total signaling cost, respectively, in HMIPv6.

2.1 User Mobility Model

For the analysis of an MN's movement behavior, we assume a simple fluid flow mobility model. The model assumes that MNs are moving at an average speed of v, and their movement direction is uniformly distributed over $[0, 2\pi]$, and that all the subnets are of the same rectangular shape and size, and form together a contiguous area.

The parameters used in our user mobility model are summarized as follows.

- γ : the border crossing rate for an MN out of a subnet

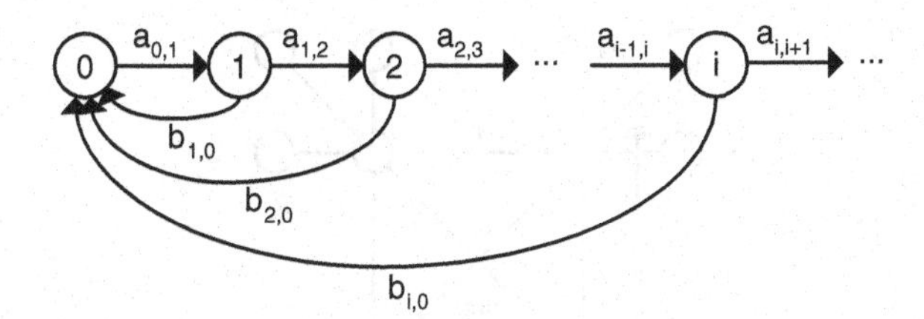

Fig. 1. State transition diagram for an imbedded Markov chain

- λ : the border crossing rate for which an MN still stays in the same domain
- μ : the border crossing rate for an MN out of a MAP domain

From [8], the border crossing rate γ for an MN out of a subnet is derived as

$$\gamma = \frac{4v}{\pi\sqrt{S}} \tag{1}$$

where S is the subnet area. We assume that a MAP domain is composed of N equally large subnets. Therefore, the border crossing rate μ for an MN out of a MAP domain is

$$\mu = \frac{4v}{\pi\sqrt{NS}} \tag{2}$$

Note that an MN that crosses a MAP domain will also cross a subnet. So, the border crossing rate λ for which the MN still stays in the same MAP domain is obtained from Eq.(1) and (2):

$$\lambda = \gamma - \mu = (1 - \frac{1}{\sqrt{N}})\gamma \tag{3}$$

Figure 1 shows an imbedded Markov chain model, which describes the binding update process of an MN, where $a_{i,i+1} = \lambda$ and $b_{i,0} = \mu$. The state of an imbedded Markov chain, i $(i \geq 0)$, is defined as the number of subnets in the same MAP domain that an MN has passed by. The state transition $a_{i,i+1}(i \geq 0)$ represents an MN's movement rate to an adjacent subnet in the same MAP domain, and the state transition $b_{i,0}(i \geq 1)$ represents an MN's movement rate (from state i to state 0) to another subnet out of the MAP domain. We assume π_i to be the equilibrium state probability of state i. Thus, we can obtain

$$\lambda\pi_i = (\lambda + \mu)\pi_{i+1} \tag{4}$$

$$\lambda\pi_0 = \mu\sum_{i=1}^{\infty}\pi_i \tag{5}$$

Using the Eq.(4) and (5), π_i can be expressed in terms of the equilibrium state probability π_0 as

$$\pi_i = (\frac{\lambda}{\lambda + \mu})^i\pi_0 = (1 - \pi_0)^i\pi_0 \tag{6}$$

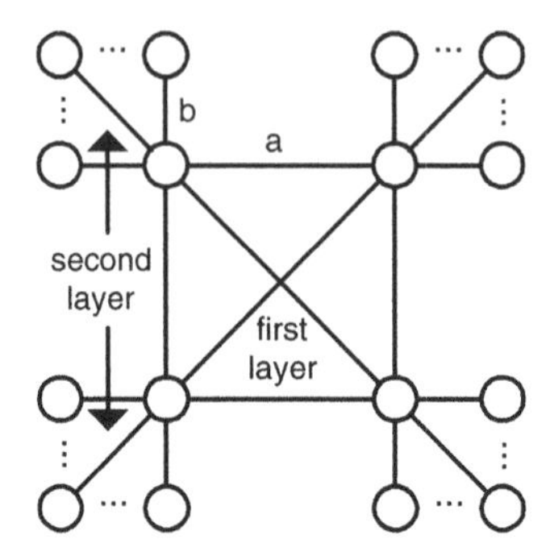

Fig. 2. Network Model

where π_0 is the equilibrium state probability of state 0. By using the law of total probability, π_0 can be obtained as

$$\pi_0 = \frac{\mu}{\lambda + \mu} = 1 - \frac{\lambda}{\lambda + \mu} = 1 - \theta \tag{7}$$

where $\theta = \frac{\lambda}{\lambda+\mu}$.

2.2 Network Model

Similar to [9], we consider a simplistic two-layer hierarchical network model given in Fig.2. The first layer has a mesh topology, which consists of M nodes. Each first layer node is a root of a N-ary tree with depth of 1. We assume that the HA and all the ARs are all second layer nodes, and that each MAP domain is composed of all the second layer nodes under the same first layer node. In addition, the functionality of the MAP is placed on the first layer node.

For the simplicity, the CN, the MN and the HA are assumed to be located in different domain. Also, we define the *domain size(N)* as the number of all the second layer nodes under the same MAP domain. In this model, the link hops between the first layer nodes are a, and the link hops between the first and second layer nodes within the same MAP domain are b, respectively. On the other hand, we assume that the link hops between the CN and the CN's default AR, and the transmission cost over the wireless link are all zero.

2.3 Cost Analysis

In this subsection, according to the user mobility model and network model given in Sect.2.1 and 2.2, in order to evaluate the performance of HMIPv6, we analytically derive the binding update cost, binding renewal cost, packet tunnelling cost, and total signaling cost generated by an MN during its average MAP domain residence time. There are two kinds of binding update messages in MIPv6 and HMIPv6. That is, the one results from the MN's subnet crossing, and the other results from the expiration of the binding lifetime.

In this paper, we use *binding update message* to refer the former, and *binding renewal message* to refer the latter to differentiate these two kinds of messages. For the analysis, several parameters and assumptions mentioned in Sect.2.1 and 2.2 are used. According to our mobility model given in Sect.2.1, the average binding update cost in HMIPv6 (U_{HMIPv6}) can be formulated as

$$\begin{aligned} U_{HMIPv6} &= \pi_0(U_m + U_h + \delta U_c) + U_m \sum_{i=1}^{\infty} i\pi_i \\ &= (1-\theta)(U_m + U_h + \delta U_c) + \frac{\theta}{1-\theta} U_m \end{aligned} \quad (8)$$

where U_m, U_h and U_c are the binding update costs to register with the MAP, the HA and the CN, respectively. Equation $\sum_{i=1}^{\infty} i\pi_i$ means the average number of subnet crossing in the MAP domain, which is derived from the imbedded Markov chain in Fig.1. Based on our network model given in Sect.2.2, U_m, U_h and U_c are as follows:

$$U_m = 2M_{BU}b \quad (9)$$

$$U_h = 2M_{BU}(a + 2b) \quad (10)$$

$$U_c = M_{BU}(a + 2b) \quad (11)$$

where M_{BU} means the bandwidth consumption caused by a binding update message. Note that the HA and the MAP must return a binding acknowledgement message, but the CNs may return it or not. For the analysis, we only consider the binding acknowledgement from the HA and the MAP, and we assume that the binding related messages are only sent alone in a separate packet without being piggybacked.

Let the binding lifetimes for the MAP, the HA and the CNs in HMIPv6 be T_M, T_H and T_C, respectively. On the other hand, from the Eq.(1) and (2), the average subnet residence time and the average MAP domain residence time of an MN are $\frac{\pi\sqrt{S}}{4v}$ and $\frac{\pi\sqrt{NS}}{4v}$, respectively. Thus, the binding renewal rate to the MAP in HMIPv6 while an MN stays in a subnet is $\lfloor \frac{\pi\sqrt{S}}{4vT_M} \rfloor$. Similarly, the binding renewal rates to the HA and the CNs in HMIPv6 while an MN stays in a MAP domain become $\lfloor \frac{\pi\sqrt{NS}}{4vT_H} \rfloor$ and $\lfloor \frac{\pi\sqrt{NS}}{4vT_C} \rfloor$, respectively.

On the other hand, for the calculation of the signaling costs generated to perform location update with the CNs, we roughly define *ratio of average binding time for the CNs to an MN's average MAP domain residence time*, δ as the following:

$$\delta = \frac{\sum_{i=1}^{n} C_i}{n\Delta} \quad (12)$$

where C_i and n represent the binding time for each CN and the number of the CNs recorded in the MN's binding update list during the MN's average MAP domain residence time, respectively. Δ represents the MN's average MAP domain residence time. Consequently, the average binding renewal cost in HMIPv6 (R_{HMIPv6}) can be formulated as follows:

$$R_{HMIPv6} = \{(1-\theta) + \frac{\theta}{1-\theta}\}\{\lfloor\frac{\pi\sqrt{S}}{4vT_M}\rfloor U_m + \frac{1}{\sqrt{N}}(\lfloor\frac{\pi\sqrt{NS}}{4vT_H}\rfloor U_h + \delta\lfloor\frac{\pi\sqrt{NS}}{4vT_C}\rfloor U_c)\} \quad (13)$$

Therefore, the average location update cost (L_{HMIPv6}) incurred by the binding update messages and binding renewal messages is

$$L_{HMIPv6} = U_{HMIPv6} + R_{HMIPv6} \quad (14)$$

Let the probability that the CN has a binding cache entry for an MN be q. Then, the average packet tunnelling cost in HMIPv6 (D_{HMIPv6}) can be formulated as follows:

$$D_{HMIPv6} = \frac{\pi p\sqrt{NS}}{4v}\{qD_{direct} + (1-q)D_{indirect}\} \quad (15)$$

where D_{direct} and $D_{indirect}$ are the tunnelling costs for a direct packet delivery (not intercepted by the HA) and the tunnelling cost for a packet routed indirectly through the HA in HMIPv6, respectively. And, p and q are the average packet arrival rate for an MN, and the probability that the CN has a binding cache for an MN, respectively.
According to our network model, D_{direct} and $D_{indirect}$ are as follows:

$$D_{direct} = M_{PD}b \quad (16)$$

$$D_{indirect} = M_{PD}(a+3b) \quad (17)$$

where M_{PD} represents the bandwidth consumption generated by tunnelling per packet. Finally, the total signaling cost (C_{HMIPv6}) generated by an MN during its average MAP domain residence time in HMIPv6 can be formulated as follows:

$$C_{HMIPv6} = L_{HMIPv6} + D_{HMIPv6} \quad (18)$$

3 Numerical Results

In this section, based on the analytical modelling of HMIPv6 in the previous section, we investigate the effects of various parameters such as the speed of an MN, binding lifetime and its packet arrival rate on the total signaling cost. The performance measure used is the signaling bandwidth consumption per packet multiplied by the number of link hops that the packet traverses during an MN's average MAP domain residence time (i.e., Bytes × Link hops / MAP domain residence time).

Table 1. Parameter value

Parameter	Type	Value
N	Domain size	64
S	Subnet area	5 Km^2
q	Probability that the CN has a binding cache for an MN	0.7
δ	Ratio of avg. binding time for CNs to avg. MAP domain residence time	0.1
a	Link hops between the first layer nodes	15
b	Link hops between the first and the second layer nodes	3
M_{BU}	The bandwidth consumption generated by a binding update message	68 *byte*
M_{PD}	The bandwidth consumption generated by tunnelling per packet	40 *byte*

The parameter values given in Table 1 are used as default values for the performance analysis. Most parameters used in this analysis are set to typical values found in [5,6,10]. The size of a binding update message is equal to the size of an IPv6 header (40 bytes) plus the size of a binding update extension header (28 bytes), so 68 bytes. In addition, the additional bandwidth consumption caused by tunnelling per packet is equal to the size of IPv6 header, so 40 bytes. According to [1,2], we set the binding lifetimes (T_H, T_M, T_C) in HMIPv6 to be the same, and denominated as T.

3.1 Location Update Cost

Figure 3(a) shows the effect of v on the average location update cost. As already mentioned, location update cost consists of both the binding update cost generated by an MN's mobility and the periodic binding renewal cost. These results presented in Fig.3(a) show that the location update cost in HMIPv6 decreases as v increases. When an MN is not moving fast, most of the signaling traffic is generated by the periodic binding renewal messages. However, as the speed of an MN increases, the periodic binding renewal messages decrease and the binding update messages generated by an MN's mobility dominate most of the signaling traffic. Note here that the location update cost in HMIPv6 remains the same when v exceeds 40 km/hour for $T = 0.2$ hour or when it exceeds 30 km/hour for $T = 0.5$ and $T = 0.8$ hour in Fig.3(a). This is due to the fact that as the speed of an MN increases, it moves to an adjacent subnet before the additional binding renewal message occurs.

Figure 3(b) shows the effect of T on the location update cost. As T gets larger, the location update cost decreases. Note that T is relatively sensitive to a slowly moving MN. This is due to the fact that the longer an MN resides in the MAP domain, the much the periodic binding renewal messages are generated. Generally, the shorter the binding lifetime is set, the much the signaling traffic is generated. Therefore, too much short binding lifetime may result in significant signaling load throughout the networks. On the other hand, the longer the binding lifetime is set, the longer the binding cache entry size of the mobility agent gets. Thus, this may result in an increase of the binding cache lookup time and memory consumption in the mobility agent. In practice, the value of binding lifetime must be specified in the implementation of HMIPv6. Therefore, further

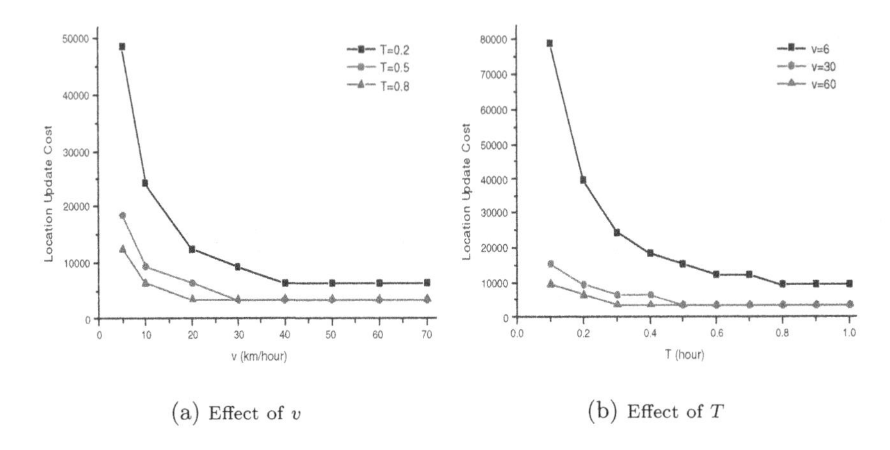

(a) Effect of v (b) Effect of T

Fig. 3. Location Update Cost

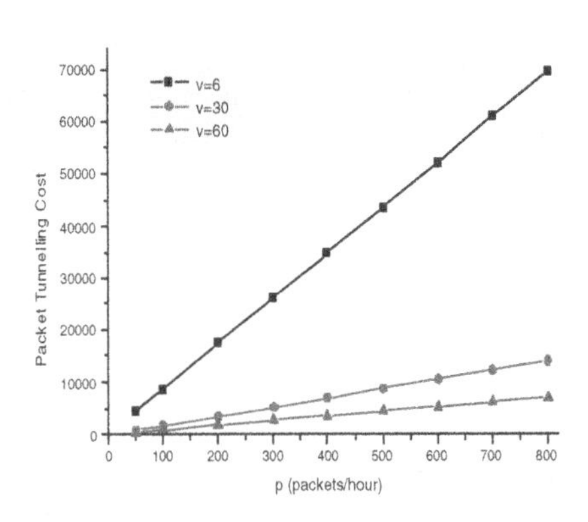

Fig. 4. Packet Tunnelling Cost

study on the effects of binding lifetime needs to be investigated to achieve the best performance.

3.2 Packet Tunnelling Cost

Figure 4 shows the effects of p on the average packet tunnelling cost. The result shown in Fig.4 indicates that the packet tunnelling cost is linearly increased as p increases. This result also indicates that a slowly moving MN is more affected by the packet arrival rate if the packet arrival rates are the same. This is due to the fact that the MAP domain residence time of a slowly moving MN is longer than that of a fast moving MN.

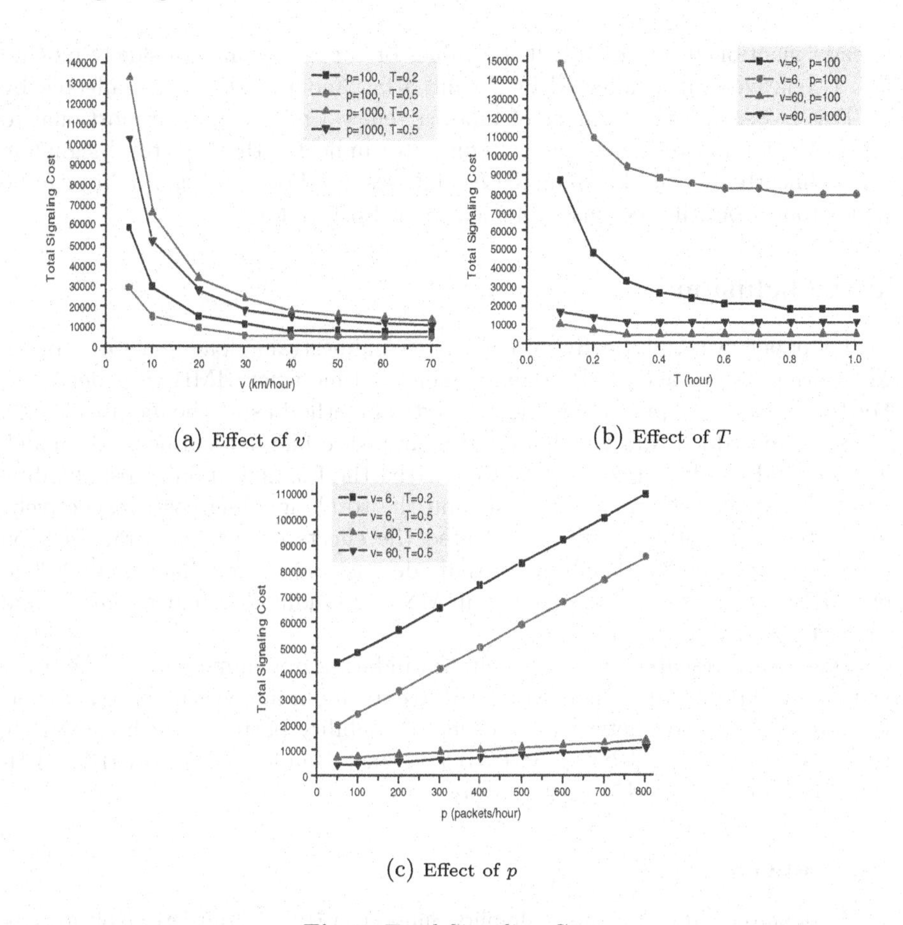

(a) Effect of v

(b) Effect of T

(c) Effect of p

Fig. 5. Total Signaling Cost

3.3 Total Signaling Cost

As already introduced and studied in location management for PCS networks, in order to evaluate the efficiency of IP mobility management protocol, the tradeoff relationship between the location update cost (i.e., binding update cost plus binding renewal cost) and the packet tunnelling cost also has to be taken into consideration [4]. That is, minimizing the total signaling cost of the location update cost and the packet tunnelling cost is the key focus when we evaluate and design the IP mobility protocol. In addition, it is essential to investigate the effects of various mobility and traffic related parameters from the viewpoint of the total mobility management cost.

We consider the four sets of parameters to show the variation in the total signaling costs in HMIPv6 according to the change of v, T, and p. Figure 5(a) indicates that the total signaling cost in HMIPv6 gets smaller as v increases, and this phenomenon becomes more prominent as T gets larger and p gets smaller. Figure 5(b) indicates that the total signaling cost gets smaller as T increases. As already mentioned in Sect.3.1, if the values of T are the same, a slowly moving MN is relatively more affected by T, and a fast moving MN is less affected by T. This is because the number of binding renewal messages is proportional to the MAP domain residence time. Figure 5(c) indicates that the total signaling cost is linearly increased as p increases. The results shown in Fig.5(a)-(c) can be understood from the reasons explained in Sect.3.1 and 3.2

4 Conclusion

In this paper, we analytically derived the signaling traffic load generated by an MN during its average MAP domain residence time when HMIPv6 is deployed. For the analysis, we presented a new analytical method using the mobility model based on imbedded Markov chain and a simplistic hierarchical network model. Then, based on these two models, we derived the binding update cost, binding renewal cost, packet tunnelling cost, and the total signaling cost, respectively, in HMIPv6. In addition, we investigated the effects of various parameters on each derived costs. The analytical results demonstrated that the signaling load of HMIPv6 decreases as the speed of an MN and binding lifetime get larger, and its packet arrival rate gets smaller.

Our future research subjects include validating our numerical results using simulation experiments. Then, we intend to extend our analytical analysis to performance comparison between the other IP mobility protocols such as MIPv6, Fast Handovers for MIPv6 (FMIPv6) and Fast Handovers for HMIPv6 (FH-MIPv6), and these works are underway.

References

1. D. Johnson and C. Perkins, "Mobility Support in IPv6," draft-ietf-mobileip-ipv6-24.txt, June 2003.
2. H. Soliman, C. Castelluccia, K. Malki, L. Bellier, "Hierarchical Mobile IPv6 Mobility Management (HMIPv6)," draft-ietf-mipshop-hmipv6-02.txt, June 2004.
3. C. Castelluccia, "HMIPv6: A Hierarchical Mobile IPv6 Proposal," ACM Mobile Computing and Communications Review, vol.4, no.1, pp.48-59, Jan. 2000
4. I. F. Akyildiz, et al., "Mobility Management in Next-Generation Wireless Systems," Proceedings of the IEEE, Aug. 1999.
5. J. Xie, and Ian F. Akildiz, "A Novel Distributed Dynamic Location Management Scheme for Minimizing Signaling Costs in Mobile IP," IEEE Trans. on Mobilecom, vol.1, no.3, Jul-Sep. 2002.
6. M. Woo, "Performance Analysis of Mobile IP Regional Registration," IEICE Trans. on Commun, vol.E86-B, no.2, pp.472-478, Feb. 2003.
7. S. Pack and Y. Choi, "Performance Analysis of Hierarchical Mobile IPv6 in IP-based Cellular Networks," PIMRC 2003, Sep. 2003.
8. F. Baumann and I. Niemegeers, "An Evaluation of Location Management Procedures," Proc. UPC'94, pp.359-364, Sep. 1994.
9. T. Ihara, H. Ohnishi, and Y. Takagi, "Mobile IP Route Optimization Method for a Carrier-Scale IP Network," ICECCS 2000, pp.11-14 Sep. 2000.
10. R. Ramjee, K. Varadhan, L. Salgarelli, S. Thuel, W. Yuan, T. Porta, "HAWAII: A Domain-Based Approach for Supporting Mobility in Wide-Area Wireless Networks," IEEE/ACM Trans. Networking, vol.10, no.3, pp.396-410, Jun. 2002.

Low-Latency Non-predictive Handover Scheme in Mobile IPv6 Environments

Geunhyung Kim[1] and Cheeha Kim[2]

[1] Technology Network Laboratory, Korea Telecom (KT),
463-1 Jeonmin-dong, Yusung-gu, Daejeon, 305-811, Korea,
geunkim@kt.co.kr
[2] Department of Computer Science and Engineering,
Pohang University of Science and Technology(POSTECH),
San 31 HyoJa-Dong, Nam-Gu, Pohang, 790-784, Korea,
chkim@postech.ac.kr

Abstract. In IP-based wireless/mobile networks, minimizing handover latency is one of the most important issues. Fast handover mobile IPv6 (FMIPv6) reduces the handover latency by handover prediction with link-layer information. It requires additional overhead in terms of signaling and may waste resources by incorrect prediction. In this paper, we introduce a new non-predictive handover scheme which employs a handover agent to speedup handover. Signaling and packet delivery costs associated with handover are introduced to compare the performance of the proposed scheme with FMIPv6 and MIPv6. As a result, we found that the proposed scheme guarantees lower handover latency as well as fewer signaling messages than FMIPv6 regardless of prediction accuracy.

1 Introduction

Mobile IPv6 (MIPv6) [1] is a protocol aimed at maintaining network connectivity for hosts roaming across the Internet. This protocol provides unbroken connectivity to IPv6 mobile hosts (MHs) using a Care-of-Address (CoA) specific to the point of network attachment, when MHs move from one access point (AP) to another in a different subnet. In MIPv6, although an MH configures its CoA when it roams to a new subnet, the MH cannot receive IP packets on its current point of attachment until the handover procedures finish. These handover procedures include a movement detection, an address configuration, and a location update. These procedures have been shown to result in a long handover latency.

In most cases, long handover latency strongly degrades the IPv6 packet stream of MHs. Two ways of overcoming this problem are to use hierarchical schemes and to apply techniques known as fast handover. Hierarchical schemes focus on reducing the latency caused by the location update procedure. As an example of hierarchical schemes, hierarchical MIPv6 (HMIPv6) [2] reduces the handover latency by minimizing the location update signaling depth. Fast handover schemes focus on reducing the latency caused by movement detection or address configuration procedures. For example, FMIPv6 [3] reduces the handover

I. Niemegeers and S. Heemstra de Groot (Eds.): PWC 2004, LNCS 3260, pp. 451–465, 2004.

latency by allowing an MH to pre-configure a new CoA before it moves to a new network.

Although FMIPv6 can provide fast handover in wireless IP networks by handover prediction based on link-layer information, it requires a higher signaling overhead than MIPv6. Also, FMIPv6 is based on the prediction made by using link-layer information of future events. As a result, this prediction may sometimes be wrong. In addition, the allocated buffer space constitutes useless overhead if the prediction is wrong since packet forwarding for smooth handover is performed in accordance with prediction.

To overcome these problems of FMIPv6, we propose a non-predictive handover scheme that can reduce the handover latency of MHs by minimizing movement detection and address configuration latencies. This paper focuses on the AP assisted handover scheme to reduce the latency caused by movement detection and duplicate address detection (DAD) procedures without handover prediction. To perform movement detection as soon as L2 handover finishes, our scheme uses link-layer state information that represents the establishment and release of L2 association between an MH and an AP. In addition, to perform movement detection procedure and DAD procedure at once, we introduce a new functional entity named as the fast handover agent (FHA) which retains a *Router Advertisement (RA) cache* and a *neighbor list* of neighbors on the link. The *RA cache* and the *neighbor list* are used for rapid movement detection and DAD, respectively. Since the FHA holds an *RA cache* and a *neighbor list*, it gives information on the default router and whether MH's link-local address to be used on the link conflicts with others. In the proposed scheme, we define a DAD flag in an RA message to indicate if MH's new CoA to be generated conflicts with others. Therefore, an MH can detect if its address conflicts with others when it receives an RA message from the FHA.

In this paper, we analyze the handover latency with a timing diagram, and formulate the signaling cost and the packet delivery cost, that consists of forwarding and loss cost, to compare the proposed scheme with FMIPv6 and MIPv6. In addition, we investigate the impact of incorrect handover prediction on these costs as well.

The remainder of this paper is organized as follows. In the next section, we investigate the characteristics of MIPv6 handover and extensions for the handover latency reduction. In section 3, we present the proposed handover scheme to reduce the handover latency without handover prediction. We give numerical analysis and numerical results in section 4 and 5, respectively. Section 6 concludes this paper.

2 Related Work

A successful handover of an MH from one point of attachment to another includes new link establishment(L2 handover), IP connectivity establishment, and new route establishment when the handover leads to subnet change (L3 handover).

We show an example of L2 and L3 handover in Fig. 1. As shown in Fig. 1, L2 handover does not always go with L3 handover. For example, L2 handover from SA 1 to SA 2 does not follow L3 handover, since SA 1 and SA 2 are in the same subnet A. However, L2 handover from SA 2 to SA 3 follows L3 handover, since the subnet of SA 2 differs from that of SA 3.

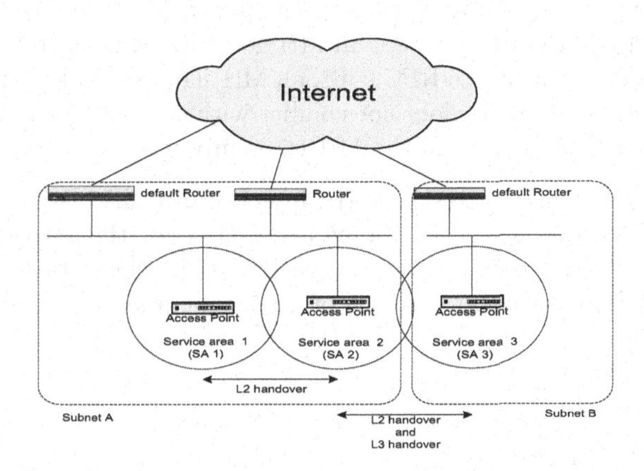

Fig. 1. An example of L2 and L3 handover

The MIPv6 is capable of handling L3 handover between different subnets, in a transparent manner for upper-layer protocol sessions. However, since L3 handover almost inevitably affects the transport protocol performance in terms of latency, mechanisms to smoothen handover, such as FMIPv6 and HMIPv6, are of interest.

For movement detection in MIPv6, an MH detects its movement by missing RA messages from the configured default router (RA assisted L3 handover) as shown in Fig. 2 (b) [1]. Therefore, an MH must wait until it receives an RA message to know the presence of a new access router (NAR) and decides that it loses reachability with the configured default router.

After detecting L3 handover, an MH performs DAD for link-local address, selects a new router, performs a prefix discovery for new router to form a new CoA and performs binding update and route optimization. Fig. 2 shows handover procedure of both RA and RS assisted L3 handover. Handover procedure can be partitioned into three steps; movement detection, address configuration, and location update.

The movement detection time (t_{MD}) is the interval from when an MH is under the coverage of an NAR to the instant it detects its movement to the NAR by receiving an RA message from the NAR. When an MH is under the coverage of new network, it can detect its movement by an RA response to an RS message from the MH as shown in Fig. 2 (a). In this case, RFC2461 specifies that RA response to RS messages must be randomly delayed by 0-500 msec. This may be a burden to provide real-time communications.

Alternatively, it may wait for an RA message. According to the latest specification, to reduce movement detection latency in MIPv6 environments, the RA interval can be a minimum of 30 msec to a maximum of 70 msec [1]. However, this reduced interval leads to bandwidth consumption of a wireless link.

The address configuration time (t_{AC}) is the interval from the time an MH detects its movement to the time it assigns its interface with a new CoA based on the prefix of the NAR. In this step, an MH generates a new CoA and performs DAD procedure. According to RFC 2461, an MH should wait 1 sec to a minimum to determine if its new CoA does not conflict with others. It is another obstacle to real-time service provision in the MIPv6 environment.

The location update time (t_U) is the interval from the instant an MH sends a binding update message to the home agent (HA) or the correspondent node (CN) to the instant it receives the first packet at the new attachment point. In general, the handover latency consists of t_{MD}, t_{AC}, and t_U.

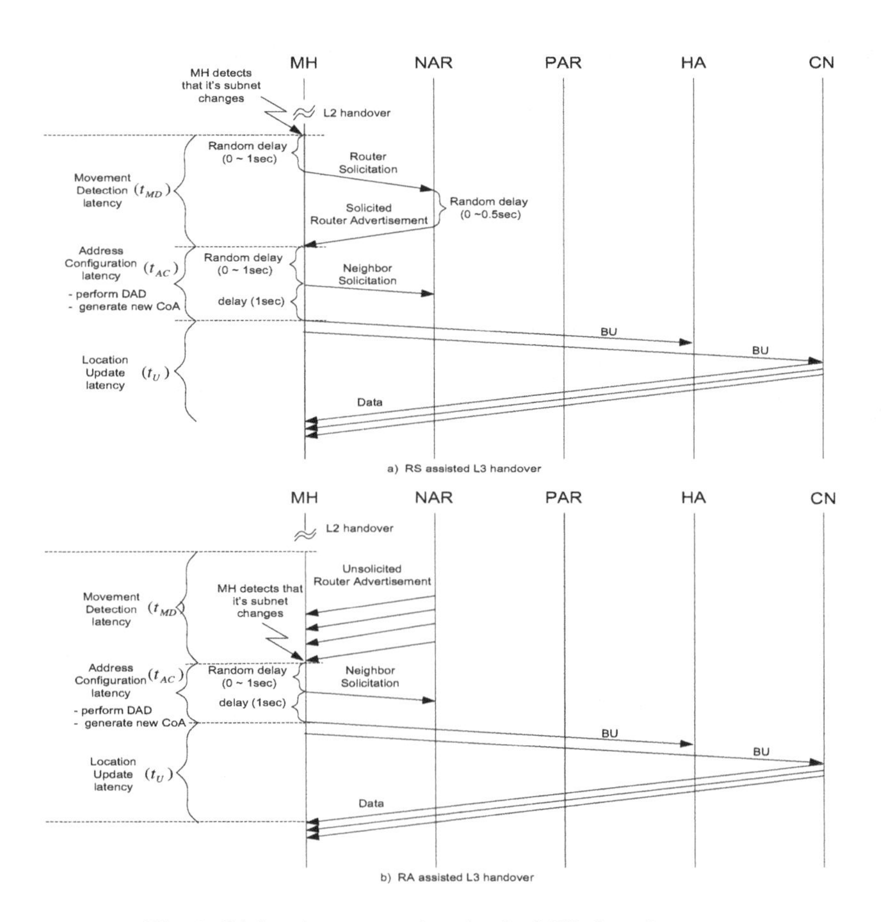

Fig. 2. L3 handover procedure in the MIPv6 environment

FMIPv6 allows an MH to predict its attachment with a prospective default router behind a new link, by helping to prepare new IP address configuration in advance. FMIPv6 assumes that this new IP address configuration is to be received through the currently used network interface and FMIPv6 requires adding additional support to IPv6 implementation in routers, which already deployed IPv6 infrastructure may not be ready to afford.

The handover prediction executed in FMIPv6 bears the conceptual uncertainty. As neither the exact moment of L2 roaming nor the definite next AP can be foreseen and even flickering may occur, extra damage is done to the traffic, which may be larger in the case of misleading anticipation. The predictions in FMIPv6 are classified into temporal (as it foresees the moment of start and completion of the handover) and spatial (as it predicts on where the handover will lead). Both predictions may be wrong; the temporal prediction can be too early or too late or an MH decides to move into a different subnet from the predicted one or not to move at all.

Two recent proposals that provide movement detection optimization are a Fast Router Advertisement (FastRA) [4] and a Router Advertisement caching (FastRD) on wireless APs [5]. In a FastRD scheme, an AP scans L2 frame for an unsolicited RA message and stores it, and sends cached RA to a new MH as soon as L2 handover finishes. However, since an unsolicited RA message is not periodic [6], the AP should scan every incoming L2 frame to look up an unsolicited RA message. It may bring about too much cost. Our proposed scheme is similar to the FastRD scheme in terms of RA caching. However, our scheme does not need to scan every L2 frame and can be used in the case when multiple routers are on the link, while FastRD scheme does not consider the case when multiple routers are on the link. Both FastRA and FastRD are concerned with only movement detection. In both FastRA and FastRD, fast DAD procedure is required. On the contrary, our proposed scheme performs movement detection and DAD procedure are performed at once to obtain low handover latency.

In this paper, we present a non-predictive handover scheme to reduce handover latency with fewer signaling than FMIPv6. In the next section, we will describe the proposed scheme in detail.

3 Proposed Scheme

From the analysis of handover procedures, movement detection and DAD procedures occupy a major part of the handover latency. Our proposed scheme aims at performing movement detection and DAD procedures quickly without handover prediction. Our proposed scheme, to accelerate L3 handover procedure, uses L2 events corresponding to establishment and release of L2 association between an MH and an AP. Nevertheless, we do not use any event that predicts L2 handover unlike FMIPv6.

In this paper, we assume that an MH generates its link-local address and a new CoA using an appropriate network prefix and its interface identifier such as link-layer address. In such a case, if link-layer address of an MH is ascertained unique on the link, we can say that link-local address and new CoA of the MH are unique on the link. In the reverse order, given a network prefix and a link-layer address, the link-local address and the CoA can be generated easily. We apply this fact to our proposed scheme.

Fig. 3 shows the proposed architecture and non-predictive handover scheme to perform movement detection and DAD procedure swiftly in MIPv6 environment. In the proposed scheme, we introduce an FHA as an assistant to perform movement detection and DAD procedures quickly. As shown in Fig. 3, movement detection and DAD procedures are performed at once.

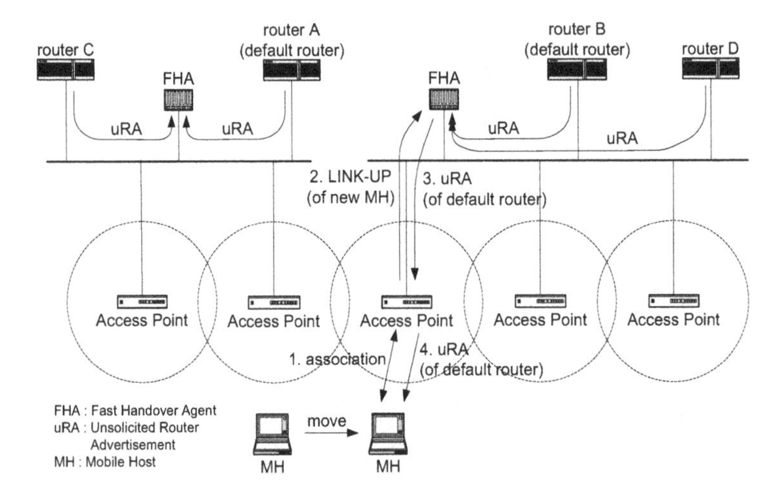

Fig. 3. Proposed architecture and non-predictive handover scheme

In the proposed scheme, FHA-based RA caching mechanism is used for default router discovery. Based on IPv6 specification, every router advertises its presence with an RA message to all node multicast address group. Therefore, the FHA receives all unsolicited RA messages from routers on the link and keeps the most recent default router's advertisement in its RA cache.

To speed up DAD procedure, we use a *neighbor list* in the FHA. The *neighbor list* stores link-layer addresses or link-local addresses of neighbors on the link. The FHA performs DAD procedure of its own accord by comparison the link-local address generated by using link-layer address of an MH with the contents of the *neighbor list*, whenever MH's handover is notified to the FHA. In addition, there may be fixed hosts on the link. Thus the FHA should have link-local addresses of fixed hosts as well as MHs in order to perform DAD procedure internally. Since fixed nodes perform DAD procedure using Neighbor Solicitation (NS) message when they detect the attachment to the network [6], the FHA performs filtering of NS messages for DAD procedure on the link and updates the *neighbor list* with link-local address or discards filtered message.

The filtering scheme of NS messages for DAD procedure is as follows. IPv6 hosts send an NS message to solicited-node multicast address group on the link. Since the solicited-node multicast address has the prefix FF:02::1:FE/104 concatenated with 24 low-order bits of IPv6 unicast address, and the NS message for DAD procedure does not include a link-layer address of the sender, the FHA can obtain NS messages for DAD procedure by filtering packets whose destination is the solicited-node multicast group and whose link-layer address option is empty.

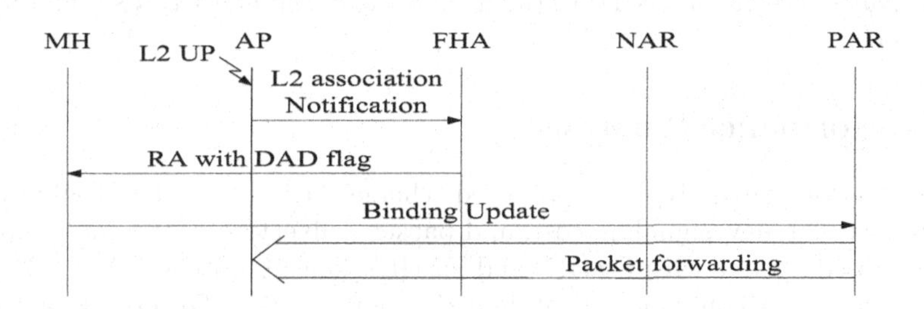

Fig. 4. The message interaction of proposed scheme

Fig. 3 and 4 represent message flows of the proposed handover scheme. Whenever an AP detects L2 handover of new MH (step 1), the AP notifies new MH's L2 handover to the FHA with LINK-UP event containing MH's link-layer address and previous AP's address (step 2). The FHA, upon receiving a notification of new MH's L2 handover, sends the cached advertisement of a default router to the MH (step 3 & 4). Then the MH uses this advertisement as a means of movement detection and DAD procedure. Based on the above procedure, the MH, upon finishing L2 handover, can discover the default router promptly by receiving an RA message and can decide whether its subnet is changed or not. To indicate whether MH's new CoA conflicts with others by using an RA message, we define a DAD flag (D bit) in the RA message as shown in Fig. 5. If the MH's new CoA conflicts with others, D bit is set to 1, otherwise, D bit is set to 0. When the MH knows that its new CoA does not conflict with others, it sends immediately binding update message to the PAR in order to receive buffered packets at the PAR.

Type	Code				Check sum
Cur Hop Limit	M	O	D	reserved	Router Lifetime
Reachable Time					
...					

Fig. 5. Modified RA message for DAD procedure

When the proposed scheme is applied to DAD procedure, there is a problem in that neighbors cached in the FHA's *neighbor list* may differ from actual neighbors on the link. This situation comes from the fact that DAD procedure is performed only when the host configures its network interface with newly generated address and there is no mechanism to notify moving out the link or disconnection of a host to the FHA.

In our proposed scheme, to maintain more accurate information on neighbors on the link in the FHA's *neighbor list*, the FHA performs periodically *Neighbor Unreachable Discovery* (NUD) to the fixed hosts in the *neighbor list* and updates the state.

4 Performance Analysis

In this section, we analyze the proposed scheme, MIPv6, and FMIPv6 in terms of handover latency, signaling cost, and packet delivery cost. We don't compare the proposed scheme with FastRD and FastRA, since both FastRA and FastRD are concerned with only movement detection optimization. For both FastRA and FastRD, fast DAD procedure is required.

4.1 Handover Latency

With handover latency, we specify three latencies such as link switching latency[1], IP connectivity latency, and location update latency similar to [10]. The link switching latency (t_L) is due to L2 handover. The IP connectivity latency (t_I) is due to new IP address configuration and movement detection after L2 handover. an MH can send packets from a new subnet link after IP connectivity latency. The location update latency (t_U) is the latency in forwarding IP packets to MH's new IP address.

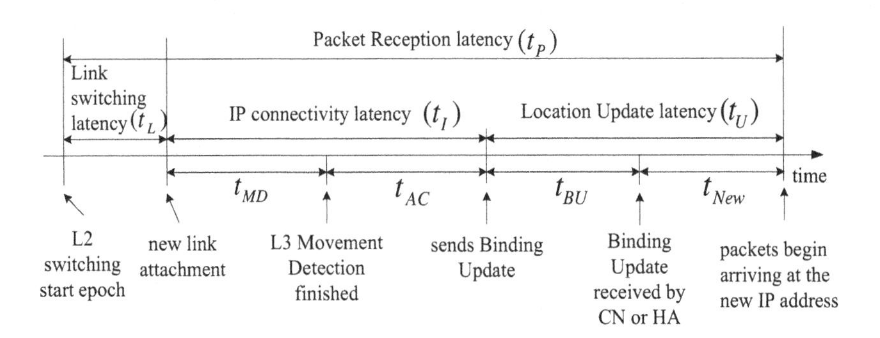

Fig. 6. Timing diagram in basic MIPv6

In discussion on handover latency, we consider two scopes; when an MH sends packets with new IP address and when the MH receives packet after L2

[1] link switching latency is the same as L2 handover latency.

handover. The IP connectivity latency reflects how quickly an MH can send IP packets after L2 handover. To consider how quickly an MH can receive IP packets after L2 handover, we specify the packet reception latency (t_P), the period from the starting point of L2 handover to when an MH receives packets for the first time after L2 handover. To represent the t_P and t_I, we define the round-trip delay (t_{MN}) between an MH and an NAR , the round-trip delay (t_{NP}) between an NAR and a PAR, the round-trip delay (t_{AF}) between an AP and an FHA, and the round-trip delay(t_{MF}) between an MH and an FHA.

Fig. 6 shows the timing diagram corresponding to basic MIPv6[2] (B-MIPv6). In basic MIPv6, the packet reception latency is $t_L + t_I + t_U$ and the IP connectivity latency is $t_{MD} + t_{AC}$.

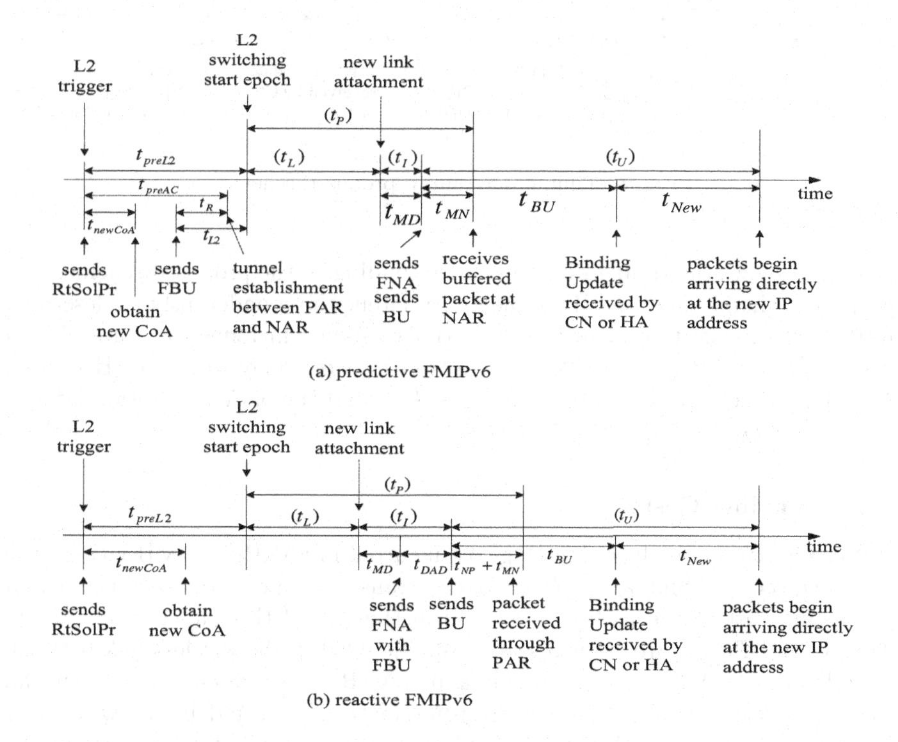

Fig. 7. Timing diagram in FMIPv6

In Fig. 7, (a) represents the timing diagram of predictive FMIPv6 (P-FMIPv6) and (b) represents that of reactive FMIPv6 (R-FMIPv6). In predictive FMIPv6, an MH can send packets as soon as movement detection, since the address configuration is finished before L2 handover finishes. In addition, packets buffered at the NAR are forwarded to an MH as soon as the NAR receives an FNA message from an MH. In R-FMIPv6, an MH can send packets after

[2] We represent MIPv6 as basic MIPv6 after this.

movement detection and DAD procedures, and can receive packets when the PAR forwards packets to an MH after it receives MH's FBU message. The IP connectivity latencies of P-FMIPv6 and R-FMIPv6 are t_{MD} and $t_{MD} + t_{DAD}$, respectively. The packet reception latencies of P-FMIPv6 and R-FMIPv6 are $t_L + t_I + t_{MN}$ and $t_L + t_I + t_{MN} + t_{NP}$, respectively.

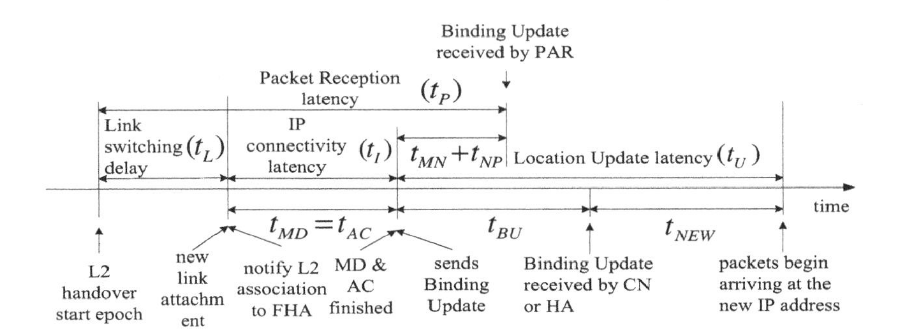

Fig. 8. Timing diagram in proposed scheme

Fig. 8 shows the timing diagram corresponding to the proposed scheme. The proposed scheme improves IP connectivity latency preponderantly as described in the previous section. The IP connectivity latency includes the transmission delay from an AP to an FHA and the transmission delay from an FHA to an MH. The IP connectivity latency is $\frac{t_{AF}}{2} + \frac{t_{MF}}{2}$ and the packet reception latency is $t_L + t_I + t_{MN} + t_{NP}$.

4.2 Signaling Cost

In this section, we analyze signaling cost incurred in FMIPv6, B-MIPv6, and the proposed scheme. Similar to [7], we define transmission cost between nodes and processing cost at the nodes to be used in the rest of this paper. The TC_{MP}, TC_{MN}, and TC_{MF} are transmission costs incurred in the wireless link between an MH and a PAR, between an MH and a NAR, and between an MH and an FHA, respectively. The TC_{NP} is transmission cost incurred in the wired link between a NAR and a PAR. The TC_{AF} is transmission cost incurred in the wired link between an AP and an FHA. The a_P, a_N, and a_{FHA} are processing costs at the PAR, the NAR, and the FHA, respectively.

We denote signaling cost of B-MIPv6, P-FMIPv6, R-FMIPv6, and the proposed scheme as CS_{basic}, CS_{pred}, CS_{react}, and CS_{prop}, respectively. In this analysis, we assume that movement detection is accomplished by receiving an RA message in B-MIPv6 and movement detection is accomplished by RS/RA message exchanges in FMIPv6. In addition, we exclude the location update cost from signaling cost.

We assume that transmission cost is proportional to the distance between the source and the destination and proportionality constant is δ_C. In addition, we assume that the transmission cost over the wireless link is ρ times higher than the unit distance wireline transmission cost, θ_S. For example, TC_{NP} can be expressed as $TC_{NP} = l_{NP}\theta_S\delta_C$, where l_{NP} is an average distance between a NAR and a PAR, and TC_{MN} can be expressed as $TC_{MN} = \rho\theta_S\delta_C$, where the average distance l_{MN} between an MH and a NAR is 1. Using the proportionality constant δ_C and weighting factor ρ for the wireless link, each signaling cost can be rewritten as follows:

$$
\begin{aligned}
CS_{basic} &= 2TC_{MN} = 2\rho\theta_S\delta_C \\
CS_{pred} &= 4TC_{MP} + 3TC_{NP} + 3TC_{MN} + 3a_P + 2a_N \\
&= (7\rho + 3l_{NP})\theta_S\delta_C + 3a_P + 2a_N \\
CS_{react} &= 2TC_{MP} + 2TC_{NP} + 3TC_{MN} + 2a_P + a_N \\
&= (5\rho + 2l_{NP})\theta_S\delta_C + 2a_P + a_N \\
CS_{prop} &= TC_{AF} + TC_{MF} + TC_{MN} + TC_{NP} + a_{FHA} + a_P \\
&= (2\rho + l_{AF} + l_{NP})\theta_S\delta_C + a_{FHA} + a_P
\end{aligned}
$$

4.3 Packet Delivery Cost

With packet delivery cost, we consider the cost associated with both forwarding packets (*forwarding cost*) and lost packets (*loss cost*). In this analysis, we consider the forwarding cost as the additional buffer space used by forwarding packets during the handover period. Packet delivery cost consisting of forwarding cost and loss cost is defined as $\alpha \cdot C_{forwarding} + \beta \cdot C_{loss}$, where α and β are weighting factors.

In B-MIPv6, only loss cost does exist in the packet delivery cost. Consequently, given the packet arrival rate λ_p, CP_{basic} can be expressed as Eq. 1. In the proposed scheme, the packet delivery cost consists of only forwarding cost. The packets, forwarded to the PAR, are buffered until MH's binding update message arrives at the PAR. As a result, the CP_{prop} can be expressed as Eq. 2.

In FMIPv6, packet buffering at the NAR, based on handover prediction, is supported to avoid packet loss and to give a smooth handover. However, because of wrong temporal prediction, some packets may be lost. t_{L2} denotes the time period from when the FBU message is sent to the starting point of L2 handover and t_R denotes the time period from when the FBU message is sent to when the tunnel is established. If t_{L2} is less than t_R, packets arriving at the PAR during $(t_R - t_{L2})$ period may be lost, since the tunnel is not yet established. Thus, the loss cost can be expressed as $\lambda_p\beta max\{(t_R - t_{L2}), 0\}$. Consequently, the packet delivery cost in P-FMIPv6 and R-FMIPv6 can be expressed as Eq. 3 and Eq. 4, respectively.

$$
CP_{basic} = \lambda_p\beta(t_L + t_I + t_U) \tag{1}
$$

$$
CP_{prop} = \lambda_p\alpha(t_L + t_I + \frac{t_{MN}}{2} + \frac{t_{NP}}{2}) \tag{2}
$$

$$CP_{pred} = \lambda_p[\alpha\{t_L + t_I + \frac{t_{MN}}{2} + max\{(t_{L2} - t_R), 0\}\} + \beta max\{(t_R - t_{L2}), 0\}] \quad (3)$$

$$CP_{react} = \lambda_p\beta(t_L + t_I + \frac{t_{MN}}{2} + \frac{t_{NP}}{2}) \quad (4)$$

In the case of a wrong spatial prediction, the forwarded packets to the wrongly predicted NAR may be lost and this packet forwarding to the wrongly predicted NAR is finished when MH's FBU message arrives at the PAR. Therefore, packet delivery cost in the wrong spatial prediction case of FMIPv6 can be expressed as Eq. 5.

$$CP_{wrong} = \lambda_p\beta[(t_L + t_I + \frac{t_{MN}}{2} + \frac{t_{NP}}{2}) + max\{(t_{L2} - t_R), 0\}] \quad (5)$$

5 Numerical Results

In this section, we demonstrate some numerical results. For the numerical analysis of handover latency and signaling cost, we set the values of parameters as shown in Table 1. These values are based on the empirical results of [8] and reference values defined in [7,9,10].

Table 1. Parameters for numerical analysis

latency analysis (in msec)						signaling cost analysis										
t_L	t_{MN}	t_{NP}	t_{AF}	t_{MF}	t_R	a_N	a_P	a_{FHA}	δ_C	l_{AF}	l_{NP}	ρ	θ_S	λ_p	α	β
50	14	6	6	14	20	5	5	10	0.1	1	5	10	1	1	0.2	0.8

In B-MIPv6 case, an MH does not use L2 trigger events and detects its movement by RA beacon message. the MH can detect its movement within 27 msec on average, and within 70 msec at worst [11]. In addition, an MH waits 1 sec after sending an NS message for DAD with random delay, to confirm that its address does not conflict with others [6]. Consequently, average t_I and t_P in B-MIPv6 are about 1527 msec and 1677 msec, respectively. In FMIPv6, an MH detects its movement by RS/RA message exchanges triggered by L2 events. Therefore, t_{MD} is 14 msec. In P-FMIPv6, t_I and t_P are 14 msec, and 78 msec, respectively. In R-FMIPv6, minimum t_I and t_P are 28 msec and 98 msec, respectively. However, if normal DAD procedure is applied to R-FMIPv6, 1 sec latency should be added to previous minimum values. In the proposed scheme, t_I is 10 msec and t_P is 83 msec. In terms of the IP connectivity latency, the proposed scheme gives the lowest latency, since movement detection and DAD procedures are performed at once. From the numerical results, we found that proposed scheme gives lower latency as well as fewer signaling messages than P-FMIPv6 in terms of the IP connectivity. Finally, we can conclude that our scheme can

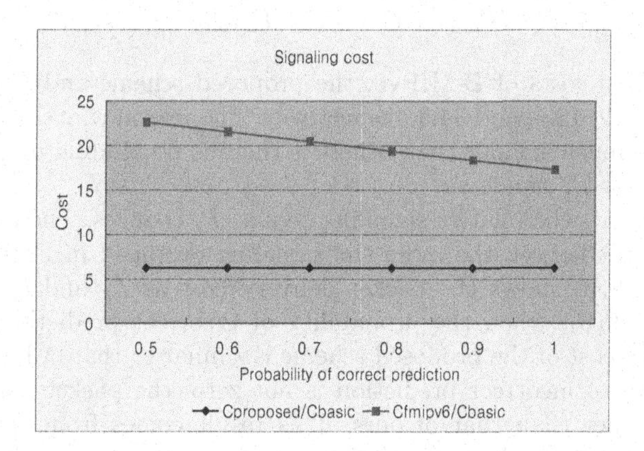

(a) Signaling cost

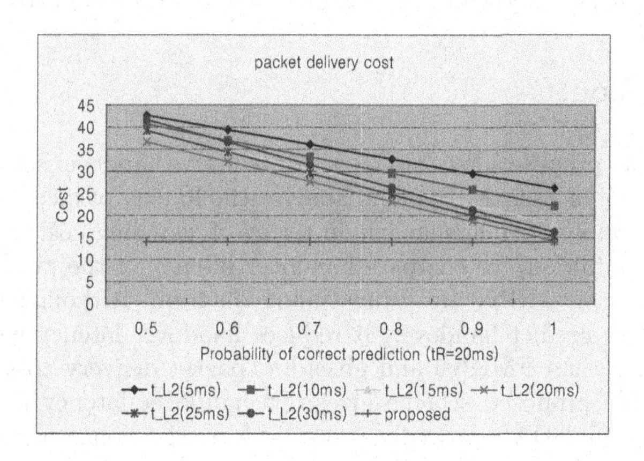

(b) Packet delivery cost

Fig. 9. Signaling and packet delivery costs

be used to support real-time applications in mobile IPv6 environment without handover prediction.

An additional signaling cost of FMIPv6 in the wrong spatial prediction case is $(3\rho+2l_{NP})\theta_S\delta_C+3a_P+2a_N$. Let P_s be the probability of prediction accuracy. Then, the signaling cost of FMIPv6 can be expressed as shown in Eq. 6. Similar to the signaling cost of FMIPv6, the packet delivery cost of FMIPv6, CP_{fmip}, can be expressed as a function of the probability of prediction accuracy as shown in Eq. 7.

$$CS_{fmip} = P_s \cdot CS_{pred} + (1 - P_s)(CS_{wrong} + CS_{react}) \quad (6)$$
$$\text{where } CS_{wrong} = (3\rho + 2l_{NP})\theta_S\delta_C + 3a_P + 2a_N.$$

$$CP_{fmip} = P_s \cdot CP_{pred} + (1 - P_s) \cdot CP_{wrong} \tag{7}$$

The signaling costs of B-MIPv6, the proposed scheme and FMIPv6 (when $P_s = 100\%$) are 2, 12.5, and 34.5, respectively. The reason why the signaling cost of B-MIPv6 is much smaller than others is there is no binding update signaling to the PAR after L2 handover.

Fig. 9(a) shows the relative signaling cost as P_s changes. The more incorrect the handover prediction, the more the signaling overhead in FMIPv6 rises. In addition, Fig. 9(b) shows the packet delivery cost as P_s and t_{L2} change. As shown in Fig. 9(b), when the probability of incorrect prediction is zero, the packet delivery cost of the proposed scheme is similar to that of FMIPv6. When the probability of incorrect prediction is not zero, the packet delivery cost of FMIPv6 is higher than that of ours. This result comes from the packet loss caused by incorrect handover prediction.

From Fig. 9, we conclude that incorrect handover prediction in FMIPv6 brings on additional overhead in terms of signaling and packet loss.

6 Conclusion

In this paper, we propose a low-latency non-predictive handover scheme to reduce handover latency in basic MIPv6 and analyze the impact of incorrect prediction in FMIPv6 on mobility management in terms of signaling cost and packet delivery cost. In addition, we compare handover latency of the proposed scheme, FMIPv6, and basic MIPv6 by using timing diagram. Although the proposed scheme does not predict handover, it reduces handover latency with fewer signaling messages than FMIPv6 and gives low packet delivery cost as well. The reasons why the proposed scheme gives low handover latency are that movement detection and DAD procedures are performed at once using FHA-based RA caching and *neighbor list* and movement detection is triggered by LINK-UP event.

In future work, we plan to extend the proposed scheme with the cases where the hierarchical MIPv6 scheme does exist. In addition, we plan to simulate the proposed scheme under several scenarios where MHs move with different velocities, incorrect movement predictions are abundant, and the movement to the same link is generated at the same time.

References

1. D. Johnson and C. Perkins, *Mobility support in IPv6, IETF drft*, March 2002.
2. H. Soliman et al., *Hierarchial MIPv6 Mobility Management IETF draft*, July 2002.
3. G. Dommety et al., *Fast Handover for Mobile IPv6, IETF draft*, March 2002.
4. J. Kempf et al., *IPv6 Fast Router Advertisement, IETF draft*, Oct. 2003.
5. J. H. Choi et al., *Fast Router Discovery with RA caching in AP, IETF draft*, Feb. 2003.
6. T. Narten et al., *RFC 2461 Neighbor Discovery for IP Version 6*, 1998.

7. J. Xie and F. Akyildiz, *A Novel Distributed Dynamic Location Management Scheme for Minimizing Signaling Costs in Mobile IP*, IEEE Trans. on Mobile Computing, 2002.
8. Mishra et al., *An Emprical Analysis of the IEEE 802.11 MAČ Layer Handff Process*, ACM SIGCOMM CCR, Oct. 2003.
9. S. Park and Y. Choi, *Performance Analysis of Fast Handover in Mobile IPv6 Networks*, IFIP PWC 2003, 2003.
10. R. Koodli and C. Perkins, *Fast Handovers and Context Transfers in Mobile Networks*, ACM SIGCOMM CCR, Oct. 2001.
11. G. Daley, B. Pentland, and R. Nelson, *Movement Detection Optimizations in Mobile IPv6*, ICON 2003, 2003.

A Framework for End-to-End QoS Context Transfer in Mobile IPv6*

Chuda Liu[1,3], Depei Qian[1,2], Yi Liu[1], and Kaiping Xiao[1]

[1] Computer Science Department, Xi'an Jiaotong University, Xi'an, P.R.C, 710049
{lcdm, depeiq, liuyi97}@263.net
[2] School of Computer, Beihang University, Beijing, P.R.C, 100083
[3] Changsha Aeronautical Vocation Technical College, Changsha, P.R.C, 410124,

Abstract. Providing Quality-of-Service (QoS) guarantees and mobility support for Internet devices has become a hot research topic in the Next Generation Internet research, since mobile computing is getting widespread. Context transfers allow better node mobility support, and avoid re-initiation of signaling to and from a Mobile Node (MN). However, Context Transfer Protocol (CTP) [1] proposed by IETF can not meet the need of end-to-end QoS mechanisms because contexts are only transferred between Access Routers (ARs). This paper presents a framework for end-to-end QoS context transfer based on the architecture of F-HMIPv6, which may provide an end-to-end QoS context transfer for real-time applications, therefore they can get promptly the same forwarding process, minimize the handover service disruption, and avoid initiating the end-to-end QoS signaling from scratch after an MN performs handovers. The Context Transfer Data message containing the QoS context information, a hop-by-hop extension IPv6 option header, is sent from the previous access router (PAR) to the next access router (NAR) via the Mobility Anchor Point (MAP) where old path and new path meet each other. The QoS entities in the nodes between MAP and NAR will be required to check the QoS context information and reserve appropriate resources for MN's sessions and update the new path data in the QoS entities. After successful context transfers, the resources reserved for MN's sessions will be released on the old path. Our scheme may also reduce the signaling overhead and handover latencies by adopting the F-HMIPv6 [2] architecture.

1 Introduction

With the rapid increase of portable devices such as laptops, PDAs, hand-held computers, and a variety of wireless devices, mobile computing applications have become more practical. Real-time services such as Internet telephony, video conferencing, and video-on-demand should be available in the mobile computing environments. It is important for the mobile Internet environment to provide QoS guarantees in the

* This work is supported by the National High-tech Research and Development Program of China, No. 2001AA112120.

I. Niemegeers and S. Heemstra de Groot (Eds.): PWC 2004, LNCS 3260, pp. 466–475, 2004.

near future. Providing QoS guarantees and mobility support for Internet devices has become a hot research topic in the next generation Internet research. MIPv6 [3] proposed by IETF allows nodes to move from one subnet to another while maintaining the connectivity and on-going connection between MN and correspondent nodes (CNs). However, the basic Mobile IPv6 protocol does not provide QoS guarantees except the best-effort services. Two different mechanisms have been proposed to provide QoS guarantees in the Internet by the IETF: the Integrated Services (IntServ) [4] based on Resource ReSeravation Protocol (RSVP) [5] and the Differentiated Services (DiffServ) [6] based on priority levels. These mechanisms may work well in the wired networks but face new challenges in the mobile networks due to the mobility of hosts. Provision of end-to-end QoS in mobile networks is more complex than in wired networks mainly due to the user mobility and the constrained bandwidth of the wireless links.

Recently, many investigators have studied the mechanisms of deploying the existing QoS architectures in the mobile wireless networks [7] ~ [10]. However these schemes provide QoS provision by re-initiating signaling from scratch after an MN performs handovers. This may introduce extra signaling overhead and complexity to the protocols, and waste the precious wireless bandwidth (i.e., MRSVP [7] will make advance resource reservations at multiple locations that the MN may possibly visit after handover), resulting in large latency and packet losses to the MN's on-going sessions.

Context transfer protocol (CTP) has been proposed by the Seamoby working group in IETF. This new concept attempts to achieve seamless handovers for Mobile IPv6. Context refers to the information about the current state of a service to be re-established on a new subnet. For example, in order to reserve the same RSVP recourses as the MN had in the previous subnet after it moves to a new subnet, the corresponding RSVP state information for obtaining the same packet forwarding treatment is called RSVP context. While context transfer refers to the movement of context from one router or network entity to another as a means of re-establishing specific services on a new subnet or collection of subnets, the main motivation of context transfer is to reduce latency and packet losses by avoiding re-initiating signaling to and from the MN [11]. Example features contained in the context are authentication, authorization, and accounting (AAA), header compression, QoS, and security. We will focus on the QoS feature of context transfer in this paper.

As described in [12], context transfer has some advantages, including seamless operation to the MN's traffic flow, bandwidth saving, and less susceptible to errors. However, CTP cannot meet the need of end-to-end QoS requirement because transferring context only at the last hop access router may be insufficient to completely re-initialize the MN's QoS treatment. Some other routers on the path between MN and CN may also need to be involved [11]. This paper proposes a scheme to provide an end-to-end QoS context transfer for real-time applications. The scheme enables the real-time applications to get promptly the same forwarding process, therefore minimizing the handover service disruption and avoiding initiating the end-to-end QoS signaling from scratch when MN performs a handover. The Context Transfer Data (CTD) message containing the QoS context information, a hop-by-hop extension to the IPv6 header, is sent from the previous access router (PAR) to the next access router (NAR) via the Mobility Anchor Point (MAP) where old path and new path meet each other.

The QoS entities in the nodes between MAP and NAR will check the QoS context information, reserve appropriate resources for MN's sessions, and update the new path data in the QoS entities. After successful context transfers, the resources reserved for MN's sessions will be released on the old path. Our scheme may also reduce the signaling overhead and handover latency by adopting the F-HMIPv6 [2] architecture.

This paper is organized as follows. In section two, we give an introduction to the related protocols. The design goals and assumptions of the proposed scheme are presented in section 3. In section 4, we propose a framework for end-to-end RSVP context transfer in Mobile IPv6, including the network model and context transfer process and the performance comparison. Finally, we conclude the paper and present the further work.

2 Related Work

2.1 Fast Handover for Hierarchical MIPv6 (F-HMIPv6) [2]

The F-HMIPv6 protocol intends to combine the Fast Handovers for Mobile IPv6 (FMIPv6) protocol [13] with the Hierarchical Mobile IPv6 Mobility Management protocol (HMIPv6) [14]. This means that the fast handover mechanism will be deployed over the HMIPv6 networks using the F-HMIPv6 protocol. Therefore, the protocol provides the advantages of both schemes, i.e., a seamless handover scheme with less signaling overhead and lower handover latencies. Moreover, the overall handover latency achieved by FMIPv6 will be further reduced because of local location updating in HMIPv6, while in the original FMIPv6, the Home Agent (HA) and CNs are usually far away.

HMIPv6 deals with reducing the amount and latency of signaling between an MN, its HA and one or more CNs by introducing the MAP which is used by the MN as a local HA. Each MN is assigned two Care-of Addresses (CoAs): RCoA (Regional CoA) and LCoA (on-Link CoA). When the MN moves around within a MAP domain, the RCoA can not be changed and only the LCoA need to be changed and registered with the MAP to bind the RCoA and LCoA. Only when the MN moves out of the MAP domain, the RCoA need to be registered to the HA and the MN sends Binding Update (BU) to CNs.

FMIPv6 reduces packet loss by providing fast IP connectivity as soon as a new link is established. It uses the Layer 2 (L2) triggers to obtain the link address and subnet prefix information of the new attachment point of an MN when it is still connected to PAR. This may reduce movement detection latency of the MN. Through the necessary messages exchanging, the MN may pre-configure the New CoA (NCoA) to reduce the NCoA configuration latency. Moreover, in FMIPv6, the data packets arriving the Previous CoA (PCoA) can be forwarded to the NAR using the tunnel established between the PAR and the NAR.

In F-HMIPv6, the MAP performs the tunneling for fast handover instead of the PAR. By the F-HMIPv6 scheme, the data packets sent by CN will be tunneled by the MAP toward the NAR during the handover.

2.2 Context Transfer Protocol (CTP) [1]

As mentioned in RFC3374 [11], CTP enables authorized context transfers. The key goals are to reduce latency, minimize packet losses and avoid re-initiation of signaling to and from MN.

The context transfer can be initiated by the mobile node (mobile-controlled) or by the network (network-initiated). The CTP typically operates between the source node (i.e., PAR) and the target node (i.e., NAR). PAR transfers contexts under two general scenarios. In the first scenario, either the MN sends the CT Activate Request (CTAR) to PAR to initiate context transfer, or a Layer 2 trigger triggers the PAR. In response PAR transmits a CTD message that contains feature contexts to NAR. In the second scenario, NAR sends a CT Request (CTR) message to PAR as a result of either receiving the CTAR message sent by the MN or an internal trigger. In response to the CTR message, PAR transmits a CTD message that includes the MN's feature contexts. When receiving a CTD message, NAR may generate a CTD Reply message to report the status of processing the received contexts.

2.3 QoS-Conditionalized Binding Update in Mobile IPv6 (QCBU) [15]

QCBU is a QoS-Conditionalized solution that is based on the HMIPv6. When a handover takes place, a QoS option, i.e., a hop-by-hop option header defined in [16], is piggybacked in the BU message and sent to the MAP. Each node between MN and MAP forwards the QoS need contained in the QoS option and checks for the resource availability in the node. If the resource is insufficient, the message is dropped and a negative acknowledgement is sent back to MN. Otherwise the message is forwarded to the next hop. When BU carrying the QoS option finally arrives MAP (the crossover point of the old path and the new path) and MAP satisfies the resource requirement, MAP sends Binding Acknowledgement (BA) back to MN.

Accordingly, a handover takes place only when sufficient resources are available in all nodes along the new transmission path. It also enables MN to choose flexibly among a set of available access points so that MN can transmit packets through a route that offers satisfied QoS. In addition, the scheme is built upon the hierarchical mobile IPv6 protocol to localize mobility management.

3 The Design Goals and Assumptions

Our scheme is based on the following goals and assumptions:

1) The key objective is to construct an effective end-to-end QoS context transfer framework, which may establish QoS context information on each new added routers on the path between the crossover router (where the old and new path meet) and NAR during the MN handover. The real-time applications can obtain the same QoS forwarding treatment as that they had before the handover.
2) The scheme should release any QoS state along the old packet path when contexts are transferred successfully after handover.

3) The scheme should achieve seamless handover and reduce the signaling overhead and handover latency to minimize service disruption of real-time applications during the handover period.
4) We assume that most handovers take(s) place locally, and we may deploy localized mobility management solution to make the MN's handover transparent to its HA and to CNs with which it is communicated.
5) The scheme operates entirely within the administrative domain of the wireless network, any handover out of the domain is beyond our scope, because the operation policies and the Service Level Agreements (SLAs) need to be re-negotiated, so re-establishment or re-negotiation of the level of service would be the preferred case proposed in RFC 3374 [12].
6) Each QoS entity in the routers between PAR-MAP-NAR (shown in Fig. 1) is required to check the QoS context information contained in the CTD message. How the QoS entity processes the QoS context information contained in CTD message is described in section 4.2.1 (QoS entity is defined in [16]).
7) MAP is the crossover point where the old path and the new path meet each other after an MN performs handovers, and can be discovered by using MAP discovery algorithm in HMIPv6.
8) QoS should be supported for both uplink (from MN) and downlink (to MN) traffic.

4 Proposed Solution

In this section, we firstly show the network reference model. Then the implementing details are described. The performance analysis of the scheme is discussed in the finally.

4.1 Network Reference Model

Fig. 1 illustrates the network reference model of the framework for the end-to-end QoS context transfer. The MAP is the crossover point where the old path and new path meet each other after MN performs handovers, which can be discovered by using MAP discovery algorithm in HMIPv6. IR is an intermediate router. When the MN moves from AR1 to AR2, the MAP is located at location (1); while from AR2 to AR3, the MAP is located at location (2).

4.2 The Framework for End-to-End QoS Context Transfer

We present an end-to-end QoS context transfer framework to overcome the drawback of the CTP in which contexts are transferred only between the PAR and NAR. In this section, we firstly give the processing rules of the CTD message in the framework, and the details of the end-to-end QoS context transfer are described successively.

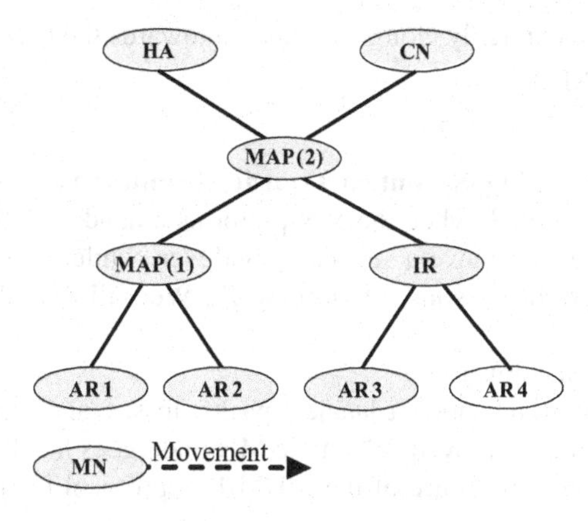

Fig. 1. Network reference model

4.2.1 The Processing Rules of the CTD Message in the Framework

In order to implement the end-to-end QoS context transfer, the CTD message is sent from PAR to NAR via MAP. The QoS entity of the router on the path between PAR and NAR is required to check the QoS context information in the CTD message based on the following rules:

1) If the QoS context information carried in the CTD message is found to be the same as the QoS states of the QoS entity in the router, the router simply forwards the CTD message to the next hop without referencing to the QoS context.
2) Otherwise, the router will extract the QoS context information from the CTD message. According to the QoS context received from the CTD message, the router will reserve corresponding resources and update the path information in the QoS entity. For example, the RSVP_HOP object of RSVP needs to be updated when the path has been changed. And the router will send a CTD Reply (CTDR) to PAR to report the status of processing the received contexts. Finally, the CTD message is forwarded to the successive router.

Therefore, when the CTD message is transmitted on the old path, that is, between PAR and MAP, routers just forward the message to the next hop. But when it is transmitted on the new path between MAP and NAR, routers on the path will extract the QoS context and reserve corresponding resources, so the corresponding QoS states are established on the new path.

Consequently, the QoS feature context information is carried in the CTD message as an IPv6 hop-by-hop option in our scheme so that each node on the path can check and process the message individually.

There are two methods to release the QoS setting in the router along the old path if contexts are transferred successfully. One is time-out, i.e., the QoS state is deleted if no refresh message arrives before the expiration of a "cleanup timeout" interval. Another way is to do it explicitly by sending a "teardown" message from MAP to PAR when the CTD Reply (CTDR) message is received by the QoS entity in MAP, for example, in RSVP, RSVP Daemon in MAP will send a "Resv_Tear" message to delete the QoS state explicitly along the old path towards the MN. The second method is adopted in this paper.

4.2.2 The End-to-End QoS Context Transfer Framework

As mentioned in section 3, when the MN performs a handover, the F-HMIPv6 architecture is employed to achieve a seamless handover, while the end-to-end QoS context transfer is performed using our framework. We will describe the details as the follows.

1) The seamless handover

In order to reduce the handover latency, packet losses, and signaling overhead, the F-HMIPv6 protocol is employed. When the MN undergoes handover, the scheme will follow the processing procedure of the F-HMIPv6 protocol [2] to achieve the seamless handover.

2) The end-to-end QoS context transfer

While the F-HMIPv6 protocol achieves the seamless handover, the end-to-end QoS context transfer is implemented as follows:

- The end-to-end QoS context transfer may be initiated by MN (MN may send the CTAR message to PAR to request contexts transfer) or by the network (PAR starts the transfer of contexts (source trigger) or NAR sends the CTR message to PAR to request contexts (target trigger)).
- The PAR sends the CTD message that includes the QoS context to the NAR via the MAP in response to the context transfer trigger. Each router on the path from PAR to NAR via MAP processes and forwards the CTD message according to the CTD message processing rules described in section 4.2.1.
  - When a router on the old path between PAR and MAP receives the CTD message, the QoS entity in the router checks the QoS context information contained in the CTD message. If the QoS context information in the CTD message is the same as that existing in the QoS entity, the router simply forwards the CTD message to the next hop;
  - When a router on the new path between MAP and NAR receives the CTD message, the QoS entity in the router checks the QoS context information carried in the CTD message. Because the QoS context information in the CTD message does not exist in the QoS entity, the router extracts the QoS context from the message. If the QoS context information is transferred successfully, the router will reserve corresponding resources and update the path information in the QoS entity. It then forwards the message to the next hop. Otherwise, a CTDR message is sent from the router to PAR to report the failure of the QoS context transfer. In failure case, the CTD message will not be forward to the next node.
  - When the NAR successfully processes the received QoS context, the "S" bit in the CTDR message sent by NAR to PAR will be set to one. This may trigger the QoS entity of the MAP to send a Teardown message toward the PAR to delete the QoS states on the old path.

4.3 The Performance Analysis and Discussion

In this section, we compare the performance of our scheme to the QoS solution requirements for Mobile IP [17].

1) Performance requirements

- Our scheme minimizes the interruption in QoS at the time of handover, because an end-to-end QoS context transfer scheme is adopted to allow better support for node mobility, and avoid re-initiating signaling to and from the MN.
- Our scheme localizes the QoS (re) programming to the affected parts of the packet path in the network, since the end-to-end QoS context transfer is performed only between the MAP and the NAR. The MAP is the cross point of the old path and the new path.
- Our scheme provides means to release any QoS state along the old path. If all routers transfer the QoS context successfully, the QoS entity of the MAP will send a Teardown message toward the PAR to delete the QoS status on the old path.

2) Interoperability requirements

- Our scheme may be better interoperable with other mobility protocols because our scheme is based on the F-HMIPv6 protocol which combines the HMIPv6 protocol and FMIPv6 protocol, two protocols that have drawn more attention in the Mipshop working group of IETF.
- Our scheme may interoperate with heterogeneous QoS paradigms such as IntServ and DiffServ, since the QoS context information retrieved from PAR can be better encoded to work with these heterogeneous QoS paradigms.

3) Miscellaneous requirements

- Our scheme can't support QoS along multiple packet paths. This is an open issue and requires further study.
- Our scheme can't provide information to link layer to support required QoS. This needs additional investigation too.

4) Standard requirements

- Scalability: With the deployment of the QoS mechanisms in the network, the QoS states are maintained in the routers by the QoS mechanism itself. Hence, this scheme does not introduce any new scalability issues.
- Security: The security issue in this scheme is as well as that of the CTP. The details are described in section 6 of the CTP [1].
- Conservation of wireless bandwidth: Re-establishing multiple contexts over an expensive, low-speed link can be avoided by relocating contexts over a potentially higher-speed wire [12]. Hence, the scheme may save the precious wireless bandwidth.
- Low processing overhead on mobile terminals: In our scheme, MN needs no additional operations.
- Providing hooks for authorization and accounting: This needs further study.
- Robustness against failures of any Mobile IP-specific QoS components in the network: This needs further study.

5 Conclusion and the Future Work

This paper proposes a framework for end-to-end QoS context transfer in Mobile IPv6 based on the F-HMIPV6 architecture, which may provide lower handover latency and packet losses and less signaling overhead handover, and implement an end-to-end QoS context transfer. The scheme overcomes the weakness of the CTP which can not meet the requirement of the end-to-end QoS mechanisms because of transferring contexts only between PAR and NAR in the CTP. We present the design goals and assumptions of the framework and also provide implementation details of the scheme in this paper. In addition, we compare the performance with scheme of the re-initiating RSVP signaling to re-establish QoS in Mobile IPv6, and show that our scheme has the less latency and packet loss than the scheme.

The simulation based on NS2 [18] platform for the scheme will be done soon to achieve the further performance analysis and discussion. And our further work will be to investigate the way of retrieving, classifying, encoding and representing of the QoS context transfer. The implementation of the end-to-end QoS context transfer framework will be based on the IntServ model, namely we will investigate the end-to-end RSVP context transfer in our further study.

References

1. J. Loughney (Ed.), M. Nakhjiri, C. Perkins, R. Koodli: Context Transfer Protocol. Internet Draft, draft-ietf-seamoby-ctp-08.txt, work in progress
2. Hee Young Jung, Seok Joo Koh, Hesham Soliman, Jun Seob Lee, Karim El-Malki, Bryan Hartwell: Fast Handover for Hierarchical MIPv6 (F-HMIPv6). Internet Draft, draft-jung-mobileip-fastho-hmipv6-02.txt, work in progress
3. D. Johnson, C. Perkins, J. Arkko: Mobility Support in IPv6. Internet Draft, draft-ietf-mobileip-ipv6-24.txt, work in progress
4. R. Braden, D. Clark, and S. Shenker: Integrated services in the Internet architecture: an overview. RFC 1633, IETF, June 1994
5. R. Braden, L. Zhang, S. Berson, S. Herzog, and S. Jamin: "Resource Reservation Protocol (RSVP) – Version 1 Functional Specification". RFC 2205, IETF, September 1997
6. S. Blake, D. Black, M. Carlson, E. Davies, Z. Wang, and W. Weiss: An architecture for differentiated services. RFC 2475, December 1998
7. Talukdar A. K., Badrinath B. R., Acharya A.: MRSVP: A Resource Reservation Protocol for An Integrated Services Network with Mobile Hosts. Wireless Networks, 2001, 7(1): 5-19
8. S. Paskalis, A. Kaloxylos, E. Zervas: An Efficient QoS Scheme for Mobile Hosts. Workshop on wireless Local Networks (Tampa Florida), November 2001
9. Q. Shen, W. Seah, A. Lo H. Zheng M. Greis: Mobility Extensions to RSVP in an RSVP-Mobile IPv6 Framework. Internet Draft, IETF, draft-shen-nsis-rsvp-mobileipv6- 00.txt, July 2002
10. Mahadevan I, Sivalingam K. M.: Architecture and Experimental Framework for Supporting QoS in Wireless Networks Using Differentiated Services. Mobile Networks and Applications, 2001, 6(4):385-395

11. J. Kempf, (Ed.): Problem Description: Reasons for Performing Context Transfers. RFC3374, IETF, September, 2002
12. R. Koodli, C. Perkins: A Context Transfer Protocol for Seamless Mobility. Internet Draft, Draft-koodli-seamoby-ct-04.txt, 2002.8
13. R. Koodli (Ed.): Fast Handovers for Mobile IPv6. Internet Draft, draft-ietf-mobileip-fast-mipv6-08.txt, work in progress
14. H. Soliman, C. Castelluccia, K. El-Malki, L. Bellier: Hierarchical Mobile IPv6 mobility management. Internet Draft, draft-ietf- mobileip-hmipv6-08.txt, work in progress
15. A. Festag, X. Fu, H. Karl, G. Schaefer, C. Fan, C. Kappler, M. Schramm, Siemens A. G.: QoS-Conditionalized Binding Update in Mobile IPv6. Internet Draft, draft-tkn-mobileip-qosbinding-mipv6-00.txt, July 2001
16. H. Chaskar, R. Koodli: A Framework for QoS Support in Mobile IPv6. Internet Draft, draft-chaskar-mobileip-qos-01.txt, March 2001
17. H. Chaskar, (Ed.): Requirements of a QoS solution for Mobile IP. RFC 3583, IETF, September 2003
18. The Network Simulator NS2: http://www.isi.edu/nsnam

Author Index

Lecture Notes in Computer Science

For information about Vols. 1–3124

please contact your bookseller or Springer

Vol. 3263: M. Weske, P. Liggesmeyer (Eds.), Object-Oriented and Internet-Based Technologies. XII, 239 pages. 2004.

Vol. 3260: I. Niemegeers, S.Heemstra de Groot (Eds.), Personal Wireless Communications. XIV, 478 pages. 2004.

Vol. 3255: A. Benczúr, J. Demetrovics, G. Gottlob (Eds.), Advances in Databases and Information Systems. XI, 423 pages. 2004.

Vol. 3254: E. Macii, V. Paliouras, O. Koufopavlou (Eds.), Integrated Circuit and System Design. XVI, 910 pages. 2004.

Vol. 3253: Y. Lakhnech, S. Yovine (Eds.), Formal Techniques in Timed, Real-Time, and Fault-Tolerant Systems. X, 397 pages. 2004.

Vol. 3249: B. Buchberger, J.A. Campbell (Eds.), Artificial Intelligence and Symbolic Computation. X, 285 pages. 2004. (Subseries LNAI).

Vol. 3246: A. Apostolico, M. Melucci (Eds.), String Processing and Information Retrieval. XIV, 316 pages. 2004.

Vol. 3241: D. Kranzlmüller, P. Kacsuk, J.J. Dongarra (Eds.), Recent Advances in Parallel Virtual Machine and Message Passing Interface. XIII, 452 pages. 2004.

Vol. 3240: I. Jonassen, J. Kim (Eds.), Algorithms in Bioinformatics. IX, 476 pages. 2004. (Subseries LNBI).

Vol. 3239: G. Nicosia, V. Cutello, P.J. Bentley, J. Timmis (Eds.), Artificial Immune Systems. XII, 444 pages. 2004.

Vol. 3238: S. Biundo, T. Frühwirth, G. Palm (Eds.), KI 2004: Advances in Artificial Intelligence. XI, 467 pages. 2004. (Subseries LNAI).

Vol. 3232: R. Heery, L. Lyon (Eds.), Research and Advanced Technology for Digital Libraries. XV, 528 pages. 2004.

Vol. 3224: E. Jonsson, A. Valdes, M. Almgren (Eds.), Recent Advances in Intrusion Detection. XII, 315 pages. 2004.

Vol. 3223: K. Slind, A. Bunker, G. Gopalakrishnan (Eds.), Theorem Proving in Higher Order Logics. VIII, 337 pages. 2004.

Vol. 3221: S. Albers, T. Radzik (Eds.), Algorithms – ESA 2004. XVIII, 836 pages. 2004.

Vol. 3220: J.C. Lester, R.M. Vicari, F. Paraguaçu (Eds.), Intelligent Tutoring Systems. XXI, 920 pages. 2004.

Vol. 3217: C. Barillot, D.R. Haynor, P. Hellier (Eds.), Medical Image Computing and Computer-Assisted Intervention – MICCAI 2004. XXXVIII, 1114 pages. 2004.

Vol. 3216: C. Barillot, D.R. Haynor, P. Hellier (Eds.), Medical Image Computing and Computer-Assisted Intervention – MICCAI 2004. XXXVIII, 930 pages. 2004.

Vol. 3210: J. Marcinkowski, A. Tarlecki (Eds.), Computer Science Logic. XI, 520 pages. 2004.

Vol. 3208: H.J. Ohlbach, S. Schaffert (Eds.), Principles and Practice of Semantic Web Reasoning. VII, 165 pages. 2004.

Vol. 3207: L.T. Yang, M. Guo, G.R. Gao, N.K. Jha (Eds.), Embedded and Ubiquitous Computing. XX, 1116 pages. 2004.

Vol. 3206: P. Sojka, I. Kopecek, K. Pala (Eds.), Text, Speech and Dialogue. XIII, 667 pages. 2004. (Subseries LNAI).

Vol. 3205: N. Davies, E. Mynatt, I. Siio (Eds.), UbiComp 2004: Ubiquitous Computing. XVI, 452 pages. 2004.

Vol. 3203: J. Becker, M. Platzner, S. Vernalde (Eds.), Field Programmable Logic and Application. XXX, 1198 pages. 2004.

Vol. 3202: J.-F. Boulicaut, F. Esposito, F. Giannotti, D. Pedreschi (Eds.), Knowledge Discovery in Databases: PKDD 2004. XIX, 560 pages. 2004. (Subseries LNAI).

Vol. 3201: J.-F. Boulicaut, F. Esposito, F. Giannotti, D. Pedreschi (Eds.), Machine Learning: ECML 2004. XVIII, 580 pages. 2004. (Subseries LNAI).

Vol. 3199: H. Schepers (Ed.), Software and Compilers for Embedded Systems. X, 259 pages. 2004.

Vol. 3198: G.-J. de Vreede, L.A. Guerrero, G. Marín Raventós (Eds.), Groupware: Design, Implementation and Use. XI, 378 pages. 2004.

Vol. 3194: R. Camacho, R. King, A. Srinivasan (Eds.), Inductive Logic Programming. XI, 361 pages. 2004. (Subseries LNAI).

Vol. 3193: P. Samarati, P. Ryan, D. Gollmann, R. Molva (Eds.), Computer Security – ESORICS 2004. X, 457 pages. 2004.

Vol. 3192: C. Bussler, D. Fensel (Eds.), Artificial Intelligence: Methodology, Systems, and Applications. XIII, 522 pages. 2004. (Subseries LNAI).

Vol. 3190: Y. Luo (Ed.), Cooperative Design, Visualization, and Engineering. IX, 248 pages. 2004.

Vol. 3189: P.-C. Yew, J. Xue (Eds.), Advances in Computer Systems Architecture. XVII, 598 pages. 2004.

Vol. 3186: Z. Bellahsène, T. Milo, M. Rys, D. Suciu, R. Unland (Eds.), Database and XML Technologies. X, 235 pages. 2004.

Vol. 3185: M. Bernardo, F. Corradini (Eds.), Formal Methods for the Design of Real-Time Systems. VII, 295 pages. 2004.

Vol. 3184: S. Katsikas, J. Lopez, G. Pernul (Eds.), Trust and Privacy in Digital Business. XI, 299 pages. 2004.

Vol. 3183: R. Traunmüller (Ed.), Electronic Government. XIX, 583 pages. 2004.

Vol. 3182: K. Bauknecht, M. Bichler, B. Pröll (Eds.), E-Commerce and Web Technologies. XI, 370 pages. 2004.

Vol. 3181: Y. Kambayashi, M. Mohania, W. Wöß (Eds.), Data Warehousing and Knowledge Discovery. XIV, 412 pages. 2004.

Vol. 3180: F. Galindo, M. Takizawa, R. Traunmüller (Eds.), Database and Expert Systems Applications. XXI, 972 pages. 2004.

Vol. 3179: F.J. Perales, B.A. Draper (Eds.), Articulated Motion and Deformable Objects. XI, 270 pages. 2004.

Vol. 3178: W. Jonker, M. Petkovic (Eds.), Secure Data Management. VIII, 219 pages. 2004.

Vol. 3177: Z.R. Yang, H. Yin, R. Everson (Eds.), Intelligent Data Engineering and Automated Learning – IDEAL 2004. XVIII, 852 pages. 2004.

Vol. 3176: O. Bousquet, U. von Luxburg, G. Rätsch (Eds.), Advanced Lectures on Machine Learning. IX, 241 pages. 2004. (Subseries LNAI).

Vol. 3175: C.E. Rasmussen, H.H. Bülthoff, B. Schölkopf, M.A. Giese (Eds.), Pattern Recognition. XVIII, 581 pages. 2004.

Vol. 3174: F. Yin, J. Wang, C. Guo (Eds.), Advances in Neural Networks - ISNN 2004. XXXV, 1021 pages. 2004.

Vol. 3173: F. Yin, J. Wang, C. Guo (Eds.), Advances in Neural Networks – ISNN 2004. XXXV, 1041 pages. 2004.

Vol. 3172: M. Dorigo, M. Birattari, C. Blum, L. M. Gambardella, F. Mondada, T. Stützle (Eds.), Ant Colony, Optimization and Swarm Intelligence. XII, 434 pages. 2004.

Vol. 3170: P. Gardner, N. Yoshida (Eds.), CONCUR 2004 - Concurrency Theory. XIII, 529 pages. 2004.

Vol. 3166: M. Rauterberg (Ed.), Entertainment Computing – ICEC 2004. XXIII, 617 pages. 2004.

Vol. 3163: S. Marinai, A. Dengel (Eds.), Document Analysis Systems VI. XI, 564 pages. 2004.

Vol. 3162: R. Downey, M. Fellows, F. Dehne (Eds.), Parameterized and Exact Computation. X, 293 pages. 2004.

Vol. 3160: S. Brewster, M. Dunlop (Eds.), Mobile Human-Computer Interaction – MobileHCI 2004. XVII, 541 pages. 2004.

Vol. 3159: U. Visser, Intelligent Information Integration for the Semantic Web. XIV, 150 pages. 2004. (Subseries LNAI).

Vol. 3158: I. Nikolaidis, M. Barbeau, E. Kranakis (Eds.), Ad-Hoc, Mobile, and Wireless Networks. IX, 344 pages. 2004.

Vol. 3157: C. Zhang, H. W. Guesgen, W.K. Yeap (Eds.), PRICAI 2004: Trends in Artificial Intelligence. XX, 1023 pages. 2004. (Subseries LNAI).

Vol. 3156: M. Joye, J.-J. Quisquater (Eds.), Cryptographic Hardware and Embedded Systems - CHES 2004. XIII, 455 pages. 2004.

Vol. 3155: P. Funk, P.A. González Calero (Eds.), Advances in Case-Based Reasoning. XIII, 822 pages. 2004. (Subseries LNAI).

Vol. 3154: R.L. Nord (Ed.), Software Product Lines. XIV, 334 pages. 2004.

Vol. 3153: J. Fiala, V. Koubek, J. Kratochvíl (Eds.), Mathematical Foundations of Computer Science 2004. XIV, 902 pages. 2004.

Vol. 3152: M. Franklin (Ed.), Advances in Cryptology – CRYPTO 2004. XI, 579 pages. 2004.

Vol. 3150: G.-Z. Yang, T. Jiang (Eds.), Medical Imaging and Augmented Reality. XII, 378 pages. 2004.

Vol. 3149: M. Danelutto, M. Vanneschi, D. Laforenza (Eds.), Euro-Par 2004 Parallel Processing. XXXIV, 1081 pages. 2004.

Vol. 3148: R. Giacobazzi (Ed.), Static Analysis. XI, 393 pages. 2004.

Vol. 3146: P. Érdi, A. Esposito, M. Marinaro, S. Scarpetta (Eds.), Computational Neuroscience: Cortical Dynamics. XI, 161 pages. 2004.

Vol. 3144: M. Papatriantafilou, P. Hunel (Eds.), Principles of Distributed Systems. XI, 246 pages. 2004.

Vol. 3143: W. Liu, Y. Shi, Q. Li (Eds.), Advances in Web-Based Learning – ICWL 2004. XIV, 459 pages. 2004.

Vol. 3142: J. Diaz, J. Karhumäki, A. Lepistö, D. Sannella (Eds.), Automata, Languages and Programming. XIX, 1253 pages. 2004.

Vol. 3140: N. Koch, P. Fraternali, M. Wirsing (Eds.), Web Engineering. XXI, 623 pages. 2004.

Vol. 3139: F. Iida, R. Pfeifer, L. Steels, Y. Kuniyoshi (Eds.), Embodied Artificial Intelligence. IX, 331 pages. 2004. (Subseries LNAI).

Vol. 3138: A. Fred, T. Caelli, R.P.W. Duin, A. Campilho, D.d. Ridder (Eds.), Structural, Syntactic, and Statistical Pattern Recognition. XXII, 1168 pages. 2004.

Vol. 3137: P. De Bra, W. Nejdl (Eds.), Adaptive Hypermedia and Adaptive Web-Based Systems. XIV, 442 pages. 2004.

Vol. 3136: F. Meziane, E. Métais (Eds.), Natural Language Processing and Information Systems. XII, 436 pages. 2004.

Vol. 3134: C. Zannier, H. Erdogmus, L. Lindstrom (Eds.), Extreme Programming and Agile Methods - XP/Agile Universe 2004. XIV, 233 pages. 2004.

Vol. 3133: A.D. Pimentel, S. Vassiliadis (Eds.), Computer Systems: Architectures, Modeling, and Simulation. XIII, 562 pages. 2004.

Vol. 3132: B. Demoen, V. Lifschitz (Eds.), Logic Programming. XII, 480 pages. 2004.

Vol. 3131: V. Torra, Y. Narukawa (Eds.), Modeling Decisions for Artificial Intelligence. XI, 327 pages. 2004. (Subseries LNAI).

Vol. 3130: A. Syropoulos, K. Berry, Y. Haralambous, B. Hughes, S. Peter, J. Plaice (Eds.), TeX, XML, and Digital Typography. VIII, 265 pages. 2004.

Vol. 3129: Q. Li, G. Wang, L. Feng (Eds.), Advances in Web-Age Information Management. XVII, 753 pages. 2004.

Vol. 3128: D. Asonov (Ed.), Querying Databases Privately. IX, 115 pages. 2004.

Vol. 3127: K.E. Wolff, H.D. Pfeiffer, H.S. Delugach (Eds.), Conceptual Structures at Work. XI, 403 pages. 2004. (Subseries LNAI).

Vol. 3126: P. Dini, P. Lorenz, J.N.d. Souza (Eds.), Service Assurance with Partial and Intermittent Resources. XI, 312 pages. 2004.

Vol. 3125: D. Kozen (Ed.), Mathematics of Program Construction. X, 401 pages. 2004.